Teaching Mathematics in Grades 6–12

P. 1-25 (Stop & Reflect p.17) due 2/8

Teaching Mathematics in Grades 6–12

Developing Research-Based Instructional Practices

Randall E. Groth

Salisbury University

Los Angeles | London | New Delhi
Singapore | Washington DC

Los Angeles | London | New Delhi
Singapore | Washington DC

FOR INFORMATION:

SAGE Publications, Inc.
2455 Teller Road
Thousand Oaks, California 91320
E-mail: order@sagepub.com

SAGE Publications Ltd.
1 Oliver's Yard
55 City Road
London EC1Y 1SP
United Kingdom

SAGE Publications India Pvt. Ltd.
B 1/I 1 Mohan Cooperative Industrial Area
Mathura Road, New Delhi 110 044
India

SAGE Publications Asia-Pacific Pte. Ltd.
3 Church Street
#10-04 Samsung Hub
Singapore 049483

Acquisitions Editor: Diane McDaniel
Editorial Assistant: Megan Koraly
Production Editor: Libby Larson
Copy Editor: Rachel Keith
Typesetter: C&M Digitals (P) Ltd.
Proofreader: Wendy Jo Dymond
Indexer: Molly Hall
Cover Designer: Anupama Krishnan
Marketing Manager: Terra Schultz
Permissions Editor Adele Hutchinson

Library of Congress Cataloging-in-Publication Data

Groth, Randall E.

Teaching mathematics in grades 6–12 : developing research-based instructional practices / Randall E. Groth.

p. cm.
Includes bibliographical references and index.

ISBN 978-1-4129-9568-9 (pbk.)

1. Mathematics--Study and teaching (Middle school)
2. Mathematics--Study and teaching (Secondary) I. Title.

QA11.2.G76 2013
510.71'2--dc23   2012016267

This book is printed on acid-free paper.

12 13 14 15 16 10 9 8 7 6 5 4 3 2 1

BRIEF CONTENTS

DETAILED CONTENTS

PREFACE

What is a mathematics teacher's main job? There are a surprisingly large number of answers to the question. Some maintain that a teacher's primary responsibility is to convey mathematics content accurately to students. From such a perspective, advanced mathematics courses are the heart of teacher preparation. Others believe that having a well-stocked "bag of tricks" for engaging and motivating students is the distinguishing mark of a well-prepared teacher. For such individuals, learning about teaching methods composes the core of mathematics teacher education. Still others assert that attending to students' needs as human beings is the fundamental mission of teachers. Hence, learning about students' social, emotional, and moral development is seen as the purpose of teacher education. Each perspective has its ardent supporters and critics.

This text is based on a premise slightly different from any of the three mentioned above. Its answer to the question about the main job of mathematics teachers is that they should be researchers who constantly experiment with methods for developing students' mathematical thinking. This does not mean that they need to publish research in academic journals, but rather that they constantly, systematically test teaching strategies. To do so, they must have strong content knowledge so they know fruitful directions in which to guide students' thinking. Content knowledge alone, however, is not sufficient. They must also have knowledge of strategies for teaching specific content. However, knowledge of teaching strategies along with knowledge of content is still not enough. To experiment intelligently with methods for developing students' mathematical thinking, teachers need knowledge of how students think about specific content areas. Such knowledge allows one to anticipate rough spots and navigate them.

The text you have in front of you is the result of my efforts to carefully comb through the mathematics education literature to highlight research and theory that is intimately related to the task of teaching mathematics in middle and high school. Of course, it was impossible to include every single study that it is useful to know. In searching through the literature and selecting studies to bring to the reader's attention, I applied my own personal lens as a secondary school teacher, researcher, university instructor, and student teaching supervisor. These various "hats" I have worn during my career informed my judgments about findings most pertinent to beginning teachers. In these roles, I have seen beginning teachers (myself included) have trouble spots that can be addressed through better knowledge of research and how it applies to practice.

A DIFFERENT APPROACH

My hope is that this will be considered a very different type of "teaching methods" text. Although it contains many specific strategies for teaching mathematics, helping the reader accumulate a random "bag of tricks" is not its primary goal. Instead, I hope that readers will come away with a deeper understanding of the types of mathematical knowledge students bring to school, and how their thinking may develop in response to different teaching strategies. It is up to the reader to experiment with the strategies and, even more important, to invent original ones on the basis of what is learned about students' thinking. Different classroom situations call for different approaches, so I hope this text helps readers develop the capacity to intelligently adapt to diverse settings. As teachers begin to view their classrooms as labs in which constant experimentation occurs, aimed at optimizing students' mathematical thinking, I believe advances can occur in mathematics education that we never dreamed possible.

Enjoy your journey into the vibrant and intriguing world of mathematics education!

ORGANIZATION

There are two parts to *Teaching Mathematics in Grades 6–12: Developing Research-Based Instructional Practices*.

Part I: Introduction to Teaching and Learning Mathematics is meant to be an induction into the field of mathematics education. Students enrolled in mathematics teaching methods courses usually have a sense of the nature of the fields of mathematics and of education, but may not have heard much about the field most relevant to their future careers—mathematics education. Chapters 1 and 2 introduce theoretical and conceptual aspects of the field, while Chapters 3, 4, and 5 deal with designing and implementing lessons and curricula. Chapter 6 is meant to help the reader make the final transition from being a university student to being a professional mathematics teacher.

Part II: Developing and Teaching Mathematical Thinking describes how students think about mathematical concepts. Borrowing an organizational strategy from the *Common Core State Standards*, standards documents of the National Council of Teachers of Mathematics, and standards documents of many states, each chapter in Part II is devoted to a specific mathematics content strand.

FEATURES

Reading a text, however, is simply not an adequate learning experience. Much of what is read is not retained if the reader does not adopt an active stance toward the text. Several pedagogical devices are included in each chapter to help readers adopt an active stance.

Vignette Analysis Activities are realistic classroom situations that highlight common pitfalls for beginning teachers and serve as the basis for analytic class discussions about teaching mathematics. One of these is included in every chapter.

Clinical Tasks engage readers in conducting small-scale research projects and bring chapter content to life. Instructors can use the tasks to provide structure for the clinical time students are required to spend in mathematics classrooms.

Implementing the Common Core marginal notes point readers toward homework and clinical tasks that relate to implementing the *Common Core State Standards*. One can glance through the text to quickly see connections between the *Common Core State Standards* and the content of each chapter.

Technology Connections help future teachers understand how today's instructional technology can enhance the teaching and learning of mathematics.

Ideas for Differentiating Instruction provide thoughts about how to structure lessons to help a wide range of students learn mathematics with understanding.

Four-Column Lesson Plans demonstrate how articles from teacher-oriented journals can be transformed into lesson plans that help develop students' mathematical thinking. Chapters 7, 8, 9, 10, and 11 each contain four sample lesson plans.

Stop to Reflect questions prompt readers to answer questions related to the reading. Instructors can also use these questions as catalysts for class discussions.

Homework Tasks extend and deepen readers' knowledge of the content and pedagogy discussed in each chapter.

Resources to Explore lists at the end of each chapter provide suggestions for print and online materials readers can explore to gain deeper understanding of the content and pedagogy discussed in each chapter.

Vocabulary Lists help highlight the important mathematics education terms used in each chapter. Common vocabulary is an essential basis for meaningful and coherent conversations about mathematics education among professionals.

INSTRUCTOR TEACHING SITE

A password-protected site, available at **www.sagepub.com/groth,** features author-provided resources that have been designed to help instructors plan and teach their courses. These resources include an extensive test bank with multiple-choice and true/false questions for each chapter as well as a lesson plan scoring rubric for instructors and a lesson plan template for students, both of which align with the lesson plans in Part II of the text.

ACKNOWLEDGMENTS

A book like this can never be attributed to the efforts of a single individual. I owe a debt of gratitude to several people who made the work possible. The mathematics educators whose writings are cited in the following chapters provided inspiration for the text, and I hope my description and synthesis of their work do justice to their efforts. I have also been fortunate to study under many talented mathematics educators at Marquette University, University of Wisconsin-Oshkosh, and Illinois State University. Their depth of knowledge and enthusiasm for the field helped me embark on the journey through the literature that provides the basis for the book. Salisbury University has provided a setting conducive to the work, providing the opportunity to teach secondary mathematics methods and supervise student teachers for several years. The staff at SAGE was a pleasure to work with throughout the process, as Diane McDaniel, Megan Koraly, Libby Larson, Brian Normoyle, and Rachel Keith were instrumental in improving the text and guiding it to completion. Most important, I owe special thanks to my family, who have been consistently encouraging and supportive of my work throughout the years.

Earlier drafts of the text were reviewed by others on several occasions. The insights of the following individuals were invaluable during the writing process:

Tracy Boone—Bald Eagle Area School District

Nancy Fardelius Fees—Northwest College

Adam P. Harbaugh—UNC Charlotte

Martha E. Hildebrandt—Chatham University

Ali Ikiz—Fayetteville State University

Suzanne C. Libfeld—Lehman College

Marta T. Magiera—Marquette University

Wendy L. McCarty—University of Nebraska at Kearney

Ann M. Rule—Saint Louis University

James Telese—University of Texas at Brownsville

Thomas P. Walsh—Kean University

Jane M. Wilburne—Penn State Harrisburg

I am indebted to these reviewers because their detailed and thoughtful analyses of the earlier chapters greatly helped me improve the finished product. Any errors or shortcomings remain my responsibility. I trust that each one will see his or her positive influence reflected at multiple points in the text.

PART I

Introduction to Teaching and Learning Mathematics

Chapter 1

Introduction to the Field of Mathematics Education

The past few decades have seen incredible growth in the study of teaching and learning mathematics. K–12 teachers, university professors, and other educators have produced standards documents, research reports, and curriculum frameworks with the potential to help improve students' learning. All of this activity makes it an exciting time to enter the profession of mathematics teaching. However, it can also be overwhelming to try to digest and reflect on everything the field has to offer. In fact, one is never really done learning about teaching mathematics. The best teachers are always learning ways to improve their practices by talking with colleagues, reading research, reading teachers' journals, carefully assessing the impact of their instructional practices on their students' thinking, and adjusting their practices to maximize students' learning.

The goal of this chapter is to provide a sense of the major issues and trends that have shaped the field of mathematics education in the recent past. By way of introduction, we will examine the standards documents published by the **National Council of Teachers of Mathematics** (NCTM), an organization with more than 90,000 members dedicated to improving mathematics education. We will then examine trends in mathematics teaching and learning around the world and the central messages of the reform movement in mathematics education. The objective is not to completely "cover" or give a comprehensive treatment of each of these topics—volumes have already been written on each of them—and resources for further study are given at the end of the chapter. Instead, the chapter provides a frame of reference for understanding the rest of the text. Remember, the best teachers are those who are always learning, and reading this chapter represents just the first step in a career-long journey of navigating the field.

A Brief History of NCTM Standards

The 1980s and 1990s marked the beginning of the "standards movement" because of the effort put into developing standards for teaching and learning in various subject areas. NCTM released three standards documents during this period: (1) *Curriculum and Evaluation Standards for School Mathematics* (1989), (2) *Professional Standards for Teaching Mathematics* (1991), and (3) *Assessment Standards for School Mathematics* (1995). The major themes from this first round of standards laid the groundwork for a fourth influential document, *Principles and Standards for School Mathematics* (NCTM, 2000). To understand the current state of the field of mathematics education, it is important to grasp the central messages conveyed by each document.

Curriculum and Evaluation Standards

NCTM's *Curriculum and Evaluation Standards for School Mathematics* (1989) described a vision for the teaching and learning of mathematics that differed sharply with much of conventional practice. For example, in regard to algebra, it called for more attention to (1) "developing an understanding of variables, expressions, and equations" (p. 70) and (2) "the use of real-world problems to motivate and apply theory" (p. 126). Less attention was to be given to (1) "manipulating symbols" (p. 70) and (2) "word problems by type, such as coin, digit, and work" (p. 127). The document contained similar direction for other mathematics content areas, including number and operations, geometry, and measurement. The recommendations sought to move school mathematics beyond an exclusive focus on the teaching and learning of procedures. A central emphasis was helping students to understand "the importance of the connections among mathematical topics and those between mathematics and other disciplines" (p. 146).

Curriculum and Evaluation Standards was also revolutionary in its call for more attention to historically neglected areas such as statistics, probability, and discrete mathematics. Although many important applications of these content areas could be found in contemporary society, they were largely absent from the school mathematics curriculum. The recommendation to give more attention to neglected areas was based on the premise that the school curriculum should change as the needs of society change. This premise also dictated that the school curriculum should take advantage of technology to help students understand the conceptual underpinnings of mathematics. In sum, *Curriculum and Evaluation Standards* recommended reform in *what* was taught as well as *how* it was taught.

Professional Standards for Teaching Mathematics

Professional Standards for Teaching Mathematics helped further clarify NCTM's vision for school mathematics reform. It recommended five major shifts in mathematics classroom environments:

- Toward classrooms as mathematical communities—away from classrooms as simply a collection of individuals;
- Toward logic and mathematical evidence as verification—away from the teacher as the sole authority for right answers;
- Toward mathematical reasoning—away from merely memorizing procedures;
- Toward conjecturing, inventing, and problem-solving—away from an emphasis on mechanistic answer-finding;

- Toward connecting mathematics, its ideas, and its applications—away from treating mathematics as a body of isolated concepts and procedures. (NCTM, 1991, p. 3)

NCTM (1991) recognized that these shifts would not occur overnight. Sustained professional development would be necessary to help teachers implement the recommendations.

NCTM's *Professional Standards for Teaching Mathematics* calls for five major shifts in the environment of mathematics classrooms (see the preceding discussion). What changes, if any, would your past mathematics teachers in Grades K–12 have needed to make to align their instruction with the five recommendations? Provide specific examples.

STOP TO REFLECT

Assessment Standards for School Mathematics

Assessment Standards for School Mathematics marked the end of the first round of NCTM standards documents. The document defined assessment broadly as "the process of gathering evidence about a student's knowledge of, ability to use, and disposition toward, mathematics and of making inferences from that evidence for a variety of purposes" (NCTM, 1995, p. 3). From this perspective, one of the primary purposes of assessment is to provide teachers information about the nature of student learning. Information about students' learning can be drawn from a variety of sources. Instead of relying solely on paper-and-pencil tests, teachers can draw information from student interviews, projects, and portfolios. Information gained about students' learning can in turn help shape future lesson plans.

Principles and Standards for School Mathematics

Principles and Standards for School Mathematics (NCTM, 2000) differed from previous standards documents in that its intent was to write standards that

- build on the foundation of the original *Standards* documents;
- integrate the classroom-related portions of *Curriculum and Evaluation Standards for School Mathematics, Professional Standards for Teaching Mathematics*, and *Assessment Standards for School Mathematics;*
- organize recommendations into four grade bands: prekindergarten through Grade 2, Grades 3–5, Grades 6–8, and Grades 9–12. (p. x)

Principles and Standards for School Mathematics organized its discussion of mathematics content around five content standards: number and operations, algebra, geometry, measurement, and data analysis and probability. The second half of this text uses a similar organizational scheme by devoting chapters to each of the content standards (with the exception that measurement is distributed among the other content strands).

As a consolidation and elaboration of the previous NCTM standards documents, *Principles and Standards for School Mathematics* represents the closest we have come to a consensus about *which* mathematical topics should be taught in school and *how* they should be taught. Teachers, university professors, mathematics supervisors, and other professionals spent three years constructing the document. As it was being written, feedback was elicited from stakeholders in mathematics education around the world. It should be noted, however, that consensus on the vision of NCTM standards has never been, and likely never will be, universal. For example, some disagree with the decreased emphasis on lecture as a teaching method (Wu, 1999b) or the manner in which technology is to be integrated into the curriculum (Askey, 1999). These kinds of conflicts have been characterized as parts of a larger "math war" over the content of the school curriculum (Schoenfeld, 2004). As with most events characterized as "wars," much of the conflict is based on the opposing sides misunderstanding each other. This book is based on the premise that one must seek to understand the NCTM standards before condemning *or* accepting them. Toward that end, in the next section, an overview of some of the most important themes in *Principles and Standards for School Mathematics* is given.

UNDERSTANDING THE PRINCIPLES AND PROCESS STANDARDS

Principles and Standards for School Mathematics goes beyond merely providing an organizational scheme for discussing mathematics content. The document also contains principles and process standards to guide the teaching of mathematics. **NCTM principles** "describe particular features of high-quality mathematics education" (NCTM, 2000, p. 11), while **NCTM process standards** describe aspects of mathematical teaching and learning that should occur in all content areas. The six principles are Equity, Curriculum, Teaching, Learning, Assessment, and Technology. The five process standards are Problem Solving, Reasoning and Proof, Communication, Connections, and Representation.

NCTM Principles

The Equity Principle states that "excellence in mathematics education requires equity—high expectations and strong support for all students" (NCTM, 2000, p. 12). This challenges the assumption that only an elite few are meant to understand mathematics. Teachers should expect *all* students to learn mathematics. The principle calls for equity of access to high-quality mathematics instruction, curriculum materials, and technology. A key to understanding the Equity Principle is the premise that "equity does not mean that every student should receive identical instruction[;] instead, it demands that reasonable and appropriate accommodations be made as needed to promote access and attainment for all students" (NCTM, 2000, p. 12). Therefore, as teachers think about how to achieve equity in their own classrooms, they need to set aside the assumption that "all students should be treated the same way."

The Curriculum Principle states that "a curriculum is more than a collection of activities: it must be coherent, focused on important mathematics, and well-articulated across the grades" (NCTM, 2000, p. 14). To understand this principle, it can be helpful to consider a counterexample: some curricula are essentially "laundry lists" of isolated "topics" to be "covered" in a prescribed order. The order in which topics are treated may or may not help students understand the fundamental concepts of the subject. Teachers in such situations may rarely, if ever, plan sequences of instruction with colleagues who teach different grade levels or subjects. The end result is that students perceive mathematics to be a disconnected body of knowledge consisting of isolated rules and procedures that do not relate to one another in a coherent fashion. To avoid this situation, it is important for teachers and instructional supervisors to communicate with one another about student learning and to modify curriculum sequences as necessary.

The Teaching Principle states that "effective mathematics teaching requires understanding what students know and need to learn and then challenging and supporting them to learn it well" (NCTM, 2000, p. 16). To pursue this principle, one must develop knowledge of students, knowledge of mathematics, and knowledge of teaching strategies (Kilpatrick, Swafford, & Findell, 2001). Developing knowledge of students' mathematical thinking allows teachers to understand the effectiveness of their instructional practices. Developing knowledge of mathematics helps teachers identify the "big ideas" in any given domain and to draw students' attention toward them. Knowledge of teaching strategies gives teachers a variety of instructional practices that can be adapted, as needed, to any given situation. A teacher is never done developing knowledge in any of the three areas. It is important to constantly seek out opportunities to develop and refine knowledge in each area.

The Learning Principle states that "students must learn mathematics with understanding, actively building new knowledge from experience and prior knowledge" (NCTM, 2000, p. 20). It is not enough for students to memorize mathematical procedures and skills in isolation from one another. While procedures and skills are important, it is unlikely that students will be able to apply them flexibly if they do not understand the big picture of why they work and how they are related. Many students have experienced "learning" a set of facts or skills needed for a test and then forgetting them shortly after taking it. Often this occurs because the facts and skills were not connected to a larger mental map of the structure of the discipline. The goal should be to help students build rich, interconnected mental roadmaps of a given content area so that they will progress beyond surface-level learning that is easily forgotten after a test has been taken.

The Assessment Principle asserts that "assessment should support the learning of important mathematics and furnish useful information for both teachers and students" (NCTM, 2000, p. 22). This principle amplifies the view of assessment given in the *Assessment Standards for School Mathematics* by emphasizing the use of assessment results to inform teaching. It also reiterates the importance of thinking beyond traditional paper-and-pencil tests for assessment. Teachers should gather evidence of students' learning from multiple sources, including "open-ended questions, constructed response tasks, selected-response items, performance tasks, observations, conversations, journals, and portfolios" (NCTM, 2000, p. 23). Insights gained from listening to students as they work with one another or reading what they write in response to a prompt from the teacher can be invaluable in improving the effectiveness of instruction.

The last of the six principles is the Technology Principle, which states that "technology is essential in teaching and learning mathematics; it influences the mathematics that is taught and enhances students' learning" (NCTM, 2000, p. 24). Properly used, technology can help students develop the type of conceptual understanding described in the Learning Principle. Instead of eroding computational skills, it can help students understand the central ideas of a given content area. Of course, if one's instructional goal is merely to teach students how to perform computations, then current technology can, in fact, seem threatening. However, since instruction needs to go beyond that sort of surface-level treatment, it is important to think about how technology can help students develop deeper understandings of important mathematical ideas. The Technology Connections included throughout this book provide specific examples of how technology can facilitate the teaching and learning of mathematics in meaningful ways.

STOP TO REFLECT

From your own experiences as a K–12 student, write an example of how schools sometimes attain each of the NCTM principles described above. Then, once again drawing on your own experiences as a K–12 student, write an example of how schools sometimes fall short of the ideals set forth in each principle. Share and compare examples with classmates.

NCTM Process Standards

Problem Solving is the first of the process standards discussed in *Principles and Standards for School Mathematics.* Several aspects of this process standard conflict with traditional notions of problem solving in mathematics. The process standard defines problem solving as "engaging in a task for which the solution method is not known in advance" (NCTM, 2000, p. 52). Under this definition, students who simply apply procedures they have been taught to produce solutions to sets of homework problems are not engaging in problem solving. In addition, problem solving should not be thought of as just doing the "application" problems that often appear at the end of a set of exercises. Students are to build mathematical knowledge *during* the process of problem solving. As they do so, they develop the attitude that mathematics makes sense and is a matter of reasoning carefully through novel situations rather than simply following teacher-prescribed recipes. Teaching via problem solving is arguably *the* central idea in reform-oriented instruction, and sometimes one of the most difficult to grasp. We will return to it at the end of this chapter and throughout the book.

Reasoning and Proof is the second process standard. While this standard sets the goal that students should be able to produce formal proofs by the end of high school, it deals with much more than formal proof. Teachers can help students develop mathematical reasoning ability simply by asking *why* a given procedure works or *how* they produced the solution to a question. Pragmatically speaking, teachers need to help students get in the habit of answering these types of questions on account of the numerous standardized tests that now include open-ended questions asking for justifications and explanations of answers. However, students' performance on tests is not the most compelling reason to help them develop the ability to reason and prove. If students do not engage in some form of reasoning, it is arguable that they are not doing mathematics at all. The goal should be to develop a classroom community in which students are curious to find out why things work as they do and to make and test their own conjectures about solutions to problems.

The third process standard is Communication. As students communicate their reasoning with peers and teachers by writing and talking, they are forced to structure their thoughts. In this process, they build and refine their mathematical knowledge. Classroom communication patterns need to go beyond teacher-student interaction alone. Student–student interaction is also important, because it gives students the opportunity to "analyze and evaluate the mathematical thinking and strategies of others" (NCTM, 2000, p. 60). The teacher's responsibility is to create a classroom environment in which students feel free to communicate their thinking and respect and value one another's contributions. This is a nontrivial task, particularly if students are not used to sharing their mathematical thinking with one another. It is, nonetheless, a valuable goal because of the opportunities it affords to help students reflect on and subsequently refine their thinking.

The fourth process standard is Connections. As stated in the Learning Principle, students should see connections among mathematical ideas rather than viewing mathematics as a subject consisting of isolated topics. For example, teachers should ask questions and present problems that help students see how fractions, decimals, and percents are related to one another. A counterexample to this would be dealing with fractions, decimals, and percents in separate chapters and never drawing students' attention to the fact that they are often used as different representations of the same quantities. A second kind of connection students should see is how the mathematics they are studying applies to contexts outside of mathematics. For example, students can be asked to solve problems that connect data analysis, probability, and mathematical modeling to making predictions about the weather or other physical phenomena. By paying careful attention to the problems they pose, teachers can systematically begin to help students see the connections among mathematical ideas as well as how those ideas connect to other fields.

The final process standard is Representation. Representation may not come to mind immediately as an important mathematical process simply because it is easy to take representations for granted. For example, when we use our place value system for representing numbers, we seldom stop to reflect on how much more efficient it is than using Roman numerals. It is important to realize, however, that any given representation is likely to be interpreted differently by individual students. Those without a deep understanding of place value may not see the efficiency in our number system. Teachers need to work to understand how their students interpret representations and design instruction to be responsive to students' needs. To facilitate learning, it is necessary to help students build bridges from their intuitive, idiosyncratic representations for concepts to conventional representations. Technology affords unique opportunities for students to explore conventional representations. For example, graphing calculators allow one to efficiently move back and forth between graphical and tabular representations of functions. Instead of simply asking students to produce graphs and tables using predetermined symbolic representations of functions, teachers can ask students to explain how specific aspects of each representation relate to one another. As students consider multiple ways to represent mathematical ideas, they can reflect on which representations are most efficient for different situations.

Common Core State Standards

Many of the recommendations from NCTM's standards documents are reflected in the *Common Core State Standards* (CCSS; National Governor's Association for Best Practices & Council of Chief State School Officers, 2010). These standards were written to provide common mathematics learning expectations across the United States. Forty-eight states participated in writing the CCSS, and the vast majority of these states also adopted them (see www.corestandards.org for information about your state). The CCSS are similar to NCTM's *Principles and Standards for School Mathematics* in that they provide standards for the content as well as the processes of learning mathematics. CCSS process and content standards relevant to middle and high school mathematics appear in Appendices A, B, and C of this text and are also available online at www.corestandards.org. Throughout this book, you will find notes in the margins pointing toward tasks to help develop your ability to implement the CCSS standards for mathematical practice and the CCSS content standards for Grades 6 through 12.

CCSS Standards for Mathematical Practice

According to the CCSS standards for mathematical practice, students should experience the following mathematical thinking processes in the classroom:

1. Make sense of problems and persevere in solving them
2. Reason abstractly and quantitatively
3. Construct viable arguments and critique the reasoning of others
4. Model with mathematics
5. Use appropriate tools strategically
6. Attend to precision
7. Look for and make use of structure
8. Look for and express regularity in repeated reasoning. (National Governor's Association for Best Practices & Council of Chief State School Officers, 2010, p. 53)

A description of each practice is provided in Appendix A.

This book includes classroom vignettes to spark thought and discussion about how each of the CCSS standards for mathematical practice can be enacted in the classroom. The vignettes do not present images of classrooms perfectly aligned with the process standards and standards for mathematical practice. Instead, they present realistic situations vividly illustrating the challenges teachers encounter on a daily basis in trying to attain the standards. Reading the vignettes and discussing them with others is a good starting point for thinking about how to overcome the challenges.

> **Implementing the Common Core**
>
> See the clinical task at the end of the chapter to analyze the extent to which a lesson aligns with all of the CCSS standards for mathematical practice.

CCSS Content Standards

The CCSS contain five content domains for middle school and six for high school. For Grades 6 through 8, the five content domains are ratios and proportional reasoning, the number system, expressions and equations, geometry, and statistics and probability. For high school level, the six content domains are number and quantity, algebra, functions, modeling, geometry, and statistics and probability. These content domains closely resemble the content strands of NCTM's (2000) *Principles and Standards for School Mathematics.* The alignment between CCSS content domains and the chapters of this book is shown in Table 1.1.

Table 1.1   Relationship Between *Common Core State Standards* Content Domains and the Chapters in This Book

Content Domains From *Common Core State Standards*	Related Chapters in *Teaching Mathematics in Grades 6–12*
Grades 6–8	
Ratios and proportional reasoning	Chapter 7: Developing Students' Thinking in Number and Operations
The number system	Chapter 7: Developing Students' Thinking in Number and Operations
Expressions and equations	Chapter 8: Developing Students' Algebraic Thinking
Geometry	Chapter 10: Developing Students' Geometric Thinking
Statistics and probability	Chapter 9: Developing Students' Statistical and Probabilistic Thinking
High School	
Number and quantity	Chapter 7: Developing Students' Thinking in Number and Operations
Algebra	Chapter 8: Developing Students' Algebraic Thinking
Functions	Chapter 8: Developing Students' Algebraic Thinking; Chapter 11: Developing Students' Thinking in Advanced Placement Courses
Modeling	Throughout Chapters 7–11
Geometry	Chapter 10: Developing Students' Geometric Thinking
Statistics and probability	Chapter 9: Developing Students' Statistical and Probabilistic Thinking; Chapter 11: Developing Students' Thinking in Advanced Placement Courses

A GLOBAL PERSPECTIVE ON MATHEMATICS EDUCATION

Most of the introductory material in this chapter has dealt with developments in mathematics education in North America. To understand more fully how mathematics education can be improved in the classroom, it is helpful to examine how mathematics is taught across the globe. Each country has a distinct culture of teaching that shapes students' experiences with mathematics. Typical lessons often vary in important ways from one country to the next. Although the mathematics students are to learn may be the same in any two given countries, the methods of teaching the content can differ sharply.

The **Third International Mathematics and Science Study (TIMSS)** was a large, comprehensive investigation of mathematics teaching and learning in different parts of the world. Conducted in 1994–1995, it tested mathematics achievement in more than 40 nations. Top-scoring countries on the assessment included Singapore, Korea, and Japan, with eighth-grade mean scores of 643, 607, and 605, respectively. Canada, England, and the United States all scored significantly lower, with eighth-grade mean scores of 527, 506, and 500, respectively (A. E. Beaton & Robitaille, 1999). Repeat administrations of the TIMSS in 1999, 2003, and 2007 showed a persistent achievement gap between U.S. students and their counterparts in other industrialized nations (National Center for Education Statistics, 2010). The TIMSS acronym now stands for "Trends in International Mathematics and Science Study."

TIMSS studies have gone beyond just measuring students' achievement. They have also uncovered factors that could explain achievement differences among countries. In a book titled *The Teaching Gap*, J. W. Stigler and Hiebert (1999) analyzed video footage of eighth-grade mathematics lessons from countries participating in the 1994–1995 TIMSS. Their analysis is especially helpful in shedding light on how mathematics is taught differently between the United States and Japan. They summarized the Japanese style of teaching in the following terms:

> In Japan, teachers appear to take a less active role, allowing their students to invent their own procedures for solving problems . . . Teachers, however, carefully design and orchestrate lessons so that students are likely to use procedures that have been developed recently in class. An appropriate motto for Japanese teaching would be "structured problem solving." (J. W. Stigler & Hiebert, 1999, p. 27)

In contrast, the typical pattern of teaching in the United States was described as follows:

> In the United States, content is not totally absent . . . but the level is less advanced and requires much less mathematical reasoning . . . Teachers present definitions of terms and demonstrate procedures for solving specific problems. Students are then asked to memorize the definitions and practice the procedures. In the United States, the motto is "learning terms and practicing procedures." (J. W. Stigler & Hiebert, 1999, p. 27)

Figures 1.1 and 1.2 show sample lessons drawn from the 1999 TIMSS video study. Figure 1.1 represents an American lesson, and Figure 1.2 represents a Japanese lesson. The American lesson illustrates "learning rules and practicing procedures," while the Japanese lesson illustrates "structured problem solving."

Figure 1.1 A typical American mathematics lesson from the TIMSS video study.

United States Public Release Lesson 4 Lesson Graph [8ᵗʰ grade]

[45 minute lesson]

4 1/2 minutes

Private Class Work: Reviewing the Previous Work

Students get their materials for the lesson ready, and then work on the following "problem of the day":

Tell about these in your own words

1) Inscribed angle theorem 2) right angle corollary 3) arc intercept corollary

26 minutes

Public Class Work

Problem of the day results

Inscribed angle theorem –Suzy: "If an angle's inscribed in a circle and it intercepts part of the circle, then the angle's measure is equal to half of the other angle."

Right angle corollary –Matt: "If an inscribed angle intercepts a semicircle, the angle is a right angle."

Arc intercept corollary – Margaret: "When you have two inscribed angles and they intercept the same arc, they have the same measure."

Review of Angles formed by Secants and Tangents

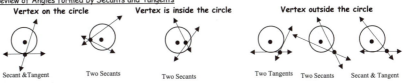

Vertex on the circle	Vertex is inside the circle	Vertex outside the circle
Secant & Tangent	Two Secants	Two Secants	Two Tangents	Two Secants	Secant & Tangent

Homework Problems

secant and tangent intersecting on a circle

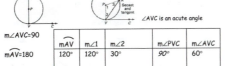
∠AVC is an acute angle

∠AVC is an obtuse angle

m∠AVC=90

m$\overset{\frown}{AV}$=180

m$\overset{\frown}{AV}$	m∠1	m∠2	m∠PVC	m∠AVC
120°	120°	30°	*90°*	60°
x°	x°	(180°-x°)÷2	90°	x°÷2

m$\overset{\frown}{AXV}$	m∠1	m∠2	m∠PVC	m∠AVC
200°	160°	10°	*90°*	100°
x°	160°-x°	(180°-m∠1)÷2	90°	x°÷2

Theorem: If a tangent and a secant (or a chord) intersect on a circle at the point of tangency, then the measure of the angle formed is half the measure of its intercepted arc – regardless of whether the angle is right, acute, or obtuse.

Two secants intersecting inside a circle

m$\overset{\frown}{AC}$	m$\overset{\frown}{BD}$	m∠1	m∠2	m∠AVC	m∠DVB
160°	40°	80°	20°	100°	100°
180°	70°	*90°*	35°	*125°*	*125°*
X₁°	X₂°	mAC÷2	mBD÷2	(mAC+ mBD) ÷2	(mAC+ mBD) ÷2

Theorem: The measure of an angle formed by two secants or chords that intersect in the interior of a circle is half the *sum* of the measures of the arcs intercepted by the angle and its vertical angle.

Two secants intersecting outside a circle

m$\overset{\frown}{BD}$	m$\overset{\frown}{AC}$	m∠1	m∠2	m∠AVC
200°	40°	100°	20°	80°

Theorem: The measure of an angle formed by two secants that intersect in the exterior of a circle is half the *difference* of the measures of the intercepted arcs.

3 1/2 minutes

Private Class Work: Example One

Given: m ∠AVC=60°, m AC=130°, find m $\overset{\frown}{BD}$

1 1/2 minutes

Public Class Work

A student puts the answer to example one on the overhead: (130°-x°)/2 = 60, 130°-x°=120°, x=10°, BD=10°

6 minutes

Private Class Work: Example Two

Given: m $\overset{\frown}{UR}$=140°, m $\overset{\frown}{RS}$=100°, m $\overset{\frown}{ST}$=30°,
Find m ∠RSU, m ∠RVU, m ∠USV, m ∠RWS
(W is the point inside the circle.)

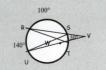

3 1/2 minutes

Public Class Work

Students give answers to example two: m ∠RSU= 70°, m ∠RVU=55°, m ∠USV=110°, m ∠RWS=95°.
The teacher reminds students to study for their final exam tomorrow.

Figure 1.2   A typical Japanese lesson from the TIMSS video study.

Japan Public Release Lesson 3 Lesson Graph [8ᵗʰ Grade]

[54 Minute Lesson]

4 Minutes

Public Class Work: Setting up the Problem– The teacher reads the following problem to his class:

It has been one month since Ichiro's mother entered the hospital. He has decided to give a prayer with his small brother at a local temple every morning so that she will be well soon. There are 18 ten-yen coins in Ichiro's wallet and just 22 five-yen coins in his smaller brother's wallet. They have decided every time to take one coin from each of them and put them in the offertory box and continue the prayer up until either wallet becomes empty. One day after they were done with their prayer, when they looked into each other's wallet the smaller brother's amount of money was bigger than Ichiro's. How many days has it been since they started the praying? That is the problem.

The teacher says, "I think that there are points hard to understand with just the sentences, so I would like to look at the figure and check it. He goes on to simulate the problem by taking coins from each wallet and putting them in the offertory box, asking how much in each wallet and which brother has more.

Private Class Work: Students work individually on the problem

13 Minutes

Midway through, the teacher says to the whole class, "If you found the answer with one method, try finding another method."

As he circulates around the room, he talks with several students about their solution methods and tells them that he is going to have them present their solution methods later on. He says to one student, "please think beforehand why you formulated an equation like this."

24 Minutes

Public Class Work: Students Presenting Solutions to Class

The teacher asks the students to present their solutions in the following order and place a pre-written title over each solution. He asks after each solution is shared, "How many others solved it in the same way?"

1. Student one (Manipulating actual objects)
Actually take one coin from each wallet and put it in offertory box until Ichiro's wallet contains less money than the brother's wallet. Or crossing out one coin from each wallet until the same condition is met. Answer: 15th day

3. Student three
(There is a difference of 5 coins per day)

$$180 - 110 = 70$$
$$70 \div (10 - 5) = 14$$
$$14 + 1 = 15$$
15日目

5. Student five (If X is the day when the brother's monetary amount exceeds Ichiro's)

$$180 - 10x < 110 - 5x$$

2. Student two
(Solving it by making a table)

Number of days	1	2	3	-----	14	15	16	17
Amount left	170	160	150	-----	40	30	20	10

Number of days	1	2	3	-----	14	15	16	17
Amount left	105	100	95	-----	40	35	30	25

Answer: 15th day

4. Student four (If X is the day when the monetary amounts become the same)

$$\begin{cases} \mathcal{P}: 180 - 10x \\ \mathcal{P}: 110 - 5x \end{cases}$$
$$(x別)\cdot(14なら)$$
15日目

Task: Complete the chart and check if the value of x holds true for this inequality. What is the relationship?

x	180-10x		110-5x
13	180-10x13=50	>	110-5x13=75
14			
15			

The teacher assigns a task and students work on it individually for six minutes.
The teacher asks a student to write her results on the chalkboard while the other students are still working.

13 Minutes

Public Class Work:

The teacher asks how many students had the same results as the last student.
He notes that x holds true for 15, 16, 17 and 18; the first one was >. The second one was equal (14) which he calls "the standard".

x	180-10x	大小	110-5x
× 13	180-10x13=50	>	110-5x13=75
× 14	180-10x14=80		110-5x14=50
○ 15	180-10x15=30	<	110-5x15=35
○ 16	180-10x16=20	<	110-5x16=30
○ 17	180-10x17=10	<	110-5x17=25

The teacher asks about the 19ᵗʰ day, and a student says Ichiro's wallet becomes empty and at that time they will end it. Students are asked to write the solution on their handouts.

The teacher presents a handout and reads the problem to the class:
The prayer was answered and their mother was able to leave the hospital safely, and that night they gave a toast with juice. At present there are 50 milliliters of juice in Ichiro's cup and 80 milliliters in the smaller brother's cup remaining. When the mother poured juice into Ichiro's cup it made Ichiro's cup reversely become more. How many milliliters had she poured?

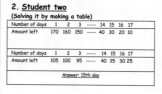

The teacher asks the students to form an inequality to solve this problem. A student comes up to the board and says that the amount needed to get up to the 80 ml should be x ml (x represents the increased amount). The teacher asks students to write the situation with the unknown x ml and the symbols (>,<). He says, "Using symbols like this, try expressing it."

After a minute, the teacher asks for the expression.
Student: 50 + x > 80. The teacher asks, "What values of x hold true for this equation?" Another student: "More than 30."
The teacher suggests using a number line and asks,
"How many numbers are there more than 30?"
A student responds "Infinite".

The teacher asks if 30 is included or not included. When the student responds that it's not included, the teacher responds yes, but if it comes over even a little, like 30.1, it becomes more than.

The TIMSS video studies reveal somewhat of an irony: although the NCTM standards documents were written in North America, lessons in Japan are generally more aligned with NCTM standards than lessons in the United States are (Jacobs et al., 2006). The U.S. goal of "learning terms and practicing procedures" conflicts with many of the ideas (discussed earlier in this chapter) in NCTM's *Principles and Standards for School Mathematics*. One of the central messages of the document is that students must learn mathematics with understanding—it is not enough to learn terms and practice procedures; students must understand *why* the procedures work and *how* they are related to one another. On the other hand, the Japanese goal of "structured problem solving" closely fits NCTM goals. According to the problem solving process standard, students should construct new mathematical knowledge in the process of solving problems. By carefully sequencing problems to capitalize on students' prior learning, Japanese teachers work toward this goal.

> **Implementing the Common Core**
>
> See Homework Task 5 to view a video of TIMSS lessons online and analyze the extent to which the lessons help students "make sense of problems and persevere in solving them" (Standard for Mathematical Practice 1).

What conclusions are to be drawn from the differences in teaching patterns between the United States and Japan? If the U.S. instructional paradigm could switch overnight from "learning terms and practicing procedures" to "structured problem solving," would the United States "catch up" to Japan on achievement tests? Of course, it is impossible to answer that question because of a host of contextual factors unique to each country (Kaiser, Luna, & Huntley, 1999). Nonetheless, there are important things to learn from the Japanese lesson pattern. In particular, the Japanese pattern illustrates how it is possible to organize mathematics instruction around problem solving. Problem solving, in contrast to memorizing disconnected bits of information, is at the heart of doing mathematics. This chapter concludes with a discussion of what it means to teach through problem solving and how one might begin to do so.

React to the statement "Problem solving, in contrast to memorizing disconnected bits of information, is at the heart of doing mathematics." Do you agree? Explain why or why not.

STOP TO REFLECT

TEACHING *THROUGH* PROBLEM SOLVING: THE CENTERPIECE OF REFORM-ORIENTED INSTRUCTION

To understand what it means to teach through problem solving, it is helpful to first consider what it does *not* mean. Schroeder and Lester (1989) distinguished among **teaching *about* problem solving**, **teaching *for* problem solving**, and **teaching *through* problem solving**. Teaching *about* problem solving involves helping students learn general problem solving strategies. For example, some curricula introduce Polya's (1945) problem solving process with the assumption that students will adopt his thinking strategies as they solve problems of their own. Teaching *for* problem solving consists of explicitly teaching students mathematical ideas that they are expected to use to solve problems later on. This is closely aligned with the pattern of "learning rules and practicing procedures" that is widespread in the United States.

Implementing the Common Core

See Homework Task 6 to analyze the thinking processes students must use to "make sense of problems and persevere in solving them" (Standard for Mathematical Practice 1).

In contrast, teaching *through* problem solving involves selecting problems containing important mathematics and helping students learn mathematical ideas as they solve the problems. Teaching *through* problem solving aligns with the typical Japanese lesson pattern described earlier as "structured problem solving."

It would be inaccurate to assume that the approach of teaching *through* problem solving can produce positive learning results only for Japanese students. Studies of curricula developed in the United States that focus on teaching *through* problem solving show that students experiencing the curricula perform as well as their counterparts from traditional classrooms on conventional tests of content, while generally outperforming them in problem solving, reasoning, and nontraditional content (Chappell, 2003; Putnam, 2003; Swafford, 2003). Therefore, teaching through problem solving is not an abstract, unrealistic, idealized notion. It is supported by careful studies of teachers and students, unlike more traditional forms of instruction that are widespread in the United States. As you prepare to teach *through* problem solving, it is important to realize that it requires a great deal of work. It is much more challenging than simply having students copy notes from a board or a screen.

Teaching through problem solving begins with the selection or design of student tasks. NCTM (1991) provides the following thoughts about **worthwhile mathematical tasks**:

> Good tasks are ones that do not separate mathematical thinking from mathematical concepts or skills, that capture students' curiosity, and that invite them to speculate and to pursue their hunches. Many such tasks can be approached in more than one interesting and legitimate way; some have more than one reasonable solution. (p. 25)

After making this statement, NCTM provided an example of one task not likely to fit the description and another that was likely to fit (Figure 1.3).

Figure 1.3   Comparison of two tasks involving area and perimeter (NCTM, 1991, p. 28).

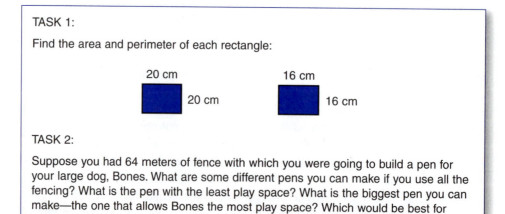

TASK 1:

Find the area and perimeter of each rectangle:

20 cm 20 cm   16 cm 16 cm

TASK 2:

Suppose you had 64 meters of fence with which you were going to build a pen for your large dog, Bones. What are some different pens you can make if you use all the fencing? What is the pen with the least play space? What is the biggest pen you can make—the one that allows Bones the most play space? Which would be best for running?

There are several important differences between Tasks 1 and 2 in Figure 1.3. Task 1 simply requires the skill of calculating area and perimeter. Task 2, in contrast, provides the opportunity to build understanding of area and perimeter as students solve a problem involving both concepts. Task 2 can be tackled in a variety of ways, ranging from trial and error to using tools from algebra and calculus. Task 2 can also be readily extended—for example, after solving the given problem successfully, students could be asked to make generalizations about which types of pen arrangements provide the most area.

Selecting and posing a problem does not mark the end of the teacher's role in the process of teaching *through* problem solving. Contrary to what some believe about this approach, the teacher does not just sit back passively as students solve problems. Instead, the teacher plays an active role in facilitating students' learning. For instance, the teacher can help students organize their thinking by having them make initial conjectures about what the answer to the problem might be (Cohen & Adams, 2004). As students work through the problem, they can then judge the reasonableness of their solution strategies by referring back to their initial predictions.

Teachers can also help draw students' attention to the structure of a problem by asking them to draw diagrams that represent the key quantities and having them "act out" the problem when possible. For example, suppose students were asked to solve the following problem:

> ### Implementing the Common Core
>
> See Homework Task 7 to think about modifying exercises in traditional textbooks to make them "worthwhile mathematical tasks" that can set the stage for students to "make sense of problems and persevere in solving them" (Standard for Mathematical Practice 1).

Fran ate one-half of a pizza. Tom ate three-fourths of another pizza that was the same size. They decided to combine the amount of pizza they had left over. What fraction of a pizza would they have if they did so?

To draw students' attention to the quantities in the problem, one could ask them to make drawings to show the relevant quantities of pizza. They might also use manipulatives such as fraction circle pieces to act out the manner in which the quantities of pizza change over time and are eventually combined. This sort of **quantitative analysis of a task** leads to much deeper learning than the traditional "key word" approach students are sometimes asked to follow (L. L. Clement & Bernhard, 2005). Notice, for instance, that students would not have a chance of being successful with the fraction problem above if they simply relied on a rule such as "The word *of* means multiply."

Imagine that you are a student who has not yet learned any of the rules for adding, subtracting, multiplying, or dividing fractions. However, you are familiar with how to represent fractions using diagrams and manipulatives. Write a solution to the Fran and Tom pizza task above from this perspective.	**STOP TO REFLECT**

As teachers use various strategies to facilitate problem solving, they need to be careful not to give away so much that the challenge of the task is taken away. Stein, Grover, and Henningsen (1996) described how teachers sometimes do this during problem-based lessons. They observed that as students press the teacher to provide solutions to problems, the teacher might show students a procedure or set of steps that will yield

Implementing the Common Core

See Homework Task 8 to reflect on how the challenge of a task is related to students' tendency to "make sense of problems and persevere in solving them" (Standard for Mathematical Practice 1).

the solution to the problem. For example, in regard to the fraction problem above involving pizzas, a teacher might reduce the challenge of the problem by telling students to find a common denominator for one-half and one-fourth and add the two fractions to obtain the solution. Notice that in doing this the teacher takes away students' opportunity to represent the quantities in the problem and think about the action in it. Although the teacher may do so with good intentions, it is important to note that what has happened, in essence, is that the lesson has reverted from "structured problem solving" to "learning rules and practicing procedures." Care must be taken to avoid this trap.

FIRST STEPS IN BEGINNING TO TEACH THROUGH PROBLEM SOLVING

Since teaching through problem solving is so central to standards-based teaching, subsequent chapters of this textbook will have much more to say about it. For now, the discussion of this topic will conclude by addressing two of the concerns teachers often have as they begin to think about teaching through problem solving. The first involves curricular resources, and the second involves time constraints.

Teachers often wonder where they will find the curricular resources to teach through problem solving. The answer to that question is influenced largely by one's school setting. Most schools have adopted textbooks that *claim* to support NCTM standards–based instruction or the *Common Core State Standards*. It is actually quite difficult to find a mathematics textbook on the market that does not make such a claim. However, many of these texts (with the notable exceptions of some of the curricula that are discussed in Chapters 4 and 5) are designed with the assumption that classrooms are set up to follow the traditional pattern of "learning rules and practicing procedures." In this case, teachers may have to modify exercises in the book to transform them into worthwhile mathematics tasks. It is also helpful to draw tasks from teachers' journals such as *Mathematics Teaching in the Middle School* and *Mathematics Teacher.* Lesson plans in Chapters 7 through 11 of this book model how one can use articles from these journals to construct lessons.

Another worry about teaching through problem solving is that it will take too much instructional time. This is an especially pressing concern in light of schools' desire to prepare students for high-stakes standardized tests. Keep in mind, however, the research evidence showing that students who experience a problem-centered curriculum generally perform as well as or better than students from traditional settings on tests of computation and higher-level thinking. It is usually more inefficient, in terms of enhancing students' achievement, to follow the pattern of learning rules and practicing procedures. If students who are used to simply learning rules and practicing procedures encounter a problem on a test for which they have forgotten the procedure, they tend to shut down, but those who have experienced teaching through problem solving have generally developed the disposition to reason through the problem even if it looks unfamiliar (Boaler, 1998). While carefully "covering" the curriculum in discrete chunks by having students learn rules and practice procedures may seem to be a commonsensical, safe way to teach, research does not support its effectiveness. Subsequent chapters of this book offer research-supported alternative approaches.

CONCLUSION

Several conclusions can be drawn from this chapter. First of all, it is an awesome responsibility to teach mathematics and a great challenge to teach it well. Anyone can teach mathematics, but it takes skill to teach it well. Good teachers seek out teaching strategies that will most effectively support their students' learning. This book introduces some of those strategies. However, it is ultimately the teacher's responsibility to study students' needs carefully to begin to understand how to apply and adapt those strategies to best facilitate learning. While some believe that the mathematics teacher's job is done once a set of notes has been transcribed onto an overhead projector or chalkboard, NCTM standards and the TIMSS studies point out that *effective* mathematics teaching is much more complex. The purpose of this textbook is to help teachers begin to understand and navigate some of those complexities, and also to help them develop the thinking skills to continue to do so throughout their teaching careers.

VOCABULARY LIST

After reading this chapter, you should be able to offer reasonable definitions for the following ideas (listed in their order of first occurrence) and describe their relevance to teaching mathematics:

National Council of Teachers of Mathematics   3

Curriculum and Evaluation Standards for School Mathematics   3

Professional Standards for Teaching Mathematics   4

Assessment Standards for School Mathematics   4

Principles and Standards for School Mathematics   4

NCTM principles   6

NCTM process standards   6

Common Core State Standards   10

Third International Mathematics and Science Study (TIMSS)   12

Teaching about problem solving   15

Teaching for problem solving   15

Teaching through problem solving   15

Worthwhile mathematical tasks   16

Quantitative analysis of a task   17

HOMEWORK TASKS

1. NCTM's *Curriculum and Evaluation Standards* called for more attention to "the use of real-world problems to motivate and apply theory" and less attention to "word problems by type, such as coin, digit, and work." Give an example of how word problems might be used for each of these purposes and then explain the difference. Draw on the *Curriculum and Evaluation Standards* as necessary in writing your response.

2. Search for the phrase "shaping the standards" on the www.nctm.org search engine (put the search phrase in quotes). Using your search results, describe how NCTM sought to have teachers contribute insights to the *Principles and Standards for School Mathematics* as they were written. Also describe

how teachers' input actually helped shape NCTM standards. Note: You will need access to NCTM's journals, either online or through your library, for this activity.

3. Visit the websites of two different sides of the "math wars": www.mathematicallycorrect.com and www.mathematicallysane.com. What is the stated purpose of each website? How well does the stated purpose align with the material posted? Provide specific examples to support your assertions.

4. Visit the TIMSS website at http://nces.ed.gov/timss/Educators.asp to see some of the items included on international assessments. Pick three items on which the United States scored poorly. For each item, form a conjecture as to why students may have scored poorly. Then, discuss how following specific aspects of the NCTM standards documents might help improve students' achievement on the problems you identified.

5. Visit the TIMSS websites http://nces.ed.gov/timss and http://timssvideo.com/ to view sample lesson video clips from the United States, Japan, and other countries. Which lessons seem to reflect the goal of "learning terms and practicing procedures"? Why? Which lessons seem to reflect the goal of "structured problem solving"? Why?

6. Find three sources that discuss the components of George Polya's problem solving process. Then describe a situation in which you solved a mathematics problem for one of your undergraduate mathematics classes. As you describe the situation, explain how the process you used to solve the problem did or did not align with Polya's ideas.

7. Find a traditional textbook or curriculum series that takes the approach of teaching *for* problem solving. Select one of the problems and rewrite it to align more closely with the guidelines for a "worthwhile mathematical task" from NCTM's (1989) *Professional Standards for Teaching Mathematics* (see pp. 25–32 in that document). Draw on the following article to stimulate your thinking, and in your write-up, explain how it helped you:

Kabiri, M. S., & Smith, N. L. (2003). Turning traditional textbook problems into open-ended problems. *Mathematics Teaching in the Middle School, 9,* 186–192.

8. Read the following article from the February 1998 issue of *Mathematics Teaching in the Middle School:*

Smith, M. S., & Stein, M. K. (1998). Selecting and creating mathematical tasks: From research to practice. *Mathematics Teaching in the Middle School, 3,* 344–350.

After reading the article, reflect on your own experiences in learning mathematics in Grades 6 through 12. Did your teachers ask you to function at high or low levels of cognitive demand? Explain.

CLINICAL TASK

Observe a full-length mathematics lesson. As you observe the lesson, put an X in the appropriate place in the following chart indicating how well the lesson is aligned with the given standard, then write an explanation justifying each rating. You should read through the description of each of the standards in Appendix A before completing this assignment. Revisit and refine your ratings and explanations as you read the rest of the chapters in this book.

	High Alignment	Some Alignment	No Alignment	Explanation
CCSS Standards for Mathematical Practice				
1. Make sense of problems and persevere in solving them				
2. Reason abstractly and quantitatively				
3. Construct viable arguments and critique the reasoning of others				
4. Model with mathematics				
5. Use appropriate tools strategically				
6. Attend to precision				
7. Look for and make use of structure				
8. Look for and express regularity in repeated reasoning				

Use the chart to write a paper that summarizes the degree of alignment observed between the lesson and each of the indicated standards. Submit your final, polished paper at the end of the course.

VIGNETTE ANALYSIS ACTIVITY **Focus on Making Sense of Problems and Persevering in Solving Them (CCSS Standard for Mathematical Practice 1)**

Items to Consider Before Reading the Vignette

Read CCSS Standard for Mathematical Practice 1 in Appendix A. Then respond to the following items:

1. Drawing on your own experiences as a mathematics student, provide specific examples of teaching practices that can help students make sense of problems and persevere in solving them.

2. Drawing on your own experiences as a mathematics student, provide specific examples of teaching practices that might prevent students from making sense of problems and persevering in solving them.

Scenario

Ms. Horton was just beginning her student teaching experience. Before assuming the role of lead classroom teacher, she observed that her mentor teacher, Mr. Sanchez, had a very good relationship with his students. Ms. Horton worried that her students might not be as accepting of her as Mr. Sanchez's students were of him and wondered how she could establish the same sort of productive relationship with her group. To help motivate students to work for her, Ms. Horton decided it would be a good idea to set the problems she posed in interesting real-world contexts. In the seventh-grade lesson described below, her goal was to help students understand the similarities and differences between circumference and diameter. She also wanted students to understand the number π as the ratio of circumference to diameter in any given circle.

The Lesson

As students entered the room, Ms. Horton had three expressions on the board that they were to simplify for a warm-up activity: (i) $7 + 3(6) \div 2$; (ii) $\frac{11 \times 8}{32 \div (5 \quad 1)}$; (iii) $\frac{2 + 3 \times 14}{17 \quad 11}$. She set a timer for five minutes and projected it on the document camera to show students how much time they had to work. Most students settled in and worked on the problems individually. Ms. Horton noticed that a few students were not done when the timer rang, so she reset it for another two minutes. The students who needed extra time kept working, while those who believed they were done began to socialize with their neighbors.

When the timer went off again, she called three students to the board to put their work up for everyone to see. As the three students put their work on the board, the others continued to fidget in their seats and talk with one another. When all three problems had been worked at the board, Ms. Horton directed everyone's attention up front. She stated that the first two were correct but that there was a problem with the third. The student assigned to post the work for the third item added $2 + 3$ to get 5 and then multiplied 5 by 14. Ms. Horton asked if students agreed that this was the correct way to simplify the problem. One student, Jamie, spoke up, saying that the multiplication in the numerator should have been done before the addition. Ms. Horton told Jamie she was correct, because multiplication comes before addition in order of operations.

The lesson then moved on to a discussion of the concepts of circumference and perimeter. Ms. Horton asked students to give examples of things that have circumference. Students quickly blurted out several examples, such as "coffee cup," "ring," "ball," "lightbulb," and "doorknob." Ms. Horton took this as enough evidence that students understood the idea of circumference and moved on to discuss how one would determine the perimeter of a rectangle. She explained that finding the perimeter of a rectangle was much like determining circumference because both involved calculating the distance around the outside of an object.

Following the discussion of circumference and perimeter, Ms. Horton announced that the students' first main problem for the day would be to determine the perimeter of a rectangular cookie using a ruler. She called on three students to distribute rulers and another three to distribute cookies. Materials were distributed efficiently and quickly. Students began to use the rulers to measure around the outside of the cookie. Most finished this task quickly and were allowed to eat their cookie. While eating, many of them talked with one another about the football playoff game the previous night. A group of students near the back of the room had difficulty reading the rulers and adding up the fractional portions of centimeters for each side length, so a teacher's assistant helped them finish the task.

Ms. Horton then moved on to what she considered to be the most challenging problem of the day. She held up a circular Oreo cookie and asked students if they could determine its perimeter. She then took out a box of Oreo cookies and had three students distribute them to the rest of the class. Students were told to use their rulers to try to determine the distance around the outside. Some became frustrated and said it was impossible to do the task using a ruler. The teacher's assistant told some of the students who became frustrated to wait a few minutes, because the teacher would be handing out string to help them measure around the outside. Other students began to roll the cookie along the ruler to measure it. Before anyone could formulate a response, Ms. Horton announced, "OK, you can see that it is really hard to do this with just a ruler. So, now I will give you a piece of string to use as well." After the string was distributed, Ms. Horton circulated about the room to direct some students to wrap the string around the cookie and then place the string on the ruler to measure the length used. Many students finished the task early and then resumed their football conversations. A few needed help measuring precisely, so Ms. Horton and the teacher's assistant helped them finish while the others waited.

After the class had finished measuring the circumference of the Oreos, Ms. Horton asked students to find other circular objects in the room and measure and record their circumferences and diameters. As students started to do so, Ms. Horton noticed that some of them were measuring the radius rather than the diameter of the circular objects. Others were measuring chords that did not pass through the center of the circle. On seeing this, Ms. Horton drew the attention of the class back to the front of the room and demonstrated how to measure the diameter of a circle.

When students had finished measuring the circumference and diameter of several circular objects, Ms. Horton once again called their attention up front. She asked students how many diameters fit into a circumference for each circular object. One student, Jessica, stated, "You need three diameters and a little bit more." Ms. Horton told Jessica she was correct, and said that the "little bit more" was the 0.14 in the 3.14 number they had been using for π.

To finish the lesson, Ms. Horton had students copy some notes from the document camera. The notes included formulas for determining the perimeter of a rectangle and the circumference of a circle. Then, Ms. Horton put a calculator on the screen and typed in $22 \div 7$. The calculator produced the decimal 3.14285714. She told students to notice that 3.14 was the number they were using for π in their circumference formula, and said it would be fine to use either 3.14 or $\frac{22}{7}$ for π when doing their homework. Students were told to write $\frac{22}{7} = \pi$ in their notebooks as a reminder.

Questions for Reflection and Discussion

1. Which aspects of the CCSS Standard for Mathematical Practice 1 did Ms. Horton's students seem to attain? Explain.

2. Which aspects of the CCSS Standard for Mathematical Practice 1 did Ms. Horton's students not seem to attain? Explain.

3. Comment on the overall value and relevance of Ms. Horton's warm-up activity to her objectives for the day.

4. Critique Ms. Horton's time management during the lesson. Identify instances of downtime and propose strategies for eliminating them.

5. How could Ms. Horton's problems for the day be enriched to offer extra challenges while still being accessible to all students?

6. There was a teacher's assistant in the classroom. How might this individual be used to effectively support the implementation of the lesson?

7. Did Ms. Horton provide appropriate support for students' problem solving activities during the lesson? Were there instances in which more support was necessary? Were there instances in which too much guidance was given? Explain.

8. Comment on the appropriateness and correctness of the statement $\frac{22}{7} = \pi$.

RESOURCES TO EXPLORE

Books

Burke, M. J., Hodgson, T., Kehle, P., & Resek, D. (2006). *Navigating through mathematical connections in Grades 9–12.* Reston, VA: NCTM.

Description: Connections can be made through a variety of different means in mathematics classes. This book provides classroom activities that build connections through mathematical models, unifying themes, multiple representations, and problem-solving processes.

Burke, M. J., Luebeck, J., Martin, T. S., McCrone, S. M., Piccolino, A. V., & Riley, K. J. (2008). *Navigating through reasoning and proof in Grades 9–12.* Reston, VA: NCTM.

Description: This book emphasizes exploration, conjecture, and justification as key to enacting the reasoning and proof process standard in high school classrooms. Reasoning and proof classroom activities related to each of the five NCTM content standards are included.

National Council of Teachers of Mathematics. (2000). *Principles and standards for school mathematics.* Reston, VA: NCTM. Available online at http://standards.nctm.org

Description: This document has been highly influential in framing debates about mathematics education and the scope of the preK–12 curriculum. Its influence can be seen across past and present state-level curriculum documents.

Pugalee, D. K., Arbaugh, F., Bay-Williams, J. M., Farrell, A., Mathews, S., & Royster, D. (2008). *Navigating through mathematical connections in Grades 6–8.* Reston, VA: NCTM.

Description: Two types of connections are emphasized in the NCTM connections process standard: connections within mathematics and connections between mathematics and other disciplines. This book contains classroom activities suitable for helping middle school students forge both types of connections.

Thompson, D. R., Battista, M. T., Mayberry, S., Yeatts, K. L., & Zawojewski, J. S. (2009). *Navigating through problem solving and reasoning in Grade 6.* Reston, VA: NCTM.

Description: This book contains examples of how teachers may enact NCTM's vision for the process standards of problem solving and reasoning in the middle grades. Specific activities involve understanding area formulas, working with scale factors, and reasoning about data.

Websites

Common Core State Standards: **http://www.corestandards.org/**

Description: Written by a consortium of 48 states, the *Common Core State Standards* represent the closest we have come to a consensus on the mathematics to be studied by students in the United States. This website contains the full text of the standards and news related to their implementation.

Official Website of TIMSS Public Use Videos: **http://timssvideo.com/**

Description: The site contains videos of mathematics lessons from seven different countries participating in the TIMSS video studies. The videos help illustrate some of the typical features of mathematics instruction in each country.

Teaching Math: A Video Library 5–8: **http://www.learner.org/resources/series33.html**

Description: These videos show middle school teachers implementing the recommendations in the NCTM standards in the classroom with their students. Lesson topics include fractions, statistics, geometry, measurement, and functions.

Teaching Math: A Video Library 9–12: **http://www.learner.org/resources/series34.html**

Description: These videos show high school teachers implementing the recommendations in the NCTM standards in the classroom with their students. Lesson topics include functions, mathematical modeling, linear programming, and probability.

Chapter 2

KEY PSYCHOLOGICAL IDEAS AND RESEARCH FINDINGS IN MATHEMATICS EDUCATION

The previous chapter began with the observation that a great deal of mathematics education research has been done over the past few decades. The TIMSS was then given as an example of a large-scale, international research study. The TIMSS, however, is just one of many studies that have helped shape the field. This chapter will offer a sense of other influential ideas and findings, focusing primarily on research that deals with the psychology of mathematics education—that is, theories and findings concerning the teaching and learning of mathematics. The goal is to give an overview of some of the most influential and useful ideas in mathematics education research without providing an exhaustive treatment. See Figure 2.1 for a map of the terrain to be explored below.

WHAT SHOULD TEACHERS EXPECT TO LEARN FROM RESEARCH?

Historically, research and practice have had a stormy relationship in the field of education. Here are just a few quotes that illustrate this point for mathematics education:

- "Within the mathematics education community, it [research] tends to be treated as a purely scientific discipline with no connection to social reality and the most urgent needs of teachers" (Malara & Zan, 2002, p. 554).
- "The manner in which research is reported in journals . . . , important as this style may be in convincing our peers that the research meets expected standards, is a turnoff for most teachers" (Sowder, 2000, pp. 3–4).
- "Research has had essentially no impact on the practice of mathematics education" (Steen, 1999, p. 240).

What, then, can teachers possibly hope to gain from studying ideas from research?

Figure 2.1 Map of ideas in Chapter 2.

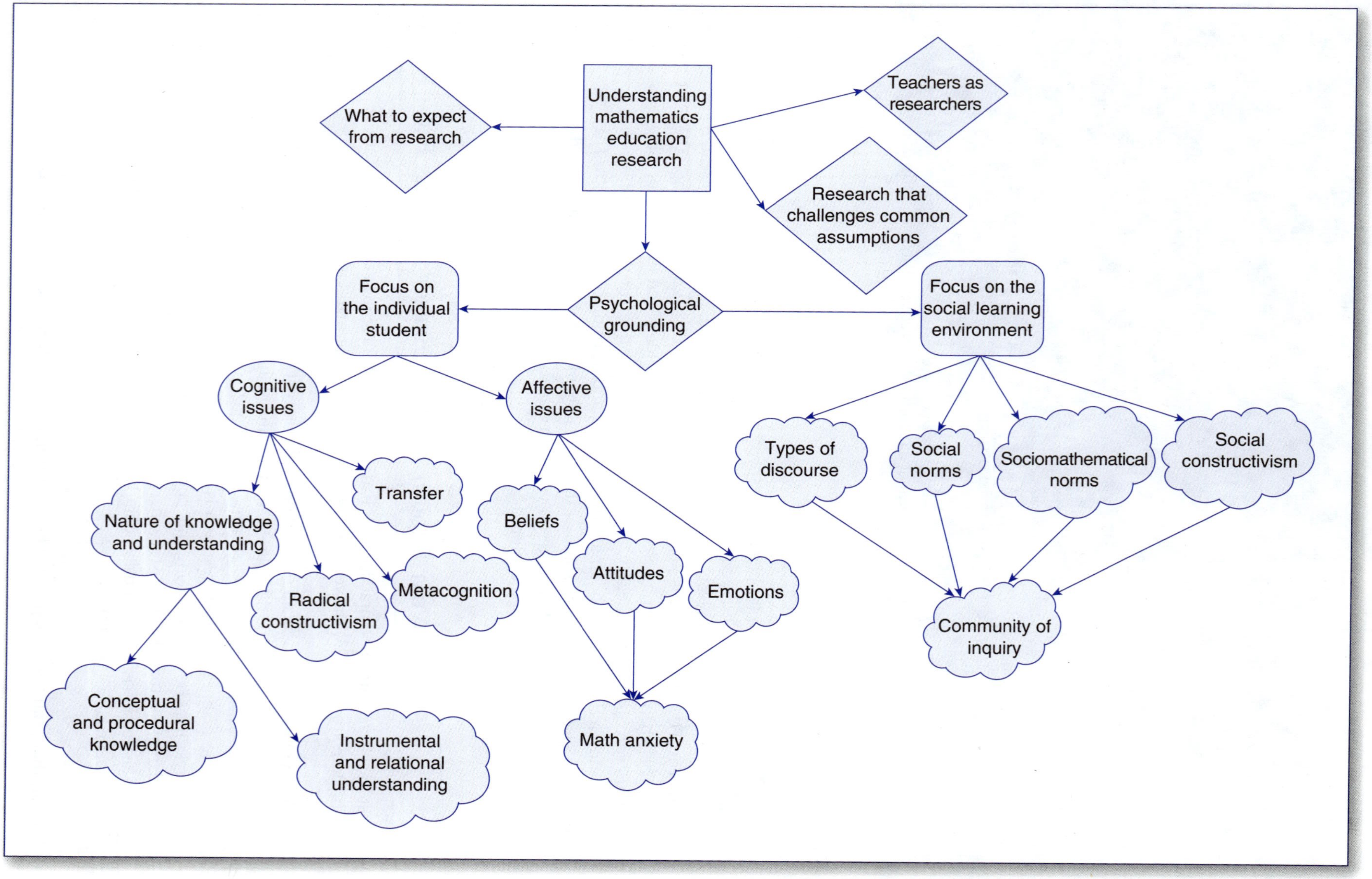

First of all, it is important to recognize that when we approach a research study or result, we all have our own preexisting notions about the purpose of research. Hammersley (2002) described three common ways of understanding the purpose of research. He called them the **engineering view**, the **strong enlightenment view**, and the **moderate enlightenment view**:

- The engineering view holds that research should provide "specific and immediately applicable technical solutions to problems, in the manner that natural science or engineering research is assumed to do" (p. 38).
- The strong enlightenment view "implies that policymakers and practitioners are normally in the dark, and that research is needed to provide the light necessary for them to see what they are doing, and/or what they ought to be doing" (p. 39).
- The moderate enlightenment view holds that "research is one among several sources of knowledge on which practice can draw. Moreover, the use made of it properly depends on practical judgments about what is appropriate and useful" (p. 42).

This chapter is based on the premise that the moderate enlightenment view is the most helpful way to view research, since it is the only one of the three that does not ascribe unrealistic, almost magical powers to research. At the same time, however, it acknowledges research findings to be one of several valuable sources of knowledge for teachers.

Hiebert (2000) provided further justification for a moderate enlightenment view of mathematics education research. In an article titled "What Can We Expect From Research," he made the following analogy to medical research:

> Health professionals have promoted the goal of a healthful life for years and have conducted a great deal of research but are still unable to specify the best way of meeting that goal. Exactly how much exercise do we need? Are seven servings of fruits and vegetables each day required, or would five be enough? What is your optimal weight? Our bodies are too complicated to specify the best path to a healthful life. The same is true in mathematics education. Teachers cannot expect to get clear and specific answers from research for exactly which textbooks or activities to use. (p. 168)

The moral of the story is that in *any* profession, one must deal with a degree of uncertainty when making important decisions. That does not mean, however, that we should just throw up our hands and disregard research that can *help* inform our decision-making. The key is to critically examine the research evidence in light of your own particular classroom situation.

THE INFLUENCE OF PIAGET AND VYGOTSKY

Jean Piaget and Lev Vygotsky are arguably the two strongest influences on the way we currently study the psychology of teaching and learning mathematics. Piaget (1896–1980) was a Swiss psychologist who studied individuals' learning and development.

Vygotsky (1896–1934) was a Russian psychologist. Woolfolk (1993) characterized the differences between the work of Piaget and Vygotsky in the following terms:

> Whereas Piaget described the child as a little scientist, constructing an understanding of the world largely alone, Vygotsky . . . suggested that cognitive development depends much more on the people in the child's world. Children's knowledge, ideas, attitudes, and values develop through interactions with others. Vygotsky also believed that language plays a very important role in cognitive development. (p. 46)

Although there is much more that could be said about the similarities and differences between the work of the two psychologists (see Jean Piaget Society, 2006, and Vygotsky Centennial Project, 2006, for helpful lists of sources), the key distinction to be explored here is that Piaget's work emphasized individual psychological processes while Vygotsky's emphasized social and cultural processes.

Piaget's and Vygotsky's differing foci have influenced mathematics education researchers. Cobb (1994) observed that two trends have emerged in mathematics education research:

> The first is the generally accepted view that students actively construct their mathematical ways of knowing as they strive to be effective by restoring coherence to the worlds of their personal experience . . . a second trend . . . emphasizes the socially and culturally situated nature of mathematical activity. (p. 13)

The first trend is very much in the tradition of Piaget's work, while the second can be linked to Vygotsky's. There is longstanding controversy over which of the two views is more reasonable, since each implicitly posits different origins of knowledge. Rather than entering that controversy, this chapter will adopt Cobb's position that "both these perspectives are of value in the current era of educational reform that stresses both students' meaningful mathematical learning and the restructuring of the school while simultaneously taking issues of diversity seriously" (p. 18). This assertion opens the door to explore psychological ideas that foreground the individual as well as those that foreground the social environment.

PSYCHOLOGICAL IDEAS FOREGROUNDING THE INDIVIDUAL

When focusing on the individual student or teacher, two domains that can be identified are **cognition** and **affect**. While the definitions and the precise boundaries of each domain have been debated by researchers, the following two definitions will be useful for beginning the conversation about them.

- Cognition: "The various ways in which people think about what they are seeing, hearing, studying, and learning" (Ormrod, 2006, p. 20).
- Affect: "A wide range of feelings and moods that are generally regarded as something different from pure cognition" (McLeod, 1989, p. 245). McLeod included beliefs, attitudes, and emotions in this category.

Given the preceding definitions, ideas that fit into both categories may come to mind. For example, one's beliefs about mathematics might fall, at least partially, into both categories. The primary purpose of this section is to draw your attention to some important ideas related to each category rather than to try to formulate precise boundaries between the two (see McLeod, 1992, for a more comprehensive treatment of this issue). Although, for the sake of convenience, the following discussion is divided into two sections, one on cognition and another on affect, note that the two domains actually have a degree of overlap.

The Cognitive Domain

Conceptual and Procedural Knowledge

In mathematics, cognitive processes can be focused on developing two different types of knowledge: **conceptual knowledge** and **procedural knowledge**. Hiebert and Lefevre (1986) offered the following definitions:

- Conceptual knowledge is characterized most clearly as knowledge that is rich in relationships. It can be thought of as a connected web of knowledge, a network in which the linking relationships are as prominent as the discrete pieces of information. Relationships pervade the individual facts and propositions so that all pieces of information are linked to some network. (p. 3)
- Procedural knowledge . . . is made up of two distinct parts. One part is composed of the formal language, or symbol representation system, of mathematics. The other part consists of the algorithms, or rules, for completing mathematical tasks. (p. 5)

These ideas may bring to mind some of the points from the first chapter. For instance, the standards developed by the National Council of Teachers of Mathematics (NCTM) assert that students need to recognize connections among mathematical ideas and how mathematics connects to other disciplines. This can be seen as a call for helping students develop conceptual knowledge. The NCTM standards state that the development of procedural knowledge is also important, but caution teachers against overemphasizing procedural aspects of mathematics and neglecting the conceptual. The typical U.S. lesson pattern of "learning rules and practicing procedures" documented by the TIMSS illustrates the danger of focusing almost exclusively on the development of procedural knowledge.

> **Implementing the Common Core**
>
> See Clinical Task 1 to examine the extent to which textbooks help students "reason abstractly and quantitatively" (Standard for Mathematical Practice 2) by supporting the development of both conceptual and procedural knowledge.

> Reflect on your past experiences as a mathematics student. Give two examples of mathematics content for which you developed procedural understanding. Give two examples of mathematics content for which you developed conceptual understanding. Explain why your examples fit each category.
>
> **STOP TO REFLECT**

Instrumental and Relational Understanding

The types of knowledge teachers focus on helping their students develop influence the understanding those students eventually attain. Again, there is a distinction useful to explore in this regard: that of **instrumental understanding** versus **relational understanding**. Skemp (1976) associated instrumental understanding with "learning rules without reasons" and relational understanding with "knowing both what to do and why." He drew an analogy between learning one's way around an unfamiliar town and learning mathematics. If someone is given a fixed set of directions from one location to the next, he or she will be able to commute between the two points. This is similar to a mathematics class in which the teacher gives a student a procedure or fixed set of steps to follow. The problem is that travelers who take a wrong turn and students who do a step incorrectly are hopelessly lost because they have developed only instrumental understanding. In contrast, if the traveler moves about the town to discover how various paths are connected, or if the student is encouraged and helped to form a mental map of the connections among the mathematical topics under study, relational understanding can be attained. With relational understanding, both the traveler and the student can construct multiple pathways to reach their respective goals and will not be lost after making a wrong turn or error.

> **Implementing the Common Core**
>
> See Homework Task 1 to reflect on the extent to which your past mathematics classes have helped you "reason abstractly and quantitatively" (Standard for Mathematical Practice 2) to attain relational understanding.

An example from algebra may help further clarify the distinction between instrumental and relational understanding. Algebra teachers frequently make use of the mnemonic FOIL (first, outside, inside, last) to teach multiplication of binomials. For example, when multiplying $(3x + 1)(x + 5)$ with this procedure, one would

1. multiply $(3x)(x)$ to obtain $3x^2$ (**F**irst),

2. multiply $(3x)(5)$ to obtain $15x$ (**O**utside),

3. multiply $(1)(x)$ to obtain x (**I**nside),

4. multiply $(1)(5)$ to obtain 5 (**L**ast), and

5. add the resulting terms to obtain the answer: $3x^2 + 16x + 5$.

Undoubtedly, FOIL is a procedure that will work to produce the product of any two binomials. The problem is that heavily emphasizing FOIL fosters mainly instrumental rather than relational understanding. Many students who learn the procedure do not understand why it works. They also often do not understand that it applies only to the multiplication of binomials. As a result, many attempt to use the technique when multiplying a trinomial by a binomial, or in some other situation in which it is not applicable. FOIL, then, often limits students' ability and inclination to form a coherent mental map of the larger domain of polynomial multiplication. Alternative approaches to polynomial multiplication that are designed to foster relational understanding will be discussed in Chapter 8.

> **Implementing the Common Core**
>
> See Clinical Task 2 to explore how a teacher's choice of teaching strategies for binomial multiplication influences students' ability to "reason abstractly and quantitatively" (Standard for Mathematical Practice 2).

Constructivism

Given that relational understanding is an important goal, the question of how one attains such understanding arises. One highly influential theory to explain how this learning process occurs is called *constructivism*. To understand the fundamental premises of constructivism, it can help to examine the work of one of its most noted theorists, Ernst von Glasersfeld, and how his theory might apply to the learning of algebra.

Von Glasersfeld (1990) endorsed a version of constructivism called **radical constructivism**, which is a theory of cognition that emphasizes the role of individual construction of knowledge in the learning process. One of the fundamental tenets of radical constructivism is that "knowledge is not passively received either through the senses or by way of communication. Knowledge is actively built up by the cognizing subject" (p. 22). An immediate implication is that it is not reasonable to think that copying notes from a teacher's presentation or a book is a sufficient condition for students to learn algebra. For that matter, one also should not expect that a student will automatically learn if simply placed in a cooperative group with others. At some point, students must perform the cognitive operations necessary to form their own meanings for the concepts, structures, and symbols that are part of school algebra. Ultimately, the individual, or "cognizing subject," must carry the load of constructing his or her own algebraic knowledge.

A second fundamental tenet of radical constructivism is that "the function of cognition is adaptive, in the biological sense of the term, tending towards fit or viability" (von Glasersfeld, 1990, p. 23). This tenet is closely related to Piaget's ideas of assimilation and accommodation (von Glasersfeld, 1990). Assimilation occurs when one's existing cognitive structures are used to make sense of experience, and accommodation occurs when one must reorganize existing cognitive structures to make sense of experience (Piaget, 1983). The FOIL example can illustrate the ideas of assimilation and accommodation. Suppose a given student's cognitive structure for dealing with polynomial multiplication consists only of the FOIL idea. When she encounters a problem involving the multiplication of a binomial by a trinomial, she may either (1) reorganize her existing cognitive structures to recognize FOIL as a technique covering a special case of polynomial multiplication (i.e., multiplication of two binomials) and not as the only way to multiply polynomials or (2) keep in place her existing cognitive structure that regards FOIL as the only way to multiply polynomials, and attempt to apply it to the problem. The first scenario illustrates accommodation and the second illustrates assimilation. In this particular situation, accommodation is needed for a successful learning outcome.

This process extends to all aspects of learning algebra in that students must often reorganize existing cognitive structures to build viable strategies for solving problems. This means that students will often necessarily experience a degree of discomfort as they are learning, which Piaget referred to as **disequilibrium**.

> **Implementing the Common Core**
>
> See Homework Tasks 2 and 4 to explore the role assimilation and accommodation play in developing the ability to "reason abstractly and quantitatively" (Standard for Mathematical Practice 2).

STOP TO REFLECT

React to the statement "Students will often necessarily experience a degree of discomfort as they are learning." What kind of "discomfort" is implied in this statement? If this statement is true, what does it mean for the often-stated goal of making mathematics "fun" for students?

A third fundamental tenet of radical constructivism is that "cognition serves the subject's organization of the experiential world, not the discovery of an objective ontological reality" (von Glasersfeld, 1990, p. 23). Von Glasersfeld contends that radical constructivism can give insight about how students learn without making any claims of an objective reality. This marks a sharp departure from philosophical schools such as Platonism, which posit the existence of an objective reality as a central tenet of their worldviews. Platonists would posit that a student who learns the solution to an equation has discovered an objectively existing truth about the world. Radical constructivists, on the other hand, would seek to avoid the issue of whether or not the student has discovered something about an objectively existing reality. They would state that if such an objective reality exists, it is beyond the comprehension of the individual, and hence it is impossible to make any kind of claim about it. This is perhaps the most subtle and philosophical of the three tenets of radical constructivism, yet it underlies a fundamental assumption the theory makes (or actually does not make) about the nature of mathematical objects. See Benacerraf and Putnam (1983) for similarly interesting issues from the philosophy of mathematics.

During the 1990s and into the 2000s, *constructivism* became somewhat of a buzzword in the world of education. The danger of forming practice around buzzwords is that the original intent and message of the theory is usually lost or misunderstood. In many circles, for instance, constructivism became synonymous with "inquiry-based teaching." The problem with uncritically linking the two is that constructivism, as described above, was originally meant to be a theory of learning rather than a theory of teaching (Simon, 1995). Although a teacher holding a constructivist view of learning may have a solid basis for justifying the use of inquiry-oriented methods, constructivism itself doesn't directly prescribe a teaching method.

STOP TO REFLECT

If one accepts the theory of radical constructivism as a good explanation of how students learn, what are the implications for teaching (i.e., are there specific strategies you would expect a teacher embracing this theory to use or avoid)? Explain.

Transfer

One of the goals of any teaching method should be teaching for **transfer**. That is, students should able to "apply what was learned in new situations and to learn related information more quickly" (Bransford, Brown, & Cocking, 1999, p. 17).

There are many cases when one might assume transfer occurs, but it does not. Carraher, Carraher, and Schliemann (1985) studied the computational ability of Brazilian schoolchildren who worked as street market vendors. They found that children were much more successful with computations in the context of selling products than they were with parallel, decontextualized computations in school. For example, one child was asked, in the context of a street vending job, how much three coconuts would cost when the price of each was 40 units of currency. The child responded by skip counting by 40s and obtaining 120. Later, when asked to solve the item 40×3 on a paper-and-pencil test in school, the same child claimed the answer was 70, since she attempted (unsuccessfully) to recall and apply a school-taught method for solving the problem. The other students in the study exhibited similar thinking patterns. The same phenomenon can be observed among people in various occupations. For instance, carpenters often do not transfer the formal geometry learned in school to their jobs, but nonetheless deal successfully with the construction of right angles and parallel lines, many times through intuitive understandings they have built on their own or have learned as "tricks of the trade" from others. One of the goals of the NCTM standards–based curricula funded by the National Science Foundation (NSF) was to facilitate transfer between in-school and out-of-school situations by teaching through problem solving with problems set in real-world contexts (these curricula will be discussed in Chapters 4 and 5).

Metacognition

Because of recent curricular emphases on teaching through problem solving, another psychological idea focused primarily on individuals' learning and development is worth considering: **metacognition**. A useful description of metacognition is that it is individuals' "awareness of their own cognitive machinery and how the machinery works" (Meichenbaum, Burland, Gruson, & Cameron, 1985, p. 5). Metacognition is vital to success in mathematical problem solving. Schoenfeld (1983) illustrated the importance of metacognition in problem solving by comparing the geometric problem solving behavior of two students to that of a mathematician. When the students were asked to determine the characteristics of the largest triangle that could be inscribed in a circle, they set about trying to calculate the area of an inscribed equilateral triangle, but could not explain how finding the area would even help to solve the problem. In contrast, when the mathematician was given an unfamiliar geometry problem to solve, he rejected unfruitful solution strategies on his way to solving the problem. The difference is that the mathematician was able to step back and critically examine his own thinking, while the students plunged straight ahead into a calculation without knowing why they were doing it. Since monitoring one's thinking is a key to solving problems successfully, it is important to teach with an eye toward helping develop students' metacognitive ability. Asking students to write about their mathematical thinking is one teaching strategy that can help address this concern, and it will be discussed in detail in Chapter 3.

> ### Implementing the Common Core
>
> See Clinical Task 3 to explore the role that metacognition plays in supporting students' ability to "make sense of problems and persevere in solving them" (Standard for Mathematical Practice 1).

The Affective Domain

Beliefs

Earlier in the chapter, it was mentioned that individuals' **beliefs** may be considered to fit the definitions given for both cognition and affect. McLeod (1992) discussed beliefs as an affective issue, but also noted that "beliefs are largely cognitive in nature, and are developed over a relatively long period of time" (p. 579). He identified several types of beliefs relevant to the learning of mathematics, including beliefs about mathematics, beliefs about self, and beliefs about mathematics teaching. Students' beliefs in all of these areas are important to understand, because they influence the manner in which students function in class. A student who believes that mathematics is essentially a body of rules that need to be memorized will be unlikely to look for connections among mathematical ideas. Students who do not believe they are "good at math" will tend to shut down when problems become difficult (unfortunately, this belief is sometimes fostered by parents who mistakenly think mathematical inability is an inherited trait). Those who believe that mathematics teaching should consist exclusively of the teacher standing at the board and writing notes will have a harder time engaging in the "structured problem solving" lesson pattern. As teachers tune in to their students' beliefs, they can begin to understand the obstacles and inroads to learning within any particular class.

Attitudes and Emotions

Attitudes and emotions are less cognitive in nature than beliefs. **Attitudes** can be thought of as "the affective responses that involve positive or negative feelings that are relatively stable. Liking geometry, disliking story problems, being curious about topology, and being bored by algebra are all examples of attitudes" (McLeod & Ortega, 1993, p. 29). Interestingly, there is not always a positive relationship between attitudes and achievement. For instance, McKnight and colleagues (1987) reported that Japanese students disliked mathematics more than students in countries that achieved at lower levels. **Emotions** "may involve little cognitive appraisal, and may appear and disappear rather quickly, as when the frustration of trying to solve a hard problem is followed by the joy of finding a solution" (McLeod, 1992, p. 579). Most of us can probably recall experiencing both positive and negative emotions in mathematics classes. Unfortunately, many can recall being made to feel foolish for not knowing the answer to a problem. On the more positive side, many may also recall experiencing an "Aha!" moment when the correct solution strategy for a problem became clear (Mason, Burton, & Stacey, 1982). Although less stable than attitudes and beliefs, emotions nonetheless can have a powerful influence on students' desire to engage in learning mathematics.

> **Implementing the Common Core**
>
> See Clinical Task 4 to analyze the beliefs, attitudes, and emotions students exhibit as they "construct viable arguments and critique the reasoning of others" (Standard for Mathematical Practice 3).

Mathematics Anxiety

No introductory discussion of the affective domain in mathematics education would be complete without mentioning **mathematics anxiety**. This is perhaps the most frequently studied issue in the affective domain related to mathematics learning.

Like many psychological constructs, it has been defined in various ways. One definition that captures much of its essence is, "the panic, helplessness, paralysis, and mental disorganization that arises among some people when they are required to solve a mathematical problem" (Tobias & Weissbrod, 1980). Individuals who experience it are often erroneously labeled by teachers and peers as being incapable of doing mathematics (Tobias, 1993). In reality, environmental factors rather than innate ability are often at the root of mathematics anxiety. Parents, for example, can cause it by having low expectations, being verbally abusive ("You're not smart enough to do well in math"), or even being physically abusive when children do not do well in school (Fiore, 1999). Teachers can also contribute to mathematics anxiety through angry behavior, unrealistic expectations, embarrassing students in front of peers, gender bias, and uncaring attitudes (Jackson & Leffingwell, 1999). While displaying respect for all students can diminish some teacher-caused mathematics anxiety, it is important to note that anxiety can also be caused by emphasizing rules in the absence of understanding (Norwood, 1994). Therefore, to keep from contributing to mathematics anxiety, teachers need to be conscious of avoiding instrumental learning (Skemp, 1976) while respecting all students.

PSYCHOLOGICAL IDEAS FOREGROUNDING THE SOCIAL ENVIRONMENT

While many research studies that investigate individuals' characteristics have been carried out in the tradition of Piaget, many of those that foreground the social environment reflect the influence of Vygotsky. One particularly influential idea stemming from Vygotsky's work is that public discourse is a key mechanism in learning. This influence can be seen throughout the NCTM standards documents. *Professional Standards for Teaching Mathematics* included special sections to describe the teacher's and student's roles in discourse. Like many other recommendations in the standards documents, these recommendations cut against the grain of traditional teaching practices. This section describes some of the elements inherent in the social environment and discourse in a standards-based classroom.

Univocal and Dialogic Discourse

A distinction that can be helpful for understanding the different types of discourse that occur in mathematics classrooms was offered by Knuth and Perressini (2001). They spoke of a continuum with **univocal discourse** on one end and **dialogic discourse** on the other. Discourse on the univocal side of the continuum generally involves teachers trying to guide students to solve a given problem in a prescribed way. The teacher places value not on trying to understand students' alternative solution strategies, but only on making sure that they received the "correct" message about the manner in which the problem should be solved. On the other hand, teachers who engage students in dialogic discourse work to understand and build on students' ways of thinking and solving problems. Dialogic discourse provides the opportunity to emphasize that the most important thing in solving a problem is to find a reasonable solution, not to try to imitate the teacher's solution. Although episodes of univocal discourse occur and are even necessary in some situations, dialogic discourse is a worthwhile goal. The give-and-take that

> **Implementing the Common Core**
>
> See Homework Task 5 to think about the roles of univocal and dialogic discourse in helping students "construct viable arguments and critique the reasoning of others" (Standard for Mathematical Practice 3).

occurs during this sort of discourse can help students take ownership of the solutions they construct for problems and can help the teacher better understand the ways in which students think.

Social Norms

In every classroom, regardless of the type of discourse that is most prevalent, interactions between students and teachers result in the negotiation of **social norms**. Cobb and Yackel (1996) described social norms as the delineation of the participation structure for classroom discourse. Norms exist for things such as justifying solutions, making sense of solution strategies offered by others in class, and indicating agreement or disagreement with a given solution. Cobb and Yackel proposed a reflexive relationship between social norms in class and students' beliefs relevant to learning mathematics—that is, students' (and teachers') beliefs help shape social norms, and the emerging norms in turn help shape students' beliefs. This being the case, teachers sometimes have to work to renegotiate social norms with students who are used to traditional classroom instruction in mathematics (characterized earlier by "learning rules and practicing procedures" and heavy emphasis on univocal discourse). While the teacher is an important figure in the classroom, he or she cannot unilaterally impose social norms on a group of students. This does not mean that a teacher is powerless to make changes to the social norms, but it points out the importance of engaging students in dialogic discourse that can help to both reveal and shape their beliefs about teaching and learning mathematics.

Sociomathematical Norms

Closely related to social norms is the concept of **sociomathematical norms**. While social norms, such as those for indicating agreement or disagreement with another student's thinking, may occur in a class for any subject, sociomathematical norms are specific to the subject of mathematics. Cobb and Yackel (1996) mentioned several types of sociomathematical norms, including what counts as a different mathematical solution, an efficient solution, or an acceptable explanation. Sociomathematical norms in each area are developed as students communicate with one another and with the teacher about the mathematics they are studying. As with social norms, teachers do not exercise complete control over the sociomathematical norms that are established. Teachers, especially those new to reform-oriented instruction, will find that their own beliefs about acceptable sociomathematical norms are shaped as they engage in dialogic discourse with students. This is yet another factor that makes the task of reform-oriented teaching both exciting and daunting.

Social Constructivism

Before closing the discussion of psychological ideas foregrounding the social environment, it should be noted that although *radical* constructivism was discussed earlier alongside theories foregrounding individual cognition, **social constructivism** foregrounds social interactions. Ernest (1996) noted that "Social constructivism regards individual subjects and the realm of the social as indissolubly connected. Human subjects are formed through their interactions with each other as well as with their individual processes" (p. 342). Social constructivism posits discourse as fundamental to students' development. It follows that the mathematical understandings they construct are shaped by the mathematical conversations they

> **Implementing the Common Core**
>
> See Clinical Task 5 to analyze how well the types of discourse and norms existing in a classroom support students' ability to "construct viable arguments and critique the reasoning of others" (Standard for Mathematical Practice 3).

are encouraged to have with others. If those conversations always involve a teacher or textbook dispensing information, students are likely to believe that mathematics has been constructed by experts and that it is simply their job to try to understand it. On the other hand, if frequent opportunities are given to discuss and debate various approaches to problems, students can see participation in a larger community as vital to the process of doing mathematics. In such classrooms, mathematics is portrayed as a dynamic, living discipline rather than a predetermined set of facts and procedures to be memorized.

The Zone of Proximal Development

To achieve the discourse goals described in this chapter, it is necessary for teachers to work to create what Goos (2004) called a "culture of inquiry." Vygotsky's notion of the **zone of proximal development** (ZPD) can be used to characterize the interactions that occur in such a classroom. The ZPD can be thought of as "a symbolic space created through the interaction of learners with more knowledgeable others and the culture that precedes them" (Goos, 2004, p. 262). In this conceptualization, the "more knowledgeable other" may be the teacher or students' peers. Goos observed that when teachers take on this role, they can **scaffold** students' thinking by conveying expectations at the beginning of the school year and then gradually withdrawing support during problem solving activities. When one teacher Goos studied withdrew his support, his students provided some of the scaffolding for their peers that he had previously provided. The teacher moved students in this direction by asking them to request help from peers and come to him only as a last resort. While some students resisted the shift toward an inquiry-based classroom, Goos's example shows how Vygotsky's ZPD characterizes the teacher's role in making the shift toward reform-oriented classroom discourse. Although teachers cannot have complete control over all aspects of the shift, they can exert a substantial influence by encouraging authentic mathematical activity.

RESEARCH THAT CHALLENGES COMMON ASSUMPTIONS ABOUT MATHEMATICS EDUCATION

When we examine public and professional discourse about mathematics education, a number of common assumptions become apparent. Often these assumptions are based on beliefs that are actually contradicted by empirical research. Examples of assumptions that are questionable, given what we know from research, include (1) procedures should be taught before students do applied mathematics, (2) teaching for relational understanding is more time-consuming than teaching for instrumental understanding, (3) students learn best when tracked into same-ability classes, and (4) extrinsic rewards help support the learning process. Examples of research studies that challenge each of these assumptions are given in the following sections.

Assumption 1: Procedures Must Be Taught Before Students Do Applied Mathematics

The assumption that procedures must be taught before applications is implicit in the instructional expositions given in many textbooks and mathematics lessons. It is common to see textbook pages with many skills exercises followed by a few word problems. In such instances, the word problems at the end usually require little to no original thought, since they require merely applying the skills practiced in the exercise set. This sequence aligns with the pattern of "learning rules and practicing procedures" that is prevalent in U.S. mathematics lessons. Given the ubiquitous nature of this instructional pattern, teachers may not even consider whether such a pattern is optimal for fostering students' learning.

Boaler (2008) presented strong evidence from her research to challenge the assumption that procedures should be taught before concepts. She described the results of in-depth case studies of two high schools using different approaches to teaching mathematics. One of the schools, Amber Hill, took a traditional approach to mathematics instruction, in that students learned procedures from the teacher and then applied them to problems. When story or word problems were presented, they generally required application of the procedures being practiced in the current lesson. Thus, when confronted with a word problem in a lesson, students found the appropriate solution strategy fairly obvious. Another school, Phoenix Park, took a very different approach. In mathematics classes, students generally completed projects requiring mathematics. The mathematics required for the projects was not usually explicitly taught before students began. Instead, students learned mathematics skills and concepts as needed. For example, in one project, students were instructed to arrange 36 sections of fence for a farmer to produce the largest possible enclosed area. Rather than teaching trigonometric ideas that could be helpful for the project at the outset, teachers waited for the need for these ideas to arise in students' work before discussing them. Although Phoenix Park and Amber Hill were very similar in terms of student and teacher demographics, the instructional approaches used in their mathematics classes differed sharply.

The sharp instructional differences between Phoenix Park and Amber Hill fostered the development of strikingly different student beliefs about mathematics. When Phoenix Park students were interviewed on their beliefs about mathematics, they described mathematical methods as useful problem solving tools. This belief coincides with the manner in which mathematical methods were portrayed by their teachers. Amber Hill students, on the other hand, had very different views. They tended to believe that mathematical methods were just rules to memorize and had no relevance to the world outside of school. Again, such beliefs coincided with the way mathematics had been portrayed in the classes they experienced. Given these differing beliefs about the subject, students at Amber Hill were much less enthusiastic about doing mathematics than were students at Phoenix Park. When students moved into the workplace, Phoenix Park students saw the relevance of high school mathematics to their work, whereas Amber Hill students still saw mathematics as something with no application outside of school.

Along with differences in beliefs about mathematics, students at Phoenix Park and Amber Hill exhibited different levels of achievement. Phoenix Park students outscored Amber Hill students on a standardized end-of-year exam taken at both schools. Although Amber Hill's instruction was carefully arranged to cover all topics that would be on the exam, the exam contained no built-in cues about which procedures should be used to solve problems. Even when students knew which procedure to use for a given item, they would sometimes carry it out incorrectly, relying purely on memorization of its steps. Phoenix Park students approached exam questions much differently. Even though they had not encountered all of the procedures necessary for solving problems during the school year, their classes had portrayed mathematics as a sense-making discipline. Therefore, when confronted with problems where the solution method was not immediately known, Phoenix Park students attempted to understand the problem based on what they knew of mathematics and the situation at hand.

When Phoenix Park and Amber Hill students were given problems requiring the application of mathematics, the difference between the two groups was even more striking. Phoenix Park students were much better equipped to use mathematics in problem situations arising outside the bounds of school, such as using a scale diagram of a house to make decisions about appropriate dimensions and design.

Assumption 2: Teaching for Relational Understanding Is Too Time-Consuming

Even if one acknowledges that developing students' relational understanding of mathematics is a worthy goal, a common worry is that teaching for relational understanding is much more time-consuming than teaching instrumentally. Boaler's (2008) study of Phoenix Park and Amber Hill provides some evidence that this worry is not well founded, since students from the school taking a relational approach outperformed those from the school taking an instrumental approach. Amber Hill students had been taught very instrumentally, rehearsing all of the procedures that would be necessary to solve problems on the end-of-year examination. However, when the time for the examination arrived, they were often unable to understand how the procedures they learned were relevant to the problems at hand. The case of Amber Hill suggests that simply "covering" all of the material students need to know at the conclusion of a course is not as safe a strategy as it may seem. Phoenix Park students spent no more time in the classroom than did their counterparts at Amber Hill, yet the relational approach taken by their mathematics teachers made them more effective problem solvers at the conclusion of the school year.

Further evidence that teaching for relational understanding is a wiser use of instructional time than teaching for instrumental understanding comes from a study done by Pesek and Kirshner (2000). They compared the performance of a group of students receiving only relational instruction (R-O) to a group receiving instrumental instruction followed by relational instruction (I-R). A key feature of the study was that students in the R-O group received only three days' worth of instruction, while those in the I-R group received five days' worth. Instruction for each group focused on area and perimeter of squares, rectangles, triangles, and parallelograms. Lessons for the I-R group included extensive review and practice of formulas for each of the geometric objects being studied. Relational lessons, which were experienced by both groups, emphasized helping students construct their own formulas for area and perimeter of objects. In relational instruction episodes, students were given problem situations requiring the use of area and perimeter, but the teacher did not specify strategies to be used to solve them. Students were to use intuitive understandings of the concept of area to arrive at solutions.

When students were given a posttest on the content of the lessons, those in the R-O group actually outperformed those in the I-R group on an achievement test, despite the fact that I-R students were given more instructional time. Student interviews indicated that the instrumental instruction the I-R group experienced actually interfered with their ability to understand the content relationally. Pesek and Kirshner (2000) identified three main types of interference: cognitive, attitudinal, and metacognitive. Students from the I-R group exhibited cognitive interference in the sense that for some applied problems (e.g., painting a wall), they did not know whether it would be appropriate to calculate area or perimeter. Because of the instructional emphasis on formulas for the I-R group, some thought that area measured length and thickness. Attitudinal interference occurred among students who preferred the instrumental mode of instruction to the relational. Such students were more receptive to instrumental instruction, and felt they learned more from it, because it more closely resembled the manner in which they were normally taught mathematics. The perception that they learned more from instrumental instruction was, of course, contradicted by their test data. Metacognitive interference was observed as students were asked to think aloud in solving problems. When I-R students explained their solution strategies, most of them applied operations such as addition and multiplication without understanding why they were doing so. When I-R students were asked to explain why formulas like $P = 2(1 + w)$ and $A = \frac{1}{2}bh$ are valid, they could not. In contrast, R-O students could explain the meanings of the formulas, even though they had not been formally introduced during lessons.

Pesek and Kirshner's (2000) research suggests that relational instruction is actually much more time efficient than instrumental instruction because of the barriers to learning and retention that occur when students are conditioned to learn rules without reasons. It is simply easier to learn and retain mathematics when it is understood relationally from the outset. The idea that instrumental instruction interferes with relational learning is supported by other studies as well. Mack (1990) found that it was very difficult to help sixth graders understand fractions relationally because they persisted in wanting to apply previously learned rote procedural knowledge to problems rather than bringing more intuitive, informal knowledge to bear. Wearne and Hiebert (1988) found that students who experienced instrumental instruction in decimals performed worse than those with no previous instruction. Kieran (1984) found that seventh graders who experienced instrumental instruction about solving equations were resistant to teaching methods that involved building on intuitive, informal knowledge. Those who did not resist the teaching methods were eventually more successful in solving equations than those who did. The overall picture painted by research is that instrumental instruction, though it may seem to be a quick, efficient way to cover material, actually prevents students from learning deeply and retaining knowledge.

Assumption 3: Students Learn Best When Tracked Into Same-Ability Groups

A common folk theory about mathematics education is that students do best academically when tracked into same-ability classes. This idea seems to be based on the perception that lower-achieving students will hold back higher-achieving ones if they are placed in the same class and that classes for higher-achieving students are too complicated and fast for lower achievers. On the surface, this may seem to be a reasonable argument. However, empirical data generally do not support it.

Some of the most compelling research on the effects of tracking on mathematics students was reported by Linchevski and Kutscher (1998). They found that placing junior high school students into mixed-ability groups did not widen the gap between higher- and lower-achieving students. This finding was particularly significant in light of earlier research stating that same-ability grouping does widen the achievement gap between higher- and lower-achieving students (Cahan & Linchevski, 1996). Furthermore, when the achievement of students placed in mixed-ability classes was contrasted to that of students placed in same-ability classes, the case against same-ability tracking became even stronger. On end-of-eighth-grade examinations, there was no significant difference between the performance of students placed in high-ability tracks and students in mixed-ability classes who would have qualified to be placed in a high-ability track. Students placed in intermediate- and low-ability tracks did significantly worse than students in mixed-ability classes who would have qualified to be placed in low and intermediate tracks. In short, lower-achieving students placed in mixed-ability classes benefitted from participating in a more rigorous curriculum, and higher-achieving students attained approximately the same academic benefits in both mixed-ability and same-ability settings.

Assumption 4: Extrinsic Rewards Support the Learning Process

Extrinsic rewards—incentives provided to induce students toward a desired behavior—are ubiquitous in education. Such rewards include praise, high grades, and certificates of merit. Some schools have programs rewarding students for doing extra academic work during the school year or over the summer. Rewards may include tickets for food and merchandise. Good behavior in class is sometimes rewarded with incentive programs. Students "caught" being on task may receive points that can accumulate and ultimately be redeemed for something of extrinsic value. Even though educators implement such programs with the best of intentions, research suggests that giving extrinsic rewards is not without cost.

Middleton and Spanias (2002) summarized much of the research literature on how extrinsic rewards can actually curtail students' learning of mathematics. One of the major costs associated with extrinsic rewards is that they diminish **intrinsic motivation** to engage in a task. Intrinsic motivation comes from pure enjoyment of an activity rather than from receiving a reward for doing it. When an extrinsic reward is removed, students are unlikely to engage in the task for the sake of learning or enjoyment. The loss of intrinsic motivation comes with heavy costs. Students who are intrinsically motivated develop greater mathematical confidence, value mathematics more highly as a discipline, and have more persistence in the face of failure. Strategies to foster intrinsic motivation include setting problems in interesting contexts and teaching for understanding. These strategies help draw students into doing mathematics without the negative effects associated with extrinsic rewards.

Of course, it is not possible to remove all extrinsic motivators from the school setting. Students must be assigned grades, and they are compelled to attend school under law. In some instances, it may even be necessary to give extrinsic rewards to get students to engage in tasks they otherwise may be hesitant to take up. In situations where the teacher is forced to provide extrinsic rewards, the negative impact can be dampened by giving group incentives rather than individual ones. Group incentives emphasize that it is important for everyone—not just select individuals—to learn. Teachers should reflect carefully, however, before giving extrinsic rewards. Questions to consider seriously before giving extrinsic rewards include the following: Could the task be made more interesting to students by setting it in a meaningful context? How long will extrinsic rewards be given? What will happen when the extrinsic reward is no longer provided? Is it the manner in which the subject is being taught, rather than the inherent properties of the subject itself, that is failing to motivate students? These can be very difficult questions to answer, but given the harmful effects associated with extrinsic rewards, they merit attention (Middleton & Spanias, 2002).

Teachers as Researchers

A major objective of this chapter has been to portray research as something closely tied to teaching practice rather than a set of ideas with no relevance to classrooms. The body of evidence available for informing the teaching and learning of mathematics has grown dramatically since the 1970s. At that time, research journals devoted exclusively to mathematics education, such as *Journal for Research in Mathematics Education* and *Educational Studies in Mathematics*, were just getting their start. While the growing amount of research available through published venues has helped move the field forward, it is important to note that effective teachers are constantly doing informal research on their own practices and their students' learning. Most of this research is never published and is usually not even thought of as "research" by the teachers doing it. Nonetheless, whenever a teacher carefully reflects on the effectiveness of a lesson or asks a question to find out how well students understand the material being studied, we could say that a form of research is taking place. When done carefully, this kind of teacher research plays a vital role in improving instruction.

Lesson study is one increasingly popular model for teacher research. Simply stated, lesson study is "collaborating with fellow teachers to plan, observe, and reflect on lessons" (Takahashi & Yoshida, 2004). J. W. Stigler and Hiebert (1999) conjectured that part of the reason for Japan's continued success on international standardized tests is its use of lesson study for teacher education. When engaged in lesson study, teachers collaboratively construct a lesson plan for a given concept based on what they know about how students will best learn it. They might draw from research literature as well as their own past experiences to identify particular difficulties students may have in comprehending the material, and then plan accordingly. One teacher in the lesson study group then teaches the collaboratively planned lesson while the others in the group observe. Following the lesson, a debriefing session is held in which group members identify the strengths and weaknesses of the lesson they observed. Responsibility for success as well as failure is shared among the group and not credited to one particular individual, since all were involved in creating the lesson plan under consideration. During the debriefing session, the group polishes the lesson and may decide to try it again. The process of designing, implementing, and polishing can be repeated until the group is satisfied with the lesson.

The idea of lesson study is deeply engrained in the culture of teaching in Japan (J. W. Stigler & Hiebert, 1999). Their success on the TIMSS is a primary motivator for the growing popularity of lesson study in other parts of the world. A difficulty with transferring the model to other countries is that time for collaboration among teachers and lesson study is built into the day of a Japanese teacher, but such time is often not given by administrators in other countries. Nonetheless, this should not deter teachers who currently do not have that time built into their schedules. A beginning step in implementing lesson study is simply to get together with one or two colleagues and carefully plan a lesson. If it is not possible to watch the lesson being carried out, it might be possible to make arrangements to videotape it. At the very least, teachers in the group can carry out the collaboratively planned lesson and then report their observations to the others. At some point, it may also be wise to invite an administrator to a meeting of the lesson study group. As administrators see the value of the lesson study model, they may begin to reprioritize the manner in which available professional development time is spent.

Another strategy for teacher research is to systematically investigate individual students' thinking. Interviews with students can reveal a great deal about their thinking because they allow teachers to ask probing questions (Buschman, 2001). The primary purpose of such interviews is to ask students questions that shed light on their thinking about key mathematical concepts taught in class. Students are sometimes asked to "think aloud" as they solve a given problem so the teacher can better understand their thought processes. Insights gained from these interviews can be used to inform instruction. If time is a factor in conducting interviews, teachers can also gain insights about students by asking them to write about their thinking in regard to a mathematics problem. L. D. Miller (1992) found that when teachers asked students to write about their understanding of concepts in algebra, they gained information useful in guiding their instruction. Specific ideas for writing prompts for students will be discussed in Chapter 3. The field experience exercises in this textbook at times involve interviewing students and having them write about their mathematical thinking processes.

Regardless of the approach taken to teacher research, it is always helpful to stimulate one's thinking by looking into published research about topics of interest. The Internet contains a wealth of options for searching out relevant literature. The Educational Resources Information Center (ERIC) database (www.eric.ed.gov) indexes many of the published journals relevant to the field of education. NCTM's website (www.nctm.org) is another valuable resource, offering online abstracts of journals and full-text access for subscribers. Other publishers' websites often contain abstracts and full-text access as well; the websites for the *Journal of Mathematical Behavior* and *Educational Studies in Mathematics* both fall into this category (enter these titles into a search engine to find their precise addresses). Some online journals, such as *Statistics Education Research Journal*, *The Mathematics Educator*, and the *International Electronic Journal of Mathematics Education*, provide full-text access to articles without a subscription. Your university library is likely to have purchased subscriptions for full-text access to many additional education-related journals. As you locate articles, be sure not to skip the reference list at the end of each one. The references often provide interesting new leads for further reading.

In reading research articles, it is important to go beyond research *results*. Although the results can be useful for guiding classroom instruction, research reports contain other valuable information for teachers. Silver (1990) observed that teachers can benefit from examining the tasks used in research studies and the researchers' theoretical perspectives. In regard to the former, the tasks used in published research studies are often far richer in their potential to reveal students' thinking than tasks published in conventional textbooks. As a result, they can be used as classroom tasks, student interview items, or writing prompts. It is often enlightening to compare students' responses to those described in published research. Second, in examining the researchers' theoretical perspectives, new ways of looking at one's classroom are revealed. For instance, the theoretical perspectives discussed in this chapter provide various ways of viewing the individual student and the collective group.

CONCLUSION

Although the relationship between research and practice is often portrayed as discordant, the field of mathematics education has a great deal to gain by bridging the two worlds. Despite the fact that research is not meant to provide precise prescriptions for the course of action to take in any given situation, it can provide evidence to inform instruction. Effective teachers are always gathering evidence from their own practice, formally or informally, on how well their students are learning. Formal research and informal observations provide a basis for informing decisions and thus can reduce, but not eliminate, some of the "guesswork" involved in teaching effectively. As one begins teaching, it is beneficial to prioritize learning from formal research as well as continuous informal classroom research.

VOCABULARY LIST

After reading this chapter, you should be able to offer reasonable definitions for the following ideas (listed in their order of first occurrence) and describe their relevance to teaching mathematics:

Engineering view of research   29

Strong enlightenment view of research   29

Moderate enlightenment view
   of research   29

Cognition   30

Affect   30

Conceptual knowledge   31

Procedural knowledge   31

Instrumental understanding   32

Relational understanding   32

Radical constructivism   33

Disequilibrium   33

Transfer   34

Metacognition   35

Beliefs   36

Attitudes   36

Emotions   36

Mathematics anxiety   36

Univocal discourse   37

Dialogic discourse   37

Social norms   38

Sociomathematical norms   38

Social constructivism   39

Zone of proximal development   39

Scaffolding   39

Extrinsic rewards   43

Intrinsic motivation   44

Lesson study   45

HOMEWORK TASKS

1. Think back to a mathematics course you took in high school or college. How would you characterize your learning during the course: mainly instrumental or mainly relational? Explain.

2. Discuss the processes of assimilation and accommodation in relation to learning polynomial multiplication. Specifically, consider the case of a student who encounters a problem involving multiplication of trinomials but has only learned the FOIL method for multiplication. What sort of assimilation and/ or accommodation would need to take place for the student to multiply the trinomials correctly?

3. Find three references that discuss Platonism and its role in the philosophy of mathematics. Then describe three actions a teacher holding Platonist beliefs about the nature of mathematics would be likely to take in teaching mathematics. Also briefly discuss how a teacher rejecting Platonist beliefs about mathematics might go about teaching. Support your answers by citing your references.

4. Find three definitions for the concept of *negative transfer*. After reading the definitions, formulate your own definition for the concept by combining common elements from all three. Finally, relate the discussion of teaching and learning FOIL to the concept of negative transfer.

5. Describe a hypothetical classroom situation in which students are working in small groups and mostly univocal discourse is taking place between the students and the teacher. Then describe a hypothetical

situation where mostly dialogic discourse is taking place between the students and the teacher, but the class is not broken into small groups. To what extent do you feel the classroom setting (small groups versus whole class) generally influences the type of discourse that takes place? Explain.

6. Read the following article:

Evitts, T. A. (2004). Action research: A tool for exploring change. *Mathematics Teacher, 97,* 366–370.

As you read, list the critical elements of action research that are identified by the author. After comparing your list with those of your peers and your instructor, create a proposal for an action research project you could carry out when you begin teaching. Locate and summarize three research articles with information relevant to your project. Describe how you will draw on information in the articles as you carry out your research, specifically, how will you draw on the research results? On the tasks used in the research? On the researcher's theoretical perspective?

7. Visit the What Works Clearinghouse online (www.whatworks.ed.gov/). What are the strengths of the website for informing mathematics teaching? What are its limitations?

CLINICAL TASKS

1. Examine the textbook used for one of the mathematics classes at the school in which you observe. Does the book support the development of both procedural and conceptual knowledge? Provide specific examples from the text to support your position.

2. Visit an algebra class in which the multiplication of binomials has already been introduced. Interview the teacher to determine the approach used to teach the subject and why he or she used it. Also interview three to four students. During the interviews, ask the students to (1) multiply $(3x + 1)$ $(2x +3)$ and (2) multiply $(3x^2 + 2x + 1)(2x + 5)$. Then write a paper in which you address the following two questions: (1) Did the teacher focus on developing students' procedural *and* conceptual knowledge in introducing binomial multiplication? (2) Did the students seem to have instrumental or relational knowledge of polynomial multiplication?

3. Choose a task from *Mathematics Teaching in the Middle School* or *Mathematics Teacher* that relates to the mathematical content being studied by a class you are observing. Check with the teacher to make sure the task is not one for which students have already learned a procedure. Then ask the teacher to identify three students: one who generally gets high grades, one who generally gets average grades, and one who generally gets low grades. Interview the students individually. During the interview, ask them to solve the task you chose. Also ask them to think aloud as they solve it, explaining both what they are doing and why they are doing it. Write a report that summarizes the metacognitive ability exhibited by each of the three students.

4. Observe a mathematics class with an eye toward the beliefs, attitudes, and emotions exhibited by the students in the class as they interact with one another and with the teacher. Make a three-column list to record observations fitting each category. Justify the manner in which you categorized your observations. Then describe how you would use the information you gathered to guide instruction if you were the teacher for this group of students.

5. Observe a mathematics classroom with an eye toward classroom discourse. Is the discourse primarily univocal or dialogic? What kinds of social norms have been negotiated by the students and teacher? What kinds of sociomathematical norms are apparent? Answer each question and provide classroom examples to support your analysis.

VIGNETTE ANALYSIS ACTIVITY	**Focus on Reasoning Abstractly and Quantitatively (CCSS Standard for Mathematical Practice 2)**

Items to Consider Before Reading the Vignette

Read CCSS Standard for Mathematical Practice 2 in Appendix A. Then respond to the following items:

1. Drawing on your own experiences as a mathematics student, provide specific examples of teaching practices that can help students reason abstractly and quantitatively.

2. Drawing on your own experiences as a mathematics student, provide specific examples of teaching practices that might prevent students from reasoning abstractly and quantitatively.

Scenario

Mr. Nelson was a first-year mathematics teacher. He was excited to have the opportunity to teach a high school geometry class because he believed the content could be made interesting and lively through the use of concrete materials. In the summer leading up to his first class, he spent a great deal of time gathering hands-on activities and manipulatives to bring geometry to life for his students. Although Mr. Nelson frequently used these activities and manipulatives during his classes, he had a vague feeling that students were not learning as much from them as he had anticipated they would. On one occasion, Mr. Nelson tried to help students examine the properties of the diagonals of several different quadrilaterals.

The Lesson

One of the first things Mr. Nelson wanted students to understand was that if two equidistant, parallel lines are intersected by another pair of equidistant, parallel lines, a rhombus is formed. After a warm-up activity and a discussion of the previous night's homework, Mr. Nelson distributed a piece of Patty Paper and a ruler to each student. He directed students to place the ruler on the paper and trace along the two long sides of the ruler to form parallel segments. He then instructed his students to turn the ruler any angle desired and repeat the process to form a pair of parallel segments intersecting the first. Mr. Nelson told them that the intersecting segments formed a rhombus. He directed students to measure the sides of the shape to confirm that this was true. Mr. Nelson then told them to take out their notebooks and write down the conjecture that two intersecting pairs of parallel lines form a rhombus if the lines in each pair are equidistant.

For the next part of the activity, Mr. Nelson told students to draw diagonals in the rhombi they had just produced. Next, students were told to measure the lengths of the diagonals, the angles formed by the intersection of the diagonals, and the distance from each vertex to the intersection of the diagonals. Students had trouble taking these measurements because the rhombi they produced at the beginning of class were very small. Upon seeing that students were not obtaining precise measurements because of the sizes of the rhombi they had to work with, Mr. Nelson directed students' attention back to the front of the room. He said, "OK, I see that this is not working well, so let's talk about what you *should* be seeing. You can put these conjectures in your notebook: (1) The diagonals of a rhombus are perpendicular; (2) the diagonals of a rhombus bisect one another. Does anyone have any questions about those?" Students busily and quietly wrote the two conjectures in their notebooks.

In addition to the conjectures that had already been presented, Mr. Nelson wanted students to see that the diagonals of a rhombus bisect the interior angles of the rhombus. Rather than having students go back to the rhombi that were too small to use during the last activity, he decided to try to reason verbally with them. He said, "Let's think about what happens where the diagonals meet each of the vertices of the rhombus. If we had larger rhombi to work with, we could see that the diagonals do *what* to the angles of the rhombus? It starts with a *b*. Kristen, do you know?" Kristen replied, "Bisect?" Mr. Nelson replied, "Good; that is exactly right. The diagonals of a rhombus bisect the angles of a rhombus. Does that make sense to everyone?" No one responded, so Mr. Nelson wrapped up this portion of the lesson by saying, "This is another important conjecture, so write it down in your notebooks: The diagonals of a rhombus bisect the angles of the rhombus."

After discussing properties of rhombi with his students, Mr. Nelson wanted them to explore properties of squares and rectangles. He asked them to use a ruler to draw a rectangle 10 centimeters long and 7 centimeters wide. Once they had done so, he said, "OK, this rectangle should be plenty big enough for us to investigate. I would like you to draw the diagonals in the rectangle." The students obliged. Mr. Nelson asked, "What do you notice about the diagonals?" Mark, one of the students, asked, "What do you want us to look at, the lengths of the diagonals or the angles in the drawing?" Mr. Nelson replied, "Measure the length of each diagonal, and then measure the distance from each vertex to where the diagonals intersect." After a few minutes, Mark said, "Oh, I see. They are all the same." Mr. Nelson replied, "That's right, Mark. Let's all write that down as another conjecture in our notebooks: The diagonals of a rectangle are congruent and they bisect one another."

There was one last conjecture Mr. Nelson wanted his students to see during the lesson. He asked the class, "Suppose we have a square and draw in its diagonals. What should be true?" His students were silent. A few of them drew squares and started drawing in diagonals. Noticing that only a few minutes of class time remained, Mr. Nelson said, "I'll give you a hint: One thing that happens is that the diagonals are—it starts with a *c*—Mary?" Mary replied, "Congruent!" "Great!" Mr. Nelson said, and then continued, "The diagonals are also—one word starts with *p* and the other starts with *b* . . ." Several students blurted out "perpendicular" and "bisect." Mr. Nelson replied, "Yes, right. That is true because a square is a special type of a rhombus and also a special type of rectangle. Please write that conjecture in your notebooks, and for the final five minutes of class, start your homework assignment, which will be the first 10 problems in this section of the book." Students wrote down their homework assignment, closed their notebooks, and started lining up at the door waiting for the bell.

Questions for Reflection and Discussion

1. Which aspects of CCSS Standard for Mathematical Practice 2 did Mr. Nelson's students seem to attain? Explain.

2. Which aspects of CCSS Standard for Mathematical Practice 2 did Mr. Nelson's students not seem to attain? Explain.

3. Identify specific points in the lesson where Mr. Nelson short-circuited students' reasoning opportunities by being too directive. Describe how these points could be reframed to allow for a greater degree of student reasoning.

4. Comment on the questions Mr. Nelson asked students. Are the questions likely to promote student reasoning? Why or why not?

5. Compare and contrast the conjecturing done during this lesson with the sort of conjecturing you would expect to see from mathematicians engaged in research.

6. Given that this was a high school geometry lesson, might it have been appropriate to attempt formal proofs for each of the conjectures? Why or why not?

RESOURCES TO EXPLORE

Brahier, D. J. (Ed.). (2011). *Motivation and disposition: Pathways to learning mathematics* (Seventy-third yearbook of the National Council of Teachers of Mathematics). Reston, VA: NCTM.

Description: A variety of issues related to student motivation are taken up in this collection of 21 chapters by various authors in the field. Chapters deal with issues such as students' motivation to take advanced mathematics courses, motivating students through problem solving, and how students develop productive dispositions toward mathematics.

Bransford, J. D., Brown, A. L., & Cocking, R. R. (Eds.). (1999). *How people learn: Brain, mind experience, and school.* Washington, DC: National Academy Press.

Description: The field of cognitive science has yielded valuable insights about how students learn and the structure of effective learning environments. In this book, the authors present a concise and very readable summary of these developments.

Herbel-Eisenmann, B., & Cirillo, M. (Eds.). (2009). *Promoting purposeful discourse: Teacher research in secondary mathematics classrooms.* Reston, VA: NCTM.

Description: The classroom learning environment plays an important role in mediating students' learning. This collection of articles presents essential ideas about classroom discourse and takes up practical topics such as managing discourse in technology-based lessons.

Kilpatrick, J., Martin, W. G., & Schifter, D. (Eds.). (2003). *A research companion to* Principles and Standards for School Mathematics. Reston, VA: NCTM.

Description: The origins of NCTM's *Principles and Standards for School Mathematics* lie within the field of mathematics education research. This collection of articles summarizes research related to each of the strands in the NCTM standards in a manner accessible to a wide audience.

Lester, F. (Ed.). (2010). *Teaching and learning mathematics: Translating research for secondary school teachers.* Reston, VA: NCTM.

Description: Mathematics education research contains valuable insights for teachers in regard to a variety of content areas. The authors of this collection of articles summarize current research about a variety of topics, including proportional reasoning, geometry and proof, probability and statistics, and the influence of technology on students' learning.

Masingila, J. O. (Ed.). (2006). *Teachers engaged in research 6–8.* Reston, VA: NCTM.

Description: This volume contains research reports written by middle school teachers investigating their own classrooms. Teachers write about topics such as learning algebra, developing a mathematical community, and teaching in urban classrooms.

Sowder, J., & Schapelle, B. (Eds.). (2002). *Lessons learned from research.* Reston, VA: NCTM.

Description: The editors of this book took up the task of making research reports published in *Journal for Research in Mathematics Education* more accessible to teachers. Topics addressed in this collection of articles include student motivation, research on effectiveness of various curricula, and students' learning in content areas such as algebra, geometry, statistics, and number and operations.

Tobias, S. (1993). *Overcoming math anxiety.* New York, NY: W. W. Norton.

Description: It is not uncommon for students to panic when asked to solve a mathematics problem. In this book, the author takes up the topic of "math anxiety" and some potential remedies.

Van Zoest, L. (Ed.). (2006). *Teachers engaged in research 9–12.* Reston, VA: NCTM.

Description: This volume contains research reports written by high school teachers investigating their own classrooms. Teachers write about topics such as probability simulations, students' understanding of limits, and learning to teach through lesson study.

Chapter 3

PLANNING MATHEMATICS LESSONS

Prospective teachers frequently put lesson planning at the top of the list of things they want to learn. This is reasonable, since teachers must deliver several lessons each day. However, a cautionary note is in order. It should be understood that there is no single "best" approach to planning a lesson. Some approaches are more effective than others, but none of them is perfect. In mathematics, we often learn algorithms that provide flawless answers every time they are executed correctly. But because of the diverse and variable nature of schools, students, and teachers, no such algorithm exists for planning mathematics lessons. Even experienced and successful teachers find they must implement a lesson several times, polishing it each time. Lesson planning is therefore best viewed as a continuous improvement process rather than an activity with a definite endpoint. This chapter provides tools for starting the process.

THE FOUR-COLUMN LESSON PLANNING MODEL

The **four-column lesson plan**, described in this chapter and illustrated in subsequent chapters, is often used when teachers engage in lesson study as described in Chapter 2. Lesson study is based on the premise that teaching progressively improves through cycles of designing, implementing, and reflecting on lessons. The four-column lesson plan is designed to emphasize the idea that lesson planning should go beyond carrying out a to-do list during a class period. The steps of the lesson compose just the first column of the plan. The headings for the four columns are:

- Steps of the lesson: Learning activities and key questions
- Expected student reactions or responses
- Teacher's response to student reactions/things to remember
- Goals and methods of evaluation (Matthews, Hlas, & Finken, 2009, p. 506)

The second, third, and fourth columns emphasize that lessons should allow teachers to assess students' thinking, respond to potential student difficulties, and extend students' thinking in mathematically productive directions. A blank template for writing a four-column lesson plan is shown in Figure 3.1. Several examples of completed four-column lessons are provided in Chapters 7 through 11.

Figure 3.1 Four-column lesson plan template. (For a Word version of the four-column lesson plan template, along with detailed instructions, go to the book's website at www.sagepub.com/groth.)

Primary objective:			
Materials needed:			
Steps of the lesson	*Expected student responses*	*Teacher's responses to students*	*Assessment strategies and goals*

A Quality Control Framework for Planning Lessons

When preparing and revising a four-column plan, it is helpful to keep some general quality control principles in mind. Fuson, Kalchman, and Bransford (2005) recommend that, to create effective classroom mathematics learning environments:

- Teachers should elicit and address students' preconceptions.
- Lessons should aim to build both conceptual and procedural knowledge.
- Instruction should support metacognition.

These three principles provide criteria for examining and improving the quality of lessons.

Eliciting and Addressing Students' Preconceptions

Teachers should plan to draw out and address **student preconceptions**. That is, instruction should build on the school-based and experiential knowledge students bring to a lesson. To do this, a lesson might begin with a problem situation such as the following: "A large container in the refrigerator has $2\frac{3}{4}$ cups of yogurt in it. One serving of yogurt is $\frac{1}{2}$ cup. How many servings of yogurt are in the large container?" To get an initial grasp of the problem, students can recall their out-of-school experiences of taking servings of food out of large containers. Teachers can ask students to use school-based formal knowledge of fraction representations to draw a diagram of the situation, as shown in Figure 3.2. By encouraging the discussion and use of both informal and formal knowledge, teachers can determine the type of support students may need. Students can use the diagram to act out the process of using a scoop that holds $\frac{1}{2}$ cup to draw servings from a large container. Doing so shows that five servings can be drawn out and half of a serving is left over. This means that the solution to the problem is $5\frac{1}{2}$ servings.

Figure 3.2   Diagram for yogurt problem.

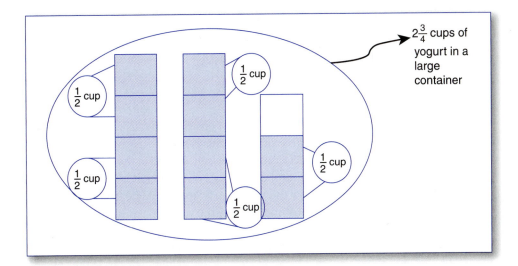

Building Conceptual and Procedural Knowledge

The distinction between conceptual and procedural knowledge was made in Chapter 2. One of the problems with traditional mathematics instruction is that it tends to overemphasize procedural knowledge and neglect conceptual knowledge. To further illustrate the distinction between conceptual and procedural knowledge, consider a more traditional approach to division of fractions. In a strictly procedural lesson, a teacher might begin by demonstrating an algorithm to compute $2\frac{3}{4} \div \frac{1}{2}$. One might show that $2\frac{3}{4}$ can be written as an improper fraction by multiplying 4 by 2, adding the result to 3, and putting the number for the finished computation in the numerator while leaving 4 in the denominator to obtain $\frac{11}{4}$. Then the teacher could demonstrate the invert-and-multiply procedure to obtain $\frac{11}{4} \times \frac{2}{1} = \frac{22}{4} = 22 \div 4 = 5\frac{1}{2}$. The solution is the same as in the diagrammatic approach to the yogurt problem described earlier, but the underlying concepts of fraction representation, the meaning of division, and measurement remain hidden. Exclusively procedural instruction would also likely not introduce the yogurt problem until after students have learned the invert-and-multiply procedure. Note, however, that the assumption that word problems involving fraction division cannot be done until after invert-and-multiply has been mastered is false. The diagrammatic solution to the yogurt problem (Figure 3.2) did not utilize the invert-and-multiply rule at any point, yet it involved fraction division.

When considering how to balance conceptual and procedural knowledge, it is important not to fall into the trap of thinking that one type of knowledge is superior to the other. Although traditional instruction has overemphasized procedural knowledge to the detriment of conceptual knowledge, it does not mean that instruction should consist only of the conceptual. Concepts and procedures complement one another in mathematics. For instance, the conceptually based diagrammatic approach to the yogurt problem could be used as a springboard for helping students develop and understand procedures for fraction division. After students solve a number of parallel tasks, they can be asked to devise solution shortcuts. To motivate the search for shortcuts, the teacher can describe a situation involving fractions that would be difficult to solve with a diagram (e.g., $9\frac{7}{8} \div 3\frac{11}{15}$). The process of looking for shortcuts for more complicated problems leads naturally to efficient fraction division procedures. By having a part in developing procedures for fraction division, students are more likely to understand and retain them.

Supporting Metacognition

Metacognition is another important psychological construct from Chapter 2. It is often described as "thinking about one's thinking." To illustrate how teachers can help students engage in metacognitive activity during a lesson, consider the yogurt problem once again. In fraction division word problems with a remainder, it is common for students to become confused about how the leftover portion should be expressed. Two of the solutions you could expect when students solve the yogurt problem are $5\frac{1}{2}$ and $5\frac{1}{4}$. The first solution, $5\frac{1}{2}$, comes from reasoning through the problem as described earlier. Students who obtain this solution could be prompted toward metacognitive activity by writing or speaking about how they obtained it. Doing so could help them consolidate their knowledge and use it again in similar situations. Students obtaining $5\frac{1}{4}$. could be asked to reflect on the reasonableness of their response. This, again, could be done by having the student write or speak about the solution process he or she used. In responding to the student's written or verbal description, the teacher should have questions ready to help the student see the mistake in his or her thinking. Some guiding questions to ask would be, What is the whole unit for the $\frac{1}{4}$ in your answer? Did you determine the number of cups or the number of servings? If you think about the process of scooping out servings with a measuring cup that will hold $\frac{1}{2}$ cup of yogurt, how much of the cup will be full on the last scoop? By asking students to explain their own thinking and reflect on the reasonableness of their solutions, teachers help students get in the habit of engaging in metacognitive activity.

> ### Implementing the Common Core
>
> See Homework Task 1 to get started on designing a lesson that helps students engage in the eight standards for mathematical practice as they "solve word problems involving division of fractions by fractions" (Content Standard 6.NS.1).

ALTERNATIVES TO THE FOUR-COLUMN LESSON PLAN MODEL

Although the lesson examples in this textbook are based on the four-column lesson plan model, it should be noted that this model is not the only viable means for constructing lessons. Some inquiry-oriented curriculum series frame lessons around three main stages: launch, explore, and summarize. This format emphasizes that inquiry-oriented lessons generally begin with an interesting task, then support students as they work on it, and finally conclude with a discussion of the mathematics needed to solve the problem. This three-stage model is described in detail on the Connected Mathematics Project website: http://connectedmath.msu.edu/. Another model for lesson planning appears on the NCTM Illuminations website: http://illuminations.nctm.org. Illuminations lessons are organized using nine headings: Learning Objectives, Materials, Instructional Plan, Questions for Students, Assessment Options, Extensions, Teacher Reflection, NCTM Standards and Expectations, and

References. The Illuminations website is an outstanding resource, as it contains hundreds of lessons organized by grade level and content strand. You should consult it frequently for sample lessons and teaching ideas.

Another influential model for planning lessons comes from outside the field of mathematics education. Tomlinson's (2005) **differentiated instruction** model has been adopted by many school districts. It emphasizes that students come to the classroom with diverse learning needs and draws attention to varying instruction along the dimensions of content, process, and product. In regard to content, teachers may offer multiple texts for students to use in studying a topic, set up interest centers in a classroom that allow students to study different content, and use any other techniques that connect with students' learning needs. Differentiating along the process dimension can be done through techniques like giving assignments at different levels of difficulty to different students and encouraging group investigations. Differentiating along the product dimension can be done by encouraging independent study and allowing students to choose assignments that meet their specific learning needs. As differentiated lessons are formed, they should address three key dimensions along which students tend to differ: readiness, interests, and learning profile.

On the surface, the four-column lesson plan appears to be substantially different from the launch-explore-summarize, NCTM Illuminations, and differentiated instruction models. However, a closer look reveals many common characteristics. All four formats prompt teachers to identify lesson objectives and necessary materials. All four emphasize the importance of assessing students' thinking and adjusting the lesson to be responsive to the assessment information gathered. Spatially, the four-column format is helpful because it prompts teachers to think about how students may respond to each portion of a lesson. Planning one's reactions to students in advance eliminates some of the "thinking on the fly" that can be difficult to do during a lesson. It also helps the differentiation process, because the teacher can think carefully about how each portion of the lesson might be altered to meet the learning needs of students with a diverse range of thinking strategies. All four lesson formats also acknowledge the importance of addressing students' preconceptions, building both conceptual and procedural knowledge, and encouraging metacognitive activity. The four-column template used in Chapters 7 through 11 of this text, however, explicitly lists these elements as a reminder to check the plan for their presence. An excellent lesson need not be written in four-column format, but it is a useful device for helping beginning teachers focus on the important elements of a lesson.

IDEA FOR DIFFERENTIATING INSTRUCTION	Open Questions and Parallel Tasks
Small (2009) discussed two core strategies for differentiation: posing **open questions** and using **parallel tasks**. Open questions are those that allow multiple solution strategies. For example, a teacher may begin a unit on linear equations by posing the following task:	A hose is running at a constant rate and filling a pool with water. The hose puts 1,250 gallons of water into the pool each hour. If the pool holds 5,000 gallons, how long will it take to be $\frac{1}{2}$ full? $\frac{3}{4}$ full? How long will it take to be completely full?

(Continued)

(Continued)

Students can approach the task in a variety of ways. Some may take an arithmetic approach, dividing 5,000 by 1,250 to obtain the answer to the final question, and then working from that answer to figure out the two questions before it. Others may construct a table to show how many gallons are in the pool after each one-hour interval. Those who have previously studied linear equations might even choose to construct graphical representations of $g = 1250h$, where g represents the number of gallons in the pool and h represents the number of hours that have passed since the hose was turned on. The pool task is an open one because it has multiple possible entry points. After students have generated individual solutions to the problem, they can be encouraged to discuss them with one another. During such discussions, students who have not yet considered an algebraic approach to the problem can begin to learn it, and those using an algebraic approach can revisit or begin to understand the arithmetic foundations of it.

Parallel tasks are those that deal with the same mathematics concepts, yet differ in ways that accommodate individual differences. In the above pool task, for example, the teacher may anticipate that some students will have difficulty with the large numbers involved. If so, it may be appropriate to pose the following task in parallel with the pool task:

A bathtub holds 20 gallons of water. A faucet fills the tub with 4 gallons per minute. How long will it take to be $\frac{1}{2}$ full? $\frac{3}{5}$ full? How long will it take to be completely full?

In the parallel task, not only were the numbers made substantially smaller—but there was also a small adjustment from $\frac{3}{4}$ to $\frac{3}{5}$ that can make a big difference in students' ability to solve the problem (try reworking the bathtub problem using $\frac{3}{4}$ instead of $\frac{3}{5}$). Students could be given the choice to solve either the pool problem or the bathtub problem, or the teacher might assign half the class to work on the pool problem and the other half to work on the bathtub problem. In the class discussion that occurs after students have worked the tasks in parallel, the teacher can still ask many of the same questions about the algebraic concepts involved, such as: What is the rate of change in amount of water? Can you construct a graph to represent the situation? Can you write an equation? Can you construct a table? Having the two tasks posed in parallel also allows for comparisons of the representations for the two different situations. For example, students might be asked to determine which situation involves a greater rate of change.

A "Top-10" List of Teaching Strategies and Tools

As you can see from the discussion so far, effective mathematics teaching is a complex endeavor. It consists of far more than simply transcribing notes on a chalkboard and hoping the students understand them. As you gain experience designing effective lessons, the various aspects surrounding them seem less daunting because they become a part of your normal mode of operation. To help you along the path to successful implementation of inquiry-oriented instruction, this section describes general strategies that have proved helpful to others. These strategies will be utilized and expanded throughout Chapters 7 through 11.

A side note: As you read about the teaching strategies described below, please notice the accompanying references, many of which refer to articles in the teacher-oriented journals *Mathematics Teaching in the Middle School* (Grades 6–8) and *Mathematics Teacher* (Grades 9–12) published by the National Council of Teachers of Mathematics (NCTM). By finding the referenced articles, you can get further details about the specific strategy being discussed. You should also keep an eye on these journals as they are published each month to continue to build your personal repertoire of teaching strategies. Do not limit yourself to just one of the two journals—high school teachers often find valuable ideas in the middle school journal, and middle school teachers often profit from reading the high school journal as well as the elementary journal *Teaching Children Mathematics.*

Reading Strategies

When teachers think about the role of reading in mathematics, the need for students to read and understand word problems often comes to mind. Pape (2004) studied middle school students' problem solving behaviors by asking them to think aloud as they solved a set of word problems. He found that

> the more successful students provided evidence that they translated and organized the given information by rewriting it on paper, and they used the context to support their solutions . . . In addition, more successful students provided explanations for their mathematical steps and/or justified their solutions. (p. 208)

On the basis of these observations, Pape suggested that teachers adopt instructional strategies that help students monitor their own problem solving behaviors. In particular, a norm for classroom discourse should be that students explain and justify their thinking. Teachers can work toward this goal simply by asking students to explain *how* and *why* they arrived at their answers. Pape also suggested having students share their solution strategies with one another so they can see a variety of ways to approach a problem. During such classroom discussions, students should be encouraged to refer back to the context of the problem to judge the reasonableness of their solutions. As they do so, they become more successful in solving word problems.

Another vital reading goal for mathematics instruction is to help students understand vocabulary. D. R. Thompson and Rubenstein (2000) identified a number of factors that impair vocabulary development in mathematics. One factor is that some words are shared by both mathematics and English, but have different meanings in each domain (e.g., *prime, origin, volume*). Another is that some words have more than one mathematical meaning (e.g., *square, base, degree*). A strategy teachers can use to enhance students' comprehension is to introduce vocabulary *after* the concept has been introduced rather than *before.* For instance, the term *slope* might be introduced only after students have worked with problems about rates of change in different situations. Visual aids can be helpful as well. Concrete visuals that show relationships among mathematical terms (such as relationships among quadrilaterals) assist students' comprehension. For example, many students confuse the terms *slant height* and *height* in relation to a pyramid. By making a concrete model of a pyramid available, teachers can have students point to where the two attributes of the pyramid are located. Finally, Seidel and McNamee (2005) recommended drawing students' attention to the fact that some words have different meanings in mathematics and in everyday usage by having them do presentations about the two different meanings. The give-and-take from such student–student interaction helps develop students' vocabulary comprehension (Borasi & Siegel, 2000).

Implementing the Common Core

See Homework Task 2 to identify potential vocabulary-related difficulties mathematics students may have as they attempt to "attend to precision" (Standard for Mathematical Practice 6).

Although the discussion of the relationship between reading and mathematics up to this point has focused on strategies for overcoming reading difficulties, it should be noted that reading can also serve as a springboard for problem solving. Such an approach is often referred to as **reading to learn mathematics**. There are many examples of how mathematics teachers have used reading in this way. Bay-Williams (2005) used poetry written by Shel Silverstein to launch problem solving investigations in a number of different content areas. In one instance, she used a poem about being 1 inch tall as inspiration for problems involving proportions. Students were asked to imagine they had shrunk to the height of 1 inch and that other things in their environment, such as their desks, had shrunk proportionally.

Students then determined the heights of various shrunken objects. Other teachers have drawn mathematical elements from popular books such as *Harry Potter* to launch problem solving investigations (T. Beaton, 2004; McShea, Vogel, & Yarnevich, 2005). Some books, such as *Alice's Adventures in Wonderland*, contain mathematics appropriate for the middle or high school that may be missed in a quick reading but can be brought out by classroom activities (Taber, 2005).

> ### Implementing the Common Core
>
> See Homework Task 3 for additional guidance on using literature to launch lessons that help students "make sense of problems and persevere in solving them" (Standard for Mathematical Practice 1).

Writing Strategies

Along with using literature to spark students' thinking, many teachers use writing to stimulate and elicit students' mathematical thinking. Aspinwall and Aspinwall (2003), for example, examined students' beliefs about mathematics by analyzing their responses to **writing prompts** such as "What is math?" "What is division?" and "What is a fraction?" Students' responses revealed their feelings about the concepts and their understandings of them. Sjoberg, Slavit, and Coon (2004) reported improving their students' understanding of mathematics by frequently asking them to respond to three prompts:

1. Explain at least one thing you learned in mathematics this week.

2. How does your answer to question 1 connect to other things in mathematics?

3. How might you use this in the real world? (p. 490)

To help students become proficient in writing about mathematics, Baxter, Woodward, Olson, and Robyns (2002) suggested that teachers begin by posing prompts that ask students to describe their attitudes and feelings toward the subject. After this, teachers can move on to prompts asking students to explain familiar mathematical ideas, and then later on to prompts dealing with advanced mathematical ideas.

> ## STOP TO REFLECT
>
> Formulate writing prompts that you could use to investigate each of the following regarding a group of students:
>
> 1. Their beliefs about mathematics
> 2. Their attitudes about mathematics
> 3. Emotional experiences they have had when studying mathematics
> 4. Their conceptual understanding of the specific mathematical topics they are studying
> 5. Their procedural understanding of the specific mathematical topics they are studying

As noted earlier, writing can be used as a tool for encouraging metacognitive activity. Goldsby and Cozza (2002) observed that "the necessity of explaining why and how a problem is solved can crystallize students' thoughts about the solution process and enhance their facility with language" (p. 520). As teachers read students' descriptions of problem solving strategies, they can adjust instruction to meet students' learning needs. Along with having students write about their thinking *as* they solve problems, some teachers find it effective to have students write about their thinking *after* they have solved a problem. Brown (2005), for example, asked students to write about their performance on an exam after it was returned to them. This gave students the opportunity to refine their thinking about problems they had missed on the exam, and it also allowed them to display understanding of topics they had expected to be on the exam but were not. Students who reflect on their problem solving strategies in writing have shown greater gains in achievement than those who do not (K. M. Williams, 2003).

When teachers have students write about mathematics, the question of how to grade the writing usually arises. N. B. Williams and Wynne (2000) described one possible approach. They began by asking students to write about their mathematical thinking twice per week. The teachers met to collaboratively decide on grading criteria for student journals, and they also exchanged graded papers to determine how consistently they were applying the criteria. This process was so time-consuming that they later reduced the frequency of writing in their mathematics classes to once per week. M. E. McIntosh and Draper (2001) discussed an alternative approach. They believed grades should not be assigned to students' responses to mathematical writing prompts. However, they did make comments on students' papers to provide feedback. They suggested that teachers who feel the need to assign grades to papers award participation points for successful completion, and they found that not assigning letter grades freed up time to implement writing several times a week in class. Whether or not teachers decide to grade student papers, they should carefully read students' written thoughts and provide feedback. This process helps students learn about mathematics and it helps teachers learn about students' thinking.

Do you believe students' writing about mathematics should be graded? Why or why not? Are there some instances in which it should be graded and others in which it should not? Support your position.

STOP TO REFLECT

Manipulatives Usage

Concrete materials are essential for helping young students learn concepts such as counting. Some teachers, however, do not believe manipulatives should be used with middle and high school students. This line of thinking seems to come from the belief that older students do not need concrete supports to understand abstract ideas. Most of the time, this belief is incorrect. For example, recall from the discussion about vocabulary development that many high school students confuse the concepts of height and slant height, and that referring to a concrete object can be helpful. This is just one example of how manipulable materials can be valuable in the mathematics instruction of older students; we will encounter many more examples in Chapters 7 through 11.

Because manipulatives play an important role in mathematics instruction, careful thought must be given to their use. Teachers sometimes believe that simply placing manipulatives in front of students will help them learn (D. L. Ball, 1992; Moyer, 2001). It is important not to fall into this trap. Lessons involving manipulatives are not all equal. In successful lessons, teachers provide students sufficient time to work through problems using manipulatives and support students without taking over the thinking involved in the problem at hand. Successful teachers also plan lessons carefully in advance to overcome difficulties associated with classroom management and mathematical understanding. Less successful teachers tend to show students exactly what to do with the manipulatives. At the other extreme, some less successful teachers provide virtually no support for students as they work with the manipulatives, and the students often drift off in unproductive directions (Stein & Bovalino, 2001). When using manipulatives, it is important for teachers to continuously strike a balance between letting students explore the mathematics and actively supporting their problem solving activities.

Calculator Usage

Perhaps no strategies for teaching mathematics are more debated than those that incorporate technology. Some believe that students should not use calculators at all in the classroom. Others advocate the use of calculators for all mathematical tasks. It is best to navigate a course midway between these two extreme views. The Technology Connections included throughout this text illustrate how technology can be a powerful tool for teaching specific content areas when it is used appropriately. The key is to carefully examine and understand what might be meant by the phrase "appropriate usage of technology."

A. D. Thompson and Sproule (2000) presented a useful framework for helping teachers make decisions about when it is appropriate to have students use calculators. They suggested examining students' mathematical backgrounds and determining whether calculator usage is *essential* or *nonessential* for the task at hand. For example, a calculator is essential for calculating the equation of a least-squares regression line when this activity takes place in a class of ninth graders who have not studied the mechanics of doing so. The calculator may be nonessential for a class that had studied the mechanics, but even then the teacher may decide to allow students to use a calculator for the task, since getting caught up in laborious calculations may hinder students from studying broader concepts. Teachers should also consider whether the goals they have for students' learning are *process* oriented or *product* oriented. When the goal is to have students understand the *processes* of mathematical problem solving, withholding calculators could simply result in students becoming bogged down in computations. When the goal is to arrive at a *product* (e.g., computing the mean for a small set of numbers), teachers *might* choose to withhold calculators from students if they are *nonessential* for the task.

Adding complexity to the issue of technology usage is the fact that graphing calculators are allowed on standardized tests such as the SAT. They are required for Advanced Placement tests in calculus and statistics. Teachers understandably feel obligated to help students prepare for these tests, since they exert great influence on many students' lives after 12th grade. Given this situation, it is unrealistic for most teachers to deal with the technology issue simply by avoiding it. Instead, it is necessary to consider when and how technology should be used in class. The A. D. Thompson and Sproule (2000) framework described earlier gives guidance for making such decisions, and the Technology Connections in this book will help you think through the use of calculators and specialized software for the content areas you will teach.

Implementing the Common Core

See Homework Task 4 to reflect on the role of calculators in relation to using appropriate tools strategically (Standard for Mathematical Practice 5).

Cooperative Learning

On the surface, it may seem that the most essential feature of cooperative learning is to place students in groups to work on mathematics problems with one another. As with most teaching strategies, however, it is necessary to look beneath the surface. Four essential aspects of a cooperative learning environment are the following:

- Students learn in small groups with two to six members in a group.
- The learning tasks in which students are engaged require that the students mutually and positively depend on one another and on the group's work as a whole.
- The learning environment offers all members of the group an equal opportunity to interact with one another regarding the learning tasks and encourages them to communicate their ideas in various ways, for example, verbally.
- Each member of the group has a responsibility to contribute to the group work and is accountable for the learning progress of the group. (Leikin & Zaslavsky, 1999, p. 240)

If any one of the four aspects is missing, students' potential for learning is diminished. For example, if a group grows to more than six members, it can become difficult for students to interact with one another. If individual group members are not held accountable for the learning progress of the entire group, some group members will become passive as others in the group do all the work. One mechanism for building individual accountability is letting students know that randomly selected group members must be able to explain the reasoning that occurred within their groups (Artzt, 1999).

Various decisions in setting up cooperative learning experiences influence the quality of instruction that takes place. One of the most crucial decisions is the selection of a mathematics problem. Recall the discussion of "worthwhile mathematical tasks" (NCTM, 1991) from Chapter 1. Worthwhile tasks are those that facilitate discourse among group members as opposed to requiring only recall of a specific fact or procedure.

Another crucial decision is the amount of time cooperative groups will be given to work. Not allowing enough time for the task will prematurely and artificially end students' conversations, while allowing too much time makes it easy for students to drift off task (Artzt, 1999). Experimenting with various tasks in cooperative learning settings will help you determine how much time is reasonable.

Another important decision involves assigning roles to individual group members. Some recommend assigning specific roles to students, such as having one record group interactions, another prompt others to speak, and another report the results to the rest of the class. However, when placed in groups, students will find

> **Implementing the Common Core**
>
> See Homework Task 5 to analyze a cooperative learning structure that supports students' ability to "construct viable arguments and critique the reasoning of others" (Standard for Mathematical Practice 3).

their own ways to do things, and sometimes their strategies are more effective (Artzt, 1999). Therefore, before assigning fixed roles to individual students during group work, one must consider how those roles might influence each student's opportunity to learn.

When implemented effectively, cooperative learning can help students "learn to cooperate" as well as "cooperate to learn" (Good, Mulryan, & McCaslin, 1992). The discussion up to this point has focused mostly on fostering the latter outcome, but the former also has value. Outside of school, occupations increasingly require collaborations among individuals. Therefore, schooling should help students develop the skills necessary to survive in such an environment. Additionally, mathematics itself is not created in a vacuum. Mathematicians benefit from cooperation with others, and knowledge is built as they subject their work to the scrutiny of peers. Cooperative learning helps students experience some of the aspects of this process of collective knowledge construction.

Encouraging Student-Invented Strategies

Teaching through problem solving requires a break from the traditional classroom pattern of learning rules and practicing procedures. Students are to engage in authentic problem solving and not just replicate problem solving strategies demonstrated by the teacher. Because of this, students should be encouraged to construct their own strategies for solving problems. If they do not have opportunities to do so, there is a danger that they will see the teacher, rather than logical reasoning, as the sole source of authority in determining solutions to problems. They also may develop the erroneous belief that there is only one way to solve any given mathematics problem.

Tasks that elicit multiple solution strategies from students are good catalysts for student-invented strategies. Consider the following problem:

Such as growing pattern tasks

> Members of the Sports Booster Club at Norton Middle School raised money to buy 12 basketballs at $24 each. Before buying the basketballs, they decided to spend $144 of the money raised for some soccer balls. How many basketballs can they buy? (Cai & Kenney, 2000, p. 535)

Several different correct strategies are possible. Some students might solve the problem with a series of arithmetic operations. Others might elect to assign a variable to the unknown quantity (the number of basketballs they can buy), set up an equation, and solve it. As students share different solution strategies with one another, they understand mathematics in more depth. For example, those initially using only arithmetic operations to solve the problem can begin to appreciate the power of algebraic techniques as they see other students use them. Those using algebraic techniques may come to appreciate the manner in which those techniques simplify the work of solving the problem as they consider arithmetic solutions.

As students invent their own solution strategies, it is inevitable that mistakes will be made. Teachers must carefully monitor the ways they react to mistakes. It is critical that mistakes be treated as learning sites rather than obstacles to be avoided at any cost (Carpenter et al., 1997). In the field of mathematics, making errors is an essential part of the process of knowledge development. Mathematicians continually make conjectures and then test them to determine their validity. Errors are part of this process of conjecturing and testing. Therefore, classrooms that ask students to avoid errors at all costs create an artificial separation between school mathematics and the discipline of mathematics. In the classroom, errors can provide springboards for inquiry, much in the same manner that getting lost in an unfamiliar town provides an opportunity to discover new pathways for getting from place to place (Borasi, 1996). To help students become confident about engaging in the conjecturing and testing process, which necessarily involves making errors, teachers can

- grade homework problems not for correctness but for effort, completeness, and logical thought;
- encourage students to do "test corrections," that is, correct errors they made on tests by using only their notes and textbook, for added credit;
- give ample credit on graded material for original thought that does not work out; and
- include more noncomputational how-to questions on tests and quizzes. (Izen, 1999, p. 757)

These steps can help teachers convey the message that they value students' well-formulated attempts to construct their own solution strategies for mathematics problems.

Having Students Pose Problems

Although many of the problems solved in any given classroom are posed by the teacher, there is great value in having students pose their own problems. Knuth (2002) suggested using the following prompt as one possible springboard for student problem posing:

Given the sequence, 1, 1, 2, 3, 5, 8, 13, 21, . . . , what questions and observations come to mind? (p. 126)

Knuth noted that high school students generally offer observations rather than questions in response to this problem. He believed that this type of response might be due to a lack of emphasis on problem posing in school mathematics. Because problem posing and conjecturing are fundamental aspects of the construction of mathematical knowledge, it is worth the effort to help students develop the disposition to pose their own problems rather than to just solve those given by the teacher or by a textbook.

Strategies to foster students' disposition to pose problems have been suggested by various teachers. Leung and Wu (1999) suggested asking students to spot errors in teacher-posed problems and to propose corrections. Through this process, students can partner with teachers in posing problems correctly. Georgakis (1999) suggested including student-created problems as items on worksheets and tests. Wiest (2000) had students brainstorm ideas for mathematics problems related to their school cafeteria. These problems became motivation for an extended project. The process of formulating and investigating problems helped students apply their existing knowledge, develop new mathematical knowledge, and develop social skills. Students were motivated to carry out the project because they had a hand in designing it. Therefore, posing problems can help students develop favorable attitudes toward mathematics as they develop knowledge of mathematical concepts.

Lecture and Note Taking

Perhaps no teaching strategy has been more criticized than lecturing. Some humorous thoughts on lecture include

- "A lecture is a process by which the notes of the professor become the notes of the student without passing through the minds of either." (R. K. Rathbun)
- "If teaching were just telling then we'd all be so smart we couldn't stand up." (Mark Twain)

All of us have probably had the experience of sitting in a lecture, understanding parts of it, but then later being unable to work with the problems and ideas independently. Many have also had the experience of sitting in a lecture and copying notes without really understanding the main ideas being discussed. Morris Kline (1977) sharply criticized lecture as a method for teaching mathematics, stating, "Mathematicians have a naïve idea of pedagogy. They believe that if they state a series of concepts, theorems, and proofs correctly and clearly, with plenty of symbols, they must necessarily be understood" (p. 117).

Despite the criticism lecture has received, it would be wrong to draw the conclusion that teachers should never tell students anything in mathematics classes. Appropriate occasions for "telling" include reminding students of mathematical conclusions that have previously been established in class, telling students that their ideas are unclear, and introducing conventional notation and terminology (Chazan & Ball, 1999). When the substance of what the teacher shares is shaped by knowledge of students' thinking and the focus is on assisting the development of conceptual understanding, positive student learning outcomes can occur (Lobato, Clarke, & Ellis, 2005). On the other hand, lecture is unproductive when the teacher falls into the trap Kline (1977) described: believing that just because something has been clearly stated, it must necessarily be understood. This

Implementing the Common Core

See Clinical Task 1 to investigate the extent to which a classroom lecture helps students attain the standards for mathematical practice.

implies that when teachers lecture on a topic, student note taking should go beyond merely transcribing information verbatim from the board. Walmsley and Hickman (2006), for example, recommended having students keep a log to record questions about the material as they take notes. By reading the logs, teachers can gain a sense of how well their lecture connects with students.

Another important thing to monitor when using lecture is the quality of the examples included to illustrate concepts and procedures. Zodik and Zaslavsky's (2008) study of secondary school teachers raises several considerations. They found that teachers tend not to plan to share counterexamples, even though they are an important part of proof in mathematics. Teachers in the study generated counterexamples only when students made unexpected, invalid conjectures or statements. More opportunities for student-constructed examples were needed as well, as Zodik and Zaslavsky documented 604 teacher-generated examples and only 35 student-generated examples in the lessons they observed. Another consideration is that teacher-selected examples should address common student difficulties. For instance, one teacher, when teaching about corresponding and alternate angles, deliberately included cases not involving parallel lines, since she knew that students tended to think of these concepts only in connection with parallel lines. Careful, deliberate selection of numbers to be used in examples is also important. When teaching the Pythagorean theorem, for example, it is important to have students deal with some triples that are not multiples of one another so they do not form the incorrect generalization that all triples are multiples of one another. Presenting uncommon examples can be helpful as well, as in the case of a teacher who showed students a picture of a concave kite and asked them if it fit the formal definition that had been provided for kites. Such an activity has the potential to expand and deepen students' knowledge of the definition.

> ### Implementing the Common Core
>
> See Clinical Task 2 to analyze the roles of examples and counterexamples in supporting students' ability to "reason abstractly and quantitatively" (Standard for Mathematical Practice 2) and "construct viable arguments and critique the reasoning of others" (Standard for Mathematical Practice 3).

When information is presented at the front of the classroom, the display device is an important consideration. Common display devices include document cameras, interactive boards, and chalkboards. Document cameras can do all the work once done by overhead projectors and more. One of the most powerful uses of the document camera is to have students present work they have done at their desks individually or in groups to the rest of the class. No special media, such as plastic sheets or large sheets of paper, are needed, because document cameras can project magnified images of ordinary sheets of paper onto a screen for all to see. Interactive boards perform a variety of functions, such as projecting images from a computer (e.g., PowerPoint slides) and allowing work to be saved from one class session to the next, which is convenient when the class is unable to finish solving a problem before the bell. In such cases, the written work can be saved to file and then reprojected on the screen at the beginning of the next class period.

Despite the advantages of technological display devices such as document cameras and interactive boards, there are many times when the chalkboard is actually the best display device. The TIMSS video study (J. W. Stigler & Hiebert, 1999) provided insight on how chalkboard usage can enhance lessons. Japanese teachers carefully use the chalkboard as a means of recording a lesson. The progression of ideas is displayed from left to right, and important concepts and definitions are not erased as the lesson goes along. Students as well as teachers may add their work to the running history of the lesson on the chalkboard. At the end of the lesson, the board contains a summary of the crucial elements of the lesson. In contrast, when overhead projectors, document cameras, and interactive boards are used as display devices, any given problem or concept displayed disappears once a new page is put up. It is then difficult to go back to compare and contrast concepts with one another (e.g., to observe how solving a system of equations using substitution compares to using addition or subtraction). Having the history of the lesson on the chalkboard facilitates the process of making connections among important concepts.

Metaphor

A special type of example that often occurs in mathematics lessons is **metaphor**. A metaphor is "a figure of speech in which a word or phrase literally denoting one kind of object or idea is used in place of another to suggest a likeness or analogy between them" (Mish, 1991, p. 746). Metaphors help students understand new ideas in terms of those they already know. The first object or idea is called the *source* and the second the *target* (Gentner & Holyoak, 1997). Any given metaphor will capture some aspects of the target while failing to capture or even misrepresenting other aspects. This phenomenon is illustrated by the common saying that "every metaphor [or analogy] limps" (note that this saying itself is a metaphor—a person with a limp is the source, while the idea of metaphor is the target). Similarities between the source and target are often called the *ground* of the metaphor, and dissimilarities are called the *tension* (Presmeg, 1998). When introducing students to metaphors, teachers should help them explore both the ground and the tension of the metaphor to minimize misunderstandings (Glynn, 1991).

Mower (2003) provided ideas for folding metaphors and analogies into mathematics instruction. She suggested beginning the school year by having students complete the statement "Doing math is like . . ." (technically a simile) and having peers share responses with one another to start a discussion. She also suggested having students write metaphors to describe ideas such as problem solving, factoring, and functions. In practice, teachers often introduce metaphors for these concepts even when they are not aware they are doing so. For example, it is quite common for algebra teachers to draw a comparison between a function and a machine. Helping students discuss and unpack these commonly used metaphors by exploring the ground and tension inherent in each one can help them develop deeper mathematical understanding.

> **STOP TO REFLECT**
>
> Write metaphors for each of the following geometric concepts and describe the ground and tension for each metaphor: (1) plane, (2) point, (3) ray, (4) line.

Games

Games can provide rich settings for mathematical thinking. Analyzing a chess position is closely connected to mathematical thinking because "successful chess players must continuously scrutinize problems in multiple ways, predict the outcome of their actions, plan several steps ahead, and use visual information" (Berkman, 2004, p. 247). Playing popular commercial games like *Connect Four* and *Mastermind* can help enhance students' spatial sense and algebraic reasoning (Lach & Sakshaug, 2004). Bingo provides a setting for introducing students to concepts from combinatorics and probability (Bey, Reys, Simms, & Taylor, 2000). Various games of chance provided motivation for much of the early development of the field of probability during the 17th century (S. M. Stigler, 1986). Hence, games can function as springboards for mathematical thinking rather than being mere diversions.

Online games expand the options available to teachers. NCTM hosts the site Calculation Nation (www.calculationnation.com), through which students can challenge others from around the world to competition. By playing these games—unlike some electronic games for learning mathematics—students learn content that goes beyond the procedural. In one game, for example, students must create a fraction that is either larger or smaller than one created by an opponent. This sort of task encourages students to focus on the conceptual issue of the link between the sizes of fractions and their representations. In another game, students draw on knowledge of angles, symmetry, and reflections to choose the best path for a ball. Since the games are freely available online, they can be used to help supplement classroom instruction in schools with access to the Internet.

AVOIDING COUNTERPRODUCTIVE TEACHING STRATEGIES

It should be noted that each of the 10 strategies and tools outlined previously can be used in counterproductive ways. For example, using manipulatives is counterproductive to teaching through problem solving if the teacher simply prescribes a series of steps that students are to carry out. Technology usage can be harmful if calculators are used as substitutes for computational fluency rather than tools to support its development. Lecture with student note taking is almost useless if it consists only of the student transcribing notes given by the teacher. Games can become unnecessary diversions if they lead to off-task behavior or do not connect to broader curricular goals. In general, before deciding on any given teaching strategy, ask yourself (1) What are my learning objectives for my students? and (2) How does this strategy help (or not help) my students attain these objectives, given what I know about my students' current mathematical thinking? Without reflection on these questions, your teaching strategies will probably not match your students' learning needs.

There are other commonly used strategies, not mentioned so far, that are usually counterproductive to learning mathematics. One example is the use of timed drills and tests. Although computational fluency is an important objective, imposing a time limit on students as they do mathematics usually does not help them achieve it. Burns (2000) argued that

> Children who perform well under time pressure display their skills. Children who have difficulty with skills, or who work more slowly, run the risk of reinforcing wrong learning under pressure. In addition, children can become fearful and negative toward their math learning. (p. 157)

Timed drills show how quickly a student can recall facts, but they do not serve a teaching function.

Another commonly used strategy of questionable value is teaching students the "key word" approach to solving word problems. L. L. Clement and Bernhard (2005) used the following problem to illustrate the pitfalls of training students to pick out key words:

> Susan collected 6 rocks, which were 4 more than Jan collected. How many rocks did Jan collect? (p. 360)

Implementing the Common Core

See Clinical Tasks 3 and 4 to explore how students' knowledge of the context of a word problem can support their ability to "reason abstractly and quantitatively" (Standard for Mathematical Practice 2).

A student trained in the key word approach would likely pick out the phrase "more than" as a signal to add the two quantities in the problem. Doing so would yield the incorrect answer.

Although timed drills and key word approaches are common strategies for teaching mathematics, they usually do more harm than good. These two examples highlight the need to critically analyze any teaching strategy before accepting or rejecting it, and not just implement a strategy because it is commonly used.

CONCLUSION

Hopefully, this chapter on planning lessons has whetted your appetite to learn and try out more teaching strategies. Chapters 7 through 11, which deal with specific mathematics content areas, will continue to discuss ways to plan lessons. With any given class, it may take some experimentation on your part to find the teaching strategies that fit you and your students. No one can simply hand you a set of strategies guaranteed to work well in any given situation. In making choices about what you will do in your own classroom, keep in mind the broad goal of enhancing students' conceptual understanding. Procedural understanding is also important, but if it becomes your sole focus, then strategies such as cooperative work, manipulatives, technology, and games will not help compensate for disregard of the conceptual.

VOCABULARY LIST

After reading this chapter, you should be able to offer reasonable definitions for the following ideas (listed in their order of first occurrence) and describe their relevance to teaching mathematics:

Four-column lesson plan   53

Student preconceptions   54

Differentiated instruction   57

Open questions   57

Parallel tasks   57

Reading to learn mathematics   60

Writing prompts   60

Metaphor   68

HOMEWORK TASKS

1. Write a four-column lesson plan for a lesson built around the yogurt problem discussed near the beginning of the chapter. Use the template shown in Figure 3.1 to write the lesson. Include both formative and summative assessment techniques in the fourth column. Then explain how the lesson will elicit and address students' preconceptions, build their conceptual and procedural knowledge, and encourage metacognitive activity.

2. Read the following article:

   Thompson, D. R., & Rubenstein, R. N. (2000). Learning mathematics vocabulary: Potential pitfalls and instructional strategies. *Mathematics Teacher, 93,* 568–574.

   Using a mathematics dictionary or a textbook, locate 10 mathematics vocabulary terms you believe it might be difficult for students to learn. The terms you choose should be different from those listed on p. 569 of the article. Place the 10 vocabulary terms into one or more of the categories listed on p. 569 of the article. Then describe two different teaching strategies you would use to address some of the potential difficulties in understanding that you have identified.

3. Read one of the following articles (or locate your own article from an NCTM publication that discusses using literature as a springboard for teaching mathematics):

   Bay-Williams, J. M. (2005). Poetry in motion: Using Shel Silverstein's works to engage students in mathematics. *Mathematics Teaching in the Middle School, 10,* 386–393.
   Beaton, T. (2004). Harry Potter in the mathematics classroom. *Mathematics Teaching in the Middle School, 10,* 23–25.
   Johnson, I. D. (2006). Grandfather Tang goes to high school. *Mathematics Teacher, 99,* 522–526.
   McShea, B., Vogel, J., & Yarnevich, M. (2005). Harry Potter and the magic of mathematics. *Mathematics Teaching in the Middle School, 10,* 408–414.
   Taber, S. B. (2005). The mathematics of *Alice's Adventures in Wonderland. Mathematics Teaching in the Middle School, 11,* 165–171.

   After reading one of the articles, describe how the ideas in it could help you launch a problem-centered lesson on one or more of the content areas prescribed in the *Common Core State Standards.* Specifically identify the standard(s) you are addressing.

4. Read the following article:

   Podlesni, J. (1999). A new breed of calculators: Do they change the way we teach? *Mathematics Teacher, 92,* 88–89.

   After reading the article, write a critique of it. With which of Podlesni's arguments do you agree? With which arguments do you disagree? Why?

5. Read the following article:

   Leikin, R., & Zaslavsky, O. (1999). Cooperative learning in mathematics. *Mathematics Teacher, 92,* 240–246.

   In 100 to 150 words, summarize the exchange-of-knowledge method of cooperative learning. Then write a four-column lesson plan that uses the method to teach a given topic.

6. Examine the websites of at least two manufacturers of interactive boards (two possibilities are Smart Technologies and Promethean). Choose two comparable interactive board products (one from each company chosen) and decide which board you would rather have in your classroom. Then write a hypothetical letter to your school principal explaining why you would like the school to purchase the board you selected. In your letter, explain why the board you prefer is superior to the board from the other company. Also include at least three learning activities you would do with the board in your classroom. Your learning activities should be consistent with research about how students learn mathematics.

CLINICAL TASKS

1. Observe a lecture-based mathematics lesson in which students are required to take notes. Take verbatim notes on the material. As you do so, write questions in the margin that you think students may have about the material. Following the lesson, interview two students who took notes during the lecture. Ask them which parts of the lecture they had questions about. Compare and contrast your list of questions with the students' questions.

2. Observe a mathematics lesson and focus on the types of examples used by the teacher. Refer to specific examples you saw the teacher use to support your responses to the following questions:

   a. Did the teacher include examples designed to illustrate procedures? To illustrate concepts?

   b. Were counterexamples used at any point?

   c. Were there any student-generated examples?

   d. Did any of the examples address common student misconceptions?

   e. Did any of the examples illustrate unusual instances of a concept or procedure?

3. Choose a word problem from a mathematics textbook being used by a class you observe. Describe at least two different correct strategies that might be used to solve it. Also describe at least two different incorrect strategies. Finally, describe how paying attention to the context of the problem might help students judge the reasonableness of their solution strategy and the answer they obtain.

4. Interview three students from classes you are observing and ask them to think aloud as they solve a word problem from a textbook. To what extent do they focus on the context of the problem to formulate a solution and judge the reasonableness of their answers? To what extent do they try to use key words to solve the problem?

VIGNETTE ANALYSIS ACTIVITY	**Focus on Constructing Viable Arguments and Critiquing the Reasoning of Others (CCSS Standard for Mathematical Practice 3)**

Items to Consider Before Reading the Vignette

Read CCSS Standard for Mathematical Practice 3 in Appendix A. Then respond to the following items:

1. Drawing on your own experiences as a mathematics student, provide specific examples of teaching practices that can help students construct viable arguments and critique the reasoning of others.

2. Drawing on your own experiences as a mathematics student, provide specific examples of teaching practices that might prevent students from constructing viable arguments and critiquing the reasoning of others.

Scenario

It was near the end of the school year, and Miss Fielder's freshman algebra class was beginning to get restless. As the students anticipated summer vacation, their thoughts seemed to drift away from academics more and more. This made Miss Fielder nervous as she thought about how to prepare them for an upcoming unit test. Students generally did not react well to days devoted entirely to reviewing a chapter. They tended to complain that such days were boring and got off task more than usual. Knowing this, Miss Fielder decided to try something different during the class she had scheduled for test review.

The Lesson

The lesson began the same way as most others. As students entered the room, they were given a set of warm-up exercises to do. The warm-up was quite lengthy, requiring students to solve six inequalities involving rational algebraic expressions. For the first 10 minutes of class, students appeared to work diligently and independently on the exercises. As students approached some of the more difficult problems near the end of the warm-up, hands shot up. For 15 minutes Miss Fielder went to each student who raised a hand and answered questions. As she did so, students who had completed the task or had given up on the more difficult exercises began to converse about their social networking websites and the previous night's wrestling match.

When Miss Fielder had finished answering questions, she turned off the document camera she had used to project the exercises on a screen and called six students to the board to present their work on the warm-up exercises. As these six copied the work from their papers to the chalkboard, the rest continued to socialize. After several minutes, the work for all six problems was on the board and the students returned to their seats. Miss Fielder directed the attention of the class to the chalkboard. She went through each exercise, explaining what the students had done at the board. She answered additional questions students posed as she explained the exercises at the board. About 30 minutes into the class period, the warm-up activity was finally completed.

The next activity was discussing the quiz the students had taken the previous day. Miss Fielder handed quiz papers back to students and then projected the quiz solutions on the document camera. When all of the students had received their quizzes, she asked if there were any questions. Regarding a quiz item having a fraction as its answer, a student asked, "Why is the top number 6 and the bottom number 7?" In response, Miss Fielder showed the procedure used to obtain the correct answer. In regard to another quiz question, a student asked, "Why do you have to times x plus 5 by 3 in that one?" Miss Fielder demonstrated a correct procedure for solving this problem as well. By the time all questions about the quiz were answered, the class had used up 45 of the 90 minutes available that day.

Finally, Miss Fielder introduced what she considered to be the main activity of the day. She told students they would be put in groups of three or four so that they could reteach the main ideas from the chapter to one another. Each group would be responsible for presenting a section of the text to the class. They were to write a summary of their assigned section, select two or three problems for the other students to solve, and write two or three problems, based on the ideas in their section, for the upcoming test. When they had finished working in their groups, they were to present their summary and practice problems to the rest of the class.

Students seemed intrigued by the task and started working on it. Procedural questions soon arose, such as "How hard should the problems be that we pick?" "How much time do we have to finish our group work?" "How much time do we have to present?" and "Should we pick easy questions for the quiz so everyone gets a good grade?" As Miss Fielder addressed these questions and students started working with one another, the fire alarm rang. Students filed out of the room and onto the front lawn for the fire drill. About 15 minutes later, they returned to their classroom.

When students returned to the classroom, about 25 minutes remained in the period. Miss Fielder was pleased to see that students quickly returned to the task of discussing their sections of the chapter with one another upon reentering the room. She could see them explaining problems to one another, writing problems related to each section of the text, and formulating some questions that would be easy for the class as well as some that would stump them. After working for 15 minutes, most groups felt they were ready to present.

Near the end of the period, Miss Fielder asked the first group to do their presentation. They summarized their section by saying, "An inequality is basically the same thing as an equation, so you have to pretend that it has an equals sign to find its answer." They went on to give examples of when to use a "closed dot" to graph an inequality and when to use an "open dot." Before they were able to present the problems they selected, the bell rang. As students filed out of the room, Miss Fielder told them that the remaining groups would present the next day.

Questions for Reflection and Discussion

1. Which aspects of the CCSS Standard for Mathematical Practice 3 did Miss Fielder's students seem to attain? Explain.

2. Which aspects of the CCSS Standard for Mathematical Practice 3 did Miss Fielder's students not seem to attain? Explain.

3. What strategies could Miss Fielder have used to redirect students toward mathematical discourse on occasions when they fell into off-task conversation?

4. Should every mathematics lesson have a warm-up activity? What is the purpose, in general, of having a warm-up activity?

5. How could Miss Fielder enhance student-student mathematical discourse?

6. Comment on Miss Fielder's use of time during the class. Was time well allocated for each activity, or could improvements be made? Explain.

7. Comment on students' use of mathematical language during the activity. Are there areas in which improvement is needed? Explain.

RESOURCES TO EXPLORE

Books

Borasi, R., & Siegel, M. (2000). *Reading counts: Expanding the role of reading in mathematics classrooms.* New York, NY: Teachers College Press.

Description: The role of reading in the learning of mathematics is often underemphasized. The authors of this book provide theoretical grounding for reading in mathematics education along with practical strategies for reading to learn mathematics.

Gutstein, E. (2005). *Reading and writing the world with mathematics: Toward a pedagogy of social justice.* New York, NY: Routledge.

Description: The author argues that mathematics should empower students to work for social change and address injustice. Examples of how such instruction can be carried out in a classroom setting are provided.

Merseth, K. K. (2003). *Windows on teaching math: Cases of middle and secondary classrooms.* New York, NY: Teachers College Press.

Description: This book takes the reader inside several middle and high school mathematics classrooms. Case studies present examples of student work, critical decisions made by teachers, and classroom discourse for readers to analyze.

Small, M., & Lin, A. (2010). *More good questions: Great ways to differentiate secondary mathematics instruction.* Reston, VA: NCTM.

Description: The author proposes two central strategies for differentiating mathematics instruction: posing open questions and using parallel tasks. Examples of these strategies are illustrated within the context of each of the content strands of the NCTM standards, and guidance is provided for creating additional open questions and parallel tasks.

Websites

Lesson study videos: **http://hrd.apecwiki.org/index.php/Mathematics_Education**

Description: This website is maintained by the Asian-Pacific Economic Cooperative (APEC). It contains videos of lesson study implementation by middle and high school mathematics teachers.

National Library of Virtual Manipulatives: **http://nlvm.usu.edu/en/nav/vlibrary.html**

Description: Virtually all types of mathematics manipulatives sold commercially are available on this website as free online versions. The website is organized by grade band and content area, making it possible to quickly identify suitable manipulatives for a given lesson.

NCTM Illuminations: **http://illuminations.nctm.org/**

Description: Illuminations is an extensive collection of lesson plans and other electronic resources. Lessons are searchable by grade level and content area, facilitating the process of finding ideas for carrying out NCTM standards–based instruction.

Chapter 4

MATHEMATICS CURRICULUM MODELS AND TECHNIQUES

The word *curriculum* has been defined in many different ways. A definition that encompasses many existing ideas about curriculum is

> An operational plan for instruction that details what mathematics students need to know, how students are to achieve the identified curricular goals, what teachers are to do to help students develop their mathematical knowledge, and the context in which teaching and learning occur. (National Council of Teachers of Mathematics [NCTM], 1989, p. 1)

Given this broad definition, it can be seen that previous chapters have already discussed some curricular considerations, and subsequent chapters will continue to do so. This chapter extends curricular ideas discussed so far and also sets the stage for the second portion of the text. Topics to be explored along the way include recent influential curriculum documents, prevalent curricular principles and trends, and criteria for choosing curriculum materials.

RECENT INFLUENTIAL CURRICULUM DOCUMENTS

Political Context

The number of documents containing recommendations for middle and high school mathematics curricula grew dramatically in the United States during the first decade of the 21st century. Much of this growth was due to the formulation of mathematics standards documents by individual states. Under the federal law **No Child Left Behind** (NCLB; 2002), states were prompted to identify mathematics learning expectations for students. As states did so, they formulated grade-level learning expectations with substantial differences from one another. For example, state-level documents varied a great deal on when fraction multiplication and division should be introduced. Some expected doing so as early as Grade 4, while others waited until Grade 5, 6, 7, or 8. One state did not even include fraction division as a topic of study (B. J. Reys, et al., 2006). Similar variation in learning expectations occurred in other mathematics content areas as well (B. J. Reys, 2006).

In addition to having wide variation in learning expectations, many state standards formulated in response to NCLB contained a large number of learning expectations per grade level; some states included more than 50 learning expectations for each grade (NCTM, 2006). In many states, mathematical reasoning and sense-making were not comprehensively addressed as part of the curriculum (Kim & Kasmer, 2006). The learning expectations in state standards documents generally consisted of isolated skills to be learned and were not presented as a coherent, connected body of knowledge. Because of this, many state standards documents became laundry lists of skills to "cover" rather than roadmaps for helping students develop mathematical thinking.

The large number of learning expectations per grade level and the variation among state recommendations led NCTM to produce two curriculum documents. The first, titled *Curriculum Focal Points for Prekindergarten Through Grade 8 Mathematics* (NCTM, 2006), contains recommendations for topics that students should learn in middle school and the earlier grades. *Focal Points* was followed by *Focus in High School Mathematics: Reasoning and Sense-Making* (NCTM, 2009), which addresses the substance of Grades 9 through 12. These two publications aimed to supplement and clarify the standards for Grades 6 through 8 and 9 through 12 set forth in *Principles and Standards for School Mathematics* (NCTM, 2000) while simultaneously addressing the problem of low-quality state standards produced in haste at the beginning of the decade. The documents also helped form the basis of the *Common Core State Standards*, so it is important to understand the nature and purpose of each.

NCTM Curriculum Focal Points

In defining the concept of **curriculum focal point**, NCTM (2006) stated,

> Curriculum focal points are important mathematical topics for each grade level, pre-K–8. These areas of instructional emphasis can serve as organizing structures for curriculum design and instruction at and across grade levels. The topics are central to mathematics: they convey knowledge and skills that are essential to educated citizens, and they provide foundations for further mathematical learning. (p. 5)

NCTM set three focal points for each grade level. By identifying a small number of focal points per grade level, NCTM hoped to encourage teaching for depth of understanding rather than hasty, cursory treatment of many topics each year. Summaries of the focal points for Grades 6, 7, and 8 appear in the following.

Grade 6 Focal Points

The first focal point NCTM (2006) identified for Grade 6 is "developing an understanding of and fluency with multiplication and division of fractions and decimals" (p. 18). In contrast to conventional approaches, this focal point recommends that students make sense of the computational procedures for multiplication and division of fractions and understand why they work. The second Grade 6 focal point, also in the realm of number and operations, is "connecting ratio and rate to multiplication and division" (NCTM, 2006, p. 18). This includes problems such as computing the price of an ounce of cereal using the price of a 12-ounce box and then using multiplication to determine what the price of a 17-ounce box should be. In the final Grade 6 focal point, NCTM stated that students should be involved in "writing, interpreting, and using mathematical expressions and equations" (p. 18). In doing so, sixth graders develop understanding of many of the fundamental concepts of formal algebra.

Grade 7 Focal Points

The first curriculum focal point for Grade 7 involves the mathematical content areas of number and operations, algebra, and geometry: Students should be "developing an understanding of and applying proportionality, including similarity" (NCTM, 2006, p. 19). Problems involving percentages are to be central to the seventh-grade curriculum, and students are to work with situations such as computing discounts, taxes, tips, percentages of increase and decrease, and scale factors. The second curriculum focal point draws on measurement, geometry, and algebra to recommend that seventh graders work on "developing an understanding of and using formulas to determine surface areas and volumes of three-dimensional shapes" (NCTM, 2006, p. 19). To help students develop an understanding of these formulas, NCTM recommends having students decompose three-dimensional figures into smaller parts for analysis. The final curriculum focal point for Grade 7 includes both number and operations and algebra, recommending that students spend time "developing an understanding of operations on all rational numbers and solving linear equations" (NCTM, 2006, p. 19). This focal point identifies work with negative integers as a particularly important goal.

Grade 8 Focal Points

NCTM (2006) identified "analyzing and representing linear functions and solving linear equations and systems of linear equations" (p. 20) as one of the focal points for Grade 8. Students are to begin to understand slope as a constant rate of change for a line, and also to interpret slope in context. Proportion should be understood in terms of its connection to linear relationships. The second Grade 8 focal point concerns the geometry and measurement learning goal of "analyzing two- and three-dimensional space and figures by using distance and angle" (NCTM, 2006, p. 20). Important geometry and measurement ideas to be developed in Grade 8 include understanding why the sum of the angles of a triangle is 180 degrees and why the Pythagorean theorem holds. The final eighth-grade focal point is "analyzing and summarizing data sets" (NCTM, 2006, p. 20). This final focal point recommends that students use descriptive statistics to characterize and compare data sets.

Implementing the Common Core

Compare NCTM's curriculum focal points for Grades 6–8 to the *Common Core State Standards* for the same grade levels (Appendix B). How are they similar? How are they different?

NCTM's *Focus in High School Mathematics*

The developers of *Focus in High School Mathematics: Reasoning and Sense-Making* (NCTM, 2009) took a unique approach to providing high school curriculum recommendations. Whereas *Focal Points* made recommendations for the content students should learn, **Focus in High School Mathematics** described the reasoning and sense-making processes students should develop and use when working across all mathematical content areas. Although the document provided examples of how reasoning and sense-making are developed and used in specific parts of the mathematics curriculum, it did not attempt to provide a comprehensive list of content students should master in high school.

In *Focus in High School Mathematics*, the terms **reasoning** and **sense-making** have specific meanings. NCTM (2009) stated, "In the most general terms, reasoning can be thought of as the process of drawing conclusions on the basis of evidence or stated assumptions" (p. 4). Under this characterization, the construction of a formal proof can be considered an example of a task requiring mathematical reasoning. However, reasoning is broader than formal proof—it also encompasses informal explanations and pattern observations. Reasoning is often "messy" in that a conjecture one thinks to be true may ultimately be proved false. NCTM defined sense-making as "developing understanding of a situation, context, or concept by connecting it with existing knowledge" (p. 4). Reasoning and sense-making are to be intertwined in the process of moving from informal observation to formal thinking and deductive proof. NCTM provided specific examples to illustrate how reasoning and sense-making can occur in working with number and measurement, algebraic symbols, functions, geometry, and statistics and probability. Some of the examples are described next.

Number and Measurement

One of the vignettes in *Focus in High School Mathematics* described a situation in which a teacher asked students to determine the total surface area of the earth. One student started by simply conjecturing it would be a large number, very roughly estimating it to be "a million or maybe a billion square miles." Another student suggested refining this estimate by thinking of the earth as a sphere and determining its radius. After determining that the radius of the earth is approximately 4,000 miles, students used the formula $A = 4\pi r^2$ to approximate its surface area. One student determined 4π to be approximately 12, multiplied 12 by 16 to obtain 192, and then rounded 192 up to 200 to produce an estimate of 200 million square miles. Another student objected to this method, claiming that it would be better to use the π-button on the calculator to obtain a more precise estimate of 201,061,930. This disagreement sparked discussion about several elements of the problem. Some students argued that the earth is not exactly spherical, raising doubt that using the π-button on the calculator would actually provide a better estimate. Students also found that the two estimates, 200,000,000 and 201,061,930 square miles, were within 1 percent of one another, calling into question the significance of the difference between the estimates.

Reasoning and sense-making were both evident in this vignette. Reasoning occurred as students made and tested conjectures about the total surface area. Sense-making occurred as students supported their reasoning by drawing on previous knowledge, such as the formula for the surface area of a sphere and the fact that the earth is not precisely spherical. Along the way, students began to build more formal notions about accuracy in estimation. The idea that one estimate was more accurate than another because it used a more precise value for π was challenged by drawing on the knowledge that the earth is not a perfect sphere. The difference between the two estimates was also ultimately expressed in formal terms by drawing on knowledge of percentages. In short, the vignette traced out how reasoning and sense-making supported moving from very informal conjectures to more formal notions about the concepts of approximation and error in measurement.

Algebraic Symbols

One vignette on reasoning with algebraic symbols in *Focus in High School Mathematics* began with a teacher posing the following problem:

> A slab of soap on one pan of a scale balances $\frac{3}{4}$ of a slab of soap of equal weight and a $\frac{3}{4}$-pound weight on the other pan. How much does the slab of soap weigh? (NCTM, 2009, p. 35)

Students were asked to solve the problem using an algebraic equation and also with arithmetic reasoning. One student reasoned that the weight of a full slab of soap could be represented with the symbol x. The situation was then represented with the equation $x = \frac{3}{4}x + \frac{3}{4}$, which produced $x = 3$. A second student, without using x to represent the weight of a full slab of soap, solved the problem by mentally removing $\frac{3}{4}$ of a slab from each side. This left $\frac{1}{4}$ slab of soap on one side of the scale and a $\frac{3}{4}$ pound weight on the other, meaning that $\frac{1}{4}$ slab must weigh $\frac{3}{4}$ pound. Multiplication of $\frac{3}{4}$ by 4 was then used to determine the weight of a full slab to be 3 pounds. Two different solution strategies yielded the same result.

This vignette illustrates how students can come to understand the meaning of their work with formal algebraic symbols. The reasoning processes students used were complementary. The approach using an equation was an abbreviated procedure made possible by formal algebraic symbols. The second approach provided an intuitive basis for the formal one. Understanding both approaches is important in helping students obtain a well-rounded understanding and appreciation of the power of algebraic symbols as they come to "see algebra as an extension of concrete arithmetic reasoning" (NCTM, 2009, p. 35).

> **Implementing the Common Core**
>
> See Clinical Task 3 to assess a student's ability to "make sense of problems and persevere in solving them" (Standard for Mathematical Practice 1), "reason abstractly and quantitatively" (Standard for Mathematical Practice 2), and "model with mathematics" (Standard for Mathematical Practice 4) when working the "slab of soap" problem.

Functions

The study of functions is foundational to high school algebra. One vignette in *Focus in High School Mathematics* illustrates how students may work with functions to model real-world phenomena. Students were told that high tide at a port occurred at 5 A.M. and that the next low tide occurred at 11 A.M. The water was 10.6 meters deep at high tide and 6.5 meters deep at low tide. Given this information, students were asked to develop a model to predict water depth as a function of time elapsed since midnight. One student proposed fitting a linear function through the two given points, but that idea was rejected when others pointed out that the tide would not continue to decrease forever. The observation that the tides would oscillate prompted consideration of sine and cosine functions as reasonable models. Students plotted the two given points on a coordinate grid and discussed a suitable period and amplitude for a function fitting the given information. As they worked, students checked the reasonableness of their models by using graphing technology.

Sense-making came into play when students judged an oscillating graph to be a more suitable model than a straight line. To reach this conclusion, they drew on previous knowledge that tides go up and down periodically and then made the connection that some trigonometric functions exhibit the same behavior. After deciding to use a trigonometric model, students used mathematical reasoning to make, test, and refine conjectures about the period and amplitude of the best trigonometric function for the situation. Throughout the vignette, communication among students played an important role as they pushed one another to construct the most accurate mathematical model possible by building on the information given and taking the context of the problem into account.

Geometry

NCTM (2009) noted that transformation geometry can help students understand congruence, similarity, and symmetry. One of the vignettes in *Focus in High School Mathematics* described events transpiring in a classroom when students were asked to determine which regular polygons had 80-degree rotational symmetry. After discussing the problem in small groups, students were to present their solutions to the rest of the class. The first group to present concluded that no regular polygons had 80-degree rotational symmetry, reasoning that 80 does not divide 360. The second group to present acknowledged initially using the same reasoning as the first group, but then they considered a polygon with 360 sides. They reasoned that if such a polygon were rotated 1 degree, its appearance would remain unchanged. Hence, rotating 80 times, by 1 degree each time, would have the same effect. After further discussion, students reasoned that a 180-gon would also have 80-degree rotational symmetry. At the end of the class period, students were given the homework problem of finding all regular polygons with 80-degree rotational symmetry and writing a short essay to prove they had found them all.

The progression from informal reasoning to formal proof is very evident in the rotational symmetry vignette. Students initially approached the problem by making conjectures that were ultimately incorrect. After class discussion, examples of polygons with 80-degree rotational symmetry were identified. The examples that were found suggested there might be additional instances of regular polygons with 80-degree rotational symmetry. By examining common characteristics of the 360-gon and 180-gon, students began to make more sophisticated conjectures about the types of regular polygons with 80-degree rotational symmetry. Writing an essay to formally show they had found all such polygons became the culminating assignment for an activity that began with informal reasoning and conjectures.

Statistics and Probability

Statistics and probability are becoming increasingly important in high school curricula as the amount of data and statistical studies available to the public continue to increase dramatically. One of the vignettes in *Focus in High School Mathematics* described a situation in which students were asked to make sense of a statistical study.

In the study, researchers had investigated whether it is easier to memorize lists of meaningful three-letter words (e.g., *car*, *bat*, *cat*) or lists of three-letter nonsense words (e.g., *cra*, *atb*, *tca*). They had randomly divided study participants into two groups of 15. One group had been asked to memorize lists of meaningful words, and the other had been asked to memorize lists of nonsense words. Researchers had gathered data on the number of words each subject recalled correctly. Students were shown the numerical data from the study and asked to construct displays to summarize and compare the data sets. From the data displays and numerical data, they were to come to a conclusion about whether it is easier to memorize lists of meaningful words or lists of nonsense words.

Reasoning and sense-making were evident as students analyzed the data. Students initially constructed dotplots to compare the data sets. Examining the values and clusters displayed in the dotplots, they reasoned that memorizing meaningful words was easier than memorizing nonsense words. When the teacher pushed students to assign a typical value to the number of words recalled in each group involved in the experiment, students initially suggested computing the arithmetic mean. After further discussion, some students suggested using the median to summarize and compare the data sets because extreme values in one data set exerted a great deal of influence on the mean. Eventually, students suggested determining the range and interquartile range of each data set to begin to quantify the spread in each one. Using these two measures led naturally to the construction of boxplots to help refine the initial arguments. Discussion of the problem eventually turned to considering the extent to which the findings of the study might generalize to other populations.

CURRICULAR TECHNIQUES IN MATHEMATICS

There are many curricula available that reflect NCTM's vision for the substance of middle and high school mathematics. After the NCTM (1989) standards were published, the National Science Foundation (NSF) funded a number of curriculum projects for middle and high school mathematics to meet the standards. **NSF-funded curricula** for middle school included *Mathematics in Context* (MiC), *MathScape*, *Math Thematics*, *Pathways to Algebra and Geometry*, and *Connected Mathematics Project* (CMP). NSF-funded high school curriculum projects included *Mathematics: Modeling Our World* (MMOW), *Interactive Mathematics Program* (IMP), *Core-Plus Mathematics Project* (CPMP), and *SIMSS Integrated Mathematics: A Modeling Approach Using Technology*. Along with the NSF projects, several other commercial curricula have been produced. Some of those discussed below are *Cognitive Tutor*, *College Preparatory Mathematics*, *Discovering Geometry*, and *Saxon Mathematics*.

Each curriculum listed above takes a different overall approach to helping students develop mathematical thinking. A variety of techniques can be seen within each approach. Some of the most prevalent techniques include use of real-world contexts, teaching through problem-solving, thematic units, integration of mathematical content areas, guided investigation, individualized instruction, progressive formalization, and distributed practice. Each technique is described in the following with illustrative examples. Many curricula employ several of the techniques.

Implementing the Common Core

See Clinical Task 4 to assess the techniques used by a textbook to support students' attainment of the standards for mathematical practice.

Use of Real-World Contexts

The MMOW high school curriculum exemplifies a strong commitment to the use of real-world contexts. Each unit is based on real-world situations. The first book in the MMOW four-year curriculum, for example, contains units that explore the mathematics involved in voting, cryptography, and tracking wildlife populations (Consortium for Mathematics and Its Applications [COMAP], 2010). It is important to note that these real-world settings do not serve as "applications" in the conventional sense of the word, but as sites for learning mathematics. Students do not learn decontextualized mathematics first and apply it later. Rather, mathematics concepts are learned as students engage in solving a given problem. Such problems are substantially different from "real-world" problems in conventional curricula, which tend to provide many practice problems before asking students to use the mathematics in a real-world situation. In such curricula, the real-world situations are primarily intended to help students practice mathematics they already know. In reform-oriented curricula, students acquire mathematical knowledge and skills as they work problems.

Consider the activity from MMOW shown in Figure 4.1. It appears as part of a unit in which students have previously developed a model for human height as a function of femur length. In solving the problem, students have the opportunity to learn a variety of mathematics concepts. Most notably, they learn how the parameter m, representing slope, affects the graph of a linear function. The final question in Figure 4.1 also leads students to notice that the y-intercept of each graph in the previous questions

Figure 4.1 Excerpt from *Mathematics: Modeling Our World* (Consortium for Mathematics and Its Applications, 2010, p. 218).

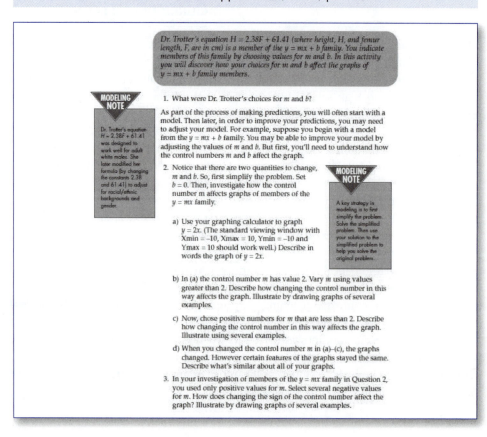

remained the same. Subsequent problems engage students further in experimenting with different values for the parameter b. This approach to learning how different values for m and b change a linear graph stands in contrast to conventional curricula that begin by teaching the ideas of slope and y-intercept in the abstract and then go on to ask students to use the concepts in application problems.

Work through the problem shown in Figure 4.1. As you work, list the mathematics concepts you would expect students to learn while doing the problem. Compare and contrast this approach to teaching those concepts to the way you initially learned the concepts embedded in the problem.

STOP TO REFLECT

Teaching Through Problem Solving

All NSF-funded mathematics curricula use the technique of teaching mathematics through problem solving. As mentioned in Chapter 1, teaching through problem solving is at the heart of NCTM standards–based instruction. Teaching through problem solving is more than just posing a set of exercises for students to solve. Instead, it involves carefully choosing mathematics problems that bring out important mathematics concepts students are to learn. The excerpt from MMOW (Figure 4.1) exemplifies this approach, because students learn about the mathematics concepts of slope and y-intercept by solving a problem. Although the MMOW excerpt involved a real-world example, interesting and worthwhile tasks can often be set in purely mathematical contexts. Consider, for example, the CMP excerpt shown in Figure 4.2. This problem prompts students to make and test a conjecture about side lengths that could form a quadrilateral. Subsequent problems in the same unit prompt students to make a generalization about segment lengths that will form a quadrilateral. Engaging in problem solving and reflecting on the strategies used to do so are pedagogical techniques used throughout the CMP curriculum.

Work through the CMP problem shown in Figure 4.2. Share your solution with a classmate and compare and contrast your strategies.

STOP TO REFLECT

Another curriculum that makes use of teaching through problem solving is *College Preparatory Mathematics* (CPM). Consider the CPM excerpt shown in Figure 4.3. It presents a scenario prompting students to investigate parabola transformations. Students are to make adjustments to the equation $y = x^2$ to change the shape, direction, and location of the graph of the corresponding parabola. As students try different strategies, they learn how graphs of parabolas are related to their symbolic representations. Each team of students is challenged to find a transformed parabola that other teams may not find. As they do so, they record their findings in writing to share with the rest of the class later on. Three discussion points are offered to help direct students toward the main mathematical objectives of the activity. The CPM excerpt exemplifies how an activity can strike a balance between providing student support for problem solving and allowing space for creative thinking.

Thematic Units

Several NSF-funded curricula use **thematic units**. A thematic unit consists of a series of problems that arise from a real-world context. Thematic units utilize the teaching through problem solving approach

Figure 4.2 Excerpt from *Connected Mathematics Project*.

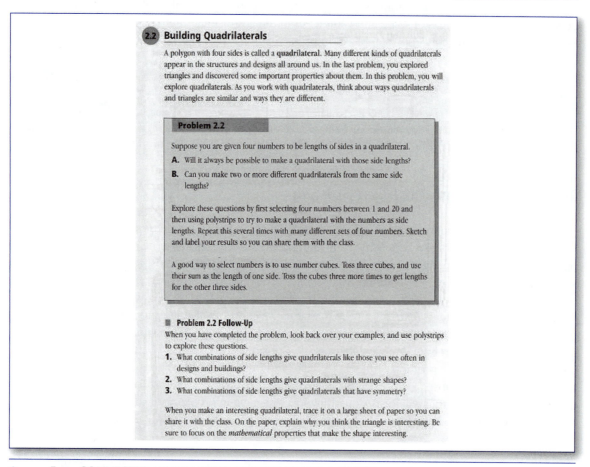

2.2 Building Quadrilaterals

A polygon with four sides is called a **quadrilateral**. Many different kinds of quadrilaterals appear in the structures and designs all around us. In the last problem, you explored triangles and discovered some important properties about them. In this problem, you will explore quadrilaterals. As you work with quadrilaterals, think about ways quadrilaterals and triangles are similar and ways they are different.

Problem 2.2

Suppose you are given four numbers to be lengths of sides in a quadrilateral.

A. Will it always be possible to make a quadrilateral with those side lengths?

B. Can you make two or more different quadrilaterals from the same side lengths?

Explore these questions by first selecting four numbers between 1 and 20 and then using polystrips to try to make a quadrilateral with the numbers as side lengths. Repeat this several times with many different sets of four numbers. Sketch and label your results so you can share them with the class.

A good way to select numbers is to use number cubes. Toss three cubes, and use their sum as the length of one side. Toss the cubes three more times to get lengths for the other three sides.

■ Problem 2.2 Follow-Up

When you have completed the problem, look back over your examples, and use polystrips to explore these questions.

1. What combinations of side lengths give quadrilaterals like those you see often in designs and buildings?

2. What combinations of side lengths give quadrilaterals with strange shapes?

3. What combinations of side lengths give quadrilaterals that have symmetry?

When you make an interesting quadrilateral, trace it on a large sheet of paper so you can share it with the class. On the paper, explain why you think the triangle is interesting. Be sure to focus on the *mathematical* properties that make the shape interesting.

Source: From CONNECTED MATHEMATICS PROJECT GRADE 6 SHAPES & DESIGNS TE © 1998 by Michigan State University, G. Lappan, J. Fey, W. Fitzgerald, S. Friel, and E. Phillips. Used by permission of Pearson Education, Inc. All Rights Reserved.

very carefully in that problems in the units are chosen not only to bring out important mathematics, but also to pertain to a given real-world situation. One of the middle school NSF projects, *MathThematics*, chose its title to emphasize its thematic curriculum. The structure of *MathThematics* can be described in the following manner:

> In a typical unit, or module, students read the section overview, which outlines the learning goals, and review the module project, which they work on throughout the unit, applying the mathematics. Each module is divided into sections. Within each section, students (1) engage in a motivating activity that sets the stage for the section, (2) work through a series of activities learning mathematical concepts and skills and solving problems, (3) review what they've learned on a summary page, and (4) practice and apply skills with a series of exercises. (American Association for the Advancement of Science, 2000)

The MathThematics curriculum contains thematic units on a variety of topics. In sixth grade, one unit is based on dinosaurs. A seventh-grade unit draws its problems

Figure 4.3 Example of teaching parabola transformations through problem solving (College Preparatory Mathematics, 2010, p. 1).

Lesson 3.1.2 How can I shift parabolas?

3-15. PARABOLA LAB, Part Two

Polly Parabola had been the manager of the parabola department of Functions of America, but she has decided to branch off and start her own company called "Professional Parabola Productions." She needs your help. See her memo below.

MEMO

To: Your Study Team
From: Ms. Polly Parabola, CEO
Re: New Parabola Possibilities

I am starting a new company specializing only in parabolas. To win over new customers, I need to be able to show them that we know more about parabolas than any of the other function factories around, since every company sells $y = x^2$ already.

My customers will need all sorts of parabolas, and we will need to have the knowledge to make them happy. I would love to offer parabolas that are completely new to them.

Please investigate all different kinds of parabolas. Determine all the ways that you can change the equation $y = x^2$ to change the shape, direction, and location of a parabola on a graph.

Remember that I'm counting on you! I need you to uncover the parabola secrets that our competitors do not know.

Sincerely,
Ms. Polly Parabola

Your Task: Work with your team to determine all the ways you can change the graph of a parabola by changing its equation. Start by choosing one transformation from the list generated by the class; then find a way to change the equation $y = x^2$ to create this transformation of the parent graph. Whenever you figure out a new transformation, record a clear summary statement on your resource page before moving on to the next transformation. Be prepared to explain your summary statement to Ms. Polly Parabola.

---

from situations faced by rescue workers. In eighth grade, a unit focusing on geometry, measurement, and algebra includes architecture problems. In any given unit, all problems are connected to a central theme.

The *Interactive Mathematics Program* (IMP), an NSF-funded high school level curriculum, also uses thematic units. Curriculum developers described their approach as follows:

> IMP units are generally structured around a complex central problem. Although each unit has a specific mathematical focus, other topics are brought in as needed to solve the central problem, rather than narrowly restricting the mathematical content. Ideas that are developed in one unit are usually revisited and deepened in one or more later units. (Interactive Mathematics Program, 2007)

Figure 4.4 shows an activity from a thematic unit on finance. It involves two of the central characters from the unit, Curtis and Hassan. As the two characters face financial choices, students begin to see the need for systems of equations and inequalities.

Implementing the Common Core

See Clinical Task 5 to assess a student's ability to interpret and analyze functions (Content Standard F-IF) when working the "Curtis and Hassan" problem shown in Figure 4.4.

Figure 4.4 Excerpt from *Interactive Mathematics Program* (Fendel, Resek, Alper, & Fraser, 2010, p. 141).

Activity

Curtis and Hassan Make Choices

1. Curtis goes into the pet store to buy a substantial supply of food for his pet. He sees that Food A costs $2 per pound and Food B costs $3 per pound. Curtis intends to vary his pet's diet from day to day, so he isn't especially concerned about how much of each type of food he buys.

   a. Suppose Curtis has $30 to spend. Find several combinations of the two foods that he might buy. Plot them on an appropriately labeled graph.

   b. Find some combinations that Curtis might buy if he were spending $50. Plot them on the same set of axes.

   c. What do you notice about your answers to parts a and b?

2. Hassan feels there will be a big demand for his work at the street fair. He is considering changing his prices so he earns a profit of $50 on each pastel and $175 on each watercolor.

   a. Based on these new profits, find some combinations of watercolors and pastels so Hassan's total profit would be $700. Plot them on a graph. (The combinations you give here don't have to fit Hassan's usual constraints.)

   b. Repeat part a for a total profit of $1,750, using the same set of axes.

   c. What do you notice about your answers to parts a and b?

Integration of Content Strands

Another common technique used by NSF-funded curricula is placing a number of mathematics content strands together in any given course. Traditionally, algebra and geometry have been considered separate courses, particularly at the high school level. The **integration of content strands** represents a sharply different approach in that students may study algebra, geometry, data analysis, probability, and other subject areas in any given year. These various content strands are often needed as students solve well-sequenced problems connected to a given theme or idea. Although such curricula can help students see the usefulness of various areas of mathematics for approaching a given problem, the integrated approach can also make educators uneasy. Those used to seeing the label "algebra" stamped on students' transcripts as a course title have to become accustomed to the idea that algebra is not contained within a single course, but is addressed in increasing depth as students progress through a four-year curriculum sequence.

IMP is one NSF-funded curriculum with an integrated approach to content. In describing the IMP approach, its developers stated, "Mathematical concepts are integrated throughout all four years of the curriculum, instead of being isolated from one another. Therefore, it is inaccurate to label any of the IMP courses with such familiar titles as Algebra, Geometry, or Trigonometry" (Interactive Mathematics Program, 2007). IMP texts blend a variety of mathematics content areas over a four-year sequence. The year 1 course includes variables, functions, graphs, geometry, and trigonometry. Year 2 includes statistics, measurement, linear programming, and maximization problems. Year 3 revisits and extends the year 1 and year 2 topics while also introducing combinatorics, derivatives, and matrices. Year 4 introduces circular functions and statistical sampling. Although the traditional labels of algebra, geometry, and precalculus are not used for IMP courses, the four-year sequence addresses topics from conventional courses while also going beyond traditional content in some cases.

CPMP is another NSF-funded curriculum that integrates content. The CPMP approach can be summarized as follows:

> Core-Plus Mathematics is a four-year unified curriculum that replaces the Algebra-Geometry-Advanced Algebra/Trigonometry-Precalculus sequence. Each course features interwoven strands of algebra and functions, geometry and trigonometry, statistics and probability, and discrete mathematics. Each of these strands is developed within coherent focused units connected by fundamental ideas such as symmetry, functions, matrices, and data analysis and curve-fitting. By actively investigating mathematics and its applications every year from an increasingly more mathematically sophisticated point of view, students' understanding of the mathematics in each strand deepens across the four year curriculum. (Hirsch, Fey, Hart, Schoen, & Watkins, 2009, p. 1)

In the CPMP curriculum, mathematical topics that arise from contemporary issues and applications of mathematics are integrated alongside more traditional topics. Some of the newer mathematical topics included in CPMP are vertex-edge graphs, mathematics of information processing and the Internet, and mathematics of democratic decision-making.

To better understand the integrated content approach, it can be helpful to trace a single content strand through a curricular sequence over a four-year time span. Consider, for example, the CPMP lesson excerpt shown in Figure 4.5. It appears in the first unit of the first year of the CPMP curriculum. Following the introductory material, the lesson goes on to introduce five-number summaries and boxplots as tools for examining variability. Ideas from data analysis and probability are further developed later in the year, when students conduct simulations to understand probability distributions. In year 2 of the curriculum, students revisit simulations and probability distributions while also learning about independent events, conditional probability, expected value, binomial distributions, and geometric distributions. Year 3 includes study of statistical surveying methods, the normal distribution, and inferential statistics. In year 4, students continue to study the binomial distribution and statistical inference. This scope of topics addressed over the four-year sequence strongly resembles that of a typical introductory college statistics course.

Implementing the Common Core

See Clinical Exercise 6 to assess the ability of a class to "draw informal comparative inferences about two populations" (Content Standard 7.SP) when completing the variability task shown in Figure 4.5.

Figure 4.5 Excerpt from *Core Plus Mathematics Project* materials.

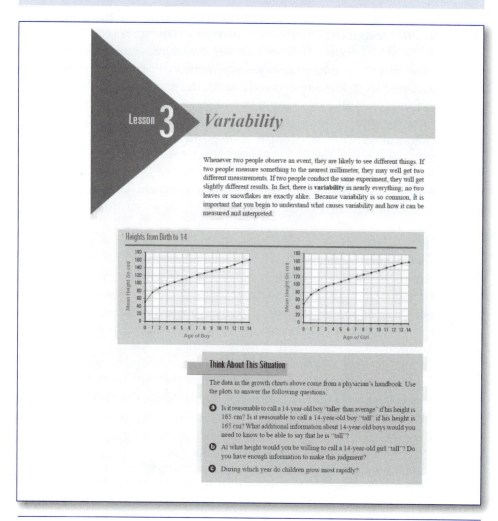

Source: Coxford, A. F., Fey, J. T., Hirsch, C. R., Schoen, H. L., Burrill, G., Hart, E. W., & Watkins, A. E., with Messenger, M. J., & Ritsema B., (1997). *Contemporary mathematics in context: A unified approach* (Course 1). Columbus, OH: Glencoe/McGraw-Hill. © The McGraw-Hill Companies, Inc.

Guided Investigation

Guided investigation is a technique employed in several NSF-funded curricula and in other mathematics curriculum sequences. Guided investigation is based on the premise that students will better retain information if they have a hand in discovering it. Curricula that utilize guided investigation do not leave students to reinvent all of the mathematics they need to learn on their own, but instead ask carefully sequenced questions to lead students in the desired direction. This sort of approach stands in contrast to curricula that primarily rely on traditional lecture to deliver content.

Discovering Geometry is one example of a curriculum that relies on guided investigation to a great extent. Consider the investigation shown in Figure 4.6. In this activity, students are not explicitly given a formula for the sum of the angle measures in a polygon, but are led to develop one. Toward the end of the activity, a transition is made from investigation of a collection of examples to formal proof. Although it would take less time to simply give students a formula and prove it for them, the guided investigation can ultimately be a wise time investment in terms of dividends paid in student understanding and retention.

> **Implementing the Common Core**
>
> See Homework Task 3 to explore how guided investigations can help students "look for and make use of structure" (Standard for Mathematical Practice 7).

Progressive Formalization

To sequence questions appropriately for guided investigation, some curricula use the principle of **progressive formalization**. Romberg (2001) explained the meaning of

Figure 4.6   Guided investigation from *Discovering Geometry.*

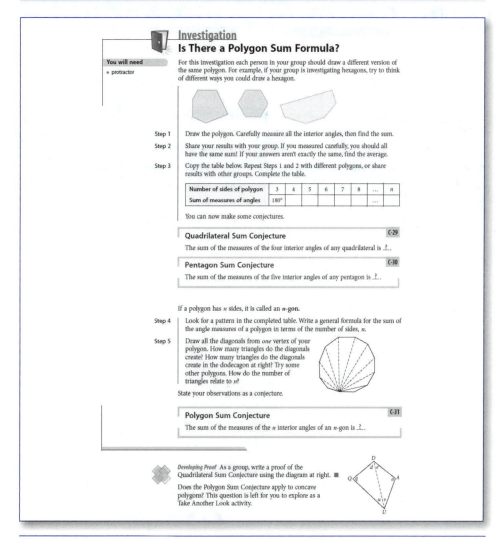

Source: From *Discovering Geometry: An Investigative Approach* by M. Serra. Copyright © 2008. Reprinted by permission of Kendall Hunt Publishing Co.

progressive formalization in discussing how *Mathematics in Context* (MiC) employs it to teach algebra:

> Progressive formalization of the mathematics involves, first, having students approach problems and acquire algebraic concepts and skills in an informal way. They use words, pictures, and/or diagrams of their own invention to describe mathematical situations, organize their own knowledge and work, solve problems, and explain their strategies. In later units, students gradually begin to use symbols to describe situations, organize their mathematical work, or express their strategies. At this level, students devise their own symbols or learn certain nonconventional notation (e.g., arrow language). Their representations of problem situations and explanations of their work are a mixture of words and symbols. (p. 5)

A fundamental characteristic of progressive formalization, in contrast to more traditional approaches, is beginning with the informal ideas students bring to the classroom. Conventional notation and language are introduced as students are led to see the need for more precise ways of expressing mathematical thinking. More traditional curricula often introduce formal notation and language prematurely, causing students to develop weak mathematical understanding.

A closer examination of how MiC approaches algebra can help clarify the idea of progressive formalization. Algebraic ideas gradually become more formalized as students move through Grades 6, 7, and 8. In one sixth-grade lesson, students use their own strategies to describe numerical patterns inherent in diagrams of V-shaped flight patterns of geese (see Figure 4.7). In seventh grade, students begin to work more with formal algebraic formulas to describe the patterns they explore in different contexts. By eighth grade, students are using multiple formal representations flexibly to approach problems. In one lesson, for example, they use tables, graphs, and equations to solve problems set in the context of hair growth. Because the curriculum gradually introduces these multiple formal algebraic representations, students have a greater opportunity to understand their meanings, interconnections, and utility.

Figure 4.7   Excerpt from *Mathematics in Context*.

Below you see the three smallest V-patterns.

10. a. Make a drawing of the next V-pattern.

  b. Is it possible for a V-pattern to have 84 dots? Why or why not?

  c. How many pairs are there in each V-pattern shown above?

  d. How many dots will there be in the sixth V-pattern?

11. Make a V-pattern with 19 dots.

12. a. Copy the table and fill in the missing values.

  b. The **V-number** tells the number of pairs of dots in a V-pattern. Describe any new patterns you see.

V-Number	Number of Dots
1	3
2	5
3	7
4	
5	
6	

13. **Reflect** You can make the row of V-patterns longer and longer. How many dots does the hundredth V-number have? How did you find your answer?

Source: Reprinted with permission from Mathematics in Context © 2006 Encyclopædia Britannica, Inc.

TECHNOLOGY CONNECTION

Computer-Assisted Individualized Instruction

A dilemma faced by teachers using any of the curricular techniques discussed in this chapter is that some students' background knowledge may prevent them from engaging deeply with the content. Furthermore, students in the same classroom generally exhibit wide variation in their level of preparedness to approach any given task. Some may have the requisite background knowledge and be prepared to exceed the requirements of the curriculum, while others may struggle with the foundational concepts they are to learn. The challenge associated with wide differences in students' background knowledge is one of the most difficult hurdles to overcome in curriculum implementation.

To help teachers manage classrooms where students have a wide variety of different learning needs, some curricula use software that provides individualized instruction. For example, the *Cognitive Tutor* curriculum (Carnegie Learning, 2009) recommends that students spend approximately 40% of their instructional time working with software that addresses individual needs. Students take a diagnostic pretest, and its results are used to set appropriate pacing for them. The software is programmed to recognize common student errors and provide hints to help put students back on the right track. Students must demonstrate mastery of a specified set of skills in order to progress through the software. This sort of immediate feedback

(Continued)

(Continued)

and support is much more difficult for a teacher to give when dealing with 30 students simultaneously in a conventional classroom setting.

Along with addressing individual students' learning needs efficiently, individualized instruction software can provide a great deal of assessment data. In the *Cognitive Tutor* software, for example, students are shown a "skillometer" to indicate their levels of proficiency with the different topics in the curriculum (Carnegie Learning, 2009). By viewing the skillometers, students and teachers alike can gain better understanding of which topics need further instructional attention. The school can use this information to customize the curriculum to address persistent areas of deficiency. The software can be adjusted to reorder, add, or delete instructional units. Hence, the software allows teachers to make optimal use of formative assessment data gathered electronically from students on a continual basis.

Follow-up questions:

1. Provide an example of a classroom situation you have observed in which students exhibited a wide range of proficiency with the concepts at hand. Would individualized computer-assisted instruction have been helpful in the situation? Why or why not?

2. What would you look for to judge the quality of any particular computer-assisted instruction software package? What kinds of features and capabilities would you expect?

Drill and Practice

A widely held belief about mathematics is that students should be given exercises to practice what has been learned in class. Some of the initial editions of NSF-funded curricula discussed previously were criticized for the small number of traditional-looking practice exercises they offered (e.g., Mathematically Correct, n.d.). This reaction was somewhat predictable in light of the fact that the new curricula sought to portray mathematics as a problem solving activity. Much of the work students were to complete consisted of analyzing novel problem situations; less emphasis was placed on traditional drill and practice exercises. Under a problem solving approach, students do often apply the same mathematical procedures and ideas within different contexts. However, since much of the practice was embedded within student problems and not explicitly identified as "practice," a quick examination would give an unfavorable impression. In response to concerns expressed by parents and teachers, subsequent editions of many NSF-funded curricula added more traditional-looking practice exercises, either within the main text of the book or in ancillary materials.

Just as some individuals are concerned about too little practice in mathematics curricula, others worry about too much of it reducing the study of mathematics to nothing more than drill of rote procedures. Some curricula, like *Saxon Mathematics* (Saxon Publishers, 2009) and popular textbooks mass produced by large publishers, have been accused of overemphasizing drill and practice. Whether or not the criticism of *Saxon* and the other texts is justified, it is important to acknowledge that mathematics curricula can and do often fall into the pattern of learning rules and practicing procedures, which is the classroom pattern the TIMSS video study associated with lower-achieving countries (Jacobs et al., 2006). Just as too little practice may impede learning, too much practice may limit students' opportunities to engage in the sort of novel problem solving central to the discipline of mathematics.

One principle about practice exercises that has been established in research literature is that mixed review is generally more effective than blocked practice (Rohrer, 2009). Mixed review, sometimes called **distributed practice**, involves having students practice skills learned in a given lesson at several different points during the school year. Distributed practice prompts students to reflect on the content that has been taught in previous lessons, not just the most recent one. This approach stands in contrast to having students extensively practice a particular set of skills only immediately after the lesson introducing them has taken place. Fairly widespread agreement on the importance of distributed practice can be seen by examining the practice problems currently included in a variety of curriculum materials. For example, distributed practice is used in NSF-funded curricula such as CPMP (Hirsch et al., 2009) and is a cornerstone of the instructional approach taken in *Saxon* texts.

Teaching for Social Justice

Despite the availability and use of curriculum materials with the potential to help promote mathematical thinking and problem solving, persistent achievement gaps remain among groups of students. Black, Hispanic, and Native American students have historically had lower mathematics scores than white students on the National Assessment of Educational Progress (NAEP). Students of lower socioeconomic status (SES) have also had lower achievement. Scores across various groups of students have steadily risen in the recent past, but differences in NAEP achievement have occurred each year the test has been administered (Lubienski & Crockett, 2007). To conclude this portion of the chapter, some of the reasons for these persistent differences, even in classrooms using reform-oriented curricula, will be considered. Recent thinking on ways to address these differences will also be discussed.

Lubienski (2000) examined seventh graders' experiences with the CMP curriculum. She studied how students of different SES levels responded to its open-ended and contextualized problems. Low-SES students tended to become frustrated with the open-ended problems. They expressed confusion about what to do with the problems because of vocabulary issues and the absence of precise problem solving instructions. Some high-SES students thought the CMP problems were easier than those found in conventional textbooks. Higher-SES students were also more likely to persist when stuck on a problem. When it came to solving problems set in real-world contexts, low-SES students were not as likely to use mathematical arguments as high-SES students. A question that asked which outcome was most likely to occur on a spinner, for example, was interpreted by a low-SES student as an opinion question requiring a guess. Generally speaking, lower-SES students had greater difficulty abstracting the mathematics from contextual situations. Lubienski's study illustrates that even though low-SES students may have a great deal to gain from innovative curriculum materials like CMP, simply providing open-ended, contextual problems will not, in and of itself, alleviate existing achievement gaps.

Gutstein (2003) reported on his efforts to implement the MiC middle school curriculum while teaching in an urban Latino school. Gutstein supplemented MiC with real-world mathematics problems that exposed inequities in society. One such problem involved analyzing a data set showing that African American and Latino drivers are pulled over in disproportionate numbers. In analyzing the problem, students had to draw on the mathematical concepts of proportionality and expected value. In another class, Gutstein helped students see the unequal distribution of wealth among continents of the world by having them stand in different parts of the room representing the continents and then distributing cookies to represent the world's wealth. When students saw that a typical person in North America had much more wealth than those in many other parts of the world, they began to discuss the idea of using one number to summarize the wealth of a typical person on the continent. Students observed that there existed a wide disparity between the richest and poorest people on the continent, making a one-number summary using the arithmetic mean of limited value.

Discussions about provocative problems exposing inequities in society often prompted Gutstein's (2003) students to go beyond the mathematics of the situation to discuss related social issues. In the process, students became better at creating mathematical generalizations and more adept at constructing their own solution strategies to nonroutine problems.

Gutstein (2005) characterized his approach to teaching in an urban Latino school as being illustrative of **social justice pedagogy**. Such an approach has three goals: to help students develop sociopolitical consciousness (which Gutstein called "reading the world with mathematics"), a sense of agency ("writing the world with mathematics"), and positive social and cultural identities. Social justice pedagogy has the potential to alleviate social inequities in at least two ways. First, Gutstein's (2003) research illustrated that students grew in their mathematical thinking ability while taking his class. This suggests that teaching for social justice can motivate students to higher levels of achievement than they would attain under more conventional approaches. Second, teaching for social justice helps students think critically about forces in society that create inequity and to become agents of change in response. Hence, teaching for social justice can address inequity in mathematics achievement while also addressing inequities outside the bounds of the mathematics classroom.

Despite emerging scholarship on ways to address inequities in the classroom, it is important to realize that many causes of inequity are deeply entrenched in the culture of schooling and not easy to uproot. Consider, for example, the practice of tracking students, as discussed in Chapter 2. Tracking often perpetuates inequity because it denies large groups of students access to the most talented teachers and most engaging problem solving modes of instruction. Students from some racial and ethnic groups are disproportionately placed into lower tracks in school. Administrators, parents, and teachers, however, often cling to the "common sense" notion that the best way to serve students is to sort them into classes by ability groups, despite research evidence contradicting this notion (Boaler & Staples, 2008; Linchevski & Kutscher, 1998).

Because forces perpetuating inequity are resistant to change, even when they are challenged by research, teachers need to prepare to be active beyond the bounds of their own classrooms if substantial progress on the issue of equity is to occur. Beginning teachers can seek out opportunities to become involved in their school district's curriculum selection process. By examining curricula being considered for adoption against guidelines discussed in this chapter, teachers can judge whether all students would be given equitable access to an engaging curriculum. Teachers can also attend and participate in local school board meetings. School board members, as individuals from the community who work in various professions, often have never studied educational research showing the ineffectiveness of practices such as tracking students. Drawing their attention to what the research has to say, in a professional manner, can help put the district along the path toward more equitable practices. Moving toward equity is a long, hard process, so teachers should not be discouraged if results are not produced overnight. However, if knowledgeable teachers are not engaged in the pursuit of equity beyond the bounds of their own classroom settings, there will be little chance to close persistent achievement gaps among groups of students.

Summary

Currently available curricula reflect a variety of techniques. NSF-funded curricula generally employ some form of teaching through problem solving by carefully sequencing problems so they build on one another and address needed mathematics content. Some use the techniques of progressive formalization and guided investigation along the way. Some also use thematic units as structures to introduce content and capture students' interest. Techniques used in NSF-funded curricula are also present to a degree in other commercially available materials. Distributed practice, for example, is a feature that appears in curricula based on various philosophical premises. Understanding these different techniques and the extent to which they are employed in different curriculum materials can facilitate the decision-making process involved in choosing materials, the topic of the next section.

CHOOSING CURRICULUM MATERIALS

The role of teachers in selecting curriculum materials varies from school to school. Some states have approved lists of textbooks that must be used. In some schools, administrators encourage teacher involvement in textbook adoption, and in others they do not seek teacher input. Beginning teachers usually inherit the curriculum materials in use when they are hired. Whatever the situation happens to be in a given school, one important principle applies: Teachers should be advocates for curricula that support student learning. Not all curricula are created equal. Available research indicates, for example, that students in classrooms using NSF-funded curricula often develop better problem solving and reasoning skills than their counterparts in traditional classrooms while acquiring approximately the same amount of computational skill (Senk & Thompson, 2003; Stein, Remillard, & Smith, 2007). If a curriculum in place in a given school is largely ineffective for fostering student learning, teachers ought to advocate curricular change.

A variety of tools can help teachers select and advocate effective mathematics curriculum materials. Release of the NSF-funded curricula discussed earlier in this chapter sparked substantial activity in the area of forming curriculum evaluation criteria. Some sets of criteria were constructed to analyze the substance of a curriculum in terms of its mathematical and/or pedagogical content; others were designed to analyze the effect of a curriculum on students' learning. Examples of both types of criteria are provided in the following.

Criteria for Examining Curricular Content

Tarr, Reys, Barker, and Billstein (2006) recommended analyzing textbooks along three dimensions: mathematics content emphasis, instructional focus, and teacher support. *Mathematics content emphasis* refers to the degree of alignment with state or district content standards. When examining texts along this dimension, teachers should not be satisfied if the text merely "covers" the standards. They should also examine how well the text motivates the study of mathematics by setting problems in context. The extent to which the content grows in sophistication from year to year should also be examined to avoid unnecessary repetition. *Instructional focus* is related to the question, "Are textbook lessons, activities, and problems likely to engage students in mathematical thinking?" (Tarr et al., 2006, p. 52). Many texts on the market begin lessons with a brief hands-on activity or a problem set in a real-world context. However, the exercises in the text are often just decontextualized practice problems. Such exercises usually do not support the goal of helping students develop conceptual understanding. Finally, *teacher support* deals with the question, "Does the textbook provide a variety of support materials to help teachers plan for instruction?" Materials helpful for planning instruction include assessment resources, descriptions of the "big ideas" embedded in units, and descriptions of common student understandings and misconceptions related to the unit. Although these three dimensions are not necessarily the only ones to consider when evaluating curriculum materials, they provide helpful starting points (Tarr et al., 2006).

Another set of textbook evaluation criteria was written by **Project 2061** of the American Association for the Advancement of Science (AAAS; Kulm, 1999). In evaluating textbooks, AAAS considered their mathematics content as well as their methods of teaching the content. To evaluate the mathematics content of a text, AAAS criteria called for examining the extent to which the content aligned with a given set of standards, such as those set forth by NCTM (2000). To be considered in alignment with a curriculum standard, activities contained in the text were to address the substance of the standard, reflect the grade level specified by the standard, and address the standard in its entirety. To evaluate the potential instructional effectiveness of a text, AAAS examined the extent to which its pedagogy reflected research describing productive mathematics classrooms. Some of the characteristics considered to be related to instructional effectiveness were: conveying a sense of purpose and direction, addressing students' ideas and potential difficulties with the content, engaging students in the process of forming generalizations, making connections among mathematical ideas, encouraging students to monitor their own thinking, and having assessments aligned with standards.

In 1999, the U.S. Department of Education released textbook evaluation criteria they had used to analyze several mathematics curricula. In analyzing each curriculum, they considered its quality, potential usefulness, educational significance, and evidence of success. Eight specific criteria were set forth:

1. The program's learning goals are challenging, clear, and appropriate for the intended student population.

2. The program's content is aligned with its learning goals, and is accurate and appropriate for the intended student population.

3. The program's instructional design is appropriate, engaging, and motivating for the intended student population.

4. The program's system of assessment is appropriate and designed to inform student learning and to guide teachers' instructional decisions.

5. The program can be successfully implemented, adopted, or adapted in multiple educational settings.

6. The program's learning goals reflect the vision promoted in national standards in mathematics education.

7. The program addresses important individual and societal needs.

8. The program makes a measureable difference in student learning. (pp. 5–7)

The eight criteria were used to identify exemplary and promising curriculum series.

The U.S. Department of Education (1999) noted that its eight evaluation criteria might have value for individual schools, districts, and boards of education. It suggested using the criteria to analyze curriculum materials being considered for adoption. It also stated that the criteria could be used as a means of improving curriculum selection processes in place in schools. By comparing and contrasting the eight criteria with existing selection processes, schools may identify features of the process that need improvement.

Examining Effects on Student Learning

The U.S. Department of Education (1999) criteria went beyond mere content analysis by examining student learning in classrooms using the curriculum materials. Criterion 8 in particular made examining impact on student learning a priority. Programs had to meet criterion 8 to be considered "exemplary," and to meet the criterion, programs had to show evidence of student gains in mathematics content knowledge or problem solving and reasoning skills. They also had to show evidence of improving students' attitudes toward learning, narrowing the achievement gap between populations of students, or improving enrollment, graduation rates, or postsecondary school attendance. The evidence of effectiveness needed to come from multiple sites with multiple populations. The U.S. Department of Education criteria marked a strong move toward the trend of demanding that curricula provide evidence of positive impact on students. When considering a curriculum for adoption, it is reasonable for teachers to demand the same sort of evidence from publishers.

Given the growing interest in evidence of the impact of curriculum materials, many publishers maintain websites with research reports about their curricula. A sampler of publishers' websites is shown in Table 4.1. Delving into the research reports, it can be seen that a wide variety of measures are used to provide evidence of curricular effectiveness. Many reports on the NSF-funded curricula use conventional measures of achievement, such as SAT, ACT, and state standardized tests, to demonstrate that their approach does not erode students' ability to pass high-stakes tests (Kilpatrick, 2003). Along with student performance on standardized tests of mathematics content, teachers should examine whether or not the curriculum shows evidence of producing gains in students' problem solving and reasoning skills. In many cases, standardized tests do not measure these areas, but qualitative research reports may describe students' problem solving activities during their engagement with a curriculum.

Table 4.1 Websites containing research on effectiveness of curricula.

Curriculum	Website Summarizing Research on Effectiveness
Connected Mathematics Project	http://www.connectedmath.msu.edu/rne/lit.shtml
College Preparatory Mathematics	http://www.cpm.org/teachers/research.htm
Cognitive Tutor	http://www.carnegielearning.com/approach.cfm
Core-Plus Mathematics	http://www.wmich.edu/cpmp/evaluation.html
Interactive Mathematics Program	http://www.mathimp.org/research/index.html
Saxon Mathematics	http://saxonpublishers.hmhco.com/en/sxnm_research.htm

Publisher-independent curriculum evaluation websites are also available, the most prominent being the **What Works Clearinghouse** (WWC; U.S. Department of Education, Institute of Education Sciences, 2009). WWC aims to produce user-friendly guides for educators about the effectiveness of curricula, assess research evidence on curricular effectiveness, and develop and implement standards for the review and synthesis of education research. Although WWC provides an independent perspective, its standards for reviewing educational research have been questioned. One problem comes from the narrowness of WWC criteria. WWC considered randomized controlled trials to be the gold standard of research design, even though such trials are limited in their feasibility, suitability, and insightfulness in the context of education. Additionally, the WWC may characterize curricula as "effective" on the basis of skill development alone, not taking into account the impact of the curriculum on students' problem solving ability (Schoenfeld, 2006). Despite its flaws, the WWC is emblematic of a shift toward demanding that publishers provide evidence of improved student learning in classrooms where their curricular materials are used. WWC is worth consulting in the curriculum evaluation process but should not be used as the sole basis for decision making.

CONCLUSION

Recent developments in mathematics curriculum make this an exciting and challenging time to become a teacher. NCTM has released landmark documents to help guide efforts to improve the coherence of mathematics curriculum standards. NSF-funded curricula facilitate the task of implementing NCTM pedagogical recommendations by providing problems and other materials to teach mathematics through problem solving. Various innovative techniques can be seen in existing curriculum series, such as the use of real-world contexts for problems, thematic units, guided investigation, and individualized instruction. Tools also exist to help teachers judge the quality of curricula. Criteria to evaluate textbooks have been formulated by agencies such as the U.S. Department of Education, and research on students' learning in classrooms using different kinds of curriculum materials is available online. After selecting a curriculum, teachers need to discern the authors' intent for classroom use of the materials. Care must be taken to ensure that opportunities the materials provide to engage students in doing mathematics are used, and that they do not degenerate to rote, mechanistic activity. Finally, it is important to realize that simply implementing a curriculum with fidelity will not automatically remedy all causes of inequity in learning mathematics. Ways to address inequities include teaching for social justice and becoming engaged in the larger context of one's school district. As well-prepared teachers become more engaged in curricular decisions in their school districts, progress can be made toward improving students' learning and closing persistent achievement gaps.

VOCABULARY LIST

After reading this chapter, you should be able to offer reasonable definitions for the following ideas (listed in their order of first occurrence) and describe their relevance to teaching mathematics:

No Child Left Behind   77

Curriculum focal point   78

*Focus in High School
    Mathematics*   80

Reasoning   80

Sense-making   80

NSF-funded curricula   83

Thematic unit   87

Integration of content strands   88

Guided investigation   90

Progressive formalization   91

Distributed practice   95

Social justice pedagogy   96

Project 2061   98

What Works Clearinghouse   100

HOMEWORK TASKS

1. Examine the middle or high school mathematics curriculum standards for the state in which you plan to teach. Describe three similarities to and three differences from the NCTM *Principles and Standards for School Mathematics* (http://standards.nctm.org). The similarities and differences should pertain to *which* mathematics is to be taught or *how* mathematics is to be taught.

2. Examine the middle or high school mathematics curriculum standards for the state in which you plan to teach. Describe three similarities to and three differences from the NCTM *Focal Points* (www.nctm.org/standards/content.aspx?id=270). The similarities and differences should pertain to *which* mathematics is to be taught or *how* mathematics is to be taught.

3. Using your state's mathematics curriculum document for middle and/or high school, identify a key concept students are to learn. Then design a guided investigation to help students learn it. See the *Discovering Geometry* excerpt in Figure 4.6 for a sample structure. Ask a classmate to work through the guided investigation you design and comment on its strengths and weaknesses.

4. Suppose you are working in a middle school that is in the process of selecting a new mathematics curriculum and that the principal has asked you to recommend two curriculum series for adoption. Write a letter to the principal with your top two choices. Support your choices by citing evidence of the impact of curriculum materials on students' learning. See the publishers'

websites and the What Works Clearinghouse to locate research about effects on students' learning. Describe at least two different research studies on students' learning to support each curriculum you recommend.

5. Suppose you are working in a high school that is in the process of selecting a new mathematics curriculum and that the principal has asked you to recommend two curriculum series for adoption. Write a letter to the principal with your top two choices. Support your choices by citing evidence of the impact of curriculum materials on students' learning. See the publishers' websites and the ERIC database (www.eric.ed.gov) to locate research about effects on students' learning. Describe at least two different research studies on students' learning to support each curriculum you recommend.

6. Use the NAEP data explorer (http://nces.ed.gov/nationsreportcard/naepdata/) to build a report that gives test performance by ethnicity for any mathematics content area on the main National Assessment of Educational Progress (NAEP) over the past decade. Print the report that you produced. Then describe any achievement gaps that exist between groups. Also compare the end-of-decade performance of a lower-scoring group against the beginning-of-decade performance of a higher-achieving group. Finally, prepare a report for the state in which you plan to teach that provides a picture of challenges it faces in regard to enhancing curricular equity. Describe at least three specific state-level challenges.

CLINICAL TASKS

1. Interview a teacher to determine the extent to which he or she relies on each of the following curricular materials:

   a. Textbooks

   b. School district curriculum scope and sequence

   c. State level curriculum scope and sequence

   d. NCTM's *Principles and Standards for School Mathematics*

   Ask the teacher to rank the items above in order of their importance for long-term planning. If the teacher feels an item is missing from the list, ask him or her to add it. Ask the teacher to justify the rationale that he or she gives. Then write a report summarizing how the teacher goes about long-term planning. Finally, write your own reaction to the teacher's long-term planning processes, commenting on their strengths and weaknesses.

2. Obtain a middle school mathematics textbook and compare its content against the recommendations of NCTM's *Curriculum Focal Points* (www.nctm.org/standards/content.aspx?id=270). Describe the extent to which the text aligns with the focal points for its grade level. Be sure to state which focal points, if any, are missing from the text. Also describe how the content of the text goes beyond the focal points for its grade level, if it does so at all.

3. Ask a student to think aloud while solving the following problem:

   > A slab of soap on one pan of a scale balances $\frac{3}{4}$ of a slab of soap of equal weight and a $\frac{3}{4}$-pound weight on the other pan. How much does the slab of soap weigh? (NCTM, 2009, p. 35)

   When the student has finished, ask if he or she can think of another way to solve it. Write a report giving a complete description of the thinking strategies the student used.

4. Examine the teacher's edition of a textbook from your field placement site. Complete the following chart to indicate whether or not each of the following techniques is used in the text. Indicating that a particular technique is used means that it appears in the manner described in this chapter.

Technique	Used in text? (y/n)	Explanation & supporting examples
Real-world contexts for problem solving		
Teaching through problem solving		
Thematic units		
Integration of mathematics content areas		
Guided investigation		
Individualized instruction		
Progressive formalization		
Distributed practice		
Social justice pedagogy		

5. Ask a student to write a solution to each problem shown in Figure 4.4. Describe and critique the student's solution strategy. Then interview the student and ask if he or she knows how to do the task a different way. If the student cannot think of a different way to do the problem, demonstrate an alternative strategy. Write a report describing the student's initial written solution and your interaction with him or her when discussing the solution.

6. Have an entire class of students write solutions for the CPMP problem shown in Figure 4.5. Upon receiving the students' written work, classify their thinking from "most sophisticated" to "least sophisticated," sorting the papers into several piles accordingly. What distinguishes sophisticated thinking about the problem from less sophisticated thinking in the students' work samples?

7. Examine the homework problems (or class work, if no homework is given) assigned in your field experience classroom over the course of a week. What percentage of them are practice problems? Is distributed practice used at all? Give specific examples of the types of practice problems that are used as well as any other types of homework (or class work) problems that are used.

8. Interview a teacher at your field placement site about the textbook selection process used in the school district. Ask who is involved in the process, and interview at least one individual directly involved in selecting textbooks. When interviewing an individual involved in selecting textbooks, ask which factors he or she considers in deciding whether or not to adopt a book. Then show the individual the U.S. Department of Education (1999) criteria for textbook evaluation and ask which criteria listed therein are taken into account in textbook selection. Write a report about your findings, along with recommendations for improving the school's textbook selection process. You may find it helpful to draw on the criteria offered by Tarr and colleagues (2006) and Kulm (1999) in giving recommendations for improving the textbook selection process.

9. Attend a school board meeting. Provide a summary of what happened along with an agenda for the meeting. Did any discussion of issues related to curricular equity arise during the meeting? If so, how did they unfold? Were there any missed opportunities to discuss equity issues? If so, how would you have raised equity issues if you were given the opportunity?

| VIGNETTE ANALYSIS ACTIVITY | Focus on Using Appropriate Tools Strategically (CCSS Standard for Mathematical Practice 5) |

Items to Consider Before Reading the Vignette

Read CCSS Standard for Mathematical Practice 5 in Appendix A. Then respond to the following items:

1. Drawing on your own experiences as a mathematics student, provide specific examples of teaching practices that can help students use appropriate tools strategically.

2. Drawing on your own experiences as a mathematics student, provide specific examples of teaching practices that might prevent students from using appropriate tools strategically.

Scenario

Mr. Hart was a beginning teacher assigned to teach an eighth-grade algebra class. He was enthusiastic about his work and believed that he related to middle school students well. Many students in the algebra class would eventually take calculus during high school. This motivated Mr. Hart to challenge his students to the fullest extent possible. During one lesson, he introduced the quadratic formula. Mr. Hart considered this to be one of the most important concepts in introductory algebra, so his goals were to help students memorize the formula and to help them understand its connections to factoring expressions and producing graphs.

The Lesson

Mr. Hart began the lesson with a task he believed students could perform and would provide a foundation for what they were to learn that day. He asked students to solve the equation $x^2 - 6x + 8 = 0$. They had been asked to solve several equations like this for the previous night's homework assignment. After a few minutes of work, one of the students, Jessica, volunteered a solution. She said, "I solved the equation and got $x = 2$." Mr. Hart responded, "No, that's not quite right. Amy, what did you get?" Amy replied, "I got $x = 2$ and $x = 4$." Mr. Hart responded, "Very good. That's right, because $x^2 - 6x + 8 = (x-2)(x-4)$. Are there any questions about that?" Seeing no hands raised, Mr. Hart decided to move on.

Mr. Hart directed all of the students to take out their graphing calculators and to graph $y = x^2 - 6x + 8$. He said, "We have looked at how to solve the equation $x^2 - 6x + 8 = 0$ by factoring. Now let's look at how you can solve it by using a graph." He plotted the graph on his own calculator and then displayed it for all of the students on the document camera. He said, "This is a U-shaped graph called a parabola." Then he asked, "What do you notice about the points where the graph crosses the x-axis?" Tyler raised his hand and said, "It crosses at 2 and 4." Mr. Hart replied, "Yes, you're right. Why does that happen? Tisha?" Tisha responded, "Those were the solutions to the problem when we factored it." Mr. Hart said, "Exactly. We can see the solutions for an equation like that in its graph by looking for where it crosses the x-axis."

Mr. Hart wanted to be careful not to imply that every parabola crosses the x-axis. So, the next thing he asked students to do was to graph two parabolas that did not cross. Mr. Hart said, "OK, now graph these two on your calculator: $y = x^2 - 2x + 5$ and $y = -x^2 + x - 6$." Students started entering the functions into their calculators, and as they worked, Mr. Hart put them in his as well. When he finished, he projected the result on the screen for everyone to see. He said, "See, neither one of these parabolas intersects the x-axis. Sometimes there are no roots to find. If you tried to factor either one of these, you wouldn't be able to do it."

Once students had seen the examples he selected, Mr. Hart decided it was a good time to introduce the quadratic formula. He said, "Instead of having to factor or rely on graphs, we can tell if a parabola touches the x-axis by using the quadratic formula. This is one you're going to want to memorize. If you have $ax^2 + bx + c = 0$, you can solve for x by using $x = \frac{-b \pm \sqrt{b^2\ 4ac}}{2a}$. If you try using this to solve $0 = -x^2 + x - 6.$, what happens?" One of the students, Victor, volunteered, "You get a negative square root, and you can't have one of those." Mr. Hart said, "Good job, Victor. Does everyone agree?" Another student, Megan, spoke up, saying, "My brother is in calculus, and he said once that you can take the square root of a negative and it's called an imaginary number." Mr. Hart replied, "Good, Megan, but we won't use those now."

Mr. Hart then jumped into his favorite part of the lesson. He pulled out a CD and put it into the CD player. He announced, "Last night, while I was on the Internet, I found the 'Quadratic Formula Song' and burned it to this CD. I'm going to play it once for you, and then I want you to sing along." The CD player blared a song about the quadratic formula set to the tune of "Pop Goes the Weasel." The teacher and most of the students sang the song four times.

When he was satisfied that the students had seemed to commit the quadratic formula to memory, Mr. Hart wanted to show them a problem for which it would be particularly useful. He put $x^2 - 3x + 1$ on the board and asked students to try to factor it. After a few minutes of trying, most students gave up. A few started trying using non-integer values to factor the expression. Seeing that most students were no longer working, Mr. Hart put the graph for $y = x^2 - 3x + 1$ on the board and said, "Look, it does cross the x-axis, but we can't tell exactly where. So, let's use the quadratic formula." He demonstrated how the formula could be used for the problem.

After showing students how to apply the quadratic formula, Mr. Hart asked them to solve the equations $x^2 + 5x = 5$ and $2x^2 - 5x - 3 = 0$. He told them to use the quadratic formula, and then to use the graphing capabilities of their calculators to check to see if their answers were reasonable. Most students were able to do so, but a portion of the class had trouble entering the expressions into their calculators. Circulating about the room, Mr. Hart saw that many of these students were using the "-" key when they needed to use the subtraction key, and he helped them enter the expressions correctly. As students finished the two problems, they received a worksheet with additional practice problems for homework. Most students continued to work up to the bell.

Questions for Reflection and Discussion

1. Which aspects of the CCSS Standard for Mathematical Practice 5 did Mr. Hart's students seem to attain? Explain.

2. Which aspects of the CCSS Standard for Mathematical Practice 5 did Mr. Hart's students not seem to attain? Explain.

3. Comment on the importance of memorizing the quadratic formula. Is this a worthwhile goal for introductory algebra? Explain your reasoning.

4. What might have led Jessica to think that $x = 2$ is the only solution for $x^2 - 6x + 8 = 0$? What kinds of questions could Mr. Hart have asked to probe her thinking?

5. Is it accurate to refer to a parabola as a "U-shape?" Explain.

6. Comment on Mr. Hart's efforts to help students see connections among multiple representations of quadratic functions. Do you think his efforts are likely to help support students' learning? Explain.

7. Write a word problem that would help students work with applications of quadratics in contexts outside of mathematics. Do you believe it would have been beneficial for Mr. Hart to include this type of connection in the lesson? Explain.

8. Mr. Hart did not connect the graphical and tabular representations of quadratics to one another during the lesson, even though graphing calculators can produce tables of values for functions. Would it have been helpful to try to make this connection? Why or why not?

9. Critique Mr. Hart's instruction about the discriminant in the quadratic formula. Were there additional mathematical connections that could have been made by prompting students to analyze its sign more closely? Explain.

RESOURCES TO EXPLORE

Books

Hirsch, C. (Ed.). (2007). *Perspectives on design and development of school mathematics curricula.* Reston, VA: NCTM.

Description: The chapters in this book provide an inside look at the development of innovative NCTM standards–based curricula. Readers will come away with a better understanding of the structure of the curricula and their underpinning philosophies.

National Council of Teachers of Mathematics. (2006). *Curriculum focal points for prekindergarten through Grade 8 mathematics.* Reston, VA: Author.

Description: The focal points were published to address the large amount of state-to-state variation in mathematics learning standards. They provide perspective on a handful of "big ideas" it is important for students to master at each grade level up through Grade 8.

National Council of Teachers of Mathematics. (2009). *Focus in high school mathematics: Reasoning and sense-making.* Reston, VA: Author.

Description: This book is based on the premise that reasoning and sense-making should pervade all aspects of the mathematics curriculum. Classroom examples are provided to illustrate how to foster reasoning and sense-making within each of the content strands in the NCTM standards.

Reys, B., & Reys, R. (Eds.). (2010). *Mathematics curriculum: Issues, trends, and future directions* (Seventy-second yearbook of the National Council of Teachers of Mathematics). Reston, VA: NCTM.

Description: This yearbook provides perspective on curriculum issues across the entire K–12 spectrum. Authors of chapters in the book address topics such as technology use, high-stakes testing, and textbook selection by school districts.

Senk, S. L., & Thompson, D. R. (Eds.). (2003). *Standards-based school mathematics curricula: What are they? What do students learn?* Mahwah, NJ: Erlbaum.

Description: As NSF-funded NCTM standards–based curricula were initially implemented, questions were asked about their influence on student learning. Authors of the reports in this book summarize some of the early research on students' learning in classrooms using the new curricula.

Website

Modeling Middle School Mathematics: **http://www.mmmproject.org/**

Description: As the NSF-funded, NCTM standards-based curricula were released, teachers needed models to illustrate the logistics of their implementation. This website addresses that need by providing video of classroom lessons exemplifying the principles of five innovative middle school curricula.

Chapter 5

IMPLEMENTING AND ASSESSING MATHEMATICS LESSONS AND CURRICULA

Written lessons and curriculum materials do not implement themselves. Because individual teachers bring their own interpretations to bear, the manner in which a written curriculum or lesson is implemented inevitably varies from classroom to classroom. To emphasize this point, Stein, Remillard, and Smith (2007) distinguished the **written curriculum** from the **intended curriculum** and the **enacted curriculum**. The written curriculum is that which exists in print on the pages of textbooks and curriculum materials. The intended curriculum refers to a teacher's instructional plans related to a given curriculum. The enacted curriculum is that which actually unfolds in the teacher's classroom setting. There is seldom a seamless transition from the written to the enacted curriculum. Teachers' beliefs and knowledge, along with a variety of constraints present in the school setting and other factors, virtually ensure that the written curriculum will be modified in a manner that conflicts with the intentions of the curriculum writers. Implementing written lessons and curriculum materials and assessing the impact of the enacted curriculum on students' mathematical understanding will be the focus of this chapter.

IMPLEMENTING NSF-FUNDED CURRICULA: INSIGHTS FROM RESEARCH

As NSF-funded mathematics curricula were introduced to schools, a substantial amount of research was done to determine how teachers used them in their classrooms. The findings of the research support the theory that teachers transform the written curriculum in unique ways when implementing it. Two illustrative studies are described below. The first discusses high school teachers' implementation of the *Core-Plus Mathematics Project* (CPMP) curriculum (Lloyd, 1999), and the second discusses middle school teachers' implementation of the *Connected Mathematics Project* (CMP curriculum; Lambdin & Preston, 1995).

Implementing CPMP

Lloyd (1999) described two teachers' implementation of CPMP. In both cases, implementation of the curriculum was influenced by the teacher's beliefs about the amount of structure provided in the materials. One teacher believed that CPMP problems were open ended and challenging, and the other believed that they were too structured. The first teacher, Mr. Allen, believed that CPMP problems contrasted sharply with those found in traditional texts because CPMP problems required student analysis and thinking rather than performance of the same steps over and over again. Although he valued the open-ended nature of CPMP problems, Mr. Allen was also concerned that students lacked confidence in their thinking while solving them. This concern led him to give more direction than recommended in the curriculum materials. Mr. Allen did not rewrite problems in CPMP, but he did provide sets of review problems and used teacher-directed instruction extensively after being overwhelmed with students' questions when they worked in cooperative groups. Mr. Allen recognized that his use of teacher-directed instruction conflicted with the written CPMP curriculum because he often ended up doing much of the work rather than allowing students to struggle with it.

The second teacher in Lloyd's (1999) study, Ms. Fay, viewed CPMP materials quite differently. She believed that there was little opportunity for student exploration because the questions in the CPMP activities tended to lead students down a specific, predefined path. She also believed that many of the CPMP problems did not effectively encourage group work because students tended to solve them individually, even when placed in cooperative groups. Ms. Fay's frustration with the perceived lack of open-ended problems was compounded by the fact that her mathematics department colleagues pushed one another to move quickly and efficiently through the text. This meant that some of the extended, in-depth projects included in CPMP were skipped. To compensate for this lack of problems encouraging in-depth cooperative work, Ms. Fay frequently asked students to work together, tried to prompt students to discuss problems with their groups, and encouraged them to value the opinions of their group members. Ms. Fay chose not to rewrite any of the CPMP problems largely because of unfamiliarity with the contexts and mathematical content of many of the problems.

The contrasting cases of Mr. Allen and Ms. Fay illustrate how different beliefs about curriculum materials may lead to different types of student–teacher interactions even in classrooms where the same written curriculum is used.

STOP TO REFLECT

What are some of your strongest beliefs about the components of effective mathematics instruction? How might these beliefs influence the manner in which you implement a curriculum adopted by a school at which you are employed?

Implementing CMP

Lambdin and Preston (1995) described three different ways teachers responded to CMP curriculum materials when using them for the first time. Some found the curriculum to be a natural match with their existing teaching philosophies, others found the adjustment to CMP difficult but rewarding, and still others struggled and ultimately did not implement the curriculum. Teachers in the first group were called "standards bearers," those in the second group were called "teachers on the grow," and those in the third group were called "frustrated methodologists."

Teachers in the "standards bearers" group generally had strong knowledge of mathematics content and pedagogy. Their preferred lesson structure was to pose a problem, have students investigate it, and then ask students to share the results of their investigations with the rest of the class. This would often be followed by exploration of related problems and then return to the original problem. Such an instructional pattern aligned well with the pedagogical approach of CMP. Also in line with the philosophy of CMP, standards bearers generally had the goal of helping students function mathematically without the teacher. Students came to expect help from the teacher when working on a problem but would not expect the teacher to provide answers or take over the problem solving process for them. The strong content knowledge of standards bearers enabled them to catch student errors and understand when a student's conjecture needed to be challenged with a counterexample. For instance, during one class, a student conjectured that objects with the same perimeter always have the same area. Recognizing the conjecture to be incorrect, the teacher asked students to formulate counterexamples. When students did find counterexamples, the teacher framed the lesson as an authentic instance of doing mathematics, since doing mathematics can involve being temporarily confused, making incorrect conjectures, and then modifying one's thinking.

Those labeled "teachers on the grow" generally had weaker content knowledge than did their counterparts. Some had taken few mathematics courses in college because they graduated from programs in elementary education that emphasized pedagogical approaches and teaching across all content areas in the school curriculum. At times they showed strong pedagogical knowledge by using teaching approaches aligned with those endorsed by CMP, such as group or whole-class discussion of problems. Strong pedagogical knowledge was also exhibited when teachers on the grow did not set themselves up as the sole arbiters of mathematical truth in the classroom, but encouraged students to discuss whether or not a given conjecture was valid. However, weaknesses in content knowledge often turned up when teachers on the grow made mathematical errors in expositions of content or overlooked mathematical errors made by students. Teachers on the grow generally recognized their need for improved content knowledge, and they participated in professional development and worked with colleagues to attain it.

Unlike teachers on the grow, "frustrated methodologists" generally had very strong content knowledge in mathematics. They also had very tight control of student behavior in the classroom. However, they experienced frustration with the CMP curriculum because its recommended pedagogical approaches conflicted with their beliefs about teaching. Frustrated methodologists generally believed that mathematics instruction should emphasize procedural knowledge development, whereas CMP materials were designed to foster a balance of procedural and conceptual knowledge. Frustrated methodologists would skip unfamiliar forms of assessment recommended by CMP, such as having students write about their mathematical thinking. Such alternative forms of assessment did not translate easily into grades, so frustrated methodologists found it difficult to see their value. When CMP materials recommended activities such as experiments, frustrated methodologists modified them, in some situations doing the experiments in front of students rather than allowing them to do the activities. Taking control of such activities was often done to maintain a sense of orderliness in the classroom. Student confusion was seen as something to be avoided at all costs, and when students did exhibit confusion, frustrated methodologists would move quickly to providing the right answer rather than probing the sources of the confusion. The multiple points of conflict between the beliefs of frustrated methodologists and the beliefs of the CMP curriculum designers resulted in an enacted curriculum differing substantively from the written CMP curriculum.

IMPLEMENTING NSF-FUNDED CURRICULA: INSIGHTS FROM PRACTICE

Given the difficulties and potential pitfalls associated with implementing a curriculum, it can be helpful to examine the stories of teachers who have succeeded in putting reform-oriented curricula in place in their classrooms. As NSF-funded curricula for middle and high school were put into place, some teachers felt compelled to share how they changed from a traditional to a reform-oriented curriculum. Two such stories are summarized in the following sections.

Implementing MiC

Stevens (2001) described her experiences with implementing the *Mathematics in Context* (MiC) curriculum for the first time. Because MiC had been in place in the school district before she arrived, she found that students were much more willing to contribute ideas during classroom discussion than students who had experienced a traditional curriculum. Because of the level of student involvement, Stevens shifted her primary mode of teaching from providing direct instruction to asking questions designed to promote student-student discourse. As students discussed problems with one another, Stevens was able to assess their understanding of the mathematics at hand and make sound decisions about whether or not to move forward to the next lesson. She summarized her role as a teacher of the MiC curriculum in the following terms: "As a teacher, I decided when to provide information; when to clarify, model, and lead; and when to let the students struggle, rethink, ponder, and formulate responses" (p. 180).

This kind of teacher role is much more complex and multifaceted than simply conveying information to students and then testing to see if they have attained targeted concepts and skills.

Implementing *Math Thematics*

Van Boening (1999) described her experiences implementing the *Math Thematics* curriculum. She agreed to try the materials because she was convinced that conventional curricula were not helping her students develop strong mathematical reasoning skills. As she used the *Math Thematics* materials, colleagues and administrators observed her classes and were impressed with the amount of productive student discussion and the levels of understanding students demonstrated. Each year she used the materials, Van Boening became better acquainted with what to expect from students and how to support their learning. Students also became more comfortable with their classroom roles. Van Boening found that comparing her experiences with those of other teachers implementing the curriculum helped further improve her instructional practices. On seeing how the reform-oriented curriculum prompted improvements in students' mathematical understanding and engagement in class, she decided to shift permanently from traditional modes of instruction.

THE ROAD TO IMPLEMENTING REFORM-ORIENTED CURRICULA

Learning From Curriculum Materials

Experiences from research and practice illustrate that making the shift from traditional to reform-oriented instruction is not easy, but that doing so has great potential benefits for students. Teachers also benefit from learning along with their students. Because the curriculum developers recognized that implementing a written curriculum is not trivial, many NSF-funded curriculum materials were designed to support teacher learning along with student learning. Hence, in implementing a curriculum, it is important for teachers to take advantage of the learning opportunities it contains. According to Davis and Krajcik (2005), learning opportunities embedded in student texts and instructor materials may help teachers to

- anticipate and interpret students' thinking about mathematical concepts,
- extend their own mathematics content knowledge,
- understand connections between units students are to study, and
- make sound decisions about how, when, and why to adapt or alter the planned curriculum.

In addition, finding opportunities to share experiences with other teachers implementing the same curriculum can be quite valuable. This sort of sharing may occur among teachers within the same school building or at professional conferences in mathematics education (see Chapter 6 for further discussion of ongoing professional development).

Reconceptualizing the Teacher's Role

Perhaps the most challenging part of implementing a reform-oriented curriculum like those produced by NSF-funded projects is reconceptualizing oneself as a teacher. Traditionally, teachers have felt the most effective when simply telling students what they need to know (J. P. Smith, 1996). Although providing information is still an important teacher function in any classroom, it is not the only function, and often it is not the primary mode of instruction to be utilized. This can be difficult for teachers to accept, because they may feel less important to students if they are not involved in conveying all information directly to them. However, this feeling of reduced importance can be overcome by observing the many roles a teacher does play in a reform-oriented classroom: carefully selecting tasks, predicting students' mathematical thinking about tasks, encouraging and supporting student discourse, and judiciously deciding when to share information and when to withhold it to allow students to struggle (J. P. Smith, 1996). Mastering these four teacher roles is much more of an accomplishment and much more beneficial to students' learning than is taking on the one-dimensional role of constantly attempting to tell students everything they should know.

Long-Term Planning and Backward Design

Another key to successfully implementing a curriculum is to think beyond the bounds of individual lessons. Teachers need to have a sense of the entire scope of a unit and how units fit together. The idea of **backward design** (Wiggins & McTighe, 2005) has been influential in the approach many schools take to unit planning. Backward design consists of three stages: (1) Identify desired learning results, (2) decide on the kinds of evidence that will be acceptable for determining if students attained the desired results, and (3) plan corresponding learning experiences and select instructional strategies. The three stages emphasize "beginning with the end in mind" rather than just selecting activities that seem engaging. The pitfalls of focusing merely on activities or cursory coverage of material have been called the "twin sins" of instructional design (Wiggins & McTighe, 2005). Coherent long-term plans are structured to address specific learning goals, and long-term plans should be judged according to how well they help students attain those goals. Long-term plans, just like individual lessons, need to be reworked and polished constantly as they are tested in the classroom.

Wiggins and McTighe (2005) offered a template for planning units using backward design (see www.grantwiggins.org/documents/UbDQuikvue1005.pdf). The template is structured around the three phases of backward design. For phase 1, identifying desired results, the template prompts teachers to list the goals of the unit, the essential questions to be considered, and the understandings, knowledge, and skills students should attain. For phase 2, determining acceptable evidence, the template asks for key tasks students are to perform, evidence of student learning to be gathered, and mechanisms to help students reflect on and assess their work. For phase 3, planning learning experiences, teachers specify the precise sequence in which lessons and tasks are to be done. Although the long-term plan specifies a sequence of instructional events, it is necessary to maintain a degree of flexibility to allow necessary adjustments to the sequence if things do not work out as expected in the classroom. Backward design can complement the implementation of reform-oriented mathematics curricula by helping teachers identify key learning goals and possible mechanisms for attaining them.

TEACHING IN A SCHOOL CONTEXT LACKING A REFORM-ORIENTED CURRICULUM

Unfortunately, many schools have adopted textbooks that do little to help support teachers' movement toward practices that enhance students' mathematical learning. Although almost every textbook on the market claims to address the National Council of Teachers of Mathematics (NCTM; 2000) standards, many do so in a superficial manner that is not likely to help establish classrooms where students engage in substantive conversations about mathematics with one another and develop a balance of procedural and conceptual knowledge. When working with such curricula, it is up to the teacher to modify the material to be more responsive to students' learning needs.

Although wholesale change of the curriculum is not usually possible in a short time, small steps can be taken to improve inadequate curriculum materials.

Levels of Cognitive Demand

One of the main problems with many curriculum materials is that they contain too many routine, mechanistic exercises. Such exercises do not help students develop mature mathematical thinking strategies. M. S. Smith and Stein (1998) outlined a framework teachers can use to evaluate the **levels of cognitive demand** involved in the tasks included in curriculum materials. They can be summarized, from lowest level to highest, as the following:

- *Memorization*: Tasks involving memorization and recitation of facts, rules, or definitions
- *Procedures without connections to concepts or meaning*: Tasks using previously learned procedures
- *Procedures with connections to concepts and meaning*: Tasks requiring the use of a procedure while prompting students to engage the procedure's conceptual underpinnings
- *Doing mathematics*: Tasks having no prespecified solution method, requiring students to draw on conceptual understandings to devise solution strategies

Although it is sometimes appropriate to give students lower-level tasks, curriculum materials that focus primarily on the lower levels of cognitive demand are unlikely to help students reach their full mathematical potential. On the other hand, tasks at higher levels of cognitive demand have been associated with gains in student learning (Boaler & Staples, 2008).

To illustrate the difference between high and low levels of cognitive demand, consider the following two tasks:

Sample task 1: At Central Middle School, 42% of all students participate in athletics. If you picked a random sample of 50 students from the school, how many would you expect to be athletes?

Sample task 2: A biologist captured and tagged 40 fish from a lake. A week later, she caught 30 fish, and 12 of them were tagged. Based on her data, how many fish do you believe are in the lake? (Groth, 2010, p. 162)

In a classroom where students have already learned to compute the percentage of a number, sample task 1 would have a low level of cognitive demand because it would simply require students to practice doing such a computation. In the same classroom, however, sample task 2 would likely have a higher level of cognitive demand. Simply knowing how to compute the percentage of a number is helpful, but not adequate, for solving it. Students have to reason that the biologist likely tagged two-fifths of the fish in the lake, and then determine which number has 40 as two-fifths of its value. Although both tasks involve sampling and proportions, the type of thinking needed to solve each one is different.

Implementing the Common Core

See Homework Task 1 and Clinical Task 3 to explore how levels of cognitive demand relate to helping students "make sense of problems and persevere in solving them" (Standard for Mathematical Practice 1).

Examining tasks in curriculum materials in terms of the levels of cognitive demand required can open the door to rewriting and improving them. Caulfield, Harkness, and Riley (2003) described how to transform mundane, routine exercises into worthwhile mathematical tasks by raising the level of challenge. In one instance, they rewrote a task requiring calculation of the probability of landing on different sections of an ordinary game spinner. The task had a low level of cognitive demand because students simply had to write a fraction representing the number of successes divided by the number of possible outcomes. To make the task more challenging, the spinner was modified to have a star-shaped area in the center. Landing inside the star constituted a win. Students were asked to shrink the star-shaped area so it would fit on a smaller, blank spinner. This required students to draw on their knowledge of angle measurement, ratios, and algebra. The modified task also provided a launching point for several related tasks, such as estimating the probability of winning when using the new spinner.

Maintaining the Level of Cognitive Demand

It is important to note that even if one writes or selects a task with a high level of cognitive demand, care must be taken so that the integrity of the task is not compromised. Henningsen and Stein (1997) observed that many classroom tasks with high levels of cognitive demand deteriorated when enacted in classrooms. One type of deterioration that occurred was a decline from doing mathematics to doing procedures without connections. A second type of decline that occurred was from doing mathematics to engaging in unsystematic exploration. A third type of decline was from doing mathematics to doing no mathematical activity. To keep from lowering the level of demand, teachers need to avoid shifting the emphasis away from reasoning and sense-making and toward producing correct answers, make sure students do not have too much or too little time, and select tasks appropriate to students' background knowledge.

Henningsen and Stein were not the only researchers who observed a decline in the levels of cognitive demand associated with tasks. Similar observations were made by J. W. Stigler and Hiebert (2004), who noted that in a sample of lessons from teachers in the United States, all tasks that initially had high levels of cognitive demand declined to lower levels. Declines in level of cognitive demand seem just as likely to occur in classrooms using reform-oriented curriculum materials as in those using more traditional curricula (Tarr et al., 2008). Hence, maintaining the cognitive demands of tasks is an issue that needs attention in any classroom setting.

Stein, Remillard, and Smith (2007) identified several ways to maintain high-level cognitive demands. One key is for teachers to scaffold students' thinking as they work. This entails providing information to help students along without giving the problem away. Teachers also need to help students monitor their own progress, since metacognitive activity is a trait of successful problem solvers. As students formulate answers to tasks, they should be pressed to explain and justify their thinking. Teachers should provide feedback to students as they do so. Tasks should also be designed or selected to build on students' prior knowledge. Explicit connections between concepts should be drawn, and students need to be given adequate time to explore. Taking these steps can help ensure that high levels of cognitive demand are maintained.

Choose one of the tasks shown in the figures in Chapter 4. Explain how the level of cognitive demand for the task might be compromised by a teacher's actions. Also explain how a teacher could avoid compromising its level of cognitive demand.

STOP TO REFLECT

FORMATIVE AND SUMMATIVE ASSESSMENT STRATEGIES

The NCTM standards portray assessment as an integral part of instruction, and not just something that occurs afterward. For example, writing activities not only help students learn mathematics—they also provide teachers with valuable information about students' understanding. This information can be used to advise the path that future instruction should take. In the following discussion, the terms **formative assessment** and **summative assessment** will be used. Formative assessments are done primarily to inform the path that future instruction should take. Summative assessments measure a student's performance at the end of a lesson or unit of study. Often, summative assessments are used for the purpose of assigning grades. Achievement tests given to students at the end of the school year are also summative in nature.

*summative

POSING HIGHER-ORDER CLASSROOM QUESTIONS AND DISCUSSION PROMPTS

Perhaps the most straightforward and common type of formative assessment occurs when teachers ask students questions during mathematics lessons. To get the most from these questions, teachers should examine their questioning patterns. The form of a question influences the types of thinking and answers likely to be elicited from students. The following examples illustrate the difference between **closed-form** and **open-form questions:** "Are you clear on the differences between the methods of elimination and substitution?" as opposed to "What should you consider when deciding which method to use in solving instances of systems of linear equations?" "How do you decide which method is more efficient?" (Manouchehri & Lapp, 2003, p. 564).

All three questions above attempt to assess students' understanding of methods for solving systems of linear equations, but the first is a closed-form question because it requests a simple "yes" or "no" answer. In contrast, the last two questions, which are in open form, probe students' reasoning processes more effectively. They allow students to express diverse patterns of thinking and justification. While it may be appropriate to ask closed-form questions on some occasions, closed-form questions usually do not provide very detailed assessment information or prompt higher-order thinking from students.

One of the most frequently used frameworks for constructing higher-order thinking prompts, in all subject areas, is **Bloom's Taxonomy** (Bloom et al., 1956). This framework consists of categories that may be used to

Implementing the Common Core

See Clinical Task 4 to observe the roles that closed-form and open-form questions play in helping students "reason abstractly and quantitatively" (Standard for Mathematical Practice 2) and "construct viable arguments and critique the reasoning of others" (Standard for Mathematical Practice 3).

construct questions that elicit various cognitive levels. Here are the categories in Bloom's Taxonomy along with sample mathematics classroom questions:

- *Knowledge:* What is the quadratic formula? Can you state the Pythagorean theorem?
- *Comprehension:* For which kinds of equations do you use the quadratic formula? For which kinds of triangles do you use the Pythagorean theorem?
- *Application:* What are the solutions to $x^2 + 2x - 6 = 0$? What is the hypotenuse of a right triangle whose legs are 7 and 12 units long?
- *Analysis:* Amy was given a right triangle. One leg was labeled 5 units long, and the hypotenuse was labeled 7 units long. She used the Pythagorean theorem to determine that the other leg was $\sqrt{5^2 + 7^2} = \sqrt{74}$. Is she correct? Why or why not?
- *Synthesis:* James walks along two streets to get to school each day. One street runs north-south, and the other runs east–west. He walks 2 miles north on the north–south street and then 1 mile east on the east-west street. What is the distance (as the crow flies) from his house to school?
- *Evaluation:* Suppose you are asked to solve the equation $x^2 + 3x + 2 = 0$. Which solution strategy is most efficient: using the quadratic formula, graphing, factoring, or using a table of values? Explain and justify your opinion.

Using Bloom's Taxonomy as a framework for thinking about questions posed to students can help ensure that higher-order thinking questions requiring analysis, synthesis, and evaluation become part of classroom instruction (Kastberg, 2003).

STOP TO REFLECT

Think of a mathematical concept you will be responsible for teaching. Write one question corresponding to each level of Bloom's Taxonomy that you could use to assess students' understanding of the concept.

When posing higher-order thinking classroom questions, **wait time** becomes an important consideration. Often, teachers do not allow enough time for students to think after they have posed a question to the class. Rowe (1986) found that teachers typically wait less than one second for a student response after asking a question. Allowing such little time limits students' processing time and encourages superficial responses. Quick, superficial student responses usually do not provide meaningful formative assessment information. After posing questions, mentally count to at least three before jumping in and offering a response or asking another question. Often, it is also valuable to have students stay silent for a few seconds after a question is asked, since doing so gives everyone in class a chance to think about the question. If there is an extended period of silence and no one offers a response, it may be an opportune time in the lesson to break students into pairs or groups to talk about the question with one another. You can then gather all students back together to discuss it once everyone has had sufficient processing time. If a response is offered fairly soon after a question has been asked, the teacher should also monitor the amount of time he or she waits to comment on the response. Doing so gives students a chance to think about the reasonableness of the response given.

TECHNOLOGY CONNECTION

Classroom Response Systems

Classroom response systems (CRSs) are devices that provide efficient ways to tabulate students' responses to classroom questions and quickly gauge their understanding. For example, using presentation software, a teacher can pose a multiple-choice question such as

Which of the following word problems corresponds to $2\frac{3}{4} \div \frac{1}{2}$?

a. A large container in the refrigerator has $2\frac{3}{4}$ cups of yogurt in it. You would like to divide it among two friends so each gets half. How many cups does each person get?

b. A large container in the refrigerator has $2\frac{3}{4}$ cups of yogurt in it. One serving of yogurt is $\frac{1}{2}$ cup. How many servings of yogurt are in the large container?

c. A large container in the refrigerator has $2\frac{3}{4}$ cups of yogurt in it. Joe came along and ate $\frac{1}{2}$ of the container's contents. How many cups did Joe eat?

d. A large container in the refrigerator has $2\frac{3}{4}$ cups of yogurt in it. You want to divide it in half so that you and your friend each get half of the yogurt.

With handheld CRS response devices, commonly called "clickers," students can choose one of the four options. The CRS will tabulate the responses and then construct a data display to show how many students chose each option.

CRS technology is particularly powerful for facilitating formative assessment (Roschelle, 2003). In the sample multiple-choice question above, teachers can quickly assess how many students in class appear to understand fraction division. Each incorrect response, or **distracter**, is designed to appeal to students with a common misunderstanding. Response a is likely to be selected by students who confuse division by one-half with division by two. Response c suggests confusion between division by one-half and multiplication by one-half. Response d suggests confusion of division by one-half with both division by two and multiplication by one-half. If the teacher finds that any of these misunderstandings are prevalent, then more instructional time or a different instructional approach to the concept (or both) is warranted. If a portion of the class appears to have the correct understanding of the concept at hand and a portion does not, then conversations among peers may be a viable instructional approach. Teachers can choose to keep students' responses anonymous and simply examine aggregate data for the class to avoid embarrassing students. Students who do exhibit misunderstandings can also see that they are not alone as they look at the data display showing the distribution of class responses.

Follow-up questions:

1. An example of a suitable question for a CRS is given above. What types of questions would not be well suited to a CRS?

2. Do you believe students' responses to CRS questions should be anonymous, or should the teacher track the identities of students providing each response? Should students know how all of their classmates responded to CRS questions? Explain.

Monitoring and Steering Classroom Discourse

When posing classroom assessment questions, teachers should consider the interaction patterns that take place around the questions. Wood (1998) made a distinction between two types of interaction patterns: **funneling** and **focusing**. Funneling occurs when a teacher uses questions to walk students through a series of steps necessary to solve a problem. Focusing occurs when teachers' questions concentrate on uncovering students' thinking; the information gained about students' thinking then guides instruction. Funneling consists of a series of closed-form questions, while focusing includes open-ended questions that allow students to explain their thinking. Funneling is more common in mathematics classrooms, perhaps because teachers feel the pressure of time constraints. However, funneling can itself be wasted time, because students experiencing such lessons often are not able to work through problems on their own when the guiding closed-form questions are no longer being asked by the teacher.

> ### Implementing the Common Core
>
> See Clinical Task 5 to assess the extent to which funneling and focusing discourse patterns allow students to "construct viable arguments and critique the reasoning of others" (Standard for Mathematical Practice 3).

Observing Students at Work

Another straightforward way of carrying out formative assessment is to observe students as they work. Assessment charts are helpful for recording observations. On one side of the chart, teachers can identify what they are looking for in terms of concepts, tools, techniques, problem solving, and student collaboration. On the other side, they can list what they actually see happening during class (Cole, 1999). Such charts provide valuable written records to help teachers reflect on the effectiveness of a lesson. Teachers may also choose to take photographs, keep a journal, and video record lessons to facilitate later reflection (Sherin & Van Es, 2003). The assessment data captured through these means provide material for teacher self-assessment as well as an assessment of students' thinking. For example, in viewing a lesson video, one teacher discovered that he did not allow students enough time to think after asking questions (Breyfogle & Herbel-Eisenmann, 2004). This led him to increase the amount of wait time after asking a question.

Rubrics

Whereas classroom questioning and observations tend to be used for the purpose of formative assessment, rubrics are often used for summative assessment. Rubrics specify the criteria teachers will use to evaluate students' work. Two different types of rubrics can be identified: holistic and analytic. A focused **holistic rubric** assigns a single score to a student's work on a problem, while an **analytic rubric** assigns scores to various phases or aspects of the problem solving process (Charles, Lester, & O'Daffer, 1987). While some teachers construct rubrics before looking through students' work, it is often useful to specify your criteria for any type of rubric after you have determined the levels of response you are likely to receive from students on any given task to be scored. Taking this approach means that the rubric cannot be shared with students in advance. This can have positive as well as negative effects. Although rubrics make a teacher's grading criteria explicit to students, they can also constrain students' creativity in answering a problem. The holistic rubric shown in Figure 5.1 was developed by the State of Maryland for scoring short-response standardized test items. Figure 5.2 shows how the holistic rubric might be restructured into an analytic rubric.

Figure 5.1   Holistic rubric (Maryland State Department of Education, 2000).

Level 3

The response indicates **application** of a reasonable strategy that leads to a correct solution in the context of the problem. The **representations** are essentially correct. The **explanation** and/or **justification** is logically sound, clearly presented, fully developed, supports the solution, and does not contain significant mathematical errors. The response demonstrates a complete understanding and **analysis** of the problem.

Level 2

The response indicates **application** of a reasonable strategy that may be incomplete or undeveloped. It may or may not lead to a correct solution. The **representations** are fundamentally correct. The **explanation** and/or **justification** supports the solution and is plausible, although it may not be well developed or complete. The response demonstrates a conceptual understanding and **analysis** of the problem.

Level 1

The response indicates little or no attempt to **apply** a reasonable strategy or applies an inappropriate strategy. It may or may not have the correct answer. The **representations** are incomplete or missing. The **explanation** and/or **justification** reveals serious flaws in reasoning. The **explanation** and/or **justification** may be incomplete or missing. The response demonstrates a minimal understanding and **analysis** of the problem.

Level 0

The response is completely incorrect or irrelevant. There may be no response, or the response may state, "I don't know."

Explanation refers to the student using the language of mathematics to communicate how the student arrived at the solution.

Justification refers to the student using mathematical principles to support the reasoning used to solve the problem or to demonstrate that the solution is correct. This could include the appropriate definitions, postulates and theorems.

Essentially correct representations may contain a few minor errors such as missing labels, reversed axes, or scales that are not uniform.

Fundamentally correct representations may contain several minor errors such as missing labels, reversed axes, or scales that are not uniform.

Last Revised 8/16/00

Figure 5.2   Analytic rubric.

	Strategy application	Representation	Explanation or Justification
Level 3	The response indicates application of a reasonable strategy that leads to a correct solution in the context of the problem.	The representations are essentially correct.	The explanation and/or justification is logically sound, clearly presented, fully developed, supports the solution, and does not contain significant mathematical errors.
Level 2	The response indicates application of a reasonable strategy that may be incomplete or undeveloped.	The representations are fundamentally correct.	The explanation and/or justification supports the solution and is plausible, although it may not be well developed or complete.
Level 1	The response indicates little or no attempt to apply a reasonable strategy or applies an inappropriate strategy.	The representations are incomplete or missing.	The explanation and/or justification may be incomplete or missing.
Level 0	The response is completely incorrect or irrelevant. There may be no response, or the response may state, "I don't know."	The response is completely incorrect or irrelevant. There may be no response, or the response may state, "I don't know."	The response is completely incorrect or irrelevant. There may be no response, or the response may state, "I don't know."

Portfolios

Another form of assessment is the **portfolio**. Portfolios give students an opportunity to assemble their best mathematical work into a single collection. Teachers may ask students to reflect on their work samples and write about how the work they have selected illustrates their growth in thinking over a given period of time. This type of reflection on past problem solving supports students' metacognition. In addition, portfolios can help illustrate and summarize students' mathematical progress in more detail than a single letter grade for a semester. Because of this, Britton and Johannes (2003) used student portfolios as a mechanism for communicating with parents. In reviewing students' portfolios, parents were able to gain a greater understanding of their students' mathematical progress than they could simply by looking at a letter grade for the semester. Portfolios have become a popular means of assessment because of their potential to help students reflect on their mathematical knowledge and their ability to provide a detailed portrait of students' progress during a course.

Homework Assignments

Perhaps the most ubiquitous type of summative assessment used in classrooms is daily homework. Because it is so commonly used, it is easy to lose sight of why it is assigned. Kohn (2007) argued that research actually does not strongly support the idea that homework helps students. Although amount of time spent on homework is positively correlated with students' achievement, the conclusion that more time spent on home-work causes higher achievement does not automatically follow. The correlation may well exist because teachers tend to assign more homework to already higher-achieving classes. Data from the 2000 National Assessment of Educational Progress (NAEP) suggested that there actually may be a negative relationship between the amount of time spent on homework and student achievement. Eighth graders who reported doing between 15 and 45 minutes of homework per night scored higher than those who reported doing more than an hour per night. At the very least, these findings cast doubt on the simplistic assumption "The more homework, the better."

When teachers do decide to assign homework, they should carefully consider the manner in which they choose problems. Two common types of homework assignments are those requiring **blocked practice** and those requiring **mixed review**. Blocked practice assignments are those for which all exercises are drawn from a single lesson. They tend to be more common than mixed review. Mixed review assignments include exercises drawn from various lessons. Rohrer (2009) found that the mixed review strategy, although more demanding for students, is associated with higher achievement. Students are required to reflect on previous lessons and not focus exclusively on one problem type when doing mixed review. Rohrer also found that blocked practice involving heavy repetition of the same type of exercise is one of the least effective forms of practice. Doing several practice problems all of the same type influences achievement slightly in the short term, but very little in the long term. Based on these findings from the research, Rohrer commented that mixed review is underutilized, and that in some cases it may be beneficial to consider using a hybrid of mixed review and blocked practice.

It is also valuable for homework assignments to go beyond just practice and review. Three additional types of assignments to consider are applications, connections, and extensions (Connected Mathematics Project, 2009). Applications are the most traditional type of homework assignment. They allow students to practice skills learned in class by solving similar problems. Applications, however, may be set in contexts different from those encountered in class. Connections ask students to find and use relationships between the content of a day's lesson and what they have previously learned. Such exercises may also ask students to connect mathematical concepts to real-world contexts in which they arise. Extensions go beyond the content students encountered in class and help them learn new mathematical concepts. They can help set the stage for future units by foreshadowing and introducing ideas to be encountered. Homework problems that involve applications, connections, and extensions provide productive sites for individual learning as well as rich follow-up in-class discussions.

CONCLUSION

Teachers are mediators between written learning materials and students. Students' experiences with any given lesson or curriculum are heavily influenced by the beliefs and knowledge the teacher brings to bear. It is important for teachers to examine the extent to which they implement curriculum materials in the manner intended by the designers. When discrepancies occur, their causes should be identified. Discrepancies from implementing a well-designed curriculum with little fidelity to the original design should trigger a search for a greater degree of alignment between the written and enacted curriculum. When a curriculum is not effective in supporting students' learning, discrepancies between the written and enacted curriculum are actually desirable. In such cases, teachers can use techniques like designing and implementing tasks with higher levels of cognitive demand. As students study any curriculum, it is important to monitor their learning progress through the use of multiple formative and summative assessments. As assessment information is gathered, it can be fed back into the instructional planning process to help determine the optimal path for future instruction.

VOCABULARY LIST

After reading this chapter, you should be able to offer reasonable definitions for the following ideas (listed in their order of first occurrence) and describe their relevance to teaching mathematics:

Written curriculum   107

Intended curriculum   107

Enacted curriculum   107

Backward design   112

Levels of cognitive demand   113

Formative assessment   115

Summative assessment   115

Closed-form question   115

Open-form question   115

Bloom's Taxonomy   115

Wait time   116

Classroom response system   117

Distracter   117

Funneling   118

Focusing   118

Holistic rubric   118

Analytic rubric   118

Portfolios   120

Blocked practice   121

Mixed review   121

HOMEWORK TASKS

1. Choose a middle school grade level (6, 7, or 8). Read the NCTM (2006) focal points for that grade level. Then write at least two tasks related to each focal point. The tasks should be at a high level of cognitive demand and help students learn important mathematics content as they solve them. Explain why each task you wrote would be reflective of a high level of cognitive demand. Also describe the mathematics content you would expect students to learn in solving each problem.

2. Use the ERIC website (www.eric.ed.gov) to locate at least three articles about Bloom's Taxonomy. Then select a mathematical topic you will be responsible for teaching in the future. Write a question about the topic for each of the categories in the taxonomy. Explain how each question you write connects to the given category.

3. Examine the websites for at least two classroom response system (CRS) manufacturers (two possibilities are Turning Technologies and Texas Instruments). Describe the types of products each manufacturer has available. Choose a CRS system from one of the manufacturers and describe its capabilities. Then write a multiple-choice question you would ask students to respond to on the CRS. The multiple-choice question should have three distracters, and each distracter should reflect a possible misconception about the mathematics. Describe which misconception each distracter reflects.

4. Find a rubric that is used by your state department of education or by a testing agency (such as the Educational Testing Service: www.ets.org) to score open-ended test items. Describe the rubric and its strengths and weaknesses. Is it best described as a holistic rubric or an analytic rubric? Why? Would you use the rubric to score students' work in your own classroom? Why or why not?

CLINICAL TASKS

1. Examine a textbook used in your field placement school. Describe its overall philosophy by comparing and contrasting its approach to that recommended in either *Curriculum Focal Points* (NCTM, 2006) or *Focus in High School Mathematics* (NCTM, 2009). Then compare and contrast its approach with your own current philosophy of teaching. Finally, identify areas in which you would wish to enhance your content knowledge in order to teach effectively in a classroom utilizing the textbook.

2. Obtain the teacher's edition of a text from your field placement school. Describe the extent to which it provides the following opportunities for teacher learning (from Davis & Krajcik, 2005):

    a. Helps teachers anticipate and interpret students' thinking about mathematical concepts

    b. Extends teachers' mathematics content knowledge

    c. Helps teachers understand connections between units students are to study

    d. Helps teachers make sound decisions about how, when, and why to adapt or alter the planned curriculum

    e. Provides specific examples of how the teacher's edition does (or does not) support teacher learning in each area

3. Observe a classroom with an eye toward the levels of cognitive demand required by the tasks posed by the teacher. Give at least three examples of tasks the teacher posed during the lesson and comment on the level of cognitive demand required by each. Also describe how students generally responded to the tasks. Did it appear that most students successfully solved the task? Did it appear that students learned mathematical content in doing the tasks? Explain. Also, were there any instances in which a task initially requiring a high level of cognitive demand deteriorated into a task requiring a low level of demand? If so, describe how and why this occurred. Did the level of cognitive demand for tasks remain constant throughout the lesson? If so, describe how and why this occurred.

4. Observe a lesson with an eye toward the teacher's questioning techniques. Are most of the questions *open form* or *closed form*? Provide specific examples of each type. Describe what you learned about students' thinking by listening to the answers they provided to teacher-posed questions.

5. Observe a mathematics lesson and focus on the types of discourse that occur between the teacher and the students in the classroom. Is the discourse best described as "funneling" or "focusing"? Give sample exchanges between the teacher and students to support your response.

VIGNETTE ANALYSIS ACTIVITY **Focus on Attending to Precision (CCSS Standard for Mathematical Practice 6)**

Items to Consider Before Reading the Vignette

Read CCSS Standard for Mathematical Practice 6 in Appendix A. Then respond to the following items:

1. Drawing on your own experiences as a mathematics student, provide specific examples of teaching practices that can help students attend to precision.

2. Drawing on your own experiences as a mathematics student, provide specific examples of teaching practices that might prevent students from attending to precision.

Scenario

Mr. Smith is looking for ways to get his class to understand inequalities. He has been unhappy with his past practices, but he has not been sure what he should change to help deepen students' understanding. He has approached you, as a colleague who teaches mathematics, to observe one of his lessons and give him feedback on his teaching. You have agreed to do an observation for him. The next section describes some of the events that occur during the lesson you observe. As you read through the following set of events, write down thoughts that you might share with Mr. Smith when you sit down with him after school to discuss the lesson.

The Lesson

As Mr. Smith's students enter the room, they find a set of warm-up exercises on the board. Students are to solve the following five equations in the warm-up:

1. $3(x + 1) = 21$
2. $3(a - 3) = 2(a + 4)$
3. $3(y - 3) = 6$
4. $3(a - 1) = 4(a - 15)$
5. $2(2w + 5) + 2w = 46$

After a few minutes of class work time, Mr. Smith asks five students to come up to the board to solve each of the equations. When the students have finished writing their work on the board, Mr. Smith asks if there are any questions. When no one asks any questions, he decides it will be OK to move on to the next portion of the lesson.

Mr. Smith begins the next portion of the lesson by asking students, "What do you think of when I say the word *inequality*?" Several students shoot their hands up to volunteer responses. Responses include

- "It's a Pac-Man between two numbers."
- "It's like an equation, but it has a pointy symbol instead of an equal sign."
- "You do an inequality almost like how you do an equation."
- "You can solve an equation but not an inequality."

Unsatisfied with the set of student responses he received, Mr. Smith continues the lesson. He is especially annoyed that students have referred to the inequality sign as a "Pac-Man" rather than using its proper name. He decides to give them simple examples of using inequality signs. He says, "In the United States, you have to be at least 18 years old to vote. We could represent that with an inequality symbol by writing, $you \geq 18$." Mr. Smith goes on to give another example: "This semester, you will get at most 250 reward tickets from me for being good in class. Two hundred fifty is the maximum any one person can get. We could represent that with an inequality symbol by writing, $you \leq 250$." He then says that when a problem contains the phrase "at least," you should use the $\geq$ sign, and when a problem contains the phrase "at most," you should use the $\leq$ sign.

Next, Mr. Smith asks students to work in groups of four to solve problems. He says, "Imagine that there are 15 students who all want to earn money by putting in hours working at the school store this week. The school can pay for at most 60 total hours this week. How many hours should each person work if the hours are split fairly?" He then puts the inequality $15b \leq 60$ up on the overhead projector and asks students to solve and graph it. Students solved inequalities last year, so some of them know that graphing inequalities has something to do with open dots, closed dots, and arrows on a number line. As the groups work, Mr. Smith circulates about the room to remind them of when to use open dots and closed dots. As he circulates, he tells them that the arrow on the graph should point in the same direction as the arrow in the inequality sentence (so the arrow on the graph of $15b \leq 60$ would point to the left).

When most of the groups have finished working on the inequality displayed on the overhead projector, Mr. Smith calls the class back together so they can watch one of the students, Keisha, solve it for everyone. Keisha draws a number line on the transparency, puts a closed dot on the 4, and then draws an arrow coming from the dot and extending to the left. When Mr. Smith asks the student to explain her thinking, she says, "First, I pretended that there was an equal sign between the $15b$ and the 60 so I could treat it like an equation. Then I divided both sides by 15. I got $b \leq 4$. That means each student can work at most 4 hours each, because $\leq$ means 'at most.' Then I put a solid dot on the 4 because there is a line underneath the 'at most' symbol. Then I drew an arrow to the left because the 'at most' symbol is pointing to the left." Pleased that Keisha is able to talk so much about how she solved the problem, Mr. Smith thinks it would be good to give his students a more challenging problem.

Mr. Smith puts the inequality $\frac{x}{-3} \leq 4$ on the overhead next. He asks students to work with their groups again in order to solve and graph it. They quickly come up with the solution $x \leq -12$. Mr. Smith calls a student, Jasmine, up to the overhead projector to graph the solution. She draws a number line, puts a solid dot on the 12, and then draws an arrow to the left. After Jasmine sits back down, Mr. Smith goes up to the graph she has drawn and points to -15 on the number line. He asks students to put -15 in for x in the original inequality. When students do so, they get $5 \leq 4$. When Mr. Smith asks students if it is true that 5 is less than or equal to 4, a few say "no." He asks them to talk in their groups for a few minutes about what has happened in this situation. After students spend a few minutes squirming in their seats and talking about who was voted off *American Idol* last night, Mr. Smith asks the class to quiet down and to direct their attention back toward him. He writes on the overhead projector, "When you divide or multiply by a negative in an inequality, you need to flip the sign." He tells students to write the rule in their notebooks because they will be using it for homework. As he writes on the projector, the bell signaling the end of the class period rings. He quickly writes the homework assignment on the projector and tells students to copy it down before leaving class.

Questions for Reflection and Discussion

1. Which aspects of the CCSS Standard for Mathematical Practice 6 did Mr. Smith's students seem to attain? Explain.

2. Which aspects of the CCSS Standard for Mathematical Practice 6 did Mr. Smith's students not seem to attain? Explain.

3. Do you think that Mr. Smith's warm-up activity was a good lead-in to the concepts he was trying to teach in the main portion of the lesson? Why or why not?

4. Does Mr. Smith's rule about translating the phrases "at least" and "at most" into inequality symbols work all the time? Why or why not?

5. Was Mr. Smith's example of 15 students working in the school store a good word problem to use to illustrate $15b \leq 60$? Why or why not?

6. Critique Mr. Smith's approach to teaching (or reminding) students how to graph inequalities.

7. Critique Keisha's work on the problem she demonstrated at the front of the room.

8. Critique Mr. Smith's use of group work. Was his use of group work appropriate for this lesson? Why or why not?

9. What steps could Mr. Smith have taken to lead students toward discovering the rule for "flipping" an inequality sign when multiplying or dividing by a negative?

10. Mr. Smith made extensive use of the overhead projector during this lesson. What are the advantages and disadvantages of using an overhead projector as a teaching tool?

11. Do you think that Mr. Smith's students will retain the material they learned during this lesson for a long time? Why or why not?

12. What should students know about inequalities at different grade levels (Grades 6–12)? How can teachers build on what students learn about inequalities at each grade level?

RESOURCES TO EXPLORE

Books

Burrill, J. (Ed.). (2005). *Grades 6–8 mathematics assessment sampler*. Reston, VA: NCTM.

Description: This book contains sample assessment items aligned with the NCTM standards from a variety of tests. Open-ended and multiple-choice items are included, along with samples of representative student work for analysis.

Bush, W. S. (Ed.). (2000). *Mathematics assessment: Cases and discussion questions for Grades 6–12*. Reston, VA: NCTM.

Description: This collection of classroom assessment cases provides a catalyst for conversations among teachers. The cases address issues such as alternative forms of assessment, scoring student work, and using assessment to inform instruction.

Laud, L. (2011). *Using formative assessment to differentiate mathematics instruction: Seven practices to maximize learning*. Thousand Oaks, CA: Corwin.

Description: This book explicitly connects the ideas of differentiated instruction and formative assessment, helping the reader see how they support one another. The author explains how to support and assess both conceptual and procedural knowledge development for all learners.

National Council of Teachers of Mathematics. (1995). *Assessment standards for school mathematics*. Reston, VA: Author.

Description: This standards document provides a means for judging the quality of assessments. Six standards for assessments are described: mathematics, learning, equity, openness, inferences, and coherence.

Park, C. S., Lane, S., Silver, E. A., & Magone, M. (2003). *Using assessment to improve middle-grades mathematics teaching and learning: Suggested activities using QUASAR tasks, scoring criteria, and students' work.* Reston, VA: NCTM.

Description: This book and accompanying CD contain a wealth of conceptual tasks and students' work on them. These resources can help spark conversations about assessment among teachers as they score the work and compare their judgments with one another.

Stenmark, J. K. (1991). *Mathematics assessment: Myths, models, good questions, and practical suggestions.* Reston, VA: NCTM.

Description: The author provides a wealth of ideas and resources for assessing students' mathematical thinking. Insights about using open-ended tasks and projects as part of an overall assessment plan are provided.

Travis, B. (Ed.). (2005). *Grades 9–12 mathematics assessment sampler.* Reston, VA: NCTM.

Description: This collection of assessment items comes from state, national, and international tests and other sources. The book is intended to help teachers design their own assessment items and to develop their ability to use assessment results to guide instruction.

Website

NAEP Questions Tool: **http://nces.ed.gov/nationsreportcard/itmrlsx/landing.aspx**

Description: This website contains assessment questions and results from past administrations of the National Assessment of Educational Progress (NAEP). Teachers can access samples of student work, scoring guides, and performance data.

Chapter 6

Becoming a Professional Mathematics Teacher

Is teaching a profession? You may be surprised to find out that many believe it is not. The National Council of Teachers of Mathematics (NCTM; 2007) stated,

> At the present time, teaching as a profession does not receive the public support and esteem that it deserves. Teachers often find themselves in positions in which decisions that have great impact on their ability to teach are being made by persons who do not have the expertise that teachers have gained through their education and experience. Yet the teachers, not the decision makers, are held accountable to the public for the mathematics proficiency of their students. (p. 183)

Autonomy regarding decisions that influence practice is one of the widely accepted characteristics of a profession. Although teachers usually have a degree of autonomy in terms of the strategies they choose to implement in their classrooms, decisions such as setting the overall curriculum are often out of their control. Given this state of affairs, some scholars have argued that teachers should abandon attempts to transform teaching into a full profession (Myers, 2008).

Even though teaching is not universally recognized as a profession, this chapter is based on the premise that teachers should continue to strive toward professional ideals. This involves fostering professional relationships, participating in professional organizations, and engaging in ongoing professional development. This chapter will discuss how attending to these three key areas can help professionalize teaching. Perhaps even more important, it will discuss how striving toward these professional ideals can help teachers become more effective in supporting students' mathematical thinking.

Striving toward professional ideals is by no means a simple task. As prospective teachers finish their formal coursework, challenges associated with beginning to teach can seem daunting. Beginning teachers report many areas of concern, such as maintaining classroom discipline, motivating students, dealing with individual differences and problems, assessing students' work, forming relationships with parents, and organizing class work (Veenman, 1984). Some theorists argue that it is difficult if not impossible for teachers to perform tasks such as attending to students' thinking before baseline survival concerns like maintaining classroom control have been addressed (Fuller & Brown, 1975). Feeling overwhelmed by basic concerns can contribute to the decision to leave teaching to pursue a different career path (Zumwalt & Craig, 2005). This chapter's discussion of striving toward the ideals of professionalism is written with typical concerns of beginning teachers in mind.

FORMING PROFESSIONAL RELATIONSHIPS

Productive relationships with students, parents, and fellow educators are essential to professional practice. Professionals have a **service orientation**, meaning that self-interest must at times be secondary to serving those in need of assistance (Vollmer & Mills, 1966). In teaching, this characteristic relates most strongly to the student-teacher relationship, since the learning needs of students are to be prioritized over the convenience of the teacher. Professionals also have a function to perform, and laypeople (e.g., parents of students) can understand it (Vollmer & Mills, 1966). Often, teachers find it necessary to explain and justify to parents how they carry out their professional functions, emphasizing the importance of parent–teacher relationships. Third, professionals are responsible for upholding operational norms (Vollmer & Mills, 1966). In teaching, this relates to relationships teachers have with students, parents, and other educators.

Relationships With Students

Among all of the professional relationships teachers must foster, the student-teacher relationship is the most vital. The vast majority of time during a school day is spent with students. Not surprisingly, many of the concerns of beginning teachers relate to building and sustaining relationships with their students. This section will take up common concerns regarding the student–teacher relationship, including organization of class work, establishing routines, classroom discipline, motivating students, assessing students' work, dealing with individual differences and problems, and moral dimensions of relationships with students.

Organization of Class Work and Establishing Routines

The idea of including a "warm-up" problem at the beginning of every lesson has become pervasive in mathematics teaching practice. Students are often given a warm-up problem to do right at the beginning of a lesson. It may be written on the board, projected on a screen, or put on a handout. It is fairly apparent why warm-ups are ubiquitous in mathematics classrooms. They give students something to do immediately at the beginning of the period. In addition, they give teachers time to do administrative tasks such as taking attendance, distributing materials, and gathering homework assignments and permission slips. Because of these benefits, warm-up tasks often compose the cornerstone of teachers' overall strategies for organizing class work and establishing routines.

However, some words of caution are in order if warm-up problems are used as an essential part of one's practice. In particular, the warm-up should serve a mathematical function, not just an organizational one. Students will quickly sense that you are giving them busywork rather than something fundamentally related to the ideas to be studied that day. Try to connect your warm-up problem closely to your mathematical objectives for the lesson. Ideally, it should capture students' interests and serve as a launching pad into the mathematical ideas for the day. This means that a fair amount of advance preparation is needed. Choosing problems hastily or at random, or waiting until the last minute before class starts to decide on the warm-up, will likely result in tasks that do not capture students' interest. It is not necessarily safe to let your textbook dictate what your warm-up problems should be. Textbooks often contain suggestions for problems to use at the beginning of the class period, but you need to use your own knowledge of students' learning needs to decide on appropriate problems.

> Design two types of warm-up problems: one that engages students in learning through problem solving (see Chapter 1) and another that requires only procedural knowledge and may be viewed by students as busywork. Describe the key differences between the two problems.
>
> **STOP TO REFLECT**

Once you have used a warm-up or another activity to begin your lesson, be sure to have efficient transitions from one class activity to the next. Common transitions are (1) from individual work to group instruction, (2) from small-group work to large-group work, and (3) from problem solving to listening to other students' presentation of problems. Beginning teachers often have longer transition times between activities than experienced teachers, making lessons seem fragmented (Leinhardt, 1989). Moreover, during protracted transition times, discipline problems surface as students find their own, nonacademic, things to do to fill the time. Having smooth transitions from one activity to the next helps eliminate many of the problems that otherwise may surface during a lesson.

> **Implementing the Common Core**
>
> See Clinical Task 1 to analyze a warm-up problem and decide how well it helps students "reason abstractly and quantitatively" (Standard for Mathematical Practice 2).

To help facilitate transitions, establish **cues** that indicate to students when and how transitions are to occur. Cues can take a variety of forms. Some teachers, for example, use the verbal cue of counting down from five to indicate how much time students have left to quiet down, put away books, take new books out, and so forth. Verbal cues may also be simple and direct, such as simply asking students to stop talking and listen. Sometimes nonverbal cues are more effective than are verbal cues. When students are making the transition from small-group work to large-group work, for example, some teachers find it helpful to use a hand raising cue. The teacher will stand in front of the room, raise his or her hand, and then wait until all students in class have raised their hands. This sort of nonverbal cue is often more effective, and less stressful, than trying to shout over groups of students as they work. Other common nonverbal cues include flicking the lights on and off, setting a timer with an alarm that indicates when students are to move on to the next activity, or just standing at the front of the room and looking directly at the class to indicate you want their attention. All of these cues can help make the transition from one task to the next go more smoothly.

IDEA FOR DIFFERENTIATING INSTRUCTION Keeping All Students Engaged

Transitions are just one place where downtime (i.e., occasions when no mathematical activity is taking place in class) may occur during a lesson. It is important to work on minimizing downtime in any context to avoid discipline problems. Set a beginning-of-class activity that lets students know what they should do upon entering class. This might involve doing a warm-up problem, as discussed earlier. It might also involve having students revisit a problem from the previous day or continuing work on a problem that was not completely solved during the previous class. In any event, have students working on significant mathematics from the outset of the lesson. When students complete a task, let them know immediately what they can do next. If they finish an individual or group task well ahead of the others, have extension activities (not additional busywork) ready. Worthwhile extensions include having students write their own, related problems to pose to others students; having students prepare to present their solution to the rest of the class; asking students if they can come up with another way to solve the given problem; and pairing them up with students who have not yet finished the task. These types of extensions not only eliminate downtime, but they also help students learn mathematics more deeply by encouraging metacognitive activity.

When the lesson begins to draw to a close, make use of every minute available up until the bell. Do not expect students to begin meaningful work on a homework assignment if you give them only a few minutes at the end of class to start on it. It is usually very difficult to keep students on task in that sort of situation. If you find yourself with a few unplanned minutes at the end of a class period, your options include helping students begin one of the homework problems as a large group, giving a writing prompt that asks students to reflect on one of the concepts learned that day, posing a thought-provoking question that provides a different perspective on the mathematics learned that day, doing a short logic puzzle; taking care of class business; and making announcements related to events going on in the school building. Experiment with some of these end-of-class activities, as the need arises, to determine which ones best hold the attention of your students.

Managing Classroom Discussions

Establishing clear classroom procedures for participating in discussions can also help you maintain discipline. Although it is sometimes desirable to pose a question for the entire class and then wait for a student response, in some situations doing this will result in several students trying to speak at once. In such situations, modify your language to emphasize that just one person should be speaking at a time. For example, instead of posing the question, "What is the equation for this line?" say, "I would like one of you to raise your hand and tell me the equation for this line." When calling on a student to respond, monitor how many times you call on individuals so that one person does not dominate the entire conversation. Let students who are eager to answer every question know that you appreciate their enthusiasm, but that you also want to get others involved in the conversation. At times, you will need to avoid calling on some students in order to give others in the class a chance to think about the question you are asking. Also be conscious of the amount of time you wait for students to respond to questions. Researchers who study teachers' wait time find that they tend to wait only one to two seconds after asking a question (Rowe, 1987). This does not allow students sufficient processing time. Experiment with mentally counting to five after you ask a question to increase wait time. Increasing wait time to 3 to 5 seconds can help raise student achievement and improve classroom discourse (Tobin, 1986).

Along with monitoring the number of students answering questions and the quantity of time you allow after asking a question, be sure to monitor the quality of the discourse sparked by the questions you ask. Chapter 5 described two types of discourse: funneling and focusing. Keep track of which of these two types of discourse your questions tend to encourage. When students offer answers to questions, get in the habit of asking them how they arrived at the answer. Ask them to explain their thinking regardless of the correctness of their response, so students don't get the idea that follow-up questions are asked only when the answer is correct or only when it is incorrect. Take care not to set yourself up as the "traffic cop" or "dartboard" for classroom discourse. Aim to make student–student conversations just as prevalent as teacher–student conversations.

One way to increase student–student communication is to redirect some student questions back to the class. At times, a question that is difficult to answer may require breaking the large group into smaller groups to allow students to talk about it with one another. When a student does arrive at a solution, consider having him or her put the work for the problem up on an overhead projector, document camera, or the board. Insist that students pay close attention to the student presenting. Also try to help the student who is presenting convey ideas effectively. This can involve fairly simple steps, such as reminding the student to point to specific symbols and representations while talking about them. At times, you may need to help students with the clarity of their presentations by revoicing a point or asking them to repeat an explanation. As you establish procedures for maintaining quality discourse, your classroom becomes a site for students to discuss conjectures, form explanations and justifications, and explore alternative solution strategies.

> ### Implementing the Common Core
>
> See Clinical Task 5 to investigate how wait time is related to the quality of students' opportunities to "construct viable arguments and critique the reasoning of others" (Standard for Mathematical Practice 3).

> ### Implementing the Common Core
>
> See Clinical Task 6 to investigate how discourse patterns in a classroom affect students' opportunities to "construct viable arguments and critique the reasoning of others" (Standard for Mathematical Practice 3).

TECHNOLOGY CONNECTION

Introducing a Class to a New Piece of Technology

A final word on organizing class work relates to lessons that involve the introduction of a new piece of technology. In such situations, it is necessary to teach use of the technology alongside the mathematics it is meant to address. Try to focus on the mathematics rather than the required sequence of keystrokes or mouse clicks. Although the procedures for using the technology are important to learn, they should not be the primary objective of a lesson. Consider giving students "cheat sheets" that outline the necessary sequences of mouse clicks and keystrokes needed to accomplish common tasks such as graphing a line or constructing a triangle. Avoid standing in front of the room and trying to have everyone on the exact same keystroke at any given time. This is generally an inefficient way to teach students how to use technology because it often leads to frustration, lots of downtime, and classroom discipline problems because some

(Continued)

(Continued)

students are invariably a few keystrokes behind the others, and those students tend to halt the progress of the lesson by demanding additional one-on-one time with the teacher while the others wait for the lesson to move on. Cheat sheets allow students to work more at their own pace and free up the teacher to circulate about the room to address problems as they arise.

Follow-up questions:

1. Are there any situations in which it is important to have students memorize a set of keystrokes for accomplishing a task? If so, provide examples and explain. If not, explain why not.

2. Is it important for teachers to know all of the capabilities of a given piece of technology before teaching with it in class? Why or why not?

Implementing the Common Core

See Homework Task 1 to get a start on helping students "use appropriate tools strategically" (Standard for Mathematical Practice 5) when classroom technology is available.

Classroom Discipline

The best way to deal with discipline problems is to stop them before they occur. Doing so has been called exercising **preventive discipline** (Grossnickle & Sessno, 1990). Establishing effective classroom procedures, as just outlined, will stem many discipline problems by eliminating downtime and keeping students engaged. Posing mathematics problems that interest students is another way to increase engagement and on-task behavior. Establishing a few simple, reasonable rules at the beginning of the year and making sure all students understand them (and the consequences of violating them) is another preventive discipline strategy. Building positive relationships with students and seeking opportunities to compliment them on work well done are also essential. **"Withitness"** (Kounin, 1970)—communicating to students an awareness of what is happening in different parts of the classroom at all times—helps you address minor problems before they become major ones. Greeting students as they enter the room helps convey the message that you care about them and are aware of their presence. Developing a strong speaking voice, or "teacher voice," is a straightforward way to establish your presence as the adult in the classroom. Taken together, all of these preventive discipline measures help establish an orderly classroom.

STOP TO REFLECT

Imagine you will begin teaching an algebra class to first-year high school students tomorrow. Write a set of classroom rules you will expect them to follow. Then compare your set of rules to a classmate's. How are the two sets of rules similar and different? What changes would you like to make to your original set of rules after making the comparison? Write your finalized set of rules in a format suitable for sharing with your algebra class.

Despite your best attempts to exercise preventive discipline, there will be occasions when discipline problems surface and require you to take action. Serious infractions that entail physical or emotional harm to individuals in the classroom should be referred to school administration immediately. In the vast majority of cases, where no such imminent threat exists, the classroom teacher should intervene. If a student is blatantly breaking a classroom rule, it is usually most productive to attempt to deal with the individual on a one-on-one basis rather than confronting him or her in front of the group. Getting into an argument with a student in front of the class causes a power struggle that no one is likely to win.

As an example of a procedure for dealing with a discipline problem in the classroom, consider a situation where a student is sleeping in class. One of your classroom expectations should be that students keep their heads up. Generally speaking, soon after one student falls asleep, others will infer that such behavior is acceptable and some of them will put their heads down as well. Instead of loudly reprimanding a sleeping student in front of the entire class, go over to him or her after you set the rest of the class to work on a problem. Ask the student why he or she is sleeping. In some cases you may find out that the student is actually sick, in which case you should send him or her to the nurse. Make it clear that sleeping is not an acceptable behavior. With any misbehavior, also be sure to give a brief explanation of *why* it is not acceptable (e.g., "If you are sleeping in my class, you will not be able to learn what we are studying, and you will send the message that sleeping in class is OK"). Doing so gives you additional credibility because it emphasizes that the rules are not based on your arbitrary whims. If a student's misbehavior persists, don't make idle threats about implementing consequences. In some cases it may be appropriate to give a final warning, but after such a warning is given, there is no turning back. Depending on the student and the circumstances, a call home or a trip to the principal's office may be necessary.

On some occasions, several students in class may misbehave at once. For example, a common problem among beginning teachers is having students who talk when they are supposed to be listening to the teacher or others in class. On such occasions, it can help to invest a few minutes of class time to explicitly address the misbehavior at length with the entire class. Be ready to assertively discuss why it is unacceptable for the misbehavior to persist. Students should be told that it is unacceptable to talk out of turn, for instance, because it makes it impossible for others to listen to what is going on and engage in meaningful conversation. Yelling or appearing out of control are usually counterproductive in these kinds of situations, but being assertive and confident are essential. Nonabusively expressing anger over the situation is appropriate as long as the reasons you give for being angry relate to how students are disrupting the learning process through their misbehavior. Expressing anger should not, however, become your normal mode of operation. If a teacher constantly displays anger, students become accustomed to it and tune it out. Strategically choosing times to forcefully explain why persistent group misbehavior is not acceptable, and to clarify the consequences associated with it, can help reestablish productive classroom behavioral norms.

Describe how you, as a teacher, would deal with the following types of misbehavior in class:

- Students talk out of turn
- A student forgets a calculator or notebook on a regular basis
- A student calls a classmate a "moron"
- A student calls a classmate of a different race a racially insulting name
- A student calls a classmate of the same race a racially insulting name
- Two students get into a fistfight in class

Compare your reactions to those of a classmate. What similarities and differences do you see? After making the comparison, what changes would you like to make to your original reactions? Summarize your finalized reactions.

STOP TO REFLECT

Motivating Students

Middleton and Spanias (2002) described five themes in the research literature on factors that influence motivation:

1. Motivations are learned.

2. Motivation hinges on students' interpretation of their successes and failures.

3. Intrinsic motivation is better than engagement for a reward.

4. Inequities are influenced by how different groups are taught to view mathematics.

5. Teachers matter. (p. 9)

Each of these five themes are considered in this section, and their relevance to helping you build productive relationships with students are discussed as well.

The first predominant theme in the literature is that motivations are learned rather than innate characteristics. One way schools heavily influence motivation to learn is by sorting students into groups of "fast" and "slow" learners in the early grades. Students put in "slow" classes gradually lose their motivation to learn mathematics. The message that quickness is the most important attribute of mathematical talent is often reinforced at the classroom level when teachers reward students who are among the first to respond to a question or task. Those who need more processing time gradually disengage from the subject, and by the time they arrive at high school or middle school, they believe they are not mathematically talented. Those who have been classified as "fast" can also become disillusioned when they encounter reasoning tasks in which quality of thought is more important than speed. Even though students in either group often have an early desire to learn mathematics, their motivation may be snuffed out when school structures encourage speed rather than depth of thought.

Although it is difficult to undo the inaccurate mathematical self-images created by the emphasis on speed of computation in the early grades, middle and high school teachers can help remedy the situation. First, it is important to do no further harm. Do not give high grades or other rewards on the basis of how quickly a task is completed. Instead, focus on the quality of students' reasoning. Second, actively convey the message that mathematics is not a race. There are many examples from the history of mathematics showing that some of the world's greatest mathematical accomplishments took long periods of thought. Famous examples include Newton's and Leibniz's discoveries of calculus and Andrew Wiles's proof of Fermat's Last Theorem. Such examples help counteract the common student perception, often encouraged by the structure of school mathematics, that most mathematics problems should take five minutes or less to solve (Schoenfeld, 1992).

The second predominant theme in the literature is that "motivation hinges on students' interpretations of their successes and failures" (Middleton & Spanias, 2002, p. 9). A key distinction is whether students conceive of mathematical ability as being fluid or fixed. Those who conceive of it as being fluid tend to attribute successes and failures in mathematics to internal, changeable traits. They believe that hard work is linked to success, and lack of effort to failure. Conversely, those who conceive of mathematical ability as a fixed characteristic tend to attribute successes and failures to external, unchangeable traits. They believe that some people are naturally good at mathematics and others are not and that no amount of work will change that situation.

Given these beliefs, students who have been placed in "slow" classes often see no point in putting forth effort to attain deeper understanding of mathematics.

Students' beliefs about the sources of success in mathematics are usually quite entrenched by the time they reach middle and high school, so the mathematics teacher has a challenging job in counteracting unhelpful and inaccurate beliefs. The first step, again, is to do no further harm. Do not send the message that some students are naturally "strong" in mathematics and others are naturally "weak." This message is often conveyed in subtle yet powerful ways, such as referring to the "strong" or "fast" students in class during conversations with other teachers. Emphasize that ability in mathematics can be improved with effort. Enlist parents in conveying this message as well. Do not remain silent at parent conferences when parents make statements such as "I don't expect my child to be good at mathematics because I was never any good at it." Let them know that research does not support the idea that mathematical ability is a genetic trait. Also let them know that students of parents in higher-achieving countries tend to attribute mathematical success to effort rather than to a fixed, innate ability (Wang, 2004).

A third theme in the research literature is that "intrinsic motivation is better than engagement for a reward" (Middleton & Spanias, 2002, p. 9). Intrinsic motivation comes from pure enjoyment of an activity rather than from receiving a reward for doing it. Intrinsic motivators include enjoyment of doing an interesting mathematics task for its own sake and satisfaction over figuring out the solution to a difficult problem. Extrinsic motivation comes from receiving a reward for doing an activity. Extrinsic motivators include receiving praise from the teacher, receiving a good grade, or receiving a prize for solving a problem. The main problem with extrinsic motivators is that they tend to sap students' intrinsic motivation to do a task by making mathematics a means to obtaining some nonmathematical end. This diminishes any enjoyment that may come from doing a task, and it discourages in-depth exploration of related problems once the original problem has been completed.

In the typical classroom, extrinsic motivators are quite prevalent. Students in most classes receive letter grades. Some teachers hand out candy when students give correct answers. In some schools, students are given tickets allowing them to purchase things they want when they behave well in class. Some teachers make heavy use of praise. There is virtually no way to avoid having some extrinsic motivators in any given classroom. However, teachers should be vigilant about their use. Teachers cannot ignore policies such as assigning letter grades, but they can adjust the reward system in place in the classroom. Practices such as awarding candy and tickets for certain behaviors should be carefully scrutinized. Using extrinsic motivators unnecessarily has the potential to do permanent damage to students' desire to engage in a task once the extrinsic motivator is removed. It may appear that there are short-term benefits to using extrinsic motivators, but the long-term development of students should be teachers' primary concern. Fostering intrinsic motivation encourages students to develop higher levels of confidence in their mathematical ability and to persist even in the face of failure (Middleton & Spanias, 2002).

A fourth theme in the research literature on motivation is that "inequities are influenced by how different groups are taught to view mathematics" (Middleton & Spanias, 2002, p. 9). For instance, some research shows that males and females are treated differently in mathematics classrooms. Some studies have found that (1) teachers may show more concern for boys than for girls when they struggle, (2) boys are often called on more than are girls to answer questions, (3) boys are more often viewed as the best students in class, and (4) teachers tend to attribute girls' failures to lack of ability and boys' failures to lack of effort (Middleton & Spanias, 2002). In light of these findings, teachers need to constantly self-monitor to ensure that preferential treatment is not given on the basis of gender, race, or other demographic characteristics. Consistently unfair treatment of a group in this manner will sap motivation to learn mathematics.

A fifth major theme in the research on motivation is that teachers' actions matter. Throughout the preceding discussion on motivation, a number of ways teachers can influence students' motivation to learn, both positive and negative, have been mentioned. It is worth emphasizing some of the positive actions teachers can take. When extrinsic motivators, or incentives, are used, they should be given to groups rather than to individuals. As mentioned earlier, extrinsic motivators should be used sparingly, but the advantage of group incentives is that they encourage students to cooperate toward a goal rather than compete with one another for a prize (Slavin, 1984). Posing engaging problems based on students' interests is another concrete action that can be taken to build motivation. Problems set in contexts students can relate to not only help motivation but also stem discipline issues by keeping students engaged and help build their understanding of mathematics.

> **Implementing the Common Core**
>
> See Homework Task 2 to start thinking about how posing appropriate tasks for students can help them "make sense of problems and persevere in solving them" (Standard for Mathematical Practice 1).

Assessing Students' Work

Chapter 5 is full of suggestions for formative and summative assessment of students. Rather than revisiting all of those recommendations, we will focus here on managing one of the most common forms of assessment in mathematics classrooms: homework. When not properly managed, the task of keeping up with homework can consume and overwhelm new teachers.

A first piece of advice is to work every homework problem you assign, especially when teaching a course for the first time. You will often find subtle nuances in problems that you did not anticipate when assigning them. Uncovering these nuances before class allows you to be ready to lead productive class discussions about them, and to avoid being blindsided by student difficulties that arise. In working homework problems, you may also find that you need to go back and review some of the mathematics for yourself. It is much easier to do so before class than in the middle of discussing the problems with your students.

It is also imperative to develop efficient procedures for going over homework the day after it has been assigned. Time spent on review of homework should be kept to a minimum so that you can spend most of the class period helping students understand the concepts to be learned that day. There are several techniques for streamlining homework review. One technique is to post the numbers of the problems assigned for homework on the board and have students make tally marks next to the ones they have questions about (D. R. Johnson, 1982). You can then quickly glance at the board to see which problems are worthwhile for the whole group to discuss and which questions should be handled individually. For some assignments, it is appropriate to simply post the work for difficult problems, allow students to compare it against their own work, and then discuss it with a partner. Questions that remain after students have discussed the work posted can then be handled in a large-group discussion. In some cases, you may choose to make students accountable for posting the work for a given problem rather than doing it yourself. Students can be told, in advance, which homework problems they will be responsible for demonstrating on the board the next day. Posting a certain number of homework problems per semester can even be factored into the letter grade students receive.

Once homework has been discussed, teachers must decide how to grade it. Checking every single homework problem for correctness consumes valuable time that would be better spent planning instruction. One alternative is simply to check every paper to see if students have done all problems and to award credit accordingly. Some teachers give all of the points if the entire assignment has been completed, half of the points if half the problems are completed, and so forth. Although it saves time, this approach has disadvantages. It tells you little about students' understanding of the day's assignment. It also may award undue credit to a student who hastily wrote answers to every problem while shortchanging a student who put a great deal of thought into most of the problems. Instead of taking this approach, consider grading a representative sample of problems that capture the most important mathematical concepts students are to learn. Perhaps you will select just one or two problems to grade for correctness each day; these can be announced along with the due date for the assignment. If you wish to give students a chance to ask questions about the problems before you grade them, you can assemble all of the problems you plan to grade for correctness into a homework quiz to be given at the end of the week. Grading these quizzes will help you put a score on students' homework performance for the week.

Response to Intervention (RTI) Programs

Teaching is a unique occupation in that teachers must deal with 20 to 30 clients (students) at a time, while other professionals usually deal with clients on a one-to-one basis. This understandably creates anxiety for those just beginning to teach. Therefore, a substantial amount of attention has been given to the question of how to meet the individual needs of students within a large class setting.

Some schools have formalized procedures for dealing with individual student differences by establishing **response to intervention (RTI)** programs. RTI programs are designed to identify and remediate students in need of special learning assistance. The amount and type of assistance students are given is based on their demonstrated learning needs. Ardoin, Witt, Connell, and Koenig (2005), for example, described a three-phase RTI model. In the first phase, all students were given an assessment to identify those with skill deficits. In the second phase, an intervention occurred in the context of the regular class to see if students' needs could be met in a whole-class setting. Then, in the third phase, a few students still needing assistance were given more intensive help. Within this framework, all students involved in the study except one were able to improve. RTI is relatively new to the field of mathematics education, and it will be interesting to watch, in the coming years, the different models that are established and how they develop.

The Individualized Education Plan (IEP)

Another formal tool for dealing with individual student differences is the **individualized education plan (IEP)**. Like RTI programs, the IEP originates from federal legislation charging schools to meet students' individual learning needs. IEPs attempt to chart paths to success for special-needs students who spend time in the regular classroom. An IEP contains information about a student's academic background, a set of learning goals, a description of how goal attainment will be measured and reported, and instructional modifications to be made for the student (Sliva, 2003). The IEP is formed by a team whose members include special education teachers, regular classroom teachers, and parents. By law, teachers must follow the IEP established by the team, which makes it vital to give careful input if you are enlisted as a member of an IEP team. When part of the team, try to make sure that none of the proposed strategies conflict with learning goals for the mathematics class the student is taking. In some IEPs, for example, it is stated that the student should be allowed to use a calculator at all times. Although it is often a good strategy to use a calculator so that computational details do not obscure big ideas, there may also be times when the calculator is to be set aside. A mathematics teacher who is an IEP team member needs to be prepared to describe the specific situations where calculator usage is not beneficial and suggest that the team clarify situations in which calculator usage is helpful and situations in which it is not. Because of the binding implications of IEPs for instructional practice, intelligent participation from mathematics teachers is vital.

Moral Dimensions of Relationships With Students

While nothing strengthens the teaching profession like positive student-teacher relationships, nothing brings it down as much as inappropriate ones. Unfortunately, it seems as if every day in the news, another instance of an improper relationship between a student and teacher is reported. Stories of sexual relationships between teachers and students have become all too common. To avoid contributing to this serious problem, it is essential that you maintain a professional distance from students. Although it is essential to be friendly to students, you must avoid becoming just another one of their friends. This means you should not become "friends" on social networking sites, meet together outside of school activities, provide rides to and from school, or engage in any other activities that compromise your professional distance. Activities that lead to illegal and improper relationships with students do nothing but harm to both the student and the teacher involved. Students retain permanent emotional scars from such experiences, and teachers forfeit their position in the professional community while eroding the trust the public places in teachers.

A good thought to keep in mind while teaching is that you are the adult with the responsibility of teaching a roomful of children. A number of implications arise from this mind-set. One of them is that there is absolutely no room for abusive or sexual relationships between the teacher and the children in school (the phrase "children" here is meant to encompass all students enrolled in school, including those who are 18 years of age or older). Another implication is that it is your responsibility to teach children, not just to present material. Teachers sometimes become frustrated and say it is not their responsibility if students choose not to listen. On the contrary, when problems arise, it is the teacher's responsibility to explore all possible avenues for solving them. If the teacher's only job were presenting material, that responsibility would have been taken over by television a long time ago. Remembering that students are children in need of instruction can also help ease daily frustrations. Middle and high school students do not have life all figured out, underscoring the need for caring adults to guide them even when they think they need no guidance and rebel against it.

Relationships With Parents

As you work to establish healthy, productive relationships with students, enlist the assistance of parents when possible. The first contact you have with the parents of your students should be positive. Do not wait until a student misbehaves or forgets to hand in homework to contact them. Home visits at the beginning of the school year are very powerful means for establishing parent-teacher relationships, although they may not be possible when a teacher is responsible for teaching a great number of students. In lieu of home visits, consider mailing a letter to parents at the beginning of the school year. These letters can describe the goals of your course, relate your procedures for reporting grades, and invite parents to contact you as concerns arise. Parents should also be sent progress reports at least once before each marking period closes, so that there are no unpleasant surprises and confrontations when report cards come out. Also send written letters home when students do something well. Positive progress reports are very meaningful to most parents and students. Establishing a class website and communicating via email are also ways for keeping parents up-to-date on what is happening in your classroom. To get a better feel for how parents in your school like to receive communication from the teacher, invite them to share their preferences at the beginning of the school year. Preferred modes of communication may vary from school to school.

As a teacher, you will be required to participate in formal venues for teacher-parent communication, such as parent conferences, but do not limit your contact with parents to these settings. Some teachers have found it helpful to have family "math nights" where they engage parents in doing some of the same kinds of mathematics their students are doing. Family math nights are essential when implementing curricular ideas that parents did not encounter when they were students. Graphing calculators, new pieces of software, and reform-oriented curricula all constitute potential reasons for family math nights. Helping parents understand how these tools develop students' mathematical thinking is essential to gaining support for their use. Engaging parents in some of the same inquiry-oriented activities that are done in class can also help parents understand the mathematics more deeply so they are in better position to help their children with homework (Taylor-Cox & Oberdorf, 2006).

Relationships With Mentor Teachers and Other Teachers

A student teacher is in the unique position of guest in the classroom of another teacher. Therefore, making drastic changes to existing classroom culture is usually neither possible nor desirable. Most likely, you will need to learn to live with existing practices and organizational routines. In cases where you think it would be helpful to make a fairly major procedural change, such as how students are graded for homework assignments, discuss the situation with your mentor teacher and be ready to explain why you think the changes would be beneficial to students. Also seek input on students' possible reactions to your proposed change. Sometimes, even though a change seems as if it would improve things, it unsettles students, and the ensuing uphill struggle distracts students from what you would like them to learn mathematically. The situation will be different when you enter your own classroom to begin teaching, since you will be responsible for setting classroom norms and procedures from the outset of the year.

Learning to live with existing practices and routines in a classroom does not mean that you should turn off your creativity. Many mentor teachers look forward to taking in student teachers because of the fresh ideas they bring from their recent education. Look for opportunities to implement innovative lessons that will enhance student learning. This textbook models how to take an NCTM resource and transform it into a four-column lesson plan. During your student teaching semester, continue to seek out lesson planning resources from NCTM and other professional organizations that align with the content you are responsible for teaching. If you find an activity that seems to depart dramatically from what students in the class are used to doing, talk it over with your mentor teacher. He or she may have ideas for ironing out the logistics of the activity to make it work within the given classroom setting.

Interactions with your mentor teacher and the other teachers in the school building will be most productive if you demonstrate baseline professional behaviors. These include dressing professionally. Take your cue from the other teachers in the building. You should dress in accord with them, and perhaps even a notch higher as a beginning teacher. Student teachers should remain in the school building each day for the entire amount of time a teacher under contract is there, except in situations where the university dictates otherwise. Do not plan on getting to school right before the first bell and leaving right after the last. The best teachers are those who put in extra time before and after school to help students, make copies, grade papers, and take care of many other miscellaneous tasks that come across their plates. Along with being in school an appropriate amount of time each day, do not take unnecessary days off. Even though you are a student teacher, the students and the mentor teacher are counting on you to be there. When you are at school, be collegial with the other teachers in the building by looking for interaction opportunities. The advice to "stay out of the teachers' lounge" has some validity, particularly when teachers' lounge conversations consist of nothing but complaining about students, but don't let that deter you from seeking out productive collegial conversations.

Relationships With University Supervisors

Along with your mentor and other teachers in the school building, your university supervisor plays a key role in the student teaching experience. Communication is the key to keeping this relationship afloat. Recognize that your university supervisor is probably supervising several others and needs to know your teaching schedule each week in order to make plans to observe you. Do not put your supervisor in a bad position by being late in communicating your schedule, or by giving a test or quiz on a day he or she is scheduled to observe. Observations of your teaching practice are necessary in order to assign a grade for the semester, and they are also necessary for writing reference letters. Bear in mind that university supervisors are often asked to give opinions to the schools at which you do job interviews, whether you list them as references or not. Hiring a teacher is a huge investment, and careful administrators try to ensure that they have all bases covered. Do not force your supervisor to report that you lack communication skills or are unprofessional in other ways.

When your supervisor observes, you are likely to get a great deal of critical feedback. If you do not feel you are getting enough feedback, ask for more. The student teaching semester is the one time in your career that you will have an opportunity to have your lessons observed and critiqued carefully on a regular basis. Upon accepting a teaching position, many teachers are observed only once per semester or once per year by a principal. Many times, the principal who observes does not have any mathematics background and hence cannot provide feedback in that area. Bearing this in mind, be sure to reflect on the feedback you receive from your university supervisor and mentor teacher carefully. Having one's teaching critiqued on a constant basis can be stressful, but it is worth the stress if you come out of the experience as a better teacher. You will not always agree with the critiques, but do think about the reasons they may have been made. When you do not agree with a piece of feedback you received, try to raise the issue respectfully with your supervisor or mentor. Disagreements need not be done in a confrontational manner. Instead, genuinely try to understand why the critique was made. The reason for the critique may be something you have not previously considered.

ENGAGING IN ONGOING PROFESSIONAL DEVELOPMENT

Although the student teaching experience should be a time of great professional growth, your education as a teacher should by no means end when it is over. Imagine going to a doctor who has not learned anything about medicine since graduating from medical school. That individual would not be in position to prescribe recently developed medicine or to carry out research-based procedures that are more effective than those used in the past. Likewise, teachers need to stay current with recent findings about the nature of student thinking and how to apply them in the classroom. Professionals continuously acquire knowledge based on scientific theory, undergo a long adult socialization period to be successful in the profession, identify with the profession, build an affiliation with it, and view it as a lifelong occupation (Vollmer & Mills, 1966).

Professional Development Through Universities

Some of your professional development will likely come from university-level courses. In some states, obtaining a certain number of credits within a given time frame is a requirement for keeping one's certification. When exploring graduate study at the university level, consider available options carefully. If you need to take a certain number of university-level credits, you might as well choose credits that will contribute to a master's degree. Some states mandate the completion of a master's degree. Among the master's degrees mathematics teachers typically pursue are those in education, mathematics, mathematics education, and administration. Programs in education, mathematics education, and administration typically offer courses at times that practicing teachers can take them. Increasingly, universities offer required courses for these programs online, although when taking online courses it is important to deal with accredited, respected institutions rather than "diploma mills." Try to align the specific degree path you choose with your professional goals. Master's degrees in education provide a broad overview of issues related to teaching, perhaps with a few mathematics-specific courses. Degrees in mathematics education are just the opposite, with intensive work in mathematics education and perhaps a few broad overview courses. Obtaining a degree in mathematics or mathematics education can also be good preparation for further graduate study at the PhD level. Quality mathematics education PhD recipients are generally in high demand at institutions of higher education (R. E. Reys & Dossey, 2008).

Professional Development Situated in Practice

In addition to taking courses at the university, seek out opportunities to connect with other teachers in your school building. In Chapter 2, Japanese lesson study (JLS) was described as one model for promoting collaborative planning and analysis of lessons. This professional development structure is becoming more and more prevalent. If forming a lesson study group in your school setting is difficult, consider smaller steps that may help you move toward the goal of establishing a group. Simply meeting with one or two colleagues to plan lessons together and observe one another's classrooms can give you perspectives on students and on teaching that you might not gain otherwise. Also consider exchanging student work samples with one another and grading them blindly. This sort of activity can encourage conversations about the crucial mathematical elements of a task and the types of knowledge students need to develop. Another way to situate professional development within everyday practice is to establish a reading group. Teachers in such groups choose a book or journal to read and discuss a given number of times per week or per month. Books and articles closely related to learning goals for students can help catalyze conversations about how to improve practice by keeping teachers up-to-date on recent developments, trends, and ideas in the field of mathematics education.

Professional Development Through Conferences

Attending conferences can be another reinvigorating form of professional development. By attending conferences, you can listen to the best ideas of others and network. NCTM has a national conference each year as well as a series of regional conferences. These conferences are a combination of addresses by featured speakers on broad topics of interest and breakout sessions targeted at specific concerns that arise in mathematics classrooms. By looking through the conference program book to identify topics and speakers of interest, you can tailor the conference to meet your specific needs. Individual states and regions of the country have NCTM affiliates who hold their own local conferences as well. The College Board also provides many kinds of professional development, from workshops on Advanced Placement (AP) calculus and statistics to vertical teams at the school level. There are also conferences in education that are not specific to mathematics but may offer ideas for improving classroom practice nonetheless. These include conferences held by the Association for Supervision and Curriculum Development (ASCD) and state-level teacher conferences.

As you continue to grow and mature professionally, you may also find yourself attracted to research-oriented conferences. Some research-oriented conferences include the research presession for the NCTM annual national conference, the annual meeting of the American Educational Research Association (AERA), the annual meeting of the International Group for the Psychology of Mathematics Education (PME), the annual meeting of PME's North American chapter (PME-NA), and the meetings of the International Congress on Mathematical Education (ICME). Such conferences are valuable in that they involve rigorous discussion and critical evaluation of trends in mathematics education. Many of them are also international in scope, hence providing a perspective not attainable in the context of your own school.

PARTICIPATING IN PROFESSIONAL ORGANIZATIONS

Professional organizations like NCTM serve many important functions in addition to hosting conferences. They participate in setting standards of education and licensure for teachers, influencing legislation, and giving teachers a degree of control over the field. These functions are all vital to the growth and maturation of a profession (Vollmer & Mills, 1966). NCTM, for example, is an organization consisting of over 100,000 members. NCTM helped set standards for teacher preparation and licensure by partnering with the National Council for Accreditation of Teacher Education (NCATE). NCTM's curriculum standards heavily influenced NCTM/NCATE collaborative work. NCTM is based near Washington, D.C., and sends representatives to meet with government officials. NCTM members can receive regular updates about legislative happenings via email. Along with many of the other professional organizations mentioned so far, NCTM helps prevent politicians from completely removing control of pivotal decisions about education from the hands of educators.

Given the important functions professional organizations serve, those in the field should support them as much as possible. There are many ways to do so. When new standards documents are written, they are generally posted online for educators' review. The feedback that is given can help shape the substance of the new document. Many professional organizations also publish teachers' journals and books. These journals and books cannot survive without quality articles about ideas for teaching mathematics. As you go through your career and establish lessons that work very well for your students, consider writing about them and submitting them for inclusion in a teachers' journal or book. Even if the articles you submit are not published, you will receive valuable feedback from reviewers that can help improve your approach to teaching. Teachers' journals also need reviewers to survive, so consider volunteering for that task as well. Professional conferences need presenters in order to offer a range of breakout sessions, thus providing another venue for sharing your best teaching ideas with others. Consider putting a presentation proposal together and submitting it. Eventually, you might even consider running for office or volunteering for service in a professional organization. NCTM, for example, has a board of directors, a president, and several committees to work on current issues. Quality teacher involvement in professional organizations helps these organizations remain vibrant and useful.

CONCLUSION

Entering and navigating the field of mathematics education is a complex endeavor. New teachers rightfully have many concerns about how to succeed in doing so. The goal of this chapter has been to try to set you along the right path as you begin teaching. No single textbook or course can adequately prepare you for all of the situations you must deal with as a teacher, so it is important to seek out opportunities that will help you continue to grow. As you find suitable professional development, also be sure to bear in mind your obligations to students, parents, colleagues, and the larger educational community. Although growing as an individual is important, it is equally important to conduct yourself in a manner that brings credit to the developing profession of teaching mathematics.

VOCABULARY LIST

After reading this chapter, you should be able to offer reasonable definitions for the following ideas (listed in their order of first occurrence) and describe their relevance to teaching mathematics:

Service orientation   130

Cues   131

Downtime   132

Preventive discipline   134

Withitness   134

Response to intervention   139

Individualized education plan   139

HOMEWORK TASKS

1. Choose a calculator, piece of software, or handheld device you plan to use to help students learn mathematics. Choose a specific class you plan to use it in (e.g., Algebra 1, Geometry, Calculus). List common functions you would like students to be able to execute with the technology (e.g., graph a line, construct a circle, calculate a derivative). Then devise a cheat sheet describing how to execute each function.

2. Identify a mathematics concept you will be responsible for teaching. Write two to three parallel tasks you could use to differentiate instruction for introducing the concept. Describe how the tasks are tailored to meet the learning needs of students at different levels of thinking. Describe at least two correct strategies students might use to solve the parallel tasks. Then write three to five questions you would ask after students completed the tasks to assess their understanding of the mathematical ideas the tasks were designed to bring out.

3. Perform an Internet search for "mathematics response to intervention." Identify three websites that describe the structure of RTI programs. Compare and contrast the three programs (i.e., what do they have in common? How are they different?). Discuss possible benefits as well as drawbacks associated with each RTI program.

4. Find the website of a university in a geographic location where you would like to teach. Identify a master's degree program you would like to enter. Write up the following information:

   a. The name of the program and the university where it is offered

   b. Entrance requirements for the program

   c. Number of credits needed for completion

   d. Types of coursework required

   After writing this information, explain why you would like to enter the program, connecting the explanation to your goals for professional growth as a teacher.

5. Find the website for an upcoming professional conference you would like to attend. You might start by browsing the website of a professional organization such as NCTM (www.nctm.org). Once you have identified a conference, write up the following information:

   a. The name of the conference and where it will be held

   b. The types of sessions that will be held

   c. The cost of attending the conference

   d. Procedures for submitting a proposal to speak at the conference

After writing this information, explain why you would like to attend the conference, connecting the explanation to your goals for professional growth as a teacher.

6. Explore the Mathematics Education Trust (MET) section of NCTM's website. Find a mini-grant proposal that interests you, and describe how obtaining the grant would help you grow professionally.

7. Using the NCTM website or an Internet search engine, find the website of your nearest local NCTM affiliate. Describe the activities in which they engage (e.g., do they have an annual conference? Do they publish a journal? Do they offer other forms of professional development?). Also describe at least one way you could serve the affiliate in a way that would bring credit to the occupation of teaching.

CLINICAL TASKS

1. Observe a lesson where one or more warm-up problems are used. Look for the following things:

   a. Did students start working immediately on the warm-up problem(s)? What did the teacher do during the time the students were supposed to be working?

   b. Did the warm-up problem(s) require procedural knowledge, conceptual knowledge, or a combination of the two? Explain.

   c. Did the warm-up problem(s) introduce any of the main ideas for the mathematics to be learned in the lesson? If so, which ideas did the problem(s) introduce? If not, what function did the warm-up(s) serve?

   d. Did the teacher refer back to the warm-up problem(s) at any point during class? Did the students refer back? If so, at which points in the lesson? If not, why not?

2. Choose a lesson from a textbook from your school placement site. Describe the content of the lesson and the main ideas students should learn from it. Also describe the tasks you would assign students for the lesson and how those tasks connect to your learning goals for students. After identifying the tasks students are to do, write a menu of extensions for each activity that you would have ready to pose to students who finish the tasks earlier than others in class.

3. Observe a mathematics lesson with an eye toward the transition points in the lesson. How many transitions occurred in the lesson? What kinds of cues, if any, did the teacher use to indicate transition points to students? How efficient were the transitions? What kinds of management problems, if any, occurred at transition points? How did the teacher handle any management problems that occurred at transition points?

4. Observe a mathematics lesson with an eye toward downtime, both for individual students and for whole groups. Document all instances of downtime that you observe. Make and support a conjecture about why each instance of downtime occurred. Then make and support a conjecture about what could be done to eliminate each instance of downtime.

5. Observe a mathematics lesson with an eye toward how much wait time a teacher allows. Provide an estimate of the average wait time the teacher gives after asking a question. Also provide an estimate of the average wait time the teacher allows after a student provides an answer to a question. Describe how the teacher handles situations where students are not successful in giving a correct answer within the allowed wait time. In cases where students do give correct answers within the prescribed amount of time, do the teacher's questions require conceptual knowledge, procedural knowledge, or a combination of the two? Provide examples to support your answer.

6. Observe a mathematics lesson with an eye toward the discourse patterns that occur. In particular, focus on open lines of communication. Describe the kinds of teacher-student interaction that occur. Also describe instances of student-student interaction. To what extent is the teacher-student interaction conducive to learning? To what extent is the student-student interaction conducive to learning? Provide examples to support your answers.

7. Interview a mathematics teacher who generally maintains an orderly classroom. Try to understand the strategies the teacher uses by asking questions like

   - What kinds of classroom rules do you establish at the beginning of the year?
   - What consequences are in place for violating classroom rules?
   - How do you develop positive personal relationships with your students?
   - What strategies do you use to stem major discipline problems before they occur?

   Ask additional questions if they will help you understand the teacher's approach to classroom management. Then write a 150- to 200-word summary of the teacher's overall philosophy of classroom management.

8. Interview a student who receives high grades in mathematics. Try to understand the student's motivation for doing mathematics by asking questions like

   - Do you believe that you are good at mathematics? Why or why not?
   - Do you think getting a problem wrong says anything about how hard you tried? Explain.
   - Would you want to do mathematics even if you did not get a grade or other reward for doing it? Why or why not?
   - Describe your all-time favorite mathematics lesson. Give as much detail about what happened during the lesson as possible.

   Ask additional questions if they will help you understand the student's motivation. Then write a 150- to 200-word summary of the factors that play into the student's motivation to learn mathematics.

9. Repeat Clinical Task 8, but this time interview a student who receives low grades in mathematics.

10. Examine the IEP of a student in your field placement school. Without giving the student's name, summarize the information given about the student's academic background, the set of learning goals for the student, how attainment of the goals will be measured and reported, and instructional modifications to be made for the student.

VIGNETTE ANALYSIS ACTIVITY   **Focus on Classroom Management**

Items to Consider Before Reading the Vignette

1. Drawing on past experiences, recall a classroom that was poorly managed. List as many dysfunctional elements of the classroom as possible and make conjectures about their underlying causes.

2. Drawing on past experiences, recall a classroom that was well managed. List as many positive elements of the classroom as possible and make conjectures about their underlying causes.

Scenario

Mr. McCarthy was a first-year teacher at the time of the lesson described next. He entered the profession because he enjoyed working with adolescents. Mr. McCarthy wanted students to see him as a friend and went to great lengths to ensure that they felt comfortable in his classroom. He was soft-spoken and put a high priority on not intimidating students, since many of them were already intimidated by mathematics itself. His first-year teaching load consisted entirely of freshman-level high school mathematics classes. Each class was taught in a 90-minute block. This made Mr. McCarthy nervous because he had experienced only 45- to 50-minute classes as a high school student. At times, he struggled to find enough meaningful activities to keep students engaged for the entire class period.

The Lesson

Mr. McCarthy liked to begin each class period with a review of the previous day's homework, and this class session was no different. The previous day, students had been asked to solve 20 quadratic equations for homework. This was a new section in the textbook, and Mr. McCarthy knew that students would struggle with some of the prescribed solution techniques, such as completing the square and using the quadratic formula correctly. So, to begin the class, he asked, "Does anyone have questions on the homework assignment?" In the class of 30 students, 10 hands shot up immediately. Mr. McCarthy pointed to a student and said, "Brianna, I think you raised your hand first. What is your question?" Brianna responded, "Number 17." A murmur started in the background. Jason mumbled to a neighbor, "Man, that one is easy," and began talking with a group of classmates about their plans for Friday night. Mr. McCarthy began to work exercise 17 on the board and completed it in about three minutes. He continued to ask if there were additional questions and kept working problems for students for the next 30 minutes.

When the review of homework concluded, Mr. McCarthy announced that he had several homework papers to hand back to students. He circulated about the room for the next 10 minutes returning the papers. As students received their papers, they began comparing grades. Beth became upset that she received a lower grade than Andrea did and raised her hand to complain to Mr. McCarthy. Mr. McCarthy went over to Beth and spent five minutes telling her why her grade was not as high as Andrea's. Beth was not entirely satisfied with the explanation, but Mr. McCarthy moved on so he could start the next part of the lesson.

After handing back papers, Mr. McCarthy announced, "I see that a lot of you had a rough time solving quadratic equations last night. So, I am going to go through several more examples of how to solve these things so you can do better on tomorrow's homework assignment. Be sure to pay close attention and raise your hand if you have any questions." He then spent the next 30 minutes working examples from the book on a sheet of paper under a document camera and told students to write the worked-out examples in their notebooks so they could look back at their notes to see how to work the exercises on that night's assignment. About half of the class diligently wrote the examples in their notebooks for the entire time. The other half quietly whispered, passed notes, or took out portable electronic devices to play games and send text messages. Some of the off-task students were sitting near the back of the room and could not see the print on the document camera, others felt they did not need more examples, and still others did not see the point of trying to understand the examples.

Once the examples were finished, Mr. McCarthy felt it was time to try something new to help students understand the material. He announced, "Today we are going to try something different. I am going to put you in groups. Each group will get a set of exercises to solve. You should help each other work the exercises, because sometimes the explanations you can give each other are better than the ones I can provide." Mr. McCarthy then went on to describe the different roles students should have in their groups. He wanted one student to be the scribe, another to be the synthesizer of ideas, and others to perform various other roles. After giving all of his directions verbally, he told the students to assemble into their groups and begin work.

Upon gathering with their groups, students began to ask, "What are we supposed to do?" As Mr. McCarthy heard this question, he circulated about to the different groups to repeat the directions he had given the entire class. He told each group to hand in one paper to be graded. Some groups then decided to split up the exercises, making each group member responsible for doing two of them. These groups finished quickly and then had time to socialize with one another and with other groups. Other groups had the student they considered "smartest" solve all the exercises. Those not involved in solving the exercises took the opportunity to converse with classmates about the previous night's baseball play-off game.

When Mr. McCarthy had received a paper from each group, he announced, "OK, it is time to move on to the next part of the lesson." The noise level in the classroom had become quite high by that point, so only about half of the students heard this announcement. Mr. McCarthy said, "Hey guys, it is time to be quiet and give me your attention." A few more students started to listen when he said this, but several were still deeply engaged in discussing other things with their classmates. Nonetheless, Mr. McCarthy started giving instructions for the next portion of the class period. He said, "I want group 1 to tell me the answer they got for the first exercise." Maggie from group 1 responded, "Five." Mr. McCarthy responded, "No, sorry, that is not right. Group 2, what did you get for that one? Group 2, are you there?" Jonathan from group 2 finally responded, "Negative two." Mr. McCarthy said, "Good, that is right. Now, group 3, what did you get for exercise 2?" Mr. McCarthy continued this portion of the class until correct answers had been given for all of the exercises.

With 20 minutes left in class, Mr. McCarthy put the homework assignment on the document camera. Students were to solve 20 more quadratic equations like the ones they were assigned for homework the previous day. He told the class to make good use of the time remaining in the class and begin their homework. About 10 students began to work and managed to finish the assignment before the end of the class period. Some of these students had questions about the exercises, so Mr. McCarthy circulated about the room to help them. The other 20 students spent the time in other ways. Some pulled out homework for other classes. Others continued to chat. A few started lining up by the door with about 10 minutes left in class. The bell finally rang, and Mr. McCarthy said to his class, "Have a good day. I hope that tomorrow there will not be as many questions about the homework assignment."

Questions for Reflection and Discussion

1. Mr. McCarthy sought to be a friend to students. What is the difference between being friendly to students and being one of their friends? Which is preferable?

2. Mr. McCarthy spent a great deal of time going over homework exercises. Should homework review be part of a class period? If not, why not? If so, how can the process of reviewing homework be made more efficient?

3. Papers were handed back to students during the class period. What are some ways to streamline and simplify the process of getting papers back to students?

4. Mr. McCarthy gave directions for group work verbally, and students generally did not heed or remember them. What other communication strategies can be used for sharing this type of information with students?

5. Mr. McCarthy had trouble getting students' attention back when he wanted them to transition from group work to another activity. Name some techniques he could use to do a better job of making this transition.

6. Critique the manner in which Mr. McCarthy led the discussion of the work students had done in groups. How could this portion of the lesson be improved?

7. Students were given the final 20 minutes of the class to start on their homework. Was this a wise decision? Explain.

RESOURCES TO EXPLORE

Books

Chappell, M. F., Choppin, J., & Salls, J. (Eds.). (2004). *Empowering the beginning teacher of mathematics: High school.* Reston, VA: NCTM.

Description: This book consists of short articles written by experienced mathematics teachers to help those just entering the profession. New teachers are given insight about student assessment, classroom management, and interactions with the school and community.

Chappell, M. F., & Pateracki, T. (2004). *Empowering the beginning teacher of mathematics: Middle school.* Reston, VA: NCTM.

Description: This book consists of short articles written by experienced mathematics teachers to help those just entering the profession. New teachers are given insight about student assessment, classroom management, and interactions with the school and community.

Johnson, D. R. (1982). *Every minute counts: Making your math class work.* Parsippany, NJ: Dale Seymour.

Description: The author draws upon his experience as a mathematics teacher to provide advice for beginning teachers establishing their own mathematics classrooms. Advice is geared toward making one's mathematics class run smoothly with techniques such as asking good questions, managing homework, and establishing a notebook system.

Johnson, D. R. (1994). *Motivation counts: Teaching techniques that work.* Parsippany, NJ: Dale Seymour.

Description: This book serves as a follow-up to *Every Minute Counts.* Here the author focuses on practical techniques and strategies to motivate students.

Martin, W. G., Strutchens, M., & Elliott, P. (2007). *The learning of mathematics* (Sixty-ninth yearbook of the National Council of Teachers of Mathematics). Reston, VA: NCTM.

Description: The collection of articles in this NCTM yearbook helps teachers understand how to enact the NCTM Learning Principle. Issues considered include theories of mathematical learning, attaining classroom equity, and online learning.

National Council of Teachers of Mathematics. (2007). *Mathematics teaching today: Improving practice, improving student learning* (2nd ed.). Reston, VA: Author.

Description: This document is the second edition of NCTM's landmark document *Professional Standards for Teaching Mathematics.* It discusses key aspects of quality practice, interaction with colleagues, and career-long professional development.

Rubenstein, R., & Bright, G. (Eds.). (2004). *Perspectives on teaching mathematics* (Sixty-sixth yearbook of the National Council of Teachers of Mathematics). Reston, VA: NCTM.

Description: The collection of articles in this yearbook discusses foundational issues for teaching, implementing instruction, and ongoing professional development for teachers. These ideas are developed within the context of a variety of mathematical content strands.

Taylor-Cox, J., & Oberdorf, C. (2006). *Family math night: Middle school math standards in action.* Larchmont, NY: Eye on Education.

Description: Family math nights provide a means for teachers to establish productive working relationships with parents. This book provides practical advice for designing and running family math nights that help parents understand the purpose of the mathematics studied by their middle school students.

Websites

Key Online: **http://keyonline.keypress.com/home/login**

Description: This site provides a number of tools for ongoing professional development. Teachers can earn graduate credits for taking online courses on using the Key Curriculum Press software *Geometer's Sketchpad*, *Fathom*, and *TinkerPlots*.

T[3]—Teachers Teaching With Technology: **http://education.ti.com/calculators/pd/US/**

Description: This site provides information about professional development opportunities offered through Texas Instruments. Teachers can find and schedule professional development workshops, find information on T[3] conferences, view webinars, read tutorials, and take online courses.

PART II

Developing and Teaching Mathematical Thinking

Chapter 7

DEVELOPING STUDENTS' THINKING IN NUMBER AND OPERATIONS

Work with number and operations forms the foundation of the K–12 mathematics curriculum. Many of children's first experiences with mathematics involve counting. In early elementary school, students work on combining sets of objects and removing objects from sets. Methods for multi-digit addition and subtraction are studied in the elementary grades as well. Young students also learn how to represent fractions and begin to solve problems with them. These ideas in number and operations account for a large amount of the time students spend studying mathematics before entering middle school.

Although the study of number and operations is most intense in the early years of school, students' understanding in this area should continue to develop throughout middle and high school. While algebra receives a progressively larger amount of attention during the later years of school, the curricular strand of number and operation should not be forgotten. Among the key competencies to be developed in middle and high school are understanding rational numbers, developing a mature view of number systems, and computing with vectors and matrices. Students also need to learn to make wise decisions about when it is most appropriate to use mental computation, paper and pencil, estimation, or a calculator to solve numerical problems (National Governor's Association Center for Best Practices & Council of Chief State School Officers, 2010; National Council of Teachers of Mathematics [NCTM], 2000).

WHAT IS NUMERICAL THINKING?

Numerical thinking is often associated with performing computations, but its scope is really much broader. In their numerical work, mathematicians also engage in activities such as looking for patterns, making and testing conjectures, and visualizing quantities (Cuoco, Goldenberg, & Mark, 1996). To help students develop numerical habits of mind similar to those of mathematicians, it is imperative for the curriculum to include much more than just numerical computation. Additional aspects of numerical thinking to be explored in this chapter include number sense and estimation, proportional reasoning, understanding generalizations of arithmetic, working with number systems and number theory, understanding vectors and matrices, and combinatorial thinking.

ALGORITHMIC THINKING

Advantages of Algorithms

Usiskin (1998) defined an **algorithm** as a "finite, step-by-step procedure for accomplishing a task that we wish to complete" (p. 7). Students encounter many algorithms as they progress through school. Those for long division, multidigit multiplication, and solving linear equations are among the most common. School curricula have favored teaching algorithms largely because they are powerful, reliable, accurate, and fast (Usiskin, 1998). Once an algorithm for performing a mathematical task is known, it can be performed with minimum effort. Algorithms form an important part of the infrastructure of mathematics because they provide tools that are essential for handling tasks that would otherwise be more difficult and time-consuming (Wu, 1999a).

Pedagogical Difficulties With Algorithms

Despite the power and advantages of algorithms, it is necessary to proceed with caution when teaching them. Usiskin (1998) warned that students sometimes blindly accept the results of algorithms or apply them overzealously. Blind acceptance of results is detrimental when students think they have performed all the steps in the algorithm correctly, but have actually taken a misstep to produce an unreasonable answer. For example, students who have learned the conventional "Count the number of digits after the decimal point" algorithm for decimal multiplication may misapply it in a task such as placing the decimal point on the right side of the equation $356 \times 2.30 = 8,188$. Counting that there are two digits after the decimal point on the left side of the equation may lead students to produce the unreasonable answer 81.88. Students often trust their application of the algorithm to such an extent that they do not pause to realize that multiplying 356 by a number slightly larger than 2 should not decrease its size to 81.88. Overzealous application occurs when students use a complicated algorithm to perform a task that would be simpler to do using other means. Examples of overzealous application include using long division for a problem such as $11 \div 2$ and using invert-and-multiply for $\frac{1}{2} \div 2$. Each division problem can be solved more quickly and efficiently if students understand the sizes of the numbers involved and the meaning of performing division.

Perhaps even more serious than the problem of overzealous application is students' development of **buggy algorithms**. Buggy algorithms are persistent, flawed procedures for carrying out mathematical tasks. They often develop when students begin to blindly accept the results of school-learned algorithms or only partially memorize algorithms. Under such circumstances, students are unlikely to view incorrect answers as unreasonable, making it difficult for them to detect errors in carrying out computational algorithms. Ashlock (2005) cataloged

> **Implementing the Common Core**
>
> See Clinical Task 1 to assess students' understanding of algorithms for multiplication of decimals (Content Standard 6.NS).

various error patterns that occur in students' work. Prevalent patterns include the following:

- Adding numerators or denominators of fractions when it is not appropriate to do so (e.g., writing $\frac{1}{2}+\frac{1}{3}=\frac{1}{5}$ when the quantities refer to points on the number line)
- Adding decimals as if they were whole numbers (e.g., $3 + 0.2 = 5$)
- Inverting the first fraction in a division problem and then multiplying (e.g., $\frac{1}{2}\div\frac{1}{3}=\frac{2}{1}\times\frac{3}{1}$)
- Disregarding negative signs when they appear in a problem (e.g., $3 - (-2) = 1$)
- Misinterpreting rules such as "Two negatives make a positive" (e.g., $-5 - 2 = 7$)

> **Implementing the Common Core**
>
> See Homework Task 1 to judge whether or not a student has developed a buggy algorithm in connection with the ability to "interpret and compute quotients of fractions" (Content Standard 6. NS.1).

Such error patterns can become so deeply ingrained in students' mathematical activity that they systematically use their own personal versions of algorithms rather than conventional ones.

Connecting Algorithms to Previous Knowledge

There are several schools of thought on how pedagogical problems associated with algorithms should be addressed. Three possible approaches are described below. One approach is to teach conventional algorithms in a manner that helps students connect them to previous knowledge. A second approach involves encouraging students to invent their own computational algorithms. A third approach is to introduce algorithms that provide alternatives to conventional ones. These approaches are not necessarily in conflict with one another, as they may be employed at different times for different purposes within the same classroom. All three acknowledge that it is important for students to avoid learning algorithms by rote (Kilpatrick, Swafford, & Findell, 2001).

> **Implementing the Common Core**
>
> See Clinical Task 2 to assess students' ability to "use appropriate tools strategically" (Standard for Mathematical Practice 5) when dealing with algorithms.

Wu (1999a) advocated teaching conventional algorithms in a manner that connects with students' previous knowledge. For example, he recommended teaching the invert-and-multiply algorithm for division of fractions by connecting it to division and multiplication with whole numbers. Wu suggested beginning by reminding students that in their work with whole numbers, if $m \div n = k$, then $m = n \times k$. Students can see that this relationship holds with fractions if they understand that a fraction is simply another type of number. With this knowledge in hand, the statement $\frac{a}{b}\div\frac{c}{d}=\frac{e}{f}$ means that $\frac{e}{f}\times\frac{c}{d}=\frac{a}{b}$. Multiplying both sides of the equation by $\frac{d}{c}$ yields $\frac{e}{f}=\frac{a}{b}\times\frac{d}{c}$. In other words, the quotient $\frac{e}{f}$ can be obtained by inverting the fraction $\frac{c}{d}$ and multiplying the result by $\frac{a}{b}$. This approach provides an alternative to simply asking students to memorize the invert-and-multiply algorithm and accept on blind faith that it works. The more students understand the conceptual basis for a conventional algorithm, the less likely they will be to develop their own alternative, buggy algorithms.

> **Implementing the Common Core**
>
> See Homework Task 2 for an exercise in explaining the conceptual underpinnings for fraction operations (Content Standards 6.NS and 7.NS).

IDEA FOR DIFFERENTIATING INSTRUCTION Encouraging Student-Generated Algorithms

Although preventing students from inventing buggy algorithms is an important goal for mathematics instruction, it does not follow that students should never invent their own algorithms. Encouraging student-invented algorithms is a powerful method for giving students a stake in the mathematics done in the classroom. It also allows for a degree of differentiation in the mathematical products students produce. One way to encourage student-invented algorithms is to give mental mathematics problems to be done without calculator, paper, or pencil. R. E. Reys, Reys, Nohda, and Emori (1995) asked students to perform a number of computations mentally. Some students approached the problems by inventing their own strategies, while others simply tried to carry out school-learned procedures mentally. For example, when given the problem of determining the product of 38×50, one eighth-grade student decided to change 50 to 100, multiply 38 by 100, and then take half the result. A contrasting approach was to multiply 38 by 5 using a procedure learned in school and then annex a zero to the product. Students who tried to carry out such school-learned procedures mentally were more likely to make errors than were those who invented their own strategies. Higher-achieving students tended to prefer to invent their own computational strategies. Some lower-achieving students actually expressed the belief that school-learned procedures were the only way to carry out the computations. These results suggest that by encouraging students to invent their own algorithms, teachers can increase awareness of the fact that there is often more than one way to solve a mathematics problem while also prompting more flexible methods of computational thinking. As students share their invented algorithms with one another, they add to their computational repertoires.

Implementing the Common Core

See Homework Tasks 3 and 4 to reflect on the need for students to "use appropriate tools strategically" (Standard for Mathematical Practice 5) and "look for and make use of structure" (Standard for Mathematical Practice 7) when dealing with algorithms and numerical quantities.

Alternative Algorithms

Not all nonstandard algorithms introduced in a class necessarily need to come from students. At times, it is appropriate for teachers to introduce alternative computational algorithms. Providing students with different types of algorithms helps reduce errors by increasing available options for computation (Randolph & Sherman, 2001). Two well-known alternative algorithms appropriate for the classroom are the lattice method of multiplication and the scaffolding algorithm for division. A computation using the lattice method is shown in Figure 7.1, and a computation using the scaffolding algorithm is shown in Figure 7.2. Alternative algorithms are often more intuitive for students than are conventional ones.

STOP TO REFLECT

Examine the lattice multiplication and scaffolding division algorithms (Figures 7.1 and 7.2, respectively); compare and contrast them with conventional multidigit multiplication and long division algorithms on three aspects: understandability, efficiency, and accuracy.

Another benefit that comes with introducing alternative computational algorithms is increasing students' awareness that many different cultures have contributed to the field of mathematics. The lattice method of multiplication, for example, has its roots in 10th-century India (Randolph & Sherman, 2001). Such algorithms demonstrate that

Figure 7.1   Using the lattice method of multiplication to solve 53 × 29.

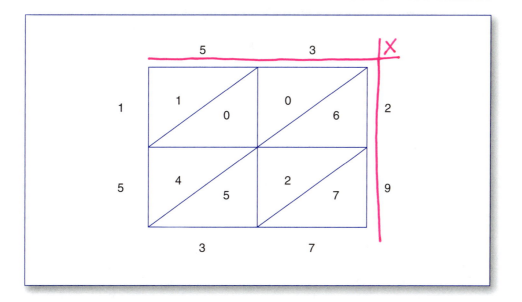

"over the generations people have grappled with many methods attempting to find those that were simpler, quicker, less wasteful of resources, or easier to communicate" (Rubenstein, 1998, p. 99). The unique mathematical histories of different cultures are reflected in present-day schools, as students from different countries often have computational methods that differ from those conventionally taught in North America (Orey, 2005). Having students share these computational algorithms, while encouraging discussion about their similarities with and differences from conventional algorithms, can help teachers show that they value a diversity of modes of mathematical thinking. This can also help spark conversations about why each procedure works to produce correct answers, which is a fundamental part of learning algorithms with understanding (Philipp, 1996).

Order of Operations

One of the most difficult algorithms for students to learn is the conventional order of operations. Teachers often resort to using a mnemonic such as "PEMDAS" (parentheses, exponents, multiplication, division, addition, and subtraction) in an attempt to help students learn. Unfortunately, this commonly used rote mnemonic device has a host of pedagogical weaknesses, and it is worthwhile to consider alternate approaches.

One of the primary weaknesses of the PEMDAS mnemonic is that its spatial ordering leads students to believe that multiplication takes precedence over division and that addition takes precedence over subtraction (Glidden, 2008). Students holding

Figure 7.2   Using a scaffolding algorithm to solve 7,884 ÷ 27.

$$
\begin{array}{r|r}
27\overline{)7884} & \\
-\ 2700 & 100 \\
\hline
5184 & \\
-\ 2700 & 100 \\
\hline
2484 & \\
-\ 1350 & 50 \\
\hline
1134 & \\
-\ 675 & 25 \\
\hline
459 & \\
-\ 270 & 10 \\
\hline
189 & \\
-\ 135 & 5 \\
\hline
54 & \\
-\ 54 & 2 \\
\hline
0 & 292 \\
\end{array}
$$

these beliefs will come to the conclusion that in a computation such as $7 - 4 + 3$, the addition of 4 and 3 must be done first. Similarly, they will conclude that $30 \div 15 \times 2$ is equal to 1 rather than 4. One way to address this common student difficulty is to teach order of operations using diagrams that place division on the same level as multiplication, and subtraction on the same level as addition (Nurnberger-Haag, 2003). The addition/subtraction level can then be placed below the multiplication/division level to show a better picture of the precedence of operations. Once students understand that the addition/subtraction level takes lowest precedence, identifying places in an expression where these operations occur outside of grouping symbols can help in parsing them into smaller, more manageable subexpressions (Rambhia, 2002). For instance, a long, unwieldy expression such as $3(2 + 1) - 8(5) + 7(9 - 4)$ can be broken into three subexpressions to evaluate—that is, $3(2 + 1)$, $8(5)$, and $7(9 - 4)$—before the subtractions and additions are done in the order they occur.

Another weakness of the PEMDAS mnemonic is that its focus on parentheses (represented by the P in the mnemonic) is misleading. Parentheses are just one sort of grouping symbol one may encounter in an expression. Other types of grouping symbols occur in complex fractions (e.g., $\frac{\frac{1}{2} + \frac{1}{2}}{\frac{3}{4}}$), radical expressions (e.g., $\sqrt[3]{7 + 20}$), and absolute value statements (e.g., $|-3 + 1|$; Nurnberger-Haag, 2003). Teaching students to concentrate on parentheses alone as grouping symbols may allow them to succeed on carefully chosen problem sets that do not include any other type of grouping symbol. However, in the long run, teachers run the risk that the P in the PEMDAS mnemonic will interfere with students' understanding of other types of grouping symbols. To avoid this difficulty, parentheses should be portrayed as just one type of grouping symbol among many.

> ### Implementing the Common Core
>
> See Clinical Task 5 to assess a student's understanding of order of operations (Content Standards 6.EE and A-SSE).

NUMBER SENSE AND ESTIMATION

Number Sense

As mentioned earlier in the chapter, numerical thinking extends beyond the use of computational algorithms. **Number sense** also plays an important role. R. E. Reys and Yang (1998) defined number sense in the following manner:

> Number sense refers to a person's general understanding of number and operations. It also includes the ability and inclination to use this understanding in flexible ways to make mathematical judgments and to develop useful strategies for handling numbers and operations. It reflects an inclination and an ability to use numbers and quantitative methods as a means of communicating, processing, and interpreting information. It results in an expectation that numbers are useful and that mathematics has a certain regularity. (pp. 225–226)

Students who are proficient in carrying out algorithms may actually have very poor number sense. For instance, when R. E. Reys and Yang asked sixth-grade students to calculate the exact result of 72 ÷ 0.025, 54% of them did so correctly. On the other hand, when they were asked to circle the best estimate for 72 ÷ 0.025, only 33% were able to do so. Of those not able to identify the best estimate, 31% thought the result would be "a lot less than 72," indicating they had the common belief that division always makes the starting quantity smaller. R. E. Reys and Yang found a pattern of consistently higher performance on algorithmic items compared to performance on items requiring number sense.

Given the results above, it is safe to conclude that number sense is very different from algorithmic thinking. Understanding the essential components in number sense is important if teachers are to help students develop it. Three components of number sense are (1) knowledge of and facility with numbers, (2) knowledge of and facility with operations, and (3) applying knowledge of and facility with numbers and operations to computational settings (A. McIntosh, Reys, & Reys, 1992). Indicators of acquisition of the first component include knowing that there are numbers between $\frac{3}{5}$ and $\frac{4}{5}$ on the number line, being able to estimate the values of points placed between integers on the number line, and recognizing that $\frac{3}{5} = 0.6$. Indicators of acquisition of the second component include knowing that the heuristics of "Multiplication makes bigger" and "Division makes smaller" are not always true (e.g., dividing a whole number by $\frac{1}{2}$ produces a larger number) and understanding differences between multiplying matrices and whole numbers (e.g., matrix multiplication is not commutative). Indicators of acquisition of the third component include having a tendency to explore a problem in different ways, using the most efficient representations, and having an inclination to check whether the result of a computation makes sense. The three components provide a working framework for teachers to assess students' acquisition of number sense.

Computational Estimation

Estimation goes hand in hand with number sense. Many of the number sense items given as examples in this chapter can be categorized as requiring **computational estimation** (Sowder, 1992), since they involve estimation of the result of a calculation (e.g., estimation of the answer to 72 ÷ 0.025). The thinking of students who are adept at computational estimation has been categorized into three processes: (1) **reformulation**, (2) **translation**, and (3) **compensation** (R. E. Reys, Rybolt, Bestgen, & Wyatt, 1982). Reformulation involves altering the numbers in a problem to make them more manageable. For instance, when estimating 2,713 ÷ 888, a student using reformulation may round 2,713 to 2,700 and then round 888 to 900 to reason that the quotient is about 3. Translation involves altering the mathematical structure of a problem. For example, a student asked to estimate $\frac{25 \times 19}{5}$ might begin by doing the division 25 ÷ 5 instead of starting with the multiplication 25 × 19. After obtaining a result of 5 from 25 ÷ 5, he or she might estimate that the answer is approximately equal to 5 × 20 = 100. Compensation occurs when students make adjustments to numbers to account for the effects of reformulation or translation. For instance, R. E. Reys and colleagues (1982) gave students a set of attendance figures from six football games (73,655; 86,421; 91,943; 96,509; 93,421; 106,409) and asked them to estimate the total attendance for all six. A ninth grader using compensation responded, "I rounded all to 100,000 except 73,000. I dropped this one to make up for rounding the others up. The numbers are so close they just make up for each other" (p. 190). Encouraging students to engage in reformulation, translation, and compensation, without reducing these three thinking processes to algorithms, may help students become better computational estimators.

Measurement Estimation

Along with computational estimation, **measurement estimation** (Sowder, 1992) is important for students to understand. Measurement estimation involves estimating geometric measurements (which will be discussed in detail in Chapter 10) or large quantities of items (numerosity estimation). Examples of numerosity tasks include estimating the number of words on a piece of paper, the number of M&Ms in a bag, the number of days one has been alive, and how many beans it would take to fill a gallon jug (Crites, 1992). When Crites presented these types of tasks to seventh graders, the most successful students used benchmark and decomposition/recomposition strategies to solve them. **Benchmark estimation** strategies involve estimating a quantity by comparing it to one that is known (e.g., estimating the number of candies in a jar by comparing its size to that of a jar holding a known number of candies). **Decomposition/recomposition estimation** strategies involve dividing a problem into parts and then recombining the parts to produce an estimate (e.g., estimating the number of words on a page by looking at the number of words in a typical line and then multiplying by the number of lines on the sheet). Students who were less successful estimators tended to prefer a **range estimation** strategy, choosing the center of an arbitrary range as their estimate (e.g., believing that the actual number of words on a page is somewhere between 100 and 150 and choosing a number in the middle of the range as an estimate).

> ### Implementing the Common Core
>
> See Clinical Tasks 6 and 7 to examine students' ability to work with estimates and numerical expressions with rational numbers (Content Standard 7.EE).

Differences between skilled and less skilled estimators manifested themselves in the Crites (1992) study in ways beyond students' preferred strategies. Skilled estimators had more success estimating large quantities. Less skilled estimators tended to guess, did not usually change their estimates in the face of contradictory evidence, and did not verbalize their thinking well. In light of these findings, Crites recommended helping students build points of reference that can be used as benchmarks when estimating. Knowing how many people a football stadium can hold, for example, can become a useful benchmark for estimating large quantities. It was also recommended that students be required to defend the reasonableness of their estimates. Students may revise their estimates when asked to reflect on how reasonable they are. Encouraging students to reason in this manner can help move their patterns of thought closer to those exhibited by successful estimators.

> **STOP TO REFLECT**
>
> Design one numerosity estimation task that deals with very large numbers (greater than 10,000), and another numerosity estimation task that deals with very small numbers (between 0 and 1). For each task, describe how benchmark, decomposition/recomposition, and range strategies might be employed by a student.

PROPORTIONAL REASONING

Proportional reasoning can be defined in the following manner:

> Proportional reasoning refers to detecting, expressing, analyzing, explaining, and providing evidence in support of assertions about proportional relationships. The word reasoning further suggests that one uses common sense, good judgment, and a thoughtful approach to problem solving, rather than plucking numbers from word problems and blindly applying rules and operations. (Lamon, 2007, p. 647)

The middle school years typically involve the most concentrated attention to developing students' proportional reasoning, but some degree of attention must be given to it throughout a student's years in school.

Operations With Rational Number Representations

One aspect of proportional reasoning is understanding representations of rational numbers and operations on them. On one famous multiple-choice National Assessment of Educational Progress (NAEP) item (Carpenter, Corbitt, Kepner, Lindquist, & Reys, 1981), when asked to choose the best estimate for $\frac{12}{13} + \frac{7}{8}$, most students answered incorrectly. Although the best choice was 2, most eighth graders chose either 19 or 21 as the best estimate. These two numbers are produced by simply adding the numerators or denominators of the fractions involved. Students also had difficulty estimating $4 + .3$. Most sixth and seventh graders believed that $.7$ was the correct response. These NAEP items have frequently been cited to make the point that students' reasoning about rational number representations and operations on them is generally quite weak.

Additional well-known examples of individuals' difficulties with operations on rational number representations come from Ma's (1999) comparison of teachers' knowledge in the United States and China. When asked to write a word problem for $1\frac{3}{4} \div \frac{1}{2}$, most U.S. teachers in Ma's study were not successful. Some of them confused division by $\frac{1}{2}$ with division by 2. One teacher making this mistake, for example, described the situation of sharing $1\frac{3}{4}$ pie fairly between two people. Another common error committed by U.S. teachers was to write a problem for multiplication by $\frac{1}{2}$ rather than division by $\frac{1}{2}$. One teacher making this mistake described the situation of having $1\frac{3}{4}$ pies, having someone steal half that amount, and then calculating $\frac{1}{2}$ of $1\frac{3}{4}$ to determine the leftover quantity of pie. These error patterns are similar to those one can expect to observe among groups of middle and high school students.

The Chinese teachers in Ma's (1999) study were much more successful in writing word problems for $1\frac{3}{4} \div \frac{1}{2}$, and they used a variety of strategies to do so. Some of them wrote word problems that fit the **measurement model of division**, which involves determining the number of groups of a certain size in a given collection. One measurement problem consisted of having $1\frac{3}{4}$ apples, knowing that $\frac{1}{2}$ apple is a serving, and determining the number of available servings. Others wrote word problems fitting the **partitive model of division**, in which the central task is to determine the number of objects per group. One partitive problem stated that $\frac{1}{2}$ of a jump rope was $1\frac{3}{4}$ meters. The task was to find the length of the whole

Implementing the Common Core

See Homework Task 5 to try your hand at creating story contexts for fraction division problems and using visual models to represent the quotient (Content Standard 6.NS.1).

rope. Finally, some teachers used the **product-and-factor model of division**. This involved finding a factor that, when multiplied by $\frac{1}{2}$, would produce $1\frac{3}{4}$. For one such problem, a teacher described the situation of knowing that the area of a rectangle was $1\frac{3}{4}$ square meters. Given that the width of the rectangle was $\frac{1}{2}$ meter, the task was to find its length.

Suggestions for Teaching Operations With Fractions

Fraction circles, manipulatives consisting of slices of a circle's interior, can help students begin to understand the meanings of fraction representations and operations on them. Cramer, Wyberg, and Leavitt (2008) reported that fraction circles were particularly helpful in building sixth-grade students' understanding of fractions as parts of a whole and also in understanding the relative sizes of fractions. With fraction circles, students observed that the more a circle is partitioned, the smaller the partitions become. An example of this is shown in Figure 7.3.

One piece of the circle diagram on the left side of Figure 7.3 can be used to represent the fraction $\frac{1}{4}$. The fraction circle on the right side was produced by cutting each of the four pieces from the diagram on the left into two equal-size pieces. Therefore, one of the pieces from the diagram on the right can be used to represent the fraction $\frac{1}{8}$. The difference in size allows students to order the two fractions by thinking about the size of each slice rather than by manipulating the conventional symbolic representations for them. Cramer, Wyberg, and Leavitt found this feature of the fraction circle model to be helpful in building students' ability to order fractions by relative magnitude.

Once students understand the sizes of fractions in comparison to one another, they can begin to better understand fraction addition and subtraction. Thinking about fractions in terms of slices of a circle enables students to form reasonable estimates of sums. Students with sound understanding of the fraction circle model, for instance, are unlikely to believe that $\frac{12}{13}+\frac{7}{8}$ is close to 19 or 21

Figure 7.3 Fraction circles with different numbers of partitions.

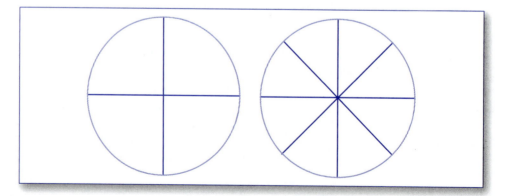

(Carpenter et al., 1981), because combining the two fraction circle representations produces only about two whole circles.

When it comes to finding the precise answer to a problem with fraction circles, students can think in terms of "trading in" sets of fraction representations for equivalent ones. For example, when faced with the problem of adding $\frac{3}{4}+\frac{5}{8}$, students can trade in three of the pieces from the circle on the left side of Figure 7.3 for six of the pieces from the circle on the right side. The fraction circle representations for each fraction are now partitioned into the same number of equal-size pieces, making it easier to combine the two fractions. The act of trading in transforms the problem into $\frac{6}{8}+\frac{5}{8}$. Combining the pieces produces one full circle and three extra pieces. The result of combining the two fractions can thus be named either $\frac{11}{8}$ or $1\frac{3}{8}$. When students are in the process of combining fraction circle pieces, it is helpful to delay the introduction of formal language such as "common denominator" and allow students to use their own informal language at first (Cramer et al., 2008). Formal, conventional language can be introduced after students have had adequate time to make sense of the process for themselves.

There should be no rush to formal language when teaching other aspects of fraction operations as well. If students have already heard formal language and practiced conventional algorithms in other classes, they can learn a great deal if they are asked to set that knowledge aside and act problems out instead. Perlwitz (2005), for example, asked students who already knew the invert-and-multiply algorithm for fraction division to devise their own strategies for determining how many pillowcases could be formed from a given amount of fabric. Students were told that each pillowcase required $\frac{3}{4}$ yard of fabric and that there were 10 yards of fabric available. Instead of jumping immediately to the invert-and-multiply algorithm, students had to attack the problem by diagramming the 10 yards of fabric and then deciding how many $\frac{3}{4}$-yard pieces could be cut from it. Students were uncomfortable and resistant to approaching the problem in this manner, but the classroom discourse sparked by setting aside formal language and procedures was exceptionally rich. In particular, the task caused a great deal of conversation about whether the correct response was 13, $13\frac{1}{3}$, or $13\frac{1}{4}$.

> ### Implementing the Common Core
>
> See Clinical Task 8 to assess a student's ability to "use visual fraction models" (Content Standard 6.NS) to represent and solve fraction division problems with fraction circles.

> **STOP TO REFLECT**
>
> Consider the following scenario: You have 10 yards of fabric, and it takes $\frac{3}{4}$ yard of fabric to make a pillowcase. How many pillowcases can you make (Perlwitz, 2005)? Is the correct response to the question 13, $13\frac{1}{3}$, or $13\frac{1}{4}$? Describe the type of thinking that would lead a student to each one of these conclusions. Then explain which answer is the most reasonable, and why.

When teaching about rational numbers, it is also important to present situations capturing the wide range of interpretations they can have. Lamon (2005) pointed out that the symbol $\frac{3}{4}$ can mean a variety of things, depending on the context in which it is used. Figure 7.4 shows possible representations for different interpretations.

Under the **part-whole interpretation**, $\frac{3}{4}$ means 3 parts out of 4 equal-size parts, as shown in the top rectangle of Figure 7.4. Under the **measure interpretation**, $\frac{3}{4}$ represents 3 lengths of size $\frac{1}{4}$ each, as shown on the number line representation in Figure 7.4. An **operator interpretation** of $\frac{3}{4}$ implies that one is recognizing $\frac{3}{4}$ of a given quantity. One can think, for example, of a copy machine operating on a setting of 75% to produce a reduced image, as shown by the sea turtle images in Figure 7.4. The **quotient interpretation** of $\frac{3}{4}$ simply means thinking of the symbol as representing 3 divided by 4, as when one considers the result of sharing three objects among four people. Finally, one may think of $\frac{3}{4}$ in terms of the **ratio interpretation**. Ratios arise in many situations, such as miles per gallon, dollars per hour, and feet per second. Colored chips are often a convenient representation to use when working with ratios, as shown in the bottom rectangle of Figure 7.4; here there are three white chips for every set of four black chips.

Figure 7.4 Representations to accompany different interpretations of $\frac{3}{4}$.

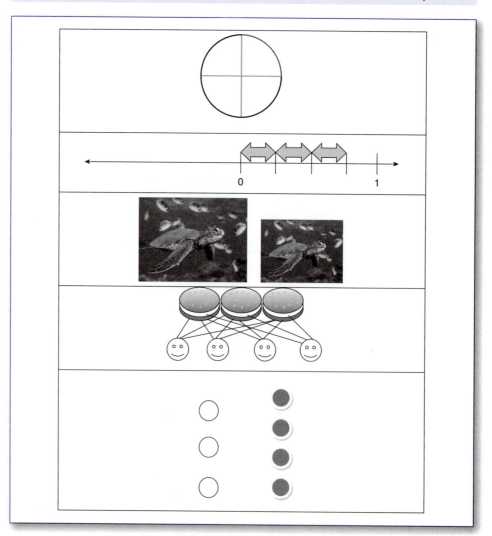

Additional Aspects of Proportional Reasoning

The variety of ways that the symbol $\frac{3}{4}$ can be interpreted brings out the idea that proportional reasoning extends beyond adding, subtracting, multiplying, and dividing fractions. According to Langrall and Swafford (2000), four essential elements of proportional reasoning are the following:

- *Recognizing the difference between **additive change** and **multiplicative change***. Additive change is simply the total raw amount of change in a quantity. Multiplicative change is the amount of change expressed as a percentage of the original quantity. Imagine that one plant started out at a height of 6 cm and grew 2 cm. Then suppose a second plant started at 2 cm and grew 1 cm. When asked which plant grew more, a student who reasons only additively will identify the first one. A student reasoning multiplicatively could argue plausibly that the second plant grew more.

- *Recognizing situations where a ratio is reasonable or appropriate*. Students with this ability understand when it makes sense to use ratios in solving problems and when it does not. For example, it makes no sense to use ratios to solve this problem: "If one man walks to the store in 20 minutes, how long will it take two to get there?" However, ratios may be useful for this task: "If one man can paint a fence in 20 minutes, how long will it take two men to paint it?"

- *Recognizing invariance in equivalent ratios*. Consider a situation in which 2 of 4 people at one table are female and 3 of 6 at another are female. Although the table compositions are additively different, the female/male ratio is invariant.

- *Thinking flexibly about units*. In proportional reasoning problems, it is often convenient to redefine the fundamental unit with which one is working. If one wants to know the cost of 24 balloons, given that 3 balloons cost $2, then it is easier to reason in terms of three-packs of balloons as a unit rather than in terms of a single balloon as a unit, since 3 divides 24. If one is working with pattern block manipulatives and needs to represent $\frac{5}{12}$, it is more convenient to define the unit as two large hexagons rather than one, since the small triangles in the manipulative set partition a hexagon into 6 pieces rather than 12.

The four elements discussed above can be considered hallmarks in the development of proportional reasoning. Teachers should regularly look for opportunities to assess students' attainment of each one, and to help support the development of each element by posing problems that elicit each type of thinking.

GENERALIZING ARITHMETIC

Just as the attainment of proportional reasoning is a significant milestone in students' mathematical development, acquiring the ability to make generalizations about arithmetic is of great importance. Algebra can in some ways be considered generalized arithmetic, so making observations about patterns that occur in work with numbers is an important foundation for algebraic thinking (Usiskin, 1988). Two key types of numerical tasks that provide rich sites for encouraging generalization of arithmetic are analyzing number sentences and analyzing numerical sequences.

> **Implementing the Common Core**
>
> See Homework Task 6 and Clinical Task 9 to create and administer an assessment to measure students' knowledge of ratios and proportional relationships (Content Standards 6.RP and 7.RP).

Analyzing Number Sentences

Open number sentences, equations with one or more blanks inserted in place of quantities, provide students with opportunities to develop understanding of the idea of equality. Surprisingly large percentages of middle school students struggle with understanding how the equal sign represents equality in number sentences. Falkner, Levi, and Carpenter (1999) reported that when 145 sixth-grade students were given the open number sentence $8 + 4 = \_\_ + 5$, none of them filled in the blank correctly. All of the students thought the correct response was either 12 or 17. Falkner and colleagues recommended having students work with these kinds of number sentences more frequently, since they strongly resemble the equations students confront when doing algebra (more will be said in Chapter 8 about how students' understanding of the equal sign affects their equation solving ability).

At times, true/false exercises can help students develop sounder conceptions of equality in number sentences. For instance, asking students to judge whether the statement $78 - 49 = 78$ is true or false has helped students analyze the meaning of the equal sign and to form generalizations such as $a - 0 = a$ and $a + b - b = a$ (Carpenter, Levi, Williams-Berman, & Pligge, 2005). Commutativity is also implicit in much of the work students do with number sentences, so teachers can draw attention toward examining whether any generalizations are suggested by statements such as $5 + 7 = 7 + 5$ or $5 \times 7 = 7 \times 5$. When students first develop the idea of commutativity, they may believe it holds only for small numbers. Hence, asking students to evaluate and discuss the truth of the statement $a + b = b + a$ can be a valuable exercise. As students come to conclusions about valid generalizations suggested by work with everyday number sentences, teachers can facilitate their learning and retention of the generalizations by posting them on large sheets of paper in the room. Doing so makes it easy for students to return to the generalizations as they need them when working or to make refinements to the generalizations as they encounter the need to be more precise in expressing them (Carpenter et al., 2005).

When helping students form generalizations, teachers should be aware that students will generally use informal language and symbols before formalizations. For instance, blanks and squares may initially occupy the positions that literal symbols representing unknowns and variables will eventually fill. One way to encourage students to move from informal notation to formal symbols is to ask them to write an open number sentence that is true for every number they can put into it (Carpenter et al., 2005). Some open number sentences, such as $\_\_ + 0 = \_\_$, lend themselves to a smooth transition to the formally stated generalization $a + 0 = a$. Other generalizations students make are initially more challenging to represent with literal symbols, such as the idea that the sum of any two even numbers is another even number. By engaging in the task of transitioning from informal notation to formal literal symbols, students have the opportunity to build meaning for the formal notation from the ground up.

In helping students make formal generalizations, an important role for the teacher to fill is facilitator of productive classroom discourse. Initially, students may come to believe that a generalization is true simply because it works for a large number of examples (Carpenter et al., 2005). After observing several examples of the commutative property of multiplication, for instance, students may be satisfied that it holds for all

Implementing the Common Core

See Homework Task 7 and Clinical Task 10 for examples of how to "reason abstractly and quantitatively" (Standard for Mathematical Practice 2), "construct viable arguments" (Standard for Mathematical Practice 3), and "look for and make use of structure" (Standard for Mathematical Practice 7) when making generalizations.

whole numbers. Although this conclusion is correct, students should understand that mathematical statements are not accepted as true simply because they hold for several examples. Teachers should look for opportunities to push students to make more general arguments that address all possible cases. Students can be encouraged to use rectangular arrays, for example, to argue that the commutative property of multiplication holds for all whole numbers. An $m \times n$ rectangular array of dots differs from an $n \times m$ array only in its orientation, and not in the total number of dots. As students realize the need for general arguments to establish mathematical proof, they develop an important idea that spans all mathematical content areas.

Analyzing Numerical Sequences

Another way to help students build a bridge from numerical to algebraic reasoning is to have them analyze numerical sequences. Examining arithmetic sequences, where there is a constant difference between terms, can lead to generalizations related to linear relationships. For instance, the arithmetic sequence 3, 6, 9, 12, . . . can be generalized to $m = 3n$, where n represents the term number and m represents its value. Similarly, the arithmetic sequence 4, 7, 10, 13, . . . can be generalized to $m = 3n + 1$, where n represents the term number and m represents its value. Although $m = 3n$ and $m = 3n + 1$ both represent linear relationships, students are likely to initially focus on differences between the two associated sequences rather than on similarities. Zazkis and Liljedahl (2002) found that students make a sharp distinction between sequences consisting of multiples of a whole number and those consisting of nonmultiples. One possible way to draw students' attention toward the similarities in the sequences rather than the differences is to encourage them to think of an arithmetic sequence such as 4, 7, 10, 13, . . . as representing "one more than a multiple of three." Doing so can help establish a link between sequences of multiples and nonmultiples.

IDEA FOR DIFFERENTIATING INSTRUCTION   Polygonal Numbers

Polygonal numbers, numbers that can be represented by dots arranged in the shapes of various polygons, provide opportunities to pose questions appropriate for students at a variety of levels of thinking.

Figure 7.5   Representations of triangular and square number sequences.

(Continued)

(Continued)

All students can be asked to determine the number of dots in the next few terms in triangular and square polygonal number sequences (Figure 7.5). Some will respond by sketching dot diagrams that extend the existing sequence. Others will search for patterns in tables showing how many dots are in each term. Teachers can try to move students' thinking forward by asking them to determine the number of dots representing a very large term. Students can be asked to determine, for example, the number of dots in the representation of the 100th triangular number. On being given this question, some may still try to draw diagrams leading up to the 100th triangular number while others will look for more efficient strat-egies. Students looking for efficient strategies can be paired with those still attempting to draw diagrams. An appropriate learning goal for all students is to devise an efficient strategy, using an algebraic formula or visual reasoning, to determine the 100th term. Students who achieve this goal in advance of others can be provided additional polygonal number sequences to explore, include pentagonal and hex-agonal numbers. Students who have worked with triangular, square, pentagonal, and hexagonal num-bers can be asked to look for patterns and generali-zations that extend across all of the number sequences and to share their discoveries in class (W. A. Miller, 1991).

| **STOP TO REFLECT** | Describe at least three thinking strategies for determining the value of the 100th triangular num-ber. Then describe at least three thinking strategies for determining the value of the 100th square number. |

Implementing the Common Core

See Clinical Task 11 to assess students' ability to "use functions to model rela-tionships between quantities" (Content Standard 8.F) and "build a function that models a relationship between two quantities" (Content Standard F-BF) when working with polygonal numbers.

Mathematical Integrity of Sequence Tasks

Although sequences provide potentially rich sites for help-ing students transition from numerical to algebraic thinking, a word of caution is in order about maintaining mathemati-cal integrity when using them. Dancis (n.d.) observed that such tasks frequently lack enough information to have a unique solution. One example he provided was the rela-tively innocent-looking task of asking students to determine the next term in the sequence 1, 2, 4, . . . Many would expect students to say that each successive term is the previous one multiplied by two. Dancis pointed out, however, that the given sequence could actually be produced by other mathematical means, such as taking the decimal expansion of $\frac{124}{999}$. This would produce the sequence 1, 2, 4, 1, 2, 4 . . . rather than 1, 2, 4, 8, 16, . . . One way to fix the original task would be to tell students the sequence is geometric, or to provide a context that describes how the numbers in the sequence are produced. Alternatively, teachers may choose to simply ask students to describe *a* pattern in the sequence (as opposed to asking for *the* pattern) and then calculate the next term (or *n*th term) in the sequence. In general, when posing tasks involving sequences, teachers need to provide enough mathematical or contextual information to ensure a unique solution, or else allow students to describe and use the patterns they detect in the sequence, which may be different from the originally intended pattern. Posing these tasks correctly is far from a trivial matter, as even some standardized test writers err when writing items involving sequences (Dancis, n.d.).

NUMBER SYSTEMS AND NUMBER THEORY

The middle and high school topics recommended in many standards documents (NCTM, 2000; National Governor's Association Center for Best Practices & Council of Chief State School Officers, 2010) provide many opportunities for the study of number systems and number theory. Studying such topics allows students to solidify, extend, and formalize existing numerical knowledge. Ideas from these areas that will be discussed below include factors and multiples, the fundamental theorem of arithmetic, divisibility, odds and evens, primes, irrationals, and negatives. Research is emerging regarding students' thinking about these ideas, providing accompanying pedagogical suggestions.

Factors and Multiples

Students often confuse the concepts of factor and multiple, so explicit attention to the distinction is necessary. Along with having difficulty distinguishing between the two, students often hold more subtle misconceptions about factors. For instance, some students believe that large numbers have more factors than small ones do. Zazkis (1999) showed students representations of two different numbers: $A = 3^2 \times 7$ and $B = 3^2 \times 17$. Even though A and B were written in their prime factorizations and each had the same number of factors, some students initially believed that B would have more factors than A because of its size. Some students ultimately rejected this belief, but they had to perform the indicated multiplications for A and B and then factor the results to be convinced that A and B had the same number of factors. Given the results of the study, Zazkis suggested that teachers expose students to examples of large composite numbers (four or five digits in length) with only four factors. Students can also be asked to find their own examples of such numbers. By studying such examples and creating their own, students can overcome the notion that large numbers always have more factors than small ones.

The Fundamental Theorem of Arithmetic

The **fundamental theorem of arithmetic** can be stated as follows: "The Fundamental Theorem of Arithmetic claims that decomposition of a number into its prime factors exists and is unique except for the order in which the prime factors appear in the product" (Zazkis & Campbell, 1996b, p. 207). To assess students' understanding of the fundamental theorem of arithmetic, Zazkis and Campbell asked a group of students to work with numbers whose prime decompositions were shown. In one such task, they asked students to determine if $M = 3^3 \times 5^2 \times 7$ is divisible by 5, 7, 9, 63, 11, 15. Some students performed the indicated multiplications for M and then checked each factor in the list. If one understands the fundamental theorem of arithmetic, this is unnecessary. The numbers 5 and 7 must divide M because they appear in its prime factorization. The numbers 9, 15, and 63 must divide M because they can be produced by combining some of the prime factors shown. The number 11 cannot be a factor, because it is prime, and the fundamental theorem of arithmetic guarantees that prime decompositions are unique. Students' responses suggest that those who know that numbers can be decomposed into primes do not necessarily know that the decompositions are unique, and hence the uniqueness aspect of the fundamental theorem of arithmetic needs explicit instructional attention.

Divisibility

Along with exploring students' understanding of the fundamental theorem of arithmetic, Zazkis and Campbell (1996a) examined their understanding of divisibility. The task of determining whether $M = 3^3 \times 5^2 \times 7$ is divisible by 5, 7, 9, 63, 11, 15 revealed that students may have procedural knowledge of how to carry out multiplication and division, but relatively weak conceptual knowledge of divisibility. Some students actually approached the task by trying to apply divisibility rules for 7 to it. Another student task was to determine if 391 is divisible by 23, and also if it is divisible by 46. In determining whether 391 is divisible by 23, performing division is warranted because there is no other obvious approach. However, once one determines that $391 = 23 \times 17$, it automatically follows that 391 cannot be divisible by 46, since 23×17 is the prime decomposition of 391. However, some of the students in the Zazkis and Campbell study who determined that $391 = 23 \times 17$ did not immediately know whether 391 was divisible by 46. They tended to carry out the unnecessary process of dividing 391 by 46. Based on the results of the study, Zazkis and Campbell recommended that teachers emphasize the inverse relationship between multiplication and division as well as postpone introduction of divisibility rules, since they might interfere with the development of conceptual understanding.

> **Implementing the Common Core**
>
> See Clinical Tasks 12 and 13 to investigate students' ability to "look for and make use of structure" (Standard for Mathematical Practice 7) when working with factors (Content Standard 6.NS) and integer exponents (Content Standard 8.EE) and when examining the structure of expressions (Content Standard A-SSE).

Odd and Even Numbers

When asked to determine whether a given number is odd or even, students apply a variety of interesting thinking strategies. Zazkis (1998) described several student strategies for this sort of task and drew implications for teaching. In one task, students were asked to determine whether 1234567 is odd or even. Most focused on the last digit of the number to conclude that it was odd. Such a strategy is commonly taught in schools and served students well on the task. Students were less successful, however, when asked whether 3^{100} is odd or even. Some applied the "Check the last digit" rule to the exponent of 100. Since the last digit of the exponent is 0, they incorrectly concluded that it is even. Others looked for a pattern in the last digits of the powers of 3 ($3^1 = 3$, $3^2 = 9$, $3^3 = 27$, $3^4 = 81$, $3^5 = 243$, . . .). Although such a strategy can lead to the correct answer, it is more time-consuming than recognizing that 3^{100} is the prime decomposition of a number, so it cannot have a factor of 2 because

2 is also prime. In general, Zazkis found that students often considered the existence of a factor of 2 to be an indicator of evenness, but the lack of a factor of 2 was not often considered an indicator of oddness. Zazkis concluded that teachers need to emphasize the equivalence of "divisibility by 2" and "evenness." She also recommended delaying exposure to the "last digit" rule for checking evenness until students understand that the presence or absence of a factor of 2 indicates whether a number is odd or even.

Determine the units digit of 33^{402}. Explain your reasoning.

STOP TO REFLECT

Prime Numbers

Many of the student difficulties described thus far relate to their understanding of prime numbers. Zazkis and Liljedahl (2004) attributed some of students' difficulties with primes to the lack of **transparent representations** for them. Transparent representations readily reveal certain characteristics of a mathematical object. Zazkis and Liljedahl noted that representing 784 as 28^2 makes transparent that 784 is a square number. At the same time, 28^2 can also be considered an **opaque representation** in that it does not readily reveal that 784 is divisible by 98. The conventional representation for prime numbers, p, is opaque in all possible ways. It does not readily reveal anything about the structure of prime numbers. The lack of a transparent representation for primes is especially troublesome when working with large primes, such as 151 and 157. Determining primality is more time-consuming in such cases, and even when students know the numbers are prime, they may encounter difficulties. When Zazkis and Liljedahl asked students if $A = 151 \times 157$ is a prime number, some responded that it was prime, stating that the product of any two primes must be another prime. This led the researchers to recommend that teachers pose tasks involving large prime numbers and help students attend to their transparent aspects. Students need to come to understand, for example, that $A = 151 \times 157$ and $B = 151^{157}$ are transparent representations of large composite numbers A and B. By understanding transparent representations of composite numbers, students can distinguish them from prime numbers, even though we lack conventional transparent representations for primes.

Implementing the Common Core

See Homework Task 8 to explore how the sieve of Eratosthenes can lead students to "look for and express regularity in repeated reasoning" (Standard for Mathematical Practice 8).

Irrational Numbers

Irrational numbers are similar to prime numbers in that they have no transparent conventional representations (Zazkis, 2005). Just as primes are best represented in terms of their relationship to composites, irrationals are best represented in terms of their relationship to rationals. Although irrationals can be represented as nonrepeating decimals, the lack of a finite representation (e.g., $\frac{a}{b}$) makes manipulation problematic (Zazkis, 2005). If students fixate on the nonrepeating decimal representation of irrationals, it can actually distract them from understanding the relationship between rationals and irrationals. Consider the case of a group of prospective teachers who were asked whether $\frac{58}{83}$ is a rational number, given the calculator output $\frac{58}{83} = 0.63855421687$ (Zazkis, 2005). Some actually doubted that it was a rational number because they could detect no repeating pattern in the calculator-provided decimal representation. Others stated that the number must be both rational and irrational, since the representation on the left side of the equation indicated rationality but the decimal representation on the right had no apparent repeating pattern. Zazkis concluded that many students rely too much on decimal representations for irrational numbers. She recommended that teachers expand students' conceptions of irrationals by providing many examples beyond conventional ones such as π and $\sqrt{2}$. Students can be encouraged to investigate, for example, whether square roots of primes beyond 2 are rational or irrational. Teachers can also give students number line intervals and have them identify several irrationals within the intervals.

Characterizing irrationals solely as nonrepeating decimals can also lead students to believe that their magnitude cannot be precisely represented. Sirotic and Zazkis (2007) asked prospective secondary school teachers to locate $\sqrt{5}$ on a number line on grid paper. They expected the prospective teachers to do the task by drawing a segment between the points (0, 0) and (2, 1), constructing a circle with a radius of that length, and then intersecting the circle with the x-axis, as shown in Figure 7.6.

Instead, the prospective teachers tended to use decimal approximations for $\sqrt{5}$. Some expressed the belief that $\sqrt{5}$ could not be placed precisely on a number line because it had no finite decimal representation. Based on the results of the study, Sirotic and Zazkis (2007) recommended that geometric representations and the Pythagorean theorem be used instead of nonrepeating decimals to introduce

Figure 7.6 Locating $\sqrt{5}$ on a number line.

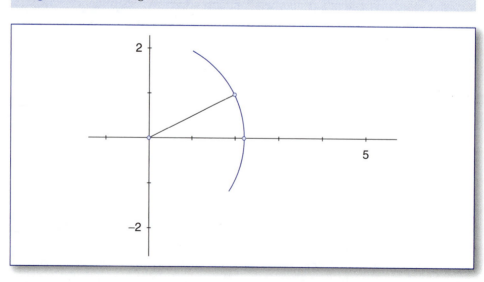

irrational numbers. They also suggested making a clear distinction between irrational numbers and common decimal approximations of them (e.g., clarifying the difference between π and $\frac{22}{7}$).

Implementing the Common Core

See Clinical Task 14 to investigate and critique a textbook's treatment of irrational numbers (Content Standards 8.NS, 8.EE, and N-RN).

Negative Numbers

Negative numbers, like irrationals, were initially difficult for some mathematicians to accept. The Pythagoreans were shocked when they discovered irrationals and tried to keep them secret. When a member of the Pythagoreans broke the secret, his death in a ship accident was considered by the other Pythagorean cult members to be punishment from God (Veljan, 2000). Gallardo (2002) discussed four levels of historical acceptance of negative numbers: (1) conception of numbers as magnitudes, making $a - b$ permissible only if $a > b$; (2) acceptance of negatives as relative numbers, capable of representing direction or situations such as debt; (3) acceptance of negatives as isolated numbers produced as a result of operations, such as $3 - 5$; and (4) acceptance of negatives as a formal part of the larger scheme of the number system. When Gallardo (2002) asked students to solve problems involving negatives, she found evidence of the first three of these historical levels in students' thinking. One of the problems students were asked to solve was "A person has a certain amount of money and receives $100.00. If he now has $50.00, how much did he have initially?" (p. 191). Some students interpreted the situation to mean that the person originally had a debt of $50. Others refused to work with negatives at all, claiming that the person had $50 to begin with. With other problems, students accepted negatives in some situations and rejected their use in others. In some respects, the patterns of thinking they exhibited mirrored philosophical dilemmas mathematicians through the ages have encountered when confronted with situations suggesting the use of negative numbers.

Although it can be difficult for students to accept negative numbers, some students construct notions about negative numbers much earlier than expected. Goldin and Shteingold (2001) reported that some second graders have awareness of characteristics of negative numbers. They investigated students' understanding of **ordinality** and **cardinality** related to negative numbers. Ordinality involves placing numbers in the appropriate order on the number line, and cardinality has to do with the number of objects in a set. A fully developed concept of negatives includes both aspects. Students may develop understanding of one aspect before the other. For instance, one of the students Goldin and Shteingold studied could place negatives correctly to the left of zero on a blank number line, but could not use negatives to keep track of the number of points in a game where one's score could dip below zero. Other students exhibited the reverse kind of understanding, being able to handle the cardinality aspects of negatives but not ordinality. Teachers should be conscious of posing tasks related to both ordinality and cardinality, since many classrooms contain students who have developed one type of understanding but not the other.

Implementing the Common Core

See Homework Task 9 to think about how to help students develop conceptual understanding of negative rational numbers (Content Standards 6.NS and 7.NS).

TECHNOLOGY CONNECTION

Operations on Integers With Virtual Manipulatives

One model for helping students make sense of operations with negative numbers involves using black chips to represent positives and red chips to represent negatives. An expression like 3 + (−2) can be represented with 3 black chips and 2 reds. Each black/red combination is a zero pair. Removing all zero pairs from the representation of the expression leaves one black chip, showing the sum to be 1. Likewise, −3 + 2 can be represented with 3 red chips and 2 black chips. Removing all zero pairs leaves one red chip, showing the sum to be (−1). The National Library of Virtual Manipulatives (http://nlvm.usu.edu/en/nav/vlibrary.html) has an applet to help extend the chip model to the subtraction of integers (Figure 7.7). Consider, for example, the case of −5−1. The number −5 can be represented with 5 red chips and a zero pair. The zero pair is included because it contains a black chip, making it possible to take a 1 away from −5. When the black chip is taken away, 6 red chips remain, showing that taking 1 away from −5 leaves −6. Users can also enter their own expressions and use the applet to evaluate them.

Figure 7.7 Chip model applet for operations with integers from the National Library of Virtual Manipulatives.

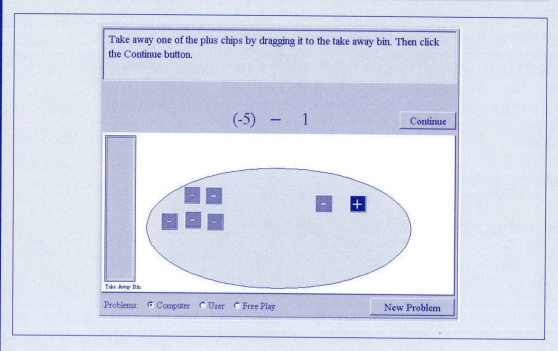

Follow-up questions:

1. How would you use the NLVM applet to evaluate (−5) − 2? How about (−5) − (−1)?

2. Are there any integer operation problems for which the NLVM applet would not be advantageous to use? Explain your position.

Complex Numbers

Complex numbers now have a prominent role in Grades 9 through 12. The *Common Core State Standards* (National Governor's Association Center for Best Practices & Council of Chief State School Officers, 2010), for example, recommend that all students learn to perform arithmetic operations with complex numbers and use them in polynomial identities and equations. This involves knowing that $i^2 = -1$ and that $a + bi$ (*a* and *b* being real numbers) is used as a general form to represent complex numbers. Students should be able to add, subtract, and multiply complex numbers and solve quadratic equations with real coefficients and complex solutions. College-bound students should also learn to represent complex numbers and their operations in the complex plane, where the horizontal is the real axis and the vertical is the imaginary. Although understanding geometric interpretations in the complex plane is not an expectation for all students in the *Common Core State Standards*, students can benefit from seeing geometric and algebraic representations developed in tandem.

As a relatively new emphasis in Grades 9 through 12, students' understanding of complex numbers has not received extensive research attention. Studies are needed regarding students' thinking while working with geometric and algebraic representations of complex numbers, and successful teaching approaches (Panaoura, Elia, Gagatsis, & Giatilis, 2006). The dearth of research provides an opportunity for teachers to do action research in their own classrooms. Lesson study, for example, can be used as a mechanism for forming and testing approaches to teaching complex number concepts in the curriculum. Initial conjectures about how students learn the concepts can be refined as they are implemented in the classroom.

UNDERSTANDING VECTORS AND MATRICES

Complex numbers can be represented with vectors and matrices. However, the usefulness of these representations is not limited to work with complex numbers. In Grades 9 through 12, students should learn several additional ideas about vectors and matrices. Students are to "understand vectors and matrices as systems that have some of the same properties of the real-number system" (NCTM, 2000, p. 290). They are also to "develop an understanding of properties of, and representations for, the addition and multiplication of matrices and vectors" (NCTM, 2000, p. 290). Students should also develop operational fluency with vectors and matrices and make sound judgments about when to use paper and pencil and when to use a calculator. In many conventional curricular sequences, students do not encounter ideas related to vectors and matrices before graduating from high school. Consequently, as with complex numbers, the research on students' understanding of the mathematical properties of vectors and matrices is in its beginning stages. The discussion in this section summarizes some of the highlights of the research available on students' thinking that can be used to inform teaching practice.

> **Implementing the Common Core**
>
> See Clinical Task 15 to investigate and critique a textbook's treatment of vectors and matrices (Content Standard N-VM).

Vectors

Vectors have so many different physical applications that it can be challenging at times for students to understand the underlying mathematics. A. Watson, Spirou, and Tall (2003) noted that in physics, vectors can represent forces, velocities, accelerations, or just a generic quantity with magnitude and direction. They claimed that different physical contexts for vectors can influence the mathematical ideas students form about them. For instance, they suggested that some physical contexts lead students naturally to use the **parallelogram rule for vector addition**, while other contexts lead more naturally to use of the **triangle rule** (Figure 7.8).

Figure 7.8   Using triangle and parallelogram rules to add vectors *a* and *b*.

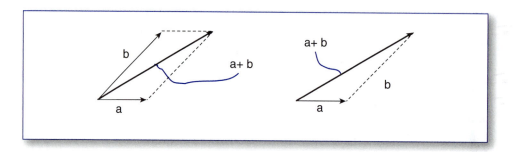

The parallelogram rule involves placing vectors together at their tails and then placing segments parallel to the two vectors to create a parallelogram. The sum of the two vectors is represented by the vector whose tail is located at the intersection of the tails of the vectors and whose tip is located at the intersection of the parallel segments. The triangle rule involves placing the vectors tip to tail. The sum of the vectors is the vector starting at the tail of the first and ending at the tip of the second. Although the two rules are mathematically equivalent, a situation such as pulling forward on two arms leads students naturally to the parallelogram rule, while situations involving successive transformations lead naturally to the triangle rule (A. Watson et al., 2003).

A. Watson and Tall (2002) had success teaching vector addition by appealing to students' sense of motion. They introduced the idea of equivalent vectors by having students move their hands from one position to another and trace out a number of alternative, equivalent pathways. This approach sparked the student conjecture that different actions can have the same net effect. This conjecture is fundamental in developing understanding of possible meanings of combining vectors. Students explored and discussed the conjecture in class. The reflective dialogue that occurred was important in helping students understand vector addition.

STOP TO REFLECT

Suppose you are teaching vector addition and students in your class are not convinced that the parallelogram and triangle rules for vector addition are equivalent. Describe how you would convince them that the two rules are mathematically equivalent.

Graphing calculators and software can also facilitate students' work with vectors. Forster (2000) described a classroom in which students used a graphing calculator function labeled ABS to determine the magnitude of a vector expressed in matrix form. Students' work in the classroom showed that magnitude can be viewed in two ways—as the result of a computation using the Pythagorean theorem and as an entity in its own right. Students need to be able to move flexibly between the two conceptions to understand the range of possible vector applications. Another study (Forster & Taylor, 2000) described students' work with the ARG function of the calculator, which computed vector direction. The students gradually developed an understanding of the ARG function, thinking of it first as a calculation process but then finally as an entity in its own right that yields direction of a vector. Being able to think of direction as an entity in its own right is important when it comes to applications such as calculation of angular velocity. To develop understanding of magnitude and direction as entities in their own right in addition to understanding them as calculations, students should reflect on what the calculator does when functions such as ABS and ARG are invoked. Writing about one's knowledge of magnitude and direction in the context of using graphing calculator functions can facilitate such reflection (Forster & Taylor, 2000).

Although vectors are commonly associated with mathematical learning achieved by advanced students, there is evidence to suggest that vector concepts are accessible to lower-achieving students. A. Watson (2002) reported her experiences as a co-teacher for lower-achieving students. These students were typically given mundane, routine, low-level tasks. In several instances, Watson tried to make the tasks richer by prompting students to think about them in different ways. In one situation, students were given a connect-the-dots exercise that simply involved plotting several points on the coordinate plane. One student decided not to simply go dot to dot. Instead, she identified all of the adjacent ordered pairs that would produce equivalent vectors when joined. Upon seeing this strategy, Watson prompted other students in the class to find similarities in lines joining adjacent points. Some students began to make generalizations about how to obtain vectors that slanted down at a certain angle. Some of the generalizations were even stated formally using coordinates—for example, coordinate (a, b) to $(a + 1, b - 2)$. Watson concluded that going "across the grain" with otherwise routine tasks can help students develop relatively sophisticated ideas about vectors and other mathematical concepts.

Matrices

Harel, Fuller, and Rabin (2008) identified several counterproductive instructional approaches to matrices. They described episodes in a single high school lesson that might have served to obscure students' understanding of matrices. At the outset of the lesson, the teacher presented the ideas of determinant and identity matrix without connecting them to anything in students' previous mathematical work. This left students wondering aloud about the point of the lesson. There was also lack of attention to the meanings of terms. At one point, a determinant was described as being "associated" with a matrix. When students asked about the sort of association being described, no satisfactory answer was provided. At another point in the lesson, a student confused the terms *determinant* and *scalar*, but the teacher did not probe the source of the confusion. When students asked about the meanings of concepts associated with matrices, such as "determinant," the teacher responded by simply restating a definition given earlier in the lesson rather than providing examples of how the concepts could be used. This lesson description brings out the idea that matrices should not be taught as an isolated unit of study. Harel and colleagues conjectured that a more meaningful approach to matrices would involve introducing terms and procedures only after students see a need for them.

One way to infuse meaning into students' work with matrices is to delve into how they can be used to represent geometric transformations. Edwards (2003) used TI-92 graphing calculators to help students see that matrix multiplication can be used to reflect, rotate, and change the size of geometric shapes in the coordinate plane. Students in his class used the TI-92s to generate and test hypotheses about matrices. Using a program for the calculator, students made conjectures about the effect that multiplication by a given matrix would have on an object represented by another matrix. Students were asked, for example, to predict the effect of the matrix $\begin{bmatrix} 1 & 0 \\ 0 & -1 \end{bmatrix}$ upon the triangle represented by $\begin{bmatrix} -1 & 0 & 4 \\ 2 & -3 & 1 \end{bmatrix}$. After considering additional specific examples of how premultiplication by $\begin{bmatrix} 1 & 0 \\ 0 & -1 \end{bmatrix}$ transforms other matrices, students made and tested conjectures about its general effect. The calculator's ability to display the results of matrix multiplication quickly in a geometric form facilitated the process of making and testing conjectures because it freed students from the time-consuming calculation and point plotting that would otherwise have been necessary.

Real-world contexts can also be used to infuse meaning into students' study of matrices. Novick (2004) argued that students should know when a real-world situation is best represented by a matrix, network, or tree diagram. She interviewed undergraduate students to determine their ability in this regard. Results of the interviews indicated that students were best at recognizing when tree diagrams would be appropriate, second best at recognizing when matrices would be appropriate, and least proficient with networks. Students' lack of success with networks was a cause for concern, given that NCTM (2000) recommended that networks be used to represent phenomena when appropriate. Novick recommended making connections between these mathematical representations and topics students study in other classes. For example, a table containing temperature data for different regions within a country can be represented in matrix form, the food web in an ecosystem can be represented using a network, and a tournament bracket can be represented using a tree diagram hierarchy. As students encounter various suitable applications for matrices, networks, and tree diagrams, they can build awareness of the range of situations they are used to represent.

Worrall and Quinn (2001) also recommended making learning of matrices meaningful by drawing on real-world situations. In one lesson with advanced algebra students, they posed a problem that prompted students to multiply two matrices containing data from a cycle factory. The first was a 3×4 matrix with rows representing the time required for parts assembly, paint, and mechanical work. The columns of the matrix indicated four kinds of cycles manufactured at the factory. The second was a 4×2 matrix. Its rows indicated the four kinds of cycles manufactured at the factory, and its columns contained the number of orders received for each type of cycle

Implementing the Common Core

See Homework Task 10 for an exercise that requires students to "perform operations on matrices and use matrices in applications" (Content Standard N-VM).

in November and December of a given year. Class discussion focused on the meaning of the resultant 3×2 matrix. Students determined that the entries in the new matrix represented the number of work hours required per month. To help students focus on the meanings of matrices produced by contextualized multiplication, Worrall and Quinn recommended that they be allowed to use a graphing calculator to automate computation after they have shown the ability to do matrix multiplication by hand.

COMBINATORIAL THINKING

Along with vectors and matrices, topics from combinatorics are important for secondary mathematics curricula. Students in Grades 9 through 12 should "develop an understanding of permutations and combinations as counting techniques" (NCTM, 2000, p. 290). Understanding these ideas is necessary in other curricular areas such as probability (National Governor's Association Center for Best Practices & Council of Chief State School Officers, 2010). It also makes it possible for students to navigate complex counting situations encountered outside of school, such as determining the odds of winning a lottery. The available research on students' combinatorial thinking describes the various types of combinatorics problems to be included in the curriculum, students' problem solving strategies, and potentially effective teaching strategies.

The Fundamental Counting Principle

Perhaps the most prevalent type of combinatorics problem is that which involves the **fundamental counting principle**—the idea that if there are x ways of doing one thing and y ways of doing another, there are xy ways of doing both together. Examples are

- How many different two-digit numbers can be formed using only the digits 1, 2, 3, and 4?
- In one restaurant, when ordering a meal, you must choose a main dish, a side dish, and a drink. Two different main dishes, three different side dishes, and four different drinks are available. How many different meals is it possible to form?
- In one state, license plates consist of six characters. The first character is one of the letters of the alphabet. The remaining five characters are each one of the digits from 1 through 9. How many different license plates can be formed?

Students use a variety of approaches when solving such problems, from randomly guessing to using systematic procedures. One of the most prevalent systematic procedures documented in the research literature is the odometer strategy, which consists of holding one term constant and then forming all possible combinations that include it. Once the possibilities have been exhausted for one constant term, students hold another term constant and then document all possible combinations that include it. English (2005) found that even elementary school students may use the odometer strategy as a natural means of solving fundamental counting principle problems. Students using the **odometer strategy** to determine how many two-digit numbers can be formed by using only the digits 1, 2, 3, and 4 might begin by listing all possible two-digit numbers that could be formed with 1 as the first digit (11, 12, 13, 14). Then, all of the two-digit numbers that could be formed with 2 as the first digit might be listed (21, 22, 23, 24). This process would continue until all 16 possible combinations were obtained.

Selection, Partition, and Distribution Problems

Three additional categories of combinatorial problems were described by Batanero, Navarro-Pelayo, and Godino (1997): **selection**, **partition**, and **distribution**. Selection problems involve drawing samples, either with or without replacement. The sample may be either ordered or unordered. Partition problems involve splitting *n* objects into *m* subsets. Distribution problems involve placing *n* objects into *m* cells. The objects may or may not be identical. The same holds for the cells. In some cases order matters, and in others it does not. There may be zero, one, or more than one object per cell. Batanero and colleagues identified these three problem categories as important because of differences in the manner students respond to them. The categories provide a working framework for formulating combinatorial problems for students to solve.

When solving selection, partition, and distribution problems, students tend to make some types of errors more frequently than others. In their study of 720 students ages 14 to 15, Batanero and colleagues (1997) found that one type of error was misinterpretation of the problem statement. Common misinterpretations included the following:

- Changing the problem from one type to another; misinterpreting the meanings of words in a problem
- Thinking that order matters in cases where it does not, and vice versa (e.g., thinking that order matters in selecting three students to clean a blackboard)
- Not understanding when repetition could or could not occur in a situation (e.g., mistaking a sampling-without-replacement situation for a sampling-with-replacement situation)
- Thinking that identical objects are distinguishable, and vice versa
- Using unsystematic strategies for listing the possible combinations in a given situation

Implementing the Common Core

See Homework Task 11 to write problems to help develop and assess students' combinatorial reasoning (Content Standard S-CP).

To help students avoid errors, Batanero and colleagues (1997) recommended engaging them in comparing and contrasting combinatorics problem types and systematically listing outcomes rather than having them simply memorize algorithms for the different problem types.

Sequencing Combinatorics Problems

Another important teaching strategy is to attend carefully to the order in which combinatorics problems are posed to students. Posing problems that have similar mathematical structures can lead students to make generalizations about types of combinatorial problems (Sriraman & English, 2004). In one study, Martino and Maher (1999) posed two similar tasks set in different contexts. The first task was to build as many towers of a given height as possible by using plastic cubes of two different colors (e.g., build as many towers as possible that are 3 cubes tall). They were also required to convince classmates that none of the towers they found were duplicates and that no possibilities had been left out. The second task involved making pizzas. Students were told that a restaurant allowed customers to select from four different toppings. They were to determine how many different pizza choices were available to the customer, list all possible choices, and convince each other that they had accounted for all possible choices. The tasks, as stated, prompted students to build justifications for their strategies and to listen to the justifications of others. As students worked, the teacher looked for opportunities to prompt students to notice connections between the mathematical structures of the two problems. As connections were noticed, students developed the ability to abstract general combinatorial principles from their work with the two specific problems.

Verifying Results of Computations

To abstract general principles from work with combinatorial problems, students must have effective strategies for reflecting on the solutions to problems. Solutions to combinatorial problems can be particularly difficult to verify, since students may often find two different, incompatible approaches to be equally convincing (Eizenberg & Zaslavsky, 2004). When Eizenberg and Zaslavsky asked 19- to 24-year-old students who had learned some combinatorics to verify the solutions they obtained for counting problems, several different strategies were apparent. The most prevalent verification strategy was to rework the solution. This could involve checking calculations or making sure all planned steps had been carried out correctly. Although reworking was the most prevalent strategy, it was the least efficient in terms of helping students obtain a correct solution. One of the more efficient verification strategies was to add justifications to the solution. Students using this strategy sometimes examined how the conditions of the problem matched the solution strategy selected, or they drew an analogy between the problem at hand and a similar, previously solved problem. Another of the more efficient strategies was to use a second method of solution and then compare its results to those of the method originally chosen. In a few cases, this verification strategy backfired, however, when students obtained the same erroneous solution using both methods. In such cases, students actually became more convinced of the correctness of the original, incorrect answer. Eizenberg and Zaslavsky concluded that it may be useful to explicitly teach verification strategies to students, and to let them explore the strengths and weaknesses of each one.

Implementing the Common Core

See Clinical Tasks 16 and 17 to investigate students' responses to combinatorial reasoning problems (Content Standard S-CP) and their strategies for verifying their computations.

CONCLUSION

The area of number and operations encompasses a wide range of ideas to be studied in middle and high school. Work with algorithms is one of the most common types of tasks associated with number and operations, but numerical thinking is needed for many other purposes as well. Number sense is necessary for estimating the results of computations and checking their reasonableness. Proportional reasoning helps students work with fractions and decimals, recognize the difference between additive and proportional change, recognize when it is reasonable to use a ratio to represent a situation, recognize invariance in equivalent ratios, and engage in the unitizing process. Along with building proportional reasoning, students should develop the ability to make generalizations about their work with numbers to lay the foundation for algebra. Generalizations can arise from work with sequences and number sentences. Students' numerical thinking should also be built by solving problems involving number theory, vectors, matrices, and combinatorics. As students develop these various facets of numerical thinking, they build a strong foundation for their work in other areas of the middle and high school mathematics curriculum.

VOCABULARY LIST

After reading this chapter, you should be able to offer reasonable definitions for the following ideas (listed in their order of first occurrence) and describe their relevance to teaching mathematics:

Algorithm   156

Buggy algorithm   156

Number sense   160

Computational estimation   161

Reformulation   161

Translation   161

Compensation   161

Measurement estimation   162

Benchmark estimation   162

Decomposition/recomposition estimation   162

Range estimation   162

Proportional reasoning   163

Measurement model of division   163

Partitive model of division   163

Product-and-factor model of division   164

Fraction circles   164

Part-whole interpretation of rational numbers   165

Measure interpretation of rational numbers   165

Operator interpretation of rational numbers   166

Quotient interpretation of rational numbers   166

Ratio interpretation of rational numbers   166

Additive change   167

Multiplicative change   167

Open number sentence   168

Polygonal numbers   169

Fundamental theorem of arithmetic   171

Transparent representation   173

Opaque representation   173

Ordinality   175

Cardinality   175

Parallelogram rule for vector addition   178

Triangle rule for vector addition   178

Fundamental counting principle   181

Odometer strategy   182

Selection combinatorics problem   182

Partition combinatorics problem   182

Distribution combinatorics problem   182

HOMEWORK TASKS

1. Suppose that you are teaching division of fractions. One of your students claims that "dividing numerators and then dividing denominators" is a better way to perform fraction division than invert-and-multiply. Respond to the following questions and provide justification.

    a. Has this student developed a buggy algorithm for fraction division?

    b. Has this student described an efficient means of dividing fractions?

2. Explain why the following commonly taught conventional algorithms for fraction operations work. Use diagrams as necessary to support your explanations.

    a. "Produce an equivalent fraction by multiplying the numerator and denominator by the same amount."

    b. "Convert a mixed number to a fraction by multiplying the denominator of the fractional portion by the whole number portion, adding the result to the numerator, and putting the sum above the original denominator."

    c. "Find the product of two fractions by multiplying numerators and multiplying denominators."

3. Perform each of the following computations mentally, without paper, pencil, or calculator.

    a. $1,001 - 998$

    b. $\frac{1}{4} \div 2$

    c. $\frac{1}{2} \div \frac{1}{4}$

    d. 38×50

    Describe the mental computation approach you took for each problem. Then describe at least one alternate approach for each one. Which approaches are most efficient? Which, if any, rely on blind or overzealous application of an algorithm? Justify your responses.

4. Visit the Algorithm Collection Project website (www.csus.edu/indiv/o/oreyd/ACP.htm_files/Alg.html). Find at least two algorithms that differ from those conventionally taught in North America. Then compare the two algorithms conceptually to their conventional counterparts. In your conceptual analysis, compare and contrast the algorithms in terms of their understandability, efficiency, and accuracy. Support your analysis with specific examples of how each algorithm can be used to solve problems.

5. Write a word problem for each of the following fraction division sentences. Make sure to draw on the measurement, partitive, and product-and-factor models in writing the problems. Identify each problem you write as one of the three models.

    a. $1\frac{1}{2} \div \frac{1}{3}$

    b. $\frac{1}{2} \div \frac{1}{3}$

    c. $\frac{1}{3} \div \frac{1}{2}$

    Explain how to solve each problem you wrote without using any conventional algorithms such as invert-and-multiply. Use diagrams to supplement your explanations as needed.

6. Design a test that could be given to students to assess their attainment of proportional reasoning. The test should contain items that assess, at minimum:

    a. addition, subtraction, multiplication, and division of fractions;

    b. additive and multiplicative reasoning;

    c. recognition of when it is reasonable or appropriate to use a ratio;

    d. recognition of invariance in equivalent ratios; and

    e. engagement in the unitizing process.

   Provide a solution to each item on the test and describe how each item assesses one or more of the above elements of proportional reasoning.

7. Write at least two general arguments that explain why each statement below is true.

    a. Two even whole numbers added together produce another even whole number.

    b. Two odd whole numbers added together produce an even number.

    c. An odd number added to an even number produces an odd number.

8. Do an Internet search on "sieve of Eratosthenes." Explain the purpose of the sieve and how it works. Give a specific example to show it in action. Then explain whether or not you would use the sieve to help students develop conceptual understanding of prime numbers. Justify your response.

9. Design at least two tasks you could use to assess students' understanding of the ordinality of negative rational numbers and at least two tasks you could use to assess students' understanding of the cardinality of negative rationals. Use a variety of rationals in your tasks (e.g., do not use only integers). Predict how the following four different types of students would respond to the tasks.

    a. Those who tend to understand both ordinality and cardinality

    b. Those who tend to understand ordinality but not cardinality

    c. Those who tend to understand cardinality but not ordinality

    d. Those who tend to understand neither ordinality nor cardinality

10. Sketch the triangle represented by $\begin{bmatrix} -1 & 0 & 4 \\ 2 & -3 & 1 \end{bmatrix}$ in the coordinate plane. Predict the nature of the geometric shapes produced by each of the following matrix multiplications:

    a. $\begin{bmatrix} 1 & 0 \\ 0 & 1 \end{bmatrix}\begin{bmatrix} -1 & 0 & 4 \\ 2 & -3 & 1 \end{bmatrix}$

    b. $\begin{bmatrix} 1 & 0 \\ 0 & -1 \end{bmatrix}\begin{bmatrix} -1 & 0 & 4 \\ 2 & -3 & 1 \end{bmatrix}$

    c. $\begin{bmatrix} 2 & 0 \\ 0 & -2 \end{bmatrix}\begin{bmatrix} -1 & 0 & 4 \\ 2 & -3 & 1 \end{bmatrix}$

    d. $\begin{bmatrix} 1 & 0 \\ 0 & -2 \end{bmatrix}\begin{bmatrix} -1 & 0 & 4 \\ 2 & -3 & 1 \end{bmatrix}$

   Check your predictions using a graphing calculator, paper and pencil, or computer software program. Compare your predictions to your results. Explain the reasons for the similarities and differences between your initial conjectures and the observed results.

11. Design two combinatorics problems for each of the following categories: selection, partition, and distribution. The problems in each category should be given the same essential mathematical structure. After writing your six problems, identify at least three errors students may make in solving each one. Then describe instructional strategies you would use to help students overcome or avoid each error.

CLINICAL TASKS

1. Interview a student and ask him or her to insert a decimal point that will make each of the following number sentences true. Do not allow the student to use a calculator or a paper-and-pencil algorithm.

    a. $356 \times 2.30 = 8,188$ (Insert a decimal point on the right side of the equation)

    b. $1.25 \times 2.2 = 275$ (Insert a decimal point on the right side of the equation)

    c. $1,376 \times 3.5 = 48.160$ (Insert a decimal point on the left side of the equation)

    After the student has finished, ask him or her to explain the strategies used. Describe the student's strategies and write about their strengths and weaknesses.

2. Write four problems for which it would be inefficient to use a conventional algorithm (e.g., $1001 - 99$). Construct one problem each for addition, subtraction, multiplication, and division. Then have a class of students solve each problem without calculators and give a written explanation of the strategies they used to solve them. Report the percentage of students who used inefficient strategies. Describe steps you would take to help more students develop efficient strategies for the problems you posed.

3. Demonstrate the lattice method of multiplication and the scaffolding algorithm for division (Figures 7.1 and 7.2, respectively) for a student. Then give the student at least two multiplication problems and two division problems to do. Ask the student to do each problem two ways—using a conventional algorithm and using the alternative algorithm—and to show all work in writing. Retain the student's written work. Ask the student to decide which algorithm he or she prefers for multiplication and which one he or she prefers for division. Also ask the student to give reasons for preferring one algorithm over the other. Report on the student's ability to use different algorithms for multiplication and division as well as on his or her reasons for preferring one algorithm over another.

4. Examine a mathematics textbook from your field experience website for evidence of alternative algorithms. Does the book contain any alternative algorithms, or does it provide only conventional ones? For each chapter of the text that contains an alternative algorithm, give at least one example of an alternative algorithm students are to study. For each chapter that does not contain an alternative algorithm, give an example of one of the conventional algorithms used in the chapter. State the title of the text, publisher, and publication date in your report.

5. Interview a student who has been taught the order of operations. Ask him or her to write at least three examples of computations for which order of operations is needed. Then ask the student to perform the computations. What do the student's responses reveal about his or her understanding of order of operations? Write your observations and then make suggestions about how you might be able to help strengthen the student's understanding.

6. Interview a student and ask him or her to estimate the following:

    a. $72 \div 0.025$

    b. $48,500 + 49,678 + 52,167 + 51,165$

    c. $\frac{25 \times 19}{5}$

    d. The average attendance for the following six football games: 73,655; 86,421; 91,943; 96,509; 93,421; 106,409

    Describe the thinking strategies the student uses for each of the tasks. Then describe each strategy as one or more of the following: reformulation, translation, or compensation. Justify your description by making reference to specific things the student said during the interview.

7. Give a class of students the following multiple-choice task (based on Carpenter et al., 1981). If possible, have them answer using a classroom response system (CRS; see Chapter 5). If no CRS is available, have them close their eyes and raise their hands to indicate which choice they believe to be most accurate. Do not allow students to use a pencil and paper or calculator for the task.

   Multiple-choice task: Choose the best estimate for $\frac{12}{13}+\frac{7}{8}$ from the choices below.

   a. 1

   b. 2

   c. 19

   d. 21

   Record how many students chose each response. Ask at least one student who chose each response to explain why. Prepare a report stating the number of students who chose each response as well as the reasons students gave for choosing each one. Also describe how you would help students who choose incorrect responses revise their thinking strategies.

8. Ask a student to solve each of the following problems using fraction circle pieces. If the student uses the conventional invert-and-multiply algorithm to solve the problems, ask him or her to do it a different way.

   a. $1\frac{1}{2} \div \frac{1}{3}$

   b. $\frac{1}{2} \div \frac{1}{3}$

   c. $\frac{1}{3} \div \frac{1}{2}$

   Describe how the student approaches each problem. Identify strong and weak points in the student's reasoning processes. Use diagrams to illustrate how he or she used the fraction circle pieces to solve the problems.

9. Administer the proportional reasoning test you created for Homework Task 6 to a class of students. Prepare a report describing students' performance on each item. For each item, include sample student work illustrating a high level of proportional reasoning and sample work illustrating a low level. List the top three instructional priorities you would set for teaching the class based on your analysis of their work. Explain why you set each priority by referring to samples of student work.

10. Give the following three true/false items to a class of students.

    a. Two even whole numbers added together produce another even whole number.

    b. Two odd whole numbers added together produce an even number.

    c. An odd number added to an even number produces an odd number.

    Tell students to write an explanation to justify their choice of true or false for each statement. Then examine the student work. Prepare a report summarizing how many students answered each item correctly. Then read the explanations students provided for choosing true or false. For each item, group similar student explanations into categories and explain which category represents the most sophisticated mathematical thinking and which category represents the least sophisticated mathematical thinking.

11. Interview at least two students individually. Show each student the sequences of triangular and square numbers in Figure 7.5. Ask them to sketch a representation of the next term in each sequence, and then ask them to determine the value of the 100th term in each sequence. Retain

any written work the students produce during the interviews, and then write a report comparing and contrasting the thinking strategies used.

12. Interview a student and ask the following three questions (from Zazkis, 1999):

   a. Which has more factors, A or B?

   $A = 3^2 \times 7$

   $B = 3^2 \times 17$

   b. Do larger numbers always have more factors than smaller ones?

   c. Can you give an example of a four-digit number that has only four factors?

   If the student has trouble getting started, review the meaning of the word *factor*. Prompt the student to explain his or her reasoning for each question, either orally or in writing. Retain any written work the student produces during the interview, and then prepare a report summarizing the thinking strategies the student used in responding to each question.

13. Interview at least three students and ask them if $M = 3^3 \times 5^2 \times 7$ is divisible by 5, 7, 9, 63, 11, and 15 (Zazkis & Campbell, 1996b). Ask the students to explain their responses completely. Also retain any written work the students produce. Allow calculator use if the students feel the need for it, but encourage them to do the task without it if possible. Write a report summarizing how each student responded to the task. Classify the responses from most sophisticated to least sophisticated.

14. Examine an algebra textbook to determine how it presents the concept of irrational numbers. How are they defined? What kinds of examples of irrationals are given? What kinds of student exercises are included? Does it provide geometric interpretations of any irrational numbers? Critique the text's treatment of irrationals using what we know from research about patterns of students' thinking. How could the textbook's treatment of irrational numbers be improved?

15. Examine a textbook from your field placement site that includes vectors and matrices. Describe the extent to which the text include tasks with the potential to help students:

   a. "understand vectors and matrices as systems that have some of the same properties of the real-number system" (NCTM, 2000, p. 290)

   b. "develop an understanding of properties of, and representations for, the addition and multiplication of matrices and vectors" (NCTM, 2000, p. 290)

   c. develop operational fluency with vectors and matrices and make sound judgments about when to use paper and pencil and when to use a calculator

   Provide specific examples from the textbook to support your description. Critique the text's treatment of the above topics using what we know from research about patterns of students' thinking. Make suggestions about how the textbook's treatment of vectors and matrices could be improved.

16. Pose the following three problems to a student and ask him or her to think aloud while solving them.

   a. How many different four-digit numbers can be formed using only the digits 1, 2, 3, and 4?

   b. In one restaurant, when ordering a meal, you must choose a main dish, a side dish, and a drink. Two different main dishes, three different side dishes, and four different drinks are available. How many different meals is it possible to form?

   c. In one state, license plates consist of six characters. The first character is one of the letters of the alphabet. Each of the remaining five characters is one of the digits from 1 to 9. How many different license plates can be formed?

Provide an analysis of the student's responses to the items. In your analysis, compare and contrast the strategies used for each problem. Were the strategies systematic or unsystematic? Describe the nature of each strategy used.

17. Pose the following three problems (from Eizenberg & Zaslavsky, 2004) to a student and ask him or her to think aloud while solving them. After the student finishes solving each problem, ask him or her to explain how he or she would verify that the solution obtained is correct.

   a. In how many ways can we choose 4 people out of 6 married couples, so that at least one married couple is chosen? (p. 35)

   b. In how many ways can we seat in one row, 2 men, 2 women, and a dog, one next to the other, so that the 2 men do not sit next to each other, and the 2 women do not sit next to each other? (p. 36)

   c Six boys and 6 girls are divided into 2 groups (not necessarily of the same size). In how many ways can they be grouped so that in each group, there is the same number of girls as boys? (p. 36)

Provide an analysis of the student's responses to the items and his or her verification strategies. In your analysis, describe the initial strategy for each problem as well as the verification strategy. Comment on the potential and limitations of each verification strategy.

VIGNETTE ANALYSIS ACTIVITY

Focus on Modeling With Mathematics (CCSS Standard for Mathematical Practice 4)

Items to Consider Before Reading the Vignette

Read CCSS Standard for Mathematical Practice 4 in Appendix A. Then respond to the following items:

1. Drawing on your own experiences as a mathematics student, provide specific examples of teaching practices that can help students model with mathematics.

2. Drawing on your own experiences as a mathematics student, provide specific examples of teaching practices that might prevent students from modeling with mathematics.

3. Draw a picture to represent $\frac{2}{3} \times 27$.

4. Solve the following problem in at least two ways.

   A pan of brownies was $\frac{2}{3}$ full. Mr. Jones bought $\frac{1}{2}$ of what was in the pan. What fraction of a pan of brownies did Mr. Jones buy?

5. Write a word problem for $\frac{5}{4} \times \frac{1}{3}$. Then solve the problem by using a diagram.

Scenario

Ms. Jennings was beginning her student teaching experience in a seventh-grade classroom. She felt fortunate to have a mentor teacher who encouraged reform-oriented instruction. There were frequent opportunities to draw class activities from innovative curriculum materials and from the NCTM website. When it was time to teach multiplication with fractions, Ms. Jennings wanted to make sure students understood what they were doing and that they were not just mindlessly learning procedures. In her teaching methods class, Ms. Jennings had the opportunity to learn with understanding by drawing diagrams to represent fractions. She wanted to provide her own students with the same sort of experience so that they would not develop strictly procedural knowledge, as she felt she had during most of her formal schooling.

The Lesson

A warm-up activity was posted for students to solve at the beginning of the class period. It stated, "Draw a picture to represent $\frac{2}{3} \times 27$. Then multiply without drawing a picture." Many of the students who had previously learned the procedure for multiplying fractions started with the second part of the task and then tried to draw a picture to show a result of 18. In many cases, the diagrams did not map the multiplication problem because they simply displayed 18 objects next to diagrammatic representations of $\frac{2}{3}$ and 27. One such student, Will, drew a circle, cut it into three parts, and shaded in two of the three parts. He then placed a multiplication symbol next to the circle, drew 27 objects after it, drew an equal sign, and then drew 18 more objects. As Ms. Jennings saw a number of students taking this approach, she stopped the activity and demonstrated her own solution to the problem at the front of the room.

After finishing her demonstration, Ms. Jennings posed another problem for students to solve. It stated, "A pan of brownies was $\frac{2}{3}$ full. Mr. Jones bought $\frac{1}{2}$ of what was in the pan. What fraction of a pan of brownies did Mr. Jones buy?" She wanted to avoid having students use the same kinds of strategies they used on the warm-up activity, so she gave some extra instructions. Ms. Jennings said, "OK, now this time I want you to actually draw the brownies and the pan before you do any of the computations. You should be able to show me the answer by using your diagram and nothing else." She went on to give further instructions. "You will be working with a partner on this problem. I am going to give each of you a number between 1 and 30. Then I will use the graphing calculator to get random numbers to determine partners." Ms. Jennings projected the random numbers generated by the calculator on a screen. As pairs of numbers were obtained, students got together to work. After about 15 minutes, everyone had a partner.

Some pairs of students simply multiplied $\frac{1}{2} \times \frac{2}{3}$ and obtained $\frac{2}{6} = \frac{1}{3}$ as their answer. As the pairs started to work, Ms. Jennings circulated among the class to push students to think about the problem as if they were cutting up a pan of brownies. She told students that an answer of $\frac{1}{3}$, without a diagram and explanation, would be marked incorrect on a test. A few of the pairs did not recognize $\frac{1}{2} \times \frac{2}{3}$ as a way to solve the brownie problem. Some of these students came to the conclusion that the solution was $\frac{2}{4}$, others believed it was $\frac{2}{6}$, and still others believed it was $\frac{1}{3}$. Ms. Jennings was surprised at this range of responses, so she once again stopped the class and demonstrated her own solution to the problem.

After completing the brownie problem, Ms. Jennings gave students an assignment from the textbook to complete during class. It consisted of 20 exercises in multiplying fractions and expressing the answer in simplest form, followed by two word problems. Ms. Jennings noted, "You should be able to do the first 20 exercises really quickly. Those of you who wanted to use procedures at the beginning of class should be really happy with those. When you get to the last two, you will have to apply some of the ideas we learned today." Students took out their books. Most worked through the first 20 problems and then started packing up their materials upon hitting the two word problems at the end.

Questions for Reflection and Discussion

1. Which aspects of the CCSS Standard for Mathematical Practice 4 did Ms. Jennings's students seem to attain? Explain.

2. Which aspects of the CCSS Standard for Mathematical Practice 4 did Ms. Jennings's students not seem to attain? Explain.

3. Comment on Will's solution to the warm-up activity. As his teacher, how would you try to elicit more conceptual thinking about the activity?

4. Do you agree with Ms. Jennings's decision to stop students' work on the warm-up activity to demonstrate her own solution to the problem? Explain.

5. Did Ms. Jennings have a good strategy for assigning partners to work on the brownie problem? Justify your position.

6. What might lead students to believe that the answer to the brownie problem is $\frac{2}{4}$? What about $\frac{2}{6}$? $\frac{1}{3}$? Draw diagrams to show how someone might arrive at each conclusion.

7. Critique the homework assignment students received. Is it likely to help Ms. Jennings attain her goal of developing students' conceptual understanding of multiplication with fractions? Why or why not?

RESOURCES TO EXPLORE

Books

Burke, M. J., Kehle, P. E., Kennedy, P. A., & St. John, D. (2006). *Navigating through number and operations in Grades 9–12.* Reston, VA: NCTM.

Description: The authors provide a collection of activities useful for supporting inquiry-oriented instruction in number and operations for high school students. Real numbers, number theory, and real-world applications are addressed in the book.

Lamon, S. J. (2011). *Teaching fractions and ratios for understanding: In-depth discussion and reasoning activities* (3rd ed.). London, England: Routledge.

Description: The author presents a careful and detailed analysis of students' proportional reasoning. Several sample problems useful for teaching and assessment are included.

Litwiller, B. (Ed.). (2002). *Making sense of fractions, ratios, and proportions* (Sixty-fourth yearbook of the National Council of Teachers of Mathematics). Reston, VA: NCTM.

Description: The chapters in this yearbook deal primarily with the development of students' proportional reasoning. Sample problems that can be used as the basis for class discussions are provided throughout the text.

Rachlin, S., Cramer, K., Finseth, C., Foreman, L. C., Geary, D., Leavitt, S., & Smith, M. S. (2006). *Navigating through number and operations in Grades 6–8.* Reston, VA: NCTM.

Description: The authors provide a collection of activities useful for supporting inquiry-oriented instruction in number and operations for middle school students. Fractions, decimals, percents, and proportional reasoning are addressed in the book.

Websites

Algorithm Collection Project: **http://www.csus.edu/indiv/o/oreyd/ACP.htm_files/Alg.html**

Description: This site contains a collection of algorithms from different cultures along with several articles about ethnomathematics. Links to resources to support classroom discussions of alternative algorithms are provided.

Learning Math: Number and Operations: **http://www.learner.org/resources/series171.html**

Description: This site contains several videos pertinent to teaching number and operations in Grades 6 through 8. Through exploring the videos, teachers can deepen their knowledge about sets, place value, number theory, and proportional reasoning.

FOUR-COLUMN LESSON PLANS TO HELP DEVELOP STUDENTS' NUMERICAL THINKING

Lesson Plan 1

Based on the following NCTM resource: Cai, J., Moyer, J. C., & Laughlin, C. (1998). Algorithms for solving nonroutine mathematical problems. In L. J. Morrow (Ed.), *The teaching and learning of algorithms in school mathematics* (pp. 218–229). Reston, VA: NCTM.

Primary objective: To engage students in inventing an algorithm to solve a problem and express it using formal notation

Materials needed: Small colored plastic chips in a variety of different sizes

Steps of the lesson	Expected student responses	Teacher's responses to students	Assessment strategies and goals
1. Pose the following problem: "Eight adults and two children need to cross a river. A small boat is available that can hold one adult or one or two children. Everyone can row the boat. What is the minimum number of trips needed for all the adults and children to cross the river? Show or explain how you got your answer" (Cai, Moyer, & Laughlin, 1998, p. 224). Students work in groups of three.	Students will become engaged in attempting solutions to the problem with colored chips or drawings. Students may not immediately have a thorough understanding of the constraints of the problem.	Encourage students to act out various different situations with the colored chips or drawings in order to understand the essence of the problem.	By listening to students as they work, determine whether or not each group of students is aware that when an adult crosses, there must always be a child waiting on the other side so that the boat may be brought back. Developing this understanding of the problem is essential in order for students to make progress on it. Guide students toward this conclusion if necessary.
2. Ask each group of students to write a description of a procedure that can be used to solve the initial problem.	Some groups will formulate a correct procedure but make errors in carrying it out.	Offer students multiple sizes of colored chips to represent the different elements in the problem: adults, children, and the boat.	The correct answer to the original problem is 33 trips. If students do not obtain this response, check to see whether their error is due to incorrect counting or incorrect execution.
3. To help students generalize their procedures, ask them to write solutions to three variations of the initial problem, where you have a. six adults and two children b. 15 adults and two children c. three adults and two children Students should devise their own ways of recording the solutions and showing all of the moves.	Students will use a variety of diagrams Some will use charts with arrows to represent river crossings. Others may use a cyclic diagram to represent the process necessary for transporting each adult across.	Encourage students to look for similarities in the procedure used across all three variations of the original problem. Ask them to make general statements about what would happen in any variation of the river problem that includes two children.	Assess whether students characterize their algorithms in terms of "chunks" of moves needed to get one adult across the river. They should begin to notice that there is a well-defined sequence of moves to get each adult across, which is repeated over and over.

Steps of the lesson	Expected student responses	Teacher's responses to students	Assessment strategies and goals
4. Ask students to write solutions to the following problem: "How many trips would you need to get everyone across the river if there were 100 adults and two children?"	Some students will attempt to apply each step of the strategy they formed, all the way up to 100. Other students will recognize that the work of obtaining the answer to the problem can be shortened by noticing patterns in the procedures they wrote.	If students want to act out river crossings for all 100 participants, ask them to examine the diagram of their procedure for a shortcut that can be used to do this variation on the task more efficiently. Help them see "chunks" or "cycles" in their diagrams that simplify the task.	Assess students' strategies for efficiency at this point in the lesson. Responses that help students them move toward mathematical generalization include those indicating that they see the need for four trips per adult and one extra trip to get the children across.
5. Ask students to write a solution to the following problem: "How many trips would you need for x adults and two children?"	Some students will arrive at the general rule $4x + 1$, reasoning that four trips are needed per adult, plus one extra to finish. Others will reason that each adult needs one trip over and two trips back, plus one extra trip, yielding $2x + 2x + 1$.	Encourage students to share solution strategies with one another. As strategies are shared, ask them to judge if they are equivalent to one another (e.g., is $4x + 1$ equivalent to $2x + 2x + 1$?).	Ask students to state what x represents in each solution strategy that is shared. This will indicate whether or not they understand how numerical thinking and generalizations can be represented formally.

How This Lesson Meets Quality Control Criteria

- *Addressing students' preconceptions:* The lesson connects to students' prior experiences in inventing strategies to win games or to carry out mathematical and nonmathematical tasks efficiently.
- *Conceptual and procedural knowledge development:* Students are encouraged to invent a general procedure to solve a problem, but also to understand the conceptual basis for the procedures by inventing them on their own.
- *Metacognition:* Students are asked to record the procedure they used to solve the problem by constructing a diagram. They are then asked to look for patterns in the diagram to describe the essential characteristics of the procedure. Finally, they use the pattern they found to approach variations of the problem and express their thinking in formal mathematical language.

Lesson Plan 2

Based on the following NCTM resource: Perlwitz, M. D. (2005). Dividing fractions: Reconciling self-generated solutions with algorithmic answers. *Mathematics Teaching in the Middle School, 10,* 278–283.

Primary objective: To develop students' conceptual understanding of division of fractions

Materials needed: Fraction circle pieces (optional), pattern blocks (optional)

Steps of the lesson	Expected student responses	Teacher's responses to students	Assessment strategies and goals
1. Pose the following task for students to solve with a partner: "In Ms. Smith's sewing class, students are making pillowcases for the open house exhibit. Ms. Smith bought 10 yards of fabric for her class project. Each pillowcase requires $\frac{3}{4}$ yard of fabric. How many pillowcases can be cut from the fabric?" (Perlwitz, 2005, p. 279).	Students may recognize the problem as a division problem and attempt to use the invert-and-multiply rule to solve it. Some may use invert-and-multiply successfully, and others may make computational errors. Some students will not recognize which operation corresponds to the problem.	Push students to explain their answers completely. For example, if they use the invert-and-multiply rule and cannot explain why it works, prompt them to use a strategy they can explain completely. It is not necessary for students to decide if the problem is addition, subtraction, multiplication, or addition at this point— they should act out the situation.	Students' written work and oral statements should be assessed at this point for quality of explanation. Blind acceptance of the invert-and-multiply rule at this point in the lesson does not constitute a high-quality explanation, since the primary objective of the lesson is to develop students' conceptual understanding of fraction division.
2. Ask each pair of students to report on their progress to the rest of the class. Each pair should give an explanation of the strategy they used to solve the problem.	Some pairs of students will be stuck in the mode of blind acceptance of invert-and-multiply; others will attempt conceptual explanations that involve acting out the situation using diagrams or manipulatives.	Encourage groups who blindly accept invert-and-multiply to revoice and explain the strategies used by groups using conceptual approaches.	In addition to pushing students toward conceptual explanations, monitor all explanations for possible flaws. Do not correct the flaws at this point, but be mindful of them in order to steer discussion in productive directions later on.
3. Ask students to choose a conceptual strategy, carry it out to completion, and then check their answer against that obtained by using invert-and-multiply.	Some student-invented strategies will align with the results of invert-and-multiply and others will not. Some common discrepant answers include 13 and $13\frac{1}{4}$. When discrepancies arise, students may begin to distrust the strategies they used if they have had extensive experience with invert-and-multiply.	Ask students to reexamine the strategies they used as discrepancies arise. Encourage them not to discard the entire strategy, but to try to identify potential problematic points within it.	Students' perseverance in revising their strategies is an important element to assess at this point. Students often discard all of the thinking they have done when an authority or an algorithm tells them they have produced an incorrect answer.
4. Pose the following problem to students to work on in pairs and then discuss as a large group: "Some believe that the answer to the initial problem is 13, others think it is $13\frac{1}{3}$, and still others think it is $13\frac{1}{4}$. Explain	While discussing the problem, students will begin to see that an answer of 13 is obtained when one does not allow for fractional portions of pillowcases, $13\frac{1}{4}$ is obtained when a yard is used as the unit instead of a pillowcase, and $13\frac{1}{3}$ is	If students have difficulty seeing why $13\frac{1}{3}$ is an acceptable answer and why $13\frac{1}{4}$ is not, focus their attention on the whole units associated with $\frac{1}{3}$ and $\frac{1}{4}$. An answer of $\frac{1}{4}$ does not make sense within the context of the problem,	Focus on the diagrams students use to represent their thinking, along with the language they use to describe their diagrams. Make sure that they thoroughly explain what each portion of the diagram represents. If they attend carefully to

Steps of the lesson	Expected student responses	Teacher's responses to students	Assessment strategies and goals
the possible thinking behind each answer. Then explain which answer is correct."	obtained when the pillowcase is the unit.	since yard produces of a pillowcase. Draw their attention to the fact that the problem asks for the number of pillowcases, not yards.	this sort of explanation, it becomes easier to convince them that $13\frac{1}{4}$ cannot be correct.
5. Ask students to write a word problem for $1\frac{3}{4} \div \frac{1}{2}$ and then solve it conceptually.	Some will write problems that confuse division by $\frac{1}{2}$ with division by 2, or with multiplication by $\frac{1}{2}$.	Ask students to exchange problems with a classmate. The classmate should solve the problem and judge whether it correctly models $1\frac{3}{4} \div \frac{1}{2}$.	As students solve a classmate's problem, have them write a solution that explains their thinking completely. Collect and analyze the written work.

How This Lesson Meets Quality Control Criteria

- *Addressing students' preconceptions:* Students are challenged throughout the lesson to explore the adequacy of the school-learned invert-and-multiply algorithm; students use out-of-school knowledge of buying and cutting fabric to act out a problem in order to invent a strategy for fraction division.
- *Conceptual and procedural knowledge development:* Students compare their procedural knowledge of the invert-and-multiply algorithm against the conceptual strategies they attempt to invent.
- *Metacognition:* Students are asked to retrace the thinking that went into producing answers of 13, $13\frac{1}{3}$, and $13\frac{1}{4}$. They are also asked to examine their conceptual strategies to explain possible discrepancies with the invert-and-multiply algorithm.

Lesson Plan 3

Based on the following NCTM resource: Quinn, A. L. (2009). Count on number theory to inspire proof. *Mathematics Teacher, 103,* 298–304.

Primary objective: To develop students' deductive proof ability by having them solve problems from number theory (discussions about the problems may occur over several class periods)			
Materials needed: Calculators (optional), paper and pencil			
Steps of the lesson	**Expected student responses**	**Teacher's responses to students**	**Assessment strategies and goals**
1. Pose the following problem: "If you take an integer that is not divisible by 3, then square the number, and then subtract 1, is your result (1) always divisible by 3, (2) never divisible by 3, or (3) sometimes divisible by 3 and sometimes not?" (Quinn, 2009, p. 299). Students can work individually or in groups.	Some students will rely on empirical examples in examining the truth of the statement. They may become convinced it is true after it works for a few specific cases. Others will attempt to reason out the situation using algebraic symbols to represent a general expression.	During whole-class discussion, elicit responses from students who were convinced by a few specific examples as well as from those who were not. In the process, attempt to raise doubt about the ability of a few examples to establish certain proof in this situation.	At this point in the lesson, students may begin to see limitations in deductive arguments. Monitor the class discussion to judge whether or not this occurs. Ask students who rely on examples if it would be possible to test every case to which the statement would apply.

(Continued)

Steps of the lesson	Expected student responses	Teacher's responses to students	Assessment strategies and goals
2. Pose the following problem: "The last digit of 2^2 is 4, the last digit of 2^3 is 8, and the last digit of 2^4 is 6 since $2^4 = 16$. Predict what the last digit of 2^{100} would be" (Quinn, 2009, p. 300). Students can work individually or in groups.	Some will reason that 2^{100} has a final digit of 4 because 2^{10} ends in 4. Others will test additional examples and find a recurring cycle in the last digits (i.e., 2, 4, 8, 6, . . .).	Try to elicit responses from students that represent a variety of different approaches. Ideally, some of these should be founded on seemingly reasonable inductive arguments, yet not adequate.	After various solution strategies are shared, give the following writing prompt: "Give specific examples of strengths and weaknesses in reasoning." Responses will indicate students' levels of acceptance of deduction.
3. Pose the following problem: "Multiply the number 142,857 by 1, then by 2, and then by 3. Predict what would happen if you multiply 142,857 by 4. Predict what would happen if you multiply by 5, 6, and 7" (Quinn, 2009, p. 300). Students can work individually or in groups.	Many students will notice that the digits 1, 2, 4, 5, 7, and 8 occur in each of the first three products in some order. This will lead them to conjecture that this holds for all cases.	Ask students to actually do the multiplications $142,857 \times 5$, $142,857 \times 6$, and $142,857 \times 7$ and see if their conjectures hold. Then ask them to make refinements to their initial conjectures as necessary.	Ask students to explain why the digits 1, 2, 4, 5, 7, and 8 do not appear in the product of $142,857 \times 7$, even though they appear in all of the products calculated up to that point.
4. Pose the following problem: "Show that any time you multiply three consecutive positive integers, the resulting product is divisible by 6" (Quinn, 2009, p. 302). Have students work individually and provide written responses that completely explain their reasoning.	Some students will continue to rely strictly on inductive reasoning, perhaps examining several specific cases such as $1 \times 2 \times 3$, $10 \times 11 \times 12$, or $100 \times 101 \times 102$. Others will look for general arguments to cover all cases, such as an explanation of why any set of three consecutive integers will contain at least one that is divisible by 2 and one that is divisible by 3.	In making comments on students' written work, acknowledge that inductive reasoning can be helpful for getting a start on the problem, but that it cannot cover all possible cases. Raise awareness of the possibility that cases not covered in the students' explanations may be problematic, as they were with the problem posed in step 3 of this lesson.	Sort students' written work into several piles. The work in each pile should represent a different level of sophistication concerning the idea of deductive proof. Levels of sophistication may include (1) no justification provided, (2) incorrect conclusion reached, (3) reliance upon inductive reasoning, (4) transition to deductive reasoning, and (5) formal deductive reasoning.
5. Select student work samples from step 4 to share with the class. Samples should represent a variety of levels of sophistication. Remove student names from the work. Ask students which arguments they find most convincing.	Some of the students who relied strictly on inductive reasoning will begin to see that deductive explanations are more convincing because they cover all possible cases and not just several examples.	Encourage students to comment on how the argument in each work sample could be strengthened. This can be done for both inductive and deductive work samples.	During class discussion, assess students' ability to distinguish between deductive and inductive reasoning by having them rank the student work samples from weakest to strongest.

How This Lesson Meets Quality Control Criteria

- *Addressing students' preconceptions:* The lesson challenges students' preconception, formed by both in-school and out-of-school experiences, that evidence from a few specific cases constitutes proof.
- *Conceptual and procedural knowledge development:* The need for the concept of deductive proof is established through experiences with problems that expose the limitations of inductive reasoning; students build computational fluency as they execute computational procedures to test conjectures.
- *Metacognition:* Students are asked to make, test, and refine conjectures in several different situations, and to explain why some conjectures prove to be true while others do not; students are asked to write about the nature of their developing understanding of the relationship between inductive and deductive reasoning.

Lesson Plan 4

Based on the following NCTM resource: Szydlik, J.E. (2000). Photographs and committees: Activities that help students discover permutations and combinations. *Mathematics Teacher, 93,* 93–99.

Primary objective: To help students distinguish between counting permutations and counting combinations

Materials needed: A classroom set of plastic chips of at least four different colors

Steps of the lesson	Expected student responses	Teacher's responses to students	Assessment strategies and goals
1. Pose the following problem: "How many ways can you arrange four people in a line to take a photograph? Each different arrangement of the four people counts as a different photograph." Students work in groups of three to four to solve the problem.	Some students may have unsystematic strategies for counting the number of possible arrangements. Others will use tree diagrams and charts to systematically track the possibilities.	Ask students using systematic strategies to share their thinking with those not doing so. In some cases, this may require moving a student from one group to another, or putting two groups together to facilitate discussion.	Assess the organizational strategies students are using at the end of this portion of the class. Each student should exhibit a feasible strategy for tracking all possible outcomes before moving on.
2. Pose the following problem as a follow-up to help students begin to generalize the situation: "Suppose another person wants to join the group of four in the photograph. How many ways are there to arrange five people in line to take a photograph?" Open the problem to whole-class discussion.	Some students will apply their tracking strategy for the first problem, from beginning to end, to this problem. Others will recognize that there are five times as many arrangements because of the new person, and then obtain a solution by multiplying the answer from the last problem by 5.	Ask students who redo the entire tracking strategy they used for the first problem to find a solution shortcut by comparing the conditions of the follow-up problem to those of the original one. Once they have done this work, push students toward generalizing the situation by asking, "How many ways are there to arrange n people in a line for a photograph?"	As students work toward seeing relationships between the original problem and the follow-up, monitor their conversations to determine if they use the idea that the nth person can stand in n different positions, the $n-1$ person could stand in $n-1$ positions, etc. Understanding this idea is key to forming a generalization.

(Continued)

(Continued)

Steps of the lesson	Expected student responses	Teacher's responses to students	Assessment strategies and goals
3. Pose one additional problem to follow up the previous one: "Suppose you have a group of n people total but you only want r of them at a time to be in your photograph. How many photographs can you make?" (Szydlik, 2000, p. 97).	Students will adjust their solution strategies from the first portion of the lesson to conclude correctly that the answer is $n(n-1)(n-2)\ldots[n-(r-1)]$.	Encourage students to compare solutions with one another. Also encourage them to try to write their ideas in as compact a form as possible.	Assess students' work to see if they arrive at the compact notation $\frac{n!}{(n-r)!}$ Also question students to determine if they see the compact form as merely a simplified form rather than an entirely new expression.
4. Have four students stand at the front of the room. Ask the class how one could determine the number of different committees of size 2 that could be formed from the group.	Students will suggest forming all possible committees of size 2 by having all of the different sets of two students step forward in a systematic manner.	As different committees step forward, record the names of the students in each respective committee on the board. Ask students how they know when all possible committees have been formed.	Assess whether or not students see the difference between this situation and the photograph situation. Ask them to write, individually, about the similarities and differences between the two situations.
5. Extend the problem of selecting committees of a given size by asking students how many committees of size 3 could be selected from the group of 4. Also ask about the number of committees of sizes 0, 1, and 4 from the group of four. Students can work in groups to determine solutions for these situations. Ask them to try to make as many conjectures as possible that pertain to a group of n people from which committees of size r are chosen.	Students' conjectures will likely include the following: - There is one committee of size n that can be formed from a group of n people. - For a group of n people, there are n committees of size 1. - For a group of n people, there is one committee of size 0. - The number of committees of size r that can be formed from a group of n people is the same as the number of committees of size $(n-r)$.	During whole-class discussion, ask students to present and defend each conjecture they make. Encourage students to try to find a general formula for counting combinations, just as they did for counting permutations. The idea that there is one committee of size 0 may be counterintuitive, so attend to this idea explicitly once a general counting formula has been found or introduced.	Ask students to revisit their initial written explanations of how the picture taking and committee selection problems are similar and different. Have them make revisions as necessary in light of what they learned during this portion of the lesson. Their ability to distinguish between the two situations can indicate whether a foundation has been established for introducing and using formal notation for permutations and combinations in subsequent lessons.

How This Lesson Meets Quality Control Criteria

- *Addressing students' preconceptions:* Students' out-of-school knowledge of making arrangements for photographs and choosing teams of individuals is engaged by the problems in the lesson; students' school-based knowledge of organizational representations such as tree diagrams is engaged as they search for systematic solution strategies.
- *Conceptual and procedural knowledge development:* The broad conceptual distinction between permutations and combinations is established as students compare the photograph and committee situations; students develop efficient procedures for counting the number of permutations and combinations arising from any given situation.
- *Metacognition:* Follow-up problems pertaining to the photograph and committee situations prompt students to revisit, clarify, and extend the solution strategies they used for simpler, related problems.

Chapter 8

DEVELOPING STUDENTS' ALGEBRAIC THINKING

Perhaps algebra receives more attention than any other area of the secondary mathematics curriculum. It serves as a foundation for higher-level mathematics, so strong conceptual and procedural knowledge of the subject are essential. Unfortunately, algebra has proven to be a gatekeeper, limiting some students' access to higher education and mathematically oriented careers (Moses & Cobb, 2001). Over the past two decades, this lack of equity in access to algebra has sparked an "**algebra-for-all**" **movement** characterized by a search for pedagogical strategies to reach all students.

A major development in the algebra-for-all movement has been the idea of centering algebra on the concept of function. A **functions-based approach to algebra** makes functions from everyday life the central objects of study in the algebra curriculum. For instance, students may examine data on the number of miles traveled in a taxi and the overall cost of the trip. Examining the data can lead to a rule describing the cost of the trip as a function of the number of miles traveled. Functions can be found in numerous other interesting situations as well, such as number of text messages sent versus total cost for the month. Possible contexts for a functions-based approach are virtually limitless, since functions arise in countless situations. In general, a functions-based approach asks students to form their own theories about how the values of quantities depend on the values of other quantities (Chazan, 1996). In doing so, they produce rules for functions. Because this approach involves engaging students in examination of an interesting context at the outset of an activity, it calls into question the common assumption that algebra is to be applied only after "basic skills" are learned. The functions-based approach engages students from the outset in abstracting rules from interesting mathematical or real-world contexts.

Although the algebra-for-all movement was interpreted by some as a call for innovative teaching ideas, such as the functions-based approach, many schools simply placed all students in algebra courses without fundamentally reconsidering the manner in which algebra was taught. The predictable consequence of this action was higher failure rates in algebra. Some sought to bring failure rates down by removing difficult

> ## Implementing the Common Core
>
> See Clinical Task 2 to evaluate the role of the concept of function (Content Standards 8.F, F-IF, and F-BF) in an algebra textbook.

concepts from the curriculum. The gradual disassembly of school algebra led the National Mathematics Advisory Panel (2008) to release a list of concepts that should be included in secondary-school algebra. The panel's list included the following categories: symbols and expressions, linear equations, quadratic equations, functions, algebra of polynomials, and combinatorics. Although specifying a list of topics to be studied in algebra may have some value, it leaves a deeper, fundamental question unanswered: How can teachers help students *learn* algebra and think algebraically? In this chapter, we explore that question, first by considering meanings ascribed to the phrase "algebraic thinking" and then by examining research about how algebraic thinking develops.

WHAT IS ALGEBRAIC THINKING?

One's definition of algebraic thinking will be strongly influenced by the aspects of algebra he or she values the most. For instance, algebraic thinking can be defined by emphasizing algebra as a generalization of arithmetic, or it can be defined in terms of functions. Those advocating a functions-based approach to algebra would likely endorse the latter approach, while those embracing a more traditional approach to algebra might favor the former. Since algebraic thinking includes multiple aspects, the formulation of overly restrictive definitions is not helpful. However, to grasp what algebraic thinking may encompass, it is useful to describe general **algebraic habits of mind**. These are thinking processes necessary for performing a variety of algebraic tasks. Three central habits of mind are doing-undoing, building rules to represent functions, and abstracting from computation (Driscoll, 1999).

The first of the three habits of mind, doing-undoing, is necessary when students are asked to execute procedures and also to do them in reverse. For instance, factoring a polynomial involves undoing the process of polynomial multiplication. Writing an equation that has specified solutions is the reverse of solving an equation. The second habit of mind, building rules to represent functions, involves recognizing algebraic patterns and writing rules that define how to move from an input variable to an output variable. It would be used in a task such as describing how x and y relate to one another in Figure 8.1.

Figure 8.1 A relationship between x and y.

x	y
1	0
2	2
3	4
4	6

In Figure 8.1, the relationship between x and y can be described by the rule $y = 2x - 2$. The third habit of mind, abstracting from computation, involves thinking about computation in the abstract, and not just in terms of particular numbers.

Consider the geometric sequence 1, 2, 4, 8, 16, ... Recursively, one can think of a given term in the sequence as being twice the previous one. Thinking about computation in the abstract, one may recognize that the nth term can be obtained in this case simply by computing 2^{n-1}. Abstracting from computation is a particularly powerful habit of mind because it allows one to perform tasks such as finding the 5,000th term in the sequence without computing each of the individual terms leading up to it.

Write your own examples of algebraic tasks that involve each of Driscoll's habits of mind. Give at least one example for each of the following habits: doing-undoing, building rules to represent functions, and abstracting from computation. Explain why each of your examples fits a particular habit of mind.

STOP TO REFLECT

Cuoco, Goldenberg, and Mark (1996) offered a different, though partially overlapping, inventory of algebraic habits of mind. The starting point for the habits of mind they discussed was not existing school curricula, but observations about algebraists at work. They considered this a valuable place to look because the activities of school algebra at times are quite detached from the types of thinking algebraists actually do in practice. The habits of mind they saw in algebraists included enjoying a good computation, using abstraction, enjoying algorithms, breaking things into parts, extending things, and representing things. Students employing these habits of mind have the opportunity to engage in some of the same thinking patterns mathematicians use. The habit of mind of using abstraction, for instance, can manifest itself in a school task such as recognizing similarities between the process of factoring a polynomial and the more concrete process of factoring a whole number. Students might also recognize similarities between long division with polynomials and long division with whole numbers. The habit of mind of breaking things into parts can turn up as students decompose integers into the products of primes. By engaging students in tasks that require the habits of mind exhibited by algebraists, teachers bring the algebra done in school closer to algebra done in practice.

The habits of mind described by Driscoll (1999) and Cuoco and colleagues (1996) surely do not exhaust all possible activities encompassed by the phrase "algebraic thinking." Nonetheless, they do provide a good backdrop for the investigation of research on students' algebraic thinking. In the research to be described in the rest of this chapter, you will see each of the habits of mind come into play. Many times, more than one habit of mind must be fostered to help students develop understanding of a particular area of algebra. Three broad areas to be considered in the following are (1) understanding symbols and conventions, (2) understanding and executing procedures, and (3) understanding functions. Research provides a great deal of insight about students' thinking in these areas and how it develops.

UNDERSTANDING ALGEBRAIC SYMBOLS AND CONVENTIONS

Interpreting Literal Symbols

To succeed in algebra, students must develop an intuitive feel for what the "letters," or **literal symbols**, in expressions and equations represent. Usiskin (1988) provided a poignant example of the complexity involved in understanding the meanings of literal symbols by asking readers to consider the equations shown in Figure 8.2. Although all five symbol strings in Figure 8.2 are equations, each uses literal symbols in different ways. In equation 2, for example, x is often referred to as an *unknown* whose value can be determined by dividing each side by five. In equation 5, however, we usually think of x as being free to vary and take on numerous different values, making x feel more like a *variable* than a specific unknown. In equation 5, k is often thought of as a *constant* that specifies the slope of a line. Therefore, this set of five equations illustrates at least three different ways literal symbols can be used in algebra—to represent unknowns, variables, and constants. In addition, the equations themselves can be used for different purposes. To illustrate this, observe that equation 1 is commonly referred to as a formula, equation 3 as an identity, and equation 4 as a property.

> **Implementing the Common Core**
>
> See Homework Task 1 to construct a task requiring students to work with literal symbols in expressions and equations (Content Standards 6.EE, 7.EE, 8.EE, and A-REI).

Figure 8.2 Examples to illustrate the complexity of literal symbols (Usiskin, 1988, p. 40).

1. $A = LW$

2. $40 = 5x$

3. $\sin x = \cos x \cdot \tan x$

4. $1 = n \cdot \left(\dfrac{1}{n}\right)$

5. $y = kx$

The difficulties students have in understanding the meanings of literal symbols in algebra have several root causes. The fact that literal symbols can represent unknowns sometimes interferes with students' ability to understand situations where they represent quantities that vary (Malisani & Spagnolo, 2009). In other words, it is more difficult to conceive of x as a variable if one is used to working only in contexts where it represents a specific unknown. This implies that teachers need to consciously pose tasks involving variables as well as unknowns, and to draw students' attention to the differences between the two situations. The many definitions that exist for the word *variable* can be another source of student difficulty (Schoenfeld & Arcavi, 1988). Some definitions include unknowns under the umbrella of variables, while others define a variable only as something that varies. No universally agreed-on

> **Implementing the Common Core**
>
> See Clinical Tasks 3 through 5 to assess students' understanding of literal symbols when they need to work with and interpret equations (Content Standards 6.EE, 7.EE, 8.EE, and A-REI) and functions (Content Standards 8.F and F-IF).

definition for the term *variable* exists, so teachers need to be aware that students ascribe many different meanings to it. Students' interpretations should be elicited, and challenged when necessary, during class discussions.

> Write your own definition of the word *variable*. Explain why you would use this notion of variable when teaching instead of other possible formulations.
>
> **STOP TO REFLECT**

Wagner (1983) gave a slightly different account of causes of students' difficulty with literal symbols. In algebra, letters are often used to indicate numerals (as in $x + 5 = 7$), but they may also be used as labels (e.g., naming a point P or a function f). Distinguishing between these two types of usage can be difficult for students who have not yet developed a feel for what letters used in algebraic expressions and equations represent. In algebra, letters are also used as placeholders in number sentences, just as pronouns are used as placeholders in grammatical sentences. In the algebraic equation $2x + 5 = 8 - x$, x is a placeholder for 1. In the sentence, "He enjoys mathematics," the word *he* is a placeholder for the individual who enjoy mathematics. This similar usage suggests a similarity between algebraic symbols and words. However, emphasizing this similarity can be misleading. Whereas x must represent the same thing throughout the number sentence $2x + 5 = 8 - x$, words can change meaning within a single sentence (e.g., "Joe sat on a tree *stump* writing a problem that would *stump* his classmates"). Developing a feel for algebra includes understanding how literal symbols in algebra compare and contrast with numerals and words. Teachers can help students develop such understanding by explicitly drawing attention to how the structure of a sentence differs from the structure of an algebraic equation.

Translating Between Words and Algebraic Symbols

Malisani and Spagnolo (2009) demonstrated that a relationship exists between students' understanding of literal symbols and their ability to form a word problem for a given equation. They asked students ages 16 to 18 to respond to the following item:

> Invent a possible situation problem that could be solved using the following relation of equality: $6x - 3y = 18$. Comment on the procedure that you have followed. (p. 25)

More than half of the students in Malisani and Spagnolo's (2009) study who responded to the question gave incorrect answers. This high failure rate prompted a follow-up investigation in which students were interviewed as they tried to form a word problem. Students' difficulties were caused by factors such as trying to use the same letter to designate two different variables and mistaking coefficients for solution pairs. Malisani and Spagnolo also hypothesized that students might have been unaccustomed to forming word problems to produce a given equation, since school mathematics usually involves teachers posing word problems and asking students to solve them. This highlights the need for teachers to have students write and pose their own problems. Doing so prompts students to think conceptually and engage more deeply with the ideas at hand.

Although students are more accustomed to translating word problems into algebraic symbols than vice versa, they are not necessarily more successful in doing so. J. Clement, Lockhead, and Monk (1981) gave a now-classic example of a translation task called the **student-professor problem** that often throws secondary-school students as well as those taking higher-level mathematics and engineering courses in college:

> Write an equation for the following statement: "There are six times as many students as professors at this university." Use S for the number of students and P for the number of professors. (p. 288)

Implementing the Common Core

See Homework Task 2 to work on anticipating students' errors when they are required to build and use functions that model a relationship between two quantities (Content Standards 8.F and F-BF) in a situation similar to the student-professor problem.

In a sample of 150 calculus students, 37% gave incorrect responses. The failure rate was higher, 57%, in a sample of college algebra students. Most of the students who gave incorrect solutions exhibited a **reversal error**, placing variables on the wrong sides of equations they wrote to represent the situation (i.e., writing the equation $6S = P$ rather than $6P = S$). The same type of reversal error turned up in problems set in different contexts.

A variety of explanations have been given for students' difficulties with translation tasks such as the student-professor problem. J. Clement and colleagues (1981) formulated two explanations for these difficulties after interviewing some of the students about their problem solving strategies. The first explanation was use of **word order matching**, meaning that students believed algebraic symbols in an equation should be written in the same order as they occur in the corresponding word problem. The second explanation was use of the **static comparison method**, which involves treating literal symbols as labels rather than variables. Students using this method realized that the number of students was greater than the number of professors, but then treated S and P as labels rather than variables. They reasoned that $6S$ should be used to indicate the larger group, and P the smaller. In contrast, successful students drew on the knowledge that S is larger than P by acting on P through multiplication to bring it to equivalence with S (i.e., $6P = S$).

It should be noted that there is not universal agreement about the predominant causes of student difficulty on the student-professor problem. MacGregor and Stacey (1993), for example, provided data contradicting the idea that word order matching is a major cause of the reversal error. They asked 281 students around age 14 to solve problems that required translation from words to symbols. Unlike the student-professor problem, some of the problems could be solved correctly using word order matching techniques. For instance, two of the problems were:

a. "z is equal to the sum of 3 and y." Write this information in mathematical symbols.

b. "The number y is 8 times the number z." Write this information in mathematical symbols. (p. 222)

Even though word order matching yields correct responses ($z = 3 + y$ and $y = 8z$), students were about equally likely to get the problems correct or incorrect. This finding suggested that many students actually do not use word order matching in approaching translation tasks. MacGregor and Stacey hypothesized instead that a predominant cause of students' translation difficulties is the presence of "cognitive models in which the numeral is associated with the larger variable" (p. 228). For instance, in problem (b), where "'y is 8 times z,' the '8 times' is associated with the perceived larger quantity y, not with z. Retrieving information directly from these models leads to the association of the number 8 with the wrong variable" (p. 222).

Students' difficulties in translating word problems to algebraic symbols have sparked interest in finding teaching strategies for such tasks. Rosnick and Clement (1980) had some success in helping students produce correct responses to tasks such as the student-professor problem by prompting them to carefully examine inputs and outputs from equations they had written (e.g., how many students does your equation say there will be for five professors? For 20 professors? Are the outputs reasonable?). However, even though these techniques helped students produce correct responses to some tasks, the reversal error sometimes resurfaced in problems set in different contexts. The resilience of this error suggests that the underlying reasons for it are not likely to be completely addressed in a single teaching intervention. Kutscher and Linchevski (1997) reported some success in helping eighth-grade students translate word problems to algebraic symbols by having them list specific numerical cases of a problem situation before attempting an algebraic equation. Considering specific cases before trying to describe the situation in general using algebraic symbols helped students avoid the reversal error. For example, students using such an approach to the student-professor problem could list several specific cases in a table: for 1 professor, there are 6 students, for 2 professors, there are 12 students, and so on. The general statement that the number of students is equal to six times the number of professors can then be abstracted from the specific cases.

> **Implementing the Common Core**
>
> See Clinical Task 6 to assess students' ability to use functions to model a relationship between two quantities (Content Standards 8.F and F-BF) within the context of the student-professor problem.

Because examining word problems numerically is a viable means of helping students translate situations into algebraic symbols, teachers sometimes need to set aside their personal preferred modes of problem solving in the classroom. Van Dooren, Verschaffel, and Onghena (2003) found that prospective secondary-school teachers strongly preferred to solve word problems by generating equations rather than by using arithmetic and reasoning through specific cases. Such a tendency is somewhat understandable, because algebraic equations are powerful tools for reasoning about problem situations. However, the prospective teachers in the study sometimes used algebraic equations for problems for which using arithmetic would have been more efficient. When asked to grade student work samples, they gave lower scores to students who used an arithmetic strategy rather than an algebraic strategy, even when the arithmetic strategy was more efficient (Van Dooren, Verschaffel, & Onghena, 2002). These findings suggest that it is necessary for teachers to reflect on their preferred strategies for solving problems so that their instruction is not constrained by their personal preferences.

> **Implementing the Common Core**
>
> See Homework Task 3 for an exercise in modeling with mathematics (Standard for Mathematical Practice 4) within the context of creating equations that describe relationships (Content Standard A-CED).

Arithmetic-Associated Difficulties

Although reasoning about arithmetic patterns can form a foundation for success in using algebraic symbols, conventions learned in arithmetic can also interfere with the development of algebraic thinking. For example, in arithmetic, two whole numbers appearing together do not represent multiplication (e.g., "56" does not mean "5 times 6"), but writing two literal symbols next to one another in algebra does indicate multiplication (e.g., xy means "x times y"). These differing conventions cause difficulty for some algebra students (Calouh & Herscovics, 1988). Once students do understand that xy means "x times y," the notation $f(x)$ can become difficult to understand. Many students then interpret $f(x)$ to mean "f times x" (Carpenter, Corbitt, Kepner, Lindquist, & Reys, 1981). These **concatenation errors** (difficulties in interpreting chains of algebraic symbols) suggest the need for explicit class discussion on the meanings of conventions that have become second nature to most secondary-school teachers. Teachers may find it difficult to think back to a time when they did not understand the conventions, so it is important not to overlook possible student difficulties in becoming acquainted with them. Asking students to explain the meanings of algebraic symbols in writing when they first encounter them can give teachers a sense of the difficulties that need to be addressed.

Along with making concatenation errors when learning algebraic conventions, students often find it difficult to interpret the symbols that populate expressions and equations. The symbol "–," for instance, can throw students when it appears in front of a literal symbol such as x. When students (and teachers) read "$-x$" aloud as "negative x," students often come to believe that $-x$ must represent a negative number, when, in fact, there are many situations where it does not. One way to address this problem is to refer to $-x$ as "the opposite of x" rather than "negative x" (D. R. Johnson, 1994). Students can then be asked to locate x and $-x$ on the number line for several different values of x (positive, negative, and zero) to reinforce the meaning of $-x$ and illustrate the symmetry of the number line.

The "=" sign can also be difficult for students to interpret because of past experiences with arithmetic. Many early experiences with arithmetic involve problems such as $2 + 4 = $ ____. After many experiences with problems like this, students begin to interpret the equal sign as a "Do something" signal. When confronted with a problem such as $2 + 4 = $ ___ $ + 5$, students with this conception will often fill in the blank with a 6, since they take the equal sign as a prompt to combine the two numbers to the left of it and write the result to the right. Behr, Erlwanger, and Nichols (1980) called this an **operational view of the equal sign** and noted that it can persist quite late into a student's schooling. They expressed the need for students to develop a **relational view of the equal sign**, meaning that they interpret it as a symbol of equivalence. The effects of the operational and relational views of the equal sign on students' ability to solve equations will be explored further in the next section.

UNDERSTANDING AND EXECUTING ALGEBRAIC PROCEDURES

Equal Sign Understanding and Solving Equations

Evidence from research suggests that students holding a relational view of the equal sign are more successful in mathematics than are students holding an operational view.

Less than half of the middle school students in a study done by Knuth, Stephens, McNeil, and Alibali (2006) held a relational view. When asked to define the equal sign, students with an operational view offered comments such as the following:

- "A sign connecting the answer to the problem." (p. 303)
- "What the problem's answer is." (p. 303)

Students with relational views offered very different definitions, such as the following:

- "It means that what is to the left and to the right of the sign mean the same thing." (p. 303)
- "The same as, the same value." (p. 303)

Students with a relational view of the equal sign tended to have higher scores on standardized mathematics tests.

As well as being associated with higher overall mathematics achievement, a relational view of the equal sign is particularly foundational to solving equations. Students in the Knuth and colleagues (2006) study who held a relational view were more likely to solve equations correctly than those holding an operational view. Those with a relational view were also more likely to use algebraic strategies for solving equations instead of inefficient strategies such as "guess and check." Students with an operational view may use such inefficient strategies when solving systems of equations (Filloy, Rojano, & Solares, 2003). The importance of helping students develop a relational view led Knuth, Alibali, Hattikudur, McNeil, and Stephens (2008) to recommend that teachers watch for incorrect uses of the equal sign, such as inappropriately stringing numbers together with it (e.g., $1 + 2 = 3 + 5 = 8 + 2 = 10$). These kinds of errors provide natural opportunities to discuss why the equal sign should not be used in such a manner and prompt explicit consideration of the meaning of the symbol.

> ### Implementing the Common Core
>
> See Clinical Task 7 to investigate students' attention to precision (Standard for Mathematical Practice 6) in regard to the use of the equal sign.

Another popular strategy aimed at helping students develop a relational view of the equal sign and solve equations is to use the metaphor of a balance scale. In the **balance scale model of equations**, equations are thought of as consisting of two sides of equal weight, meaning that the operations performed on each side of the scale must be the same to maintain the balance. A sequence of steps that could be used to solve the equation $3x + 2 = x + 4$ with a balance scale is shown in Figure 8.3. Borenson (2009) claimed that the balance scale model helps foster students' ability to solve equations even before they begin the middle grades.

Although the balance scale metaphor generates an interesting diagrammatic representation of the equation solving process, it is not a panacea. After using the balance scale model with eighth-grade students, Vlassis (2002) described errors that students make when using it. Some of the most common errors occur when students attempt to use balance scales to solve equations containing negative signs and subtraction symbols. When solving an equation such as $5 - 2x + 4 = x + 9$ with the balance scale model, students may not regard the minus sign as being associated with the $2x$ term. This "detachment" of the minus sign (Herscovics & Linchevski, 1991) may cause them to subtract x from both sides and think that doing so produces $5 - x + 4 = 9$.

Figure 8.3 Solving an equation using the balance model.

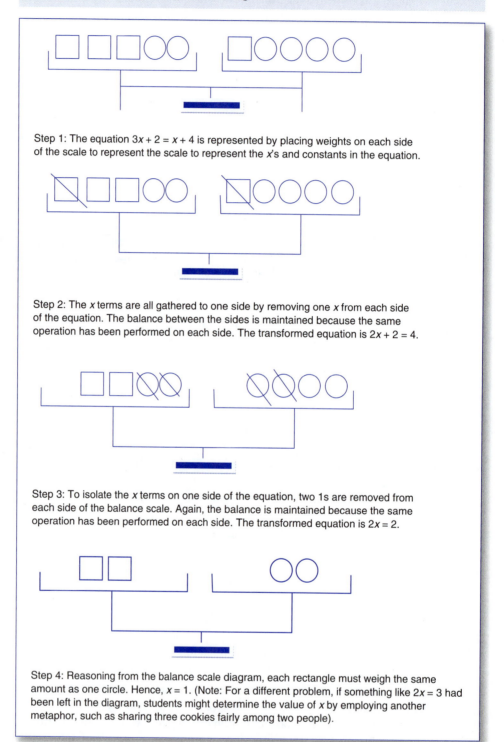

Step 1: The equation $3x + 2 = x + 4$ is represented by placing weights on each side of the scale to represent the scale to represent the x's and constants in the equation.

Step 2: The x terms are all gathered to one side by removing one x from each side of the equation. The balance between the sides is maintained because the same operation has been performed on each side. The transformed equation is $2x + 2 = 4$.

Step 3: To isolate the x terms on one side of the equation, two 1s are removed from each side of the balance scale. Again, the balance is maintained because the same operation has been performed on each side. The transformed equation is $2x = 2$.

Step 4: Reasoning from the balance scale diagram, each rectangle must weigh the same amount as one circle. Hence, $x = 1$. (Note: For a different problem, if something like $2x = 3$ had been left in the diagram, students might determine the value of x by employing another metaphor, such as sharing three cookies fairly among two people).

Students using the balance scale model may also have difficulty when confronted with an equation such as $-x = 7$. When faced with this situation, even if they recognize that $-x = -1 \cdot x$, they may attempt to subtract -1 from each side of $-x = 7$ and obtain the incorrect solution $x = 8$ or the incorrect solution $x = 6$. The balance scale may be interpreted by some students to provide a warrant for subtracting -1 from each side of the equation in this case, particularly if they have internalized the heuristic "Whatever is done to one side of the balance must be done to the other" in a superficial manner.

Despite possible student errors associated with the balance scale model, there are reasons to retain it in some form when teaching equations. Vlassis (2002) described several positive aspects. First, the model helps students learn to gather x terms to one side of an equation when they appear on both sides. Second, the balance model can help students understand the need to perform the same operation on each side of an equation. This general principle can then be carried over to various types of equations, even when students stop using the balance scale. Third, the model appears to enhance student retention of the ideas behind solving equations by providing an image of the equation solving process. Such an image places less demand on memory than a verbal description would. Students in the Vlassis study, for example, had no trouble recalling the balance model eight months after initially learning it. Vlassis concluded by noting that most criticisms of the balance model are related to uses for which the model is not well suited. Equations involving negatives and equations that can be solved more simply with arithmetic (e.g., $x + 1 = 2$, $-x = 5$) may be among those for which the balance scale model is not a useful pedagogical tool. Nonetheless, the balance scale model helps introduce essential principles and procedures that can be used in solving such equations.

> Use the balance scale model to solve the equation $3x + 5 = 2x + 7$. Show all your work and diagrams, and explain each step thoroughly. Then write three equations for which the balance scale model would be more difficult to use. Explain why it would be more difficult in each situation. Then describe the types of equations for which you would have students use a balance scale, and the types for which you would not. Justify your position.
>
> **STOP TO REFLECT**

Transition From Solving Equations to Solving Inequalities

Just as care must be taken when using the balance scale metaphor to teach equations, teachers must take care not to draw a seamless analogy from solving equations to solving inequalities. Tsamir and Bazzini (2002) found that many 16- to 17-year-old students tend to uncritically use procedures for solving equations when solving inequalities. One of the tasks they gave students was "Examine the following statement: for any $a \neq 0$ in R, $a \cdot x < 5 \Rightarrow x < 5/a$" (p. 291). Only 30% of the students in the study were able to respond correctly and provide a valid justification. Students tended to think the statement was true because division by zero was excluded, which is the same condition needed to make $5/a$ a valid solution to the equation $a \cdot x = 5$. They tended not to realize that division by a negative changes the direction of the inequality. Students frequently used the justification that it is necessary to carry out the same operation on both sides of an inequality in order to solve it, just as with an equation.

IDEA FOR DIFFERENTIATING INSTRUCTION: Comparing Multiple Strategies for Solving Inequalities

Tsamir, Almog, and Tirosh (1998) found that students who were most successful in solving linear, quadratic, rational, and square root inequalities had multiple strategies at their disposal. Students who used only algebraic manipulations to solve inequalities tended to be the least successful in producing correct answers. Hence, it can be profitable for teachers to encourage students to solve inequalities with many different thinking processes: using algebraic symbol manipulation, number lines, and function graphs. Students can be encouraged to use as many of these tools as possible and then share strategies with one another. For example, consider the inequality $-\frac{x}{3} \leq 5$. Students using only algebraic manipulation may correctly conclude that $x \geq -15$. Others will incorrectly conclude that $x \leq -15$ if they consider solving an inequality to be essentially the same as solving an equation. Students using number lines and function graphs can check to find out which of the two solutions is actually correct. A student using a number line might test points to see whether they make the inequality true. For example, after testing $x = 0, 10, -10, 20,$ and -20, a student may have the partial graph of the inequality shown in Figure 8.4.

Figure 8.4   Using a number line to reason about $-\dfrac{x}{3} \leq 5$

Testing a few points on the number line can help the student see that the inequality appears to hold for x values greater than -10, but not for those as far back as -20. This provides evidence that the correct algebraic solution must be $x \geq -15$ rather than $x \leq -15$. Students who understand functions might produce the graphs shown in Figure 8.5 to attack the problem.

Figure 8.5   Using function diagrams to reason about $-\dfrac{x}{3} \leq 5$

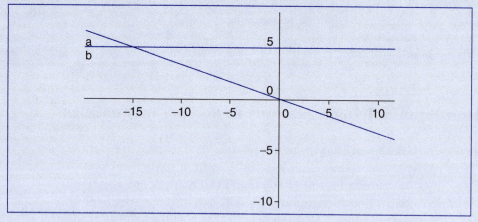

In Figure 8.5, the diagonal line represents the function $y = -\frac{x}{3}$ (function from the left side of the inequality) and the horizontal line represents $y = 5$ (function from the right side of the inequality). The two lines intersect at the point $x = -15$, showing that it is one of the solutions to the inequality. The rest of the solutions to the inequality can be found at the remaining x values where $y = -\frac{x}{3}$ is lower than $y = 5$ (to the right of the intersection point). The strategies outlined above are complementary, so it is profitable to have students discuss them alongside one another. In this case, encouraging differentiation along the dimension of the thinking processes students use sets the stage for rich classroom discussion as they start to share their different approaches.

Simplifying Expressions

Working with inequalities is not the only area in which manipulation of algebraic symbols without understanding occurs. Simplification of algebraic expressions also causes difficulty. One common student tendency is to **cojoin terms**, meaning to combine algebraic terms that should remain separate. For example, students may attempt to simplify an expression such as $3x + 4$ by writing $7x$. Booth (1988) attributed this tendency to students' previous experiences with the "+" sign. The plus sign, much like the equal sign, is often used in arithmetic as a signal to "do something." After working numerous problems involving the addition of whole numbers (e.g., $2 + 4$), students may feel a need to combine any two terms that are separated by a plus sign.

Implementing the Common Core

See Homework Task 4 to explore different methods for solving inequalities in one variable (Content Standards 7.EE and A-REI), anticipate student solution strategies, and address incorrect thinking patterns.

Teaching practices in algebra can also contribute to students' tendency to cojoin terms. Tirosh, Even, and Robinson (1998) investigated teachers' awareness of students' tendency to cojoin terms and observed the teachers' initial lessons on simplifying expressions. One of the teachers who was not aware of students' tendency to cojoin terms, Benny, taught simplification of expressions by giving students a rule of "adding numbers separately and adding letters separately" (p. 54). Students were confused by the rule, as it seemed to imply that an expression such as $2m + 3$ might simplify to $5 + m$ or $5m$, and they conjoined terms frequently throughout the lesson. Another teacher, Drora, who did not show awareness of students' tendency to cojoin expressions, used a "fruit salad" approach. That is, she explained to students that an expression such as $3t + 2$ cannot be simplified to $5t$ because combining the terms would be like adding three pears and two apples. Although the fruit salad approach works in this specific instance, students may carry it over to problems such as $(3a)(5b)$ and claim that it is not possible to multiply the two because you cannot multiply apples by bananas. The cases of Benny and Drora suggest that teaching practices, rather than innate characteristics of students, are sometimes the cause of students' difficulties. Teacher-invented "shortcuts" can cause more confusion than the formal rules themselves.

Use of formal rules is just one of many strategies students tend to use when simplifying expressions. Demby (1997) documented seven different types of procedures students follow: automization, formulas, guessing-substituting, preparatory modification, concretization, rules, and quasi-rules. Automization occurred when students saw an expression and immediately knew how to simplify it. When asked about the strategies they used, students generally could not give an explanation of how they arrived at the correct answer because the procedure they used was so deeply interiorized. In contrast, students using formulas identified specific formulas learned in school to approach tasks. For instance, a student might recall the general formula $(a + b)^2 = a^2 + 2ab + b^2$. Guessing-substituting procedures were characterized by first guessing the result of a simplification and then substituting specific numbers to check the guess. For example, a student might guess that $(a + b)^2 = a^2 + b^2$, but then substitute $a = 1$ and $b = 2$ and see that the two expressions are not equivalent. At that point, the student might begin the guessing-substituting process over again. Students using preparatory modification made adjustments to expressions to render them easier to deal with. An example of this would be changing the expression $3x - x + 1 - 2x$ to $3x + -x + 1 + -2x$ before attempting to simplify it, since changing from subtraction to addition allows more freedom to move terms. Preparatory modification could also involve expressing divisions as fractions or breaking multiplication problems into their component parts. Concretization strategies involved recasting algebraic expressions as something more concrete from everyday life—such as thinking of $3a + 2b$ as "three apples plus two bananas."

Perhaps the most interesting portion of Demby's (1997) study was the distinction between using rules to simplify expressions and using quasi-rules. Rules employed by students could lead to either correct or incorrect answers. Sometimes the rules they used were precisely those they learned in the classroom, but students often made their own modifications. Students using rules applied them consistently. If, for instance, one term in an algebraic expression was $5x^2$ and they simplified it to $25x^2$, they would consistently apply the same (in this case, incorrect) rule to obtain $4x^2$ from $2x^2$. On the other hand, students using quasi-rules did not have consistency of application. They might, for instance, believe that $2x^2$ cannot be simplified any further in one problem, and then in another problem think that $5x^2$ is the same as $25x^2$. The use of quasi-rules appeared to decline as students in the study gained more experience in algebra. Demby observed that the development of incorrect rules may be a normal stage in students' development, and recommended that teachers concentrate on helping students learn to construct correct rules for themselves rather than attempting to directly transmit rules to them.

As an example of the advice to help students construct the correct rules for simplifying expressions, consider the case of rules for exponents. In many traditional classrooms, the teacher provides a rule such as, "When you multiply and have the same bases, add the exponents," and then has students practice it on numerous exercises. Under this sort of teaching approach, many students inevitably modify the rule or get it mixed up with another. In the case of the preceding exponent rule, they might forget the part about having the same bases, or they may get confused about which situations involve addition of exponents and which involve subtraction. An alternative approach is to ask students to simplify expressions with exponents initially by using expanded form. For example, students might be asked to simplify expressions like those shown on the far left side of Figure 8.6 as far as possible by using expanded form before they have learned any of the rules for operations with exponents.

> ### Implementing the Common Core
>
> See Clinical Tasks 8 and 9 to assess students' ability to "apply the properties of operations to generate equivalent expressions" (Content Standard 6.EE) and to think about how to use assessment observations to choose instructional approaches that prevent students from cojoining terms inappropriately.

After working several problems like those shown in Figure 8.6, the teacher can ask questions to help students formulate rules for exponents. Some questions pertinent to Figure 8.6 include

- Which types of expressions reverted back to their original form after having been written in expanded form? Which did not?
- Are there any shortcuts you see for simplifying expressions? If so, what are they?
- Will the shortcut you invented always work? Why or why not?

In addition to leading students toward the correct rules for operating on expressions with exponents, this sort of activity provides students a fallback method (writing things in expanded form) if they forget one of the "shortcuts" or rules. The activity can be extended to teaching other exponent rules and other mathematical rules. The key is to activate previous knowledge that is relevant to formulating the rule, and then to ask guiding questions that will help students reflect on and consolidate their work.

Figure 8.6   Writing expressions in expanded form.

$$3^2 \cdot 3^4 = 3 \cdot 3 \cdot 3 \cdot 3 \cdot 3 \cdot 3 = 3^6$$

$$3^2 \cdot 4^4 = 3 \cdot 3 \cdot 4 \cdot 4 \cdot 4 \cdot 4 = 3^2 \cdot 4^4$$

$$x^2 \cdot x^4 = x \cdot x \cdot x \cdot x \cdot x \cdot x = x^6$$

$$x^2 \cdot y^4 = x \cdot x \cdot y \cdot y \cdot y \cdot y = x^2 \cdot y^4$$

STOP TO REFLECT

Suppose your objective for a lesson is to have students understand and use the rule $\frac{a^m}{a^n} = a^{m-n}$, $a^n \neq 0$. Design a sequence of five exercises that will help students invent this rule for themselves. Give a rationale for each exercise you include, and explain how the exercises build on the previous knowledge you expect students to bring to the lesson. Then design three problems you would ask students to work to assess whether or not they understand the rule. Give a rationale for each assessment exercise you include.

It should be noted that simplification of expressions is another area in which mathematics education researchers do not uniformly agree on the causes of student difficulties. Kirshner and Awtry (2004) gave a different type of explanation: They argued that the **visual salience** of an algebraic rule heavily influences students' thinking. To illustrate the idea of visual salience, consider the following rules:

- $x(y + z) = xy + xz$

- $(xy)^z = x^z y^z$

- $\sqrt[n]{xy} = \sqrt[n]{x}\,\sqrt[n]{y}$ (Kirshner & Awtry, 2004, p. 230)

Kirshner and Awtry argued that the three rules above have a great deal of visual salience because the left and right sides appear to be naturally related to one another. In contrast, in the three following equations, there appears to be a lesser degree of visual salience:

- $x^2 - y^2 = (x - y)(x + y)$

- $\dfrac{w}{x} \div \dfrac{y}{z} = \dfrac{wz}{xy}$

- $(x + y)^2 = x^2 + 2xy + y^2$ (Kirshner & Awtry, 2004, p. 230)

Implementing the Common Core

Complete the "Stop to reflect" task to help students "look for and make use of structure" (Standard for Mathematical Practice 7) and "look for and express regularity in repeated reasoning" (Standard for Mathematical Practice 8) in learning to "apply the properties of integer exponents to generate equivalent numerical expressions" (Content Standard 8.EE.1).

Implementing the Common Core

See Clinical Task 10 to investigate the impact of visual salience on "seeing structure in expressions" (Content Standard A-SSE) and performing "arithmetic with polynomials and rational expressions" (Content Standard A-APR).

Although visual salience is a difficult concept to define, most will admit that the first three rules seem more natural than the last three. This difference makes the second set more difficult to internalize than the first. Concrete supports (e.g., algebra tiles, to be discussed below) can be used to help students understand and retain rules with low degrees of visual salience.

STOP TO REFLECT

Give three of your own examples of algebraic rules that seem to have a high degree of visual salience. Then give three of your own examples of algebraic rules that seem to have a low degree of visual salience. Explain your reasoning for each of the six examples. Which characteristics seem to mark the difference between a rule with high visual salience and one with low visual salience?

Factoring and Multiplying Polynomials

Two of the most predominant types of symbol manipulation problems in school algebra are factoring and multiplying polynomials. As with the types of simplification problems previously discussed, students tend to develop their own sets of beliefs about how these operations are to be carried out. Students' rules for polynomial multiplication and factoring often conflict with those taught in school. Rauff (1994) found several different types of beliefs about factoring. One student definition for factoring was "unFOILING." Students with this belief factored polynomials such as $x^2 + 5x - 24$ correctly, but then factored $x^4 - 9x^2$ as $(x^2 + 3x)(x^2 - 3x)$, and believed the polynomial $3x + 6 - ax - 2a$ to be unfactorable. Another student definition was "breaking down a problem into simpler terms to the point where it can't be simplified anymore" (p. 423). This belief led some students to factor $x^2 + 5x - 24$ as $x(x + 5 + 3 \cdot 8)$. Although the student rules described above seem reasonable on the surface (and are even taught in some classrooms), they can lead to unproductive approaches to factoring expressions.

Algebra tiles are tools for helping students learn polynomial multiplication and factoring conceptually. They are designed to help students think about polynomial factoring and multiplication in terms of dimensions and areas of rectangles. The idea of thinking about algebraic properties geometrically is not a new one. Euclid, for example, gave geometric proofs of properties that we now think of as being purely algebraic, such as $a^2 - b^2 = (a - b)(a + b)$ (Boyer & Merzbach, 1989). Pieces in a typical set of tiles are shown in Figure 8.7. The label on each tile represents its area. Pieces in a typical set of tiles are shown in Figure 8.7. The area and dimensions of each tile are labeled. The three tiles on the top of Figure 8.7 have dimensions (from left to right) of $x \times x, x \times 1$, and 1×1, respectively. The three tiles on the bottom can be used to represent the opposites of the three on the top. The 1×1, tile, sometimes called the units tile, is designed so it will not measure out the length of the x tile precisely. This feature is intended to emphasize the idea that x represents a variable quantity. Some online versions of algebra tiles (Cannon, Dorward, Heal, & Edwards, 2001) do an even more effective job of portraying x as a variable quantity, with built-in sliders that allow users to vary the length of x. Plastic tiles are available through school supply shops, but teachers may also easily create laminated paper or cardboard versions (Leitze & Kitt, 2000).

Figure 8.7 Pieces available in a typical set of algebra tiles.

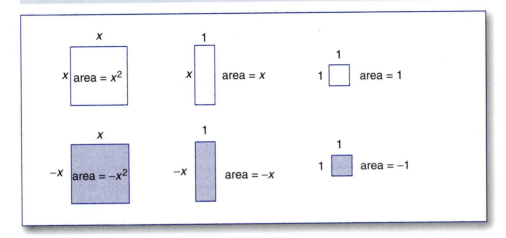

For an example of the algebra tile model in action, consider the polynomial multiplication problem $(x + 1)(x + 3)$. This problem can be interpreted as asking for the area of a rectangle whose dimensions are $(x + 1)$ by $(x + 3)$. To emphasize the connection to area, it is helpful to ask students to first recall how one determines the area of a rectangle with whole-number dimensions. The polynomial multiplication is essentially the same situation, except that it involves building a rectangle with variable rather than fixed dimensions. A diagram of the polynomial multiplication with algebra tiles is shown in Figure 8.8.

Figure 8.8 Algebra tile diagram for $(x + 1)(x + 3)$.

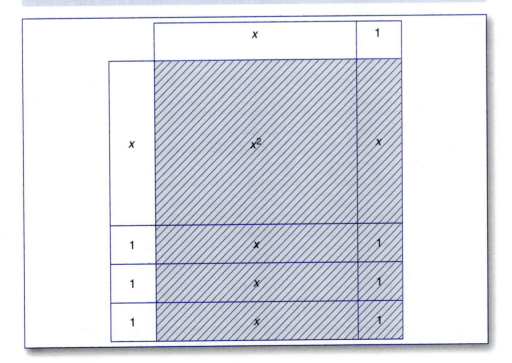

The shaded region in Figure 8.8 indicates that the area of a rectangle with dimensions $(x + 1)$ and $(x + 3)$ is $x^2 + x + x + x + x + 1 + 1 + 1$, or $x^2 + 4x + 3$. Now, imagine that students instead had the problem of factoring $x^2 + 4x + 3$. They could begin by building the rectangle shown in the shaded region. The task of factoring could then be interpreted as reading off the dimensions of a rectangle with area $x^2 + 4x + 3$. Doing so would give the factorization $(x + 1)(x + 3)$. Students need no knowledge of conventional polynomial multiplication and factoring procedures to solve these problems. The essential piece of background knowledge for performing the tasks is simply to understand the relationship between the area of a rectangle and its dimensions. In fact, the algebra tile model loses much of its power as a pedagogical device if it is only used to provide a visual representation of procedures students have already learned. If students are asked to use the tile model after they have already been taught procedures for multiplying and factoring polynomials, they are prone to interpreting the tile solution as just another procedure to memorize and carry out (Chappell & Strutchens, 2001). Eventually, students should be asked to look for ways to factor and multiply without the tiles, but teachers should not rush into this before students are ready.

STOP TO REFLECT

Write a polynomial multiplication problem that would require some shaded tiles (see Figure 8.7). Draw an algebra tile diagram for the problem, solve it, and explain how your diagram shows the solution. Repeat for a polynomial factoring problem that would require some shaded tiles.

TECHNOLOGY CONNECTION

Computer Algebra Systems

With the emergence of **computer algebra systems** (CASs), technology that can perform traditional symbol manipulation tasks such as polynomial multiplication and factoring in a few keystrokes, many questions about which algebraic procedures are still important for students to learn have arisen. For example, because of the CAS, the ability to factor or use the quadratic formula is no longer a prerequisite for studying solutions to quadratic equations (Heid, 2005). Even before the CAS became widely available, some teachers believed procedures such as polynomial factoring were arcane skills that should be removed from the curriculum. Usiskin (1980), for example, argued that trinomial factoring should not be a central part of school mathematics because it is not applicable to the vast majority of trinomials. Hence, the availability of the CAS has cast more doubt on the already hotly debated practice of spending a great deal of instructional time on execution of algebraic procedures.

Some teachers believe that deemphasizing algebraic procedures contributes to the erosion of algebraic thinking. However, strong arguments have been made that the CAS can actually prompt students to engage in deeper algebraic thinking. When one adopts a definition of algebraic thinking that extends beyond the execution of algebraic procedures, many possibilities for CAS use in algebra classrooms emerge. Heid, Blume, Hollebrands, and Piez (2002), for example, noted that understanding the distinction between the concepts of variable and parameter becomes much more important in a CAS environment than in a conventional one. They noted that on the CAS, exploring a function rule for interest, $A(t)1000(1 + i)^t$, requires knowledge of which literal symbols represent variables and which represent parameters. In this case, t can be regarded as a variable for time and i can be considered a parameter for interest rate. Developing knowledge of the meanings that literal symbols can take on in algebra is a central component of algebraic thinking, and also an issue that needs extensive attention in CAS environments.

Along with providing opportunities for considering the meanings of literal symbols used as input, a CAS provides opportunities for analysis and interpretation of output. Kieran and Saldanha (2005) asked students to interpret CAS output when using a built-in equality test. When using the test on "$(x - 3)(4x - 3) = (-x + 3)^2 + x(3x - 9)$," the CAS predictably returned the output "true." However, a "true" output was also given when the input was "$x^2 - x - 2 = \frac{(x-2)(x^2+3x+2)}{x+2}$." This presented an opportunity to discuss why the calculator took into account the restriction $x \neq 2$. Students essentially were prompted to do "reverse engineering" to explain the output. This sort of exercise can help promote flexible thinking about algebraic concepts and procedures.

A CAS also can return unexpected output when it automatically simplifies expressions put in by the user (e.g., changing $\frac{10x^2}{2}$ to $5x^2$) or when it expresses results in an unconventional form (e.g., returning $-(\sqrt{5} - 2)$ rather than $2 - \sqrt{5}$). In such cases, users need to decide if their expected output is the same as that given by the CAS. L. Ball, Pierce, and Stacey (2003) gave examples of expressions 10th grade students generally find easy to interpret as equivalent and some they are not likely to see as equivalent. Among the easier items were those involving notation (e.g., comparing $5m$ and m^5), linear factoring (e.g., comparing $-2y + 6$ and $-2(y - 3)$), and collecting like terms (e.g., comparing $2f - g + 3f - g$ to $5f - 2g$). More difficult items included expressions with fractions (e.g., comparing $\frac{12x}{2x}$ to 6x) and radicals (e.g., comparing $\sqrt{2} + \sqrt{y}$ to $\sqrt{2 + y}$). To help students overcome these difficulties, and in the process develop a deeper understanding of algebra, L. Ball and colleagues recommended giving more attention to examining a wide variety of equivalent expressions and having students compare their expectations against CAS output. Different-looking, yet equivalent, expressions produced by hand and by the CAS provide rich opportunities to discuss the meaning of algebraic equivalence.

Follow-up questions:

1. React to the claim "Trinomial factoring should not be a central part of school mathematics because it is not applicable to the vast majority of trinomials." Do you agree? Why or why not?
2. Do you consider the existence of CAS technology to be, on balance, a detriment to mathematics education or an opportunity for deeper learning? Explain.

UNDERSTANDING FUNCTIONS

The placement of this section on functions in the chapter is not meant to suggest that the study of functions should occur only after students become proficient in understanding algebraic symbols and performing procedures. Reform-oriented curricula often ask students to think about functional relationships even before rules for manipulating algebraic expressions are fully developed. For example, in the *Connected Mathematics* series (Lappan, Fey, Fitzgerald,

Implementing the Common Core

See Homework Task 5 to start thinking about how to help students "use appropriate tools strategically" (Standard for Mathematical Practice 5) when CASs are available.

Friel, & Phillips, 1998b), middle school students work with situations modeled by linear, quadratic, and exponential relationships before extensive attention is given to manipulating algebraic symbols. The *Contemporary Mathematics in Context* series (Hirsch, Coxford, Fey, & Schoen, 1998) for high school students treats algebra in a similar fashion, emphasizing mathematical modeling with functions early on and saving extensive work with complicated symbol manipulations for the final year of the sequence. The philosophy underlying a functions-based approach to algebra is that students are able to develop functional reasoning and solve challenging mathematics problems at the same time as they develop proficiency in manipulating symbolic expressions.

Reification Theory

Sfard and Linchevski (1994) hypothesized that the psychological process of **reification** plays a vital role in understanding functions. Reification occurs as students move from viewing functions as processes to viewing them as objects. As an illustration of this, consider the symbol string $y = x^2 + x + 1$. Depending on the context in which the symbol string is set, one may see it as a process or an object. A **process view of function** interprets $x^2 + x + 1$ as a procedure to be carried out by substituting a value for x and computing the result. From this perspective, the symbol y represents the finished computation. In contrast, an **object view of function** views $y = x^2 + x + 1$ as a single entity. The entity might be thought of graphically in terms of a parabola, and one may think about modifying it by altering its coefficients or constants. Depending on the mathematical task at hand, sometimes it is helpful to think of $y = x^2 + x + 1$ as a process, and other times it is more appropriate to think of it as an object. Sfard and Linchevski (1994) hypothesized that the process-oriented view emerges for students before the object-oriented view.

Many of the student difficulties described earlier in this chapter can be explained in terms of reification and process and object orientations toward algebraic symbol strings. Tirosh and colleagues (1998) attributed students' tendency to cojoin terms in algebraic expressions (e.g., thinking $5x + 1 = 6x$) to their inability to view expressions such as $5x + 1$ as objects. In such an expression, students may see the "+" sign as an indicator of an unfinished part of the process to be carried out. Likewise, students who view the "=" sign strictly as a "Do something" symbol (Behr et al., 1980) understand it only as the precursor to the end result of a process and not as a symbol that can be used to define the characteristics of an object. Students' difficulties with inequalities can also be explained in terms of reification. Goodson-Espy (1998) described the case of a student who could represent the parts of a word problem involving an inequality with algebraic symbols, but did not see the two symbol strings she generated as objects that could be compared to one another with an inequality symbol. Difficulties moving from process to object conceptions are not isolated to the students in the handful of studies mentioned here. In a review of the literature on algebraic learning, Kieran (1992) estimated that only 7% to 10% of students attain reification.

TECHNOLOGY CONNECTION

Analyzing Multiple Representations

Heid and Blume (2008) argued that having students study **multiple representations of functions**, including graphs, tables, equations, and verbal representations, can support the process of reification. To illustrate this in practice, consider the task of familiarizing students with the symbol string $y = x^2$. Students with a strictly process-oriented view will interpret this string as a set of instructions to square a numerical value for x and then call the result y. It can be very difficult for students to move beyond the process conception if the symbolic representation is the only one they encounter. Tabular and graphical representations help provide a richer picture.

Tabular representations show that x and y can take on different values in $y = x^2$, emphasizing the interpretation of x and y as variables. A tabular representation for $y = x^2$ might even be compared to one for $y = x$ to help build students' understanding (Figure 8.9).

Figure 8.9   Comparing tabular representations of $y = x$ and $y = x^2$.

x	$y = x$		x	$y = x^2$
0	0		0	0
1	1		1	1
2	2		2	4
3	3		3	9
4	4		4	16
5	5		5	25
6	6		6	36

The two tables shown in Figure 8.9 are only partial representations for $y = x$ and $y = x^2$, yet they hint at some interesting properties of each function. Calculating the differences between successive values in the right-hand column for $y = x$ yields a constant difference of 1. In the table for $y = x^2$, the difference between successive y-values is not constant, but the difference between differences is constant. This observation can spark a number of interesting questions to investigate: Why are the second differences constant in $y = x^2$? Are the second differences constant for any quadratic? As students explore questions like these, they can begin to think of $y = x$ and $y = x^2$ as objects with specific properties that can be compared to other objects. Handheld calculators and computer software programs that produce tabular representations can help automate the tedious computation that would be involved in producing several examples for comparison.

Graphical representations can also prompt students to develop object-oriented views of functions. A graphical representation for $y = x^2$ reveals that the points in its tabular representation fall along a parabolic curve. Students can be asked to predict how the shape of the parabola would change if terms were added, coefficients were changed, or constants were added or subtracted. Making all of these modifications to $y = x^2$ by hand would involve tedious hand computation and graphing that could easily be automated with a hand calculator or computer software. Automating the process allows students to reflect on how changes to the relationship influence the graphical objects that are produced. Sets of curves students might explore and compare to one another using graphing technology are shown in Figure 8.10.

(Continued)

(Continued)

Figure 8.10   Output from student exploration of parabola graphs.

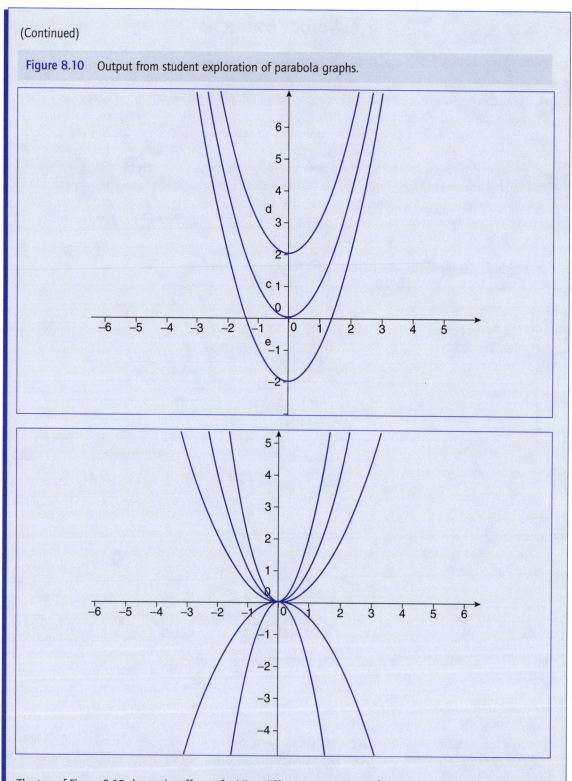

The top of Figure 8.10 shows the effects of adding different constants to x^2, and the bottom shows the effects of multiplying by different constants. After generating these types of graphs, students can be asked to explain why the graphs turned out as they did by referring to the tabular and symbolic representations. A teacher could also begin a task by providing the graphical representations and asking students to generate corresponding tables and graphs. By having students shuttle back and forth among various representations for functions, teachers can emphasize the idea that strings of symbols representing functions can be thought of as objects with several different representations that each serve to reveal different characteristics.

Follow-up questions:

1. Are there any situations in which students should graph points and functions by hand rather than using technology? If so, provide specific examples. If not, explain why not.

2. What kinds of assessment questions would you ask students as they explore the $y = x^2$ family of functions in the manner described earlier?

The Cartesian Connection

Research suggests that multiple representations of functions are not used to their fullest extent in traditional mathematics instruction. Knuth (2000) noted that traditional algebra instruction emphasizes producing graphs from symbolic representations of functions (e.g., the task, "Produce a graph of $y = x^2$"), but generally does not ask students to reason from graphs back to symbolic representations. To illustrate the detrimental effects of this practice, Knuth gave high school students tasks in which they were to determine the equation of a given linear graph. In one task, students were asked to determine the value of "?" in $?x + 3y = -6$. They were given a graph of the equation to use. Many of the students who gave a correct solution to the task used the inefficient process of calculating the slope of the graph provided, determining the y-intercept from the graph, writing the equation in slope-intercept form, and then converting it back to standard form. Students did not seem to recognize that every point on the graph represented a solution to $?x + 3y = -6$. If they understood this idea, called the **Cartesian connection**, they likely would have chosen any point (x, y) from the graph to substitute into the equation to determine the value of the "?" symbol. Although students in traditional algebra classes produce tables, graphs, and equations for functions, the act of producing these representations becomes a rote process devoid of meaning if problems that prompt them to recognize ideas such as the Cartesian connection are not included.

> **Implementing the Common Core**
>
> See Homework Task 6 to investigate a tabular perspective on quadratic models (Content Standard F-LE).

Students who do not fully grasp the Cartesian connection may also lack skill in choosing the most efficient representations for solving problems. Slavit (1998) examined the algebraic problem solving strategies of students in a precalculus course where the instructor emphasized graphical representations. Despite this emphasis, some students persisted in using equations and symbol manipulation even when it was inefficient to do so. For example, when given a task requiring a solution to $-.1x^2 + 3x + 80 = x$, one of the

> **Implementing the Common Core**
>
> See Clinical Task 11 to think about the Cartesian connection in relationship to the idea that "the graph of a function is the set of ordered pairs consisting of an input and the corresponding output" (Content Standard 8.F).

students interviewed first tried to factor. When factoring became difficult, she used the quadratic formula. Although she was prompted by the interviewer to discuss other solution strategies, approaching the problem graphically never occurred to her. A graphical approach might involve locating the roots of the parabola $y = -.1x^2 + 2x + 80$ or determining the intersection point of $y = -.1x^2 + 3x + 80$ and $y = x$. Slavit partially attributed the lack of use of graphical representations to past instruction focusing heavily on symbol manipulation. When instruction emphasizes only one representational system, some students come to see graphical and symbolic representations as two separate systems of procedures to follow rather than as representations that complement one another.

When students view function representations as complementary to one another, they develop better understanding of the attributes of functions. Schwarz and Hershkowitz (1999) found that students who explored multiple representations with technology understood functions more deeply than those who did not. Students using the technology were able to generate a broader range of examples of functions and also understood multiple representations as different descriptions of the same function. The zooming and scaling features of the technology helped students develop better part-whole reasoning about function representations. The results of the study strongly suggest that technology capable of generating multiple representations of functions should be a foundational part of algebra instruction.

Nickerson (2002) described how the multiple representations in one technology environment, SimCalc, helped students reason about solutions to systems of equations. As students worked with SimCalc, they were shown multiple representations of word problems. Students were able to describe a line graph in terms of a character's speed and direction, view slope as constant for a line, and use slope to determine if lines met. Similarly, Yerushalmy (2006) reported that numeric and tabular representations helped less successful algebra students approach word problems for systems of equations they otherwise might have given up on or carried out with little understanding. These studies support the idea that use of technology should be accompanied by class discussions where students are encouraged to explore the different representations with one another and with the teacher. During such discussions, students can focus on interpreting the representations and deciding what each one reveals.

TECHNOLOGY CONNECTION

Parabolic Area Functions

As another example of how technology can facilitate discussions of multiple representations of functions, consider Figure 8.11. It shows a Geometer's Sketchpad document exploring how the area of a rectangle with a perimeter of 15.45 cm (JHFE) is related to the length of one of its sides (FH). As point F is dragged to the left and right, rectangle JHFE changes its area, but not its perimeter. The graph of $f(x)$ shows how the area of rectangle JHFE varies with the length of segment HF. Since it is impossible for $f(x)$ to take on negative values in this situation, there is an opportunity to discuss the parts of the graph that are actually useful. The table of values to the right helps tell more of the story by showing specific inputs and outputs for the situation. In class discussions, students can be asked to examine the equation for $f(x)$ and predict where it has negative values, and where it is at a maximum. Technology supports the process of making and testing predictions by displaying multiple representations simultaneously for analysis and removing the time-consuming task of plotting and graphing every point needed.

Figure 8.11   Multiple representations for a function indicating the area of rectangle *JHFE*.

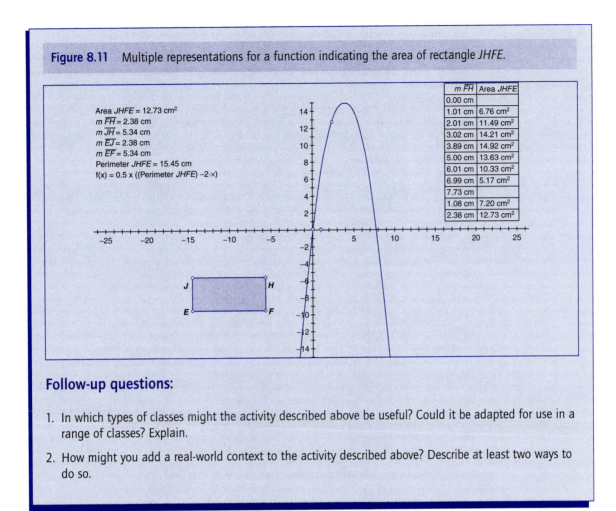

Follow-up questions:

1. In which types of classes might the activity described above be useful? Could it be adapted for use in a range of classes? Explain.

2. How might you add a real-world context to the activity described above? Describe at least two ways to do so.

Creating Meaning From Context

A longstanding assumption in mathematics instruction is that word problems are to be used as applications of procedures students have learned rather than sites for learning mathematics. Much of the research on students' learning of functions directly challenges that assumption. Confrey and Smith (1991) outlined a general approach to teaching functions using realistic contexts. Under their approach, students build understanding by producing multiple representations for situations that give rise to different types of functions. Students compare and contrast strategies for solving problems that produce different types of functions, such as linear, quadratic, trigonometric, and exponential functions. Instead of simply distinguishing among these types of functions by examining their symbolic forms, students compare and contrast the tabular and graphical representations of the functions as well. Technology allows the efficient generation of representations, which facilitates exploration of how varying the parameters involved influences each representation.

Confrey and Smith (1995) used the case of exponential functions as an extended illustration of how meaning for functions can be built from context. The following task was given as an example of a contextual problem to introduce exponential functions:

An efficiency expert works in a factory that builds robots. A robot builds another robot in 5 minutes, and then the robot walks into a crate and is shipped off. The expert comes up with a great idea for a labor-saving device. He creates a robot that is programmed to build two versions of itself in 5 minutes. It then walks into a crate and is shipped off. He goes into the supervisor's office to explain to her that he has doubled the efficiency of the plant and starts the robot off in order to be able to demonstrate his creation. She is in a meeting and he waits 3 hours. When he explains his invention, she looks up in alarm and rushes out into the factory. How many robots does she find? How many robots does the expert expect her to find? (p. 81)

Students can initially approach the problem by using a tabular representation like the one shown in Figure 8.12.

Figure 8.12   Table for robot problem.

Periods (*t*)	1	2	3	4	5	6	7	8	9
Robots (*r*)	2	4	8	16	32	64	128	256	512

The table is just an entry point to the problem, since it calls for determining the number of robots after three hours, when $t = 36$. The table indicates that a linear model will not give an accurate prediction, since the change is not constant from one cell to the next. This observation helps motivate a search for a formula to make the prediction. The creation of an explicit formula can then be done by making conjectures about its structure and testing the conjectures against the data in the table. Such a problem solving sequence demonstrates how symbolic representations for functions can grow from tabular ones.

STOP TO REFLECT

Write a complete solution for the robot task using the table of values shown in Figure 8.12. In your solution, explain how the table could be used to build an explicit formula to solve the task. Then, modify the robot problem to make it a linear situation, and explain how generating a tabular representation could lead to building an explicit formula to describe the new situation.

The approach of building tables and formulas from problem situations is not limited to the case of exponential functions. Consider the following task from Swafford and Langrall (2000):

Mary's basic wage is $20 per week. She is also paid another $2 for every hour of overtime she works.

- What would her total wage be if she worked 4 hours of overtime in one week? 10 hours of overtime?

- Describe the relationship between the amount of hours of overtime Mary works and her total wage.

- If *H* stands for the number of hours of overtime Mary works and if *W* stands for her total wage, write an equation for finding Mary's total wage.

- Can you use your equation to find out how much overtime Mary would have to work to earn a total wage of $50? One week Mary earned $36. A coworker of Mary's had worked 1 hour less overtime than Mary had worked. What would her friend's wage be for the week? (p. 93)

Sixth-grade students studied by Swafford and Langrall (2000) were able to perform the first few parts of this task successfully, even before formal instruction in algebra. The final portion of the task, where students were asked to use an equation to make a prediction, proved to be the most difficult. Nonetheless, students' ability to deal with an algebraic situation like this relatively early in their schooling suggests that it may be fruitful to have students reason about situations that give rise to different families of functions to help them gradually develop understanding of functions. Such an approach is used in reform-oriented middle school curricula such as the *Connected Mathematics Project* (CMP; Lappan et al., 1998b).

As with any teaching approach, building meaning from context is not a panacea. Lubienski (2000) described difficulties faced by students studying a pilot version of the CMP curriculum. Some students preferred the traditional approach over the more open-ended CMP approach. They complained that the text and the teacher were not directive enough. Students of lower socioeconomic status (SES), in particular, wanted to be given the "rules" for problems rather than to discover patterns through exploration. A second major difficulty was that students sometimes attended to mathematically irrelevant features of the context. Again, this problem occurred more among lower-SES students. As a result, they had more difficulty seeing connections among the mathematical structures of problems set in different contexts. This is particularly problematic when teaching functions, since one goal is to have students recognize some situations as linear, some as quadratic, others as exponential, and so on. Although engaging students in the study of functions in real-world contexts may ultimately help lower-SES students learn more deeply than they would in a more traditional curriculum, teachers should not expect the more open-ended approach to automatically solve all problems learners may face.

> **Implementing the Common Core**
>
> See Clinical Task 12 to assess a student's ability to "interpret functions that arise in applications in terms of the context" (Content Standard F-IF).

Finding Patterns and Making Generalizations

When students look for numerical patterns in contextualized situations, teachers can expect to see instances of both **recursive** and **explicit** reasoning. Recursive reasoning involves examining individual terms in a sequence to determine subsequent terms, and explicit reasoning involves formulating a rule to describe the characteristics of any given term without reference to previous terms. To illustrate the difference between the two, consider the following problem:

The Jog Phone Company is currently offering a calling plan that charges 10 cents per minute for the first five minutes of any telephone call. Any additional minutes cost only 6 cents per minute.

1. How much would a 7-minute phone call cost? A 12-minute phone call? A 32-minute phone call? A 57-minute phone call?

2. Explain how you would determine the cost for any phone call that is 5 minutes long or longer. Write a rule to explain how you would determine this cost. (Lannin, 2004, p. 219)

Students might approach the problem with a table of values like the one shown in Figure 8.13.

Figure 8.13   Table of values for telephone problem.

Minutes (m)	1	2	3	4	5	6	7	8	9
Cost (c)	10	20	30	40	50	56	62	68	74

The distinction between recursive and explicit reasoning appears in how the table may be used to perform the task. Reasoning recursively, students might notice that 10 is added from one cell to the next in the first few entries in the cost row, and then 6 is added to each value from there on out. A recursive formula to model this type of thinking for $m > 5$ would be $c_m = c_{m-1} + 6$, $c_5 = 50$. Reasoning explicitly, a student might notice that "the number of times that 6 is added is equal to the number of minutes by which the telephone call exceeds the initial five minutes" (Lannin, 2004, p. 219). This leads to the explicit rule $c = 6(m - 5) + 50$, $m > 5$.

At first glance, the explicit rule for the phone problem looks much better suited to approaching the task of determining the cost of a 57-minute phone call, so teachers may be tempted to try to get students to jump to the explicit rule quickly. However, two things should be taken into consideration. First, the formulation of the explicit rule is actually grounded in recursive reasoning in the telephone problem example, suggesting that students should be encouraged to reason recursively about it at first. Second, with the emergence of spreadsheets as commonly available tools, recursive rules are actually quite powerful for solving tasks like the telephone problem. A spreadsheet can quickly calculate successive values when the user inputs a single recursive formula. While it would be laborious for a student to compute recursively all the way out to 57 in the telephone problem, a spreadsheet can do so in an instant.

Students' choice to use either explicit and recursive reasoning is tied to a number of classroom-based factors. Among these are the mathematical structure of the task, its input values, the social influence of other students and the teacher, prior strategies, and the visual image of the task situation (Lannin, Barker, & Townsend, 2006). For instance, if part 1 of the Jog Phone Company task asked students only to determine the cost of a seven-minute call, it is likely that students would use a simple recursive strategy to determine the solution. Asking for the output for an input value of 57 is meant to prompt students toward finding an explicit formula. In addition, if only recursive strategies have previously been used in class, students will likely

Implementing the Common Core

See Homework Task 7 for an exercise in constructing a function given a visual pattern (Content Standard F-LE).

begin approaching the problem with such a strategy. The demonstration of an explicit reasoning strategy by another student or by the teacher may prompt a change. Students who focus on the tabular representation of the problem rather than the story representation may also be more prone to using recursive reasoning if they focus on how the values change from one cell to the next rather than on the overall structure of the situation.

IDEA FOR DIFFERENTIATING INSTRUCTION: TOOTHPICK TOWERS

One of the benefits of asking students to examine patterns and make generalizations is that such problems provide opportunities for communication among students employing diverse thinking strategies. For example, Tabach and Friedlander (2008) asked students to think of the illustrations shown in Figure 8.14 as two sequences of growing toothpick towers. Reasoning from the illustrations, students were to determine the fourth, fifth, tenth, and nth tower in each sequence. Students were encouraged to use their own strategies for the task—no teacher-prescribed strategies were provided.

Figure 8.14   Sequences of toothpick towers.

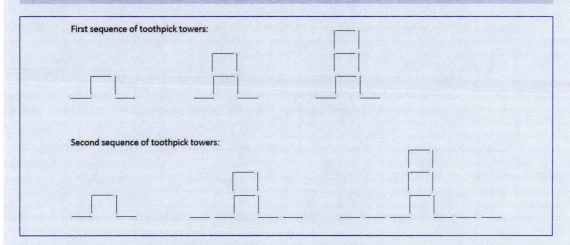

First sequence of toothpick towers:

Second sequence of toothpick towers:

Encouraging students to devise their own strategies for this task allows for differentiation of thinking processes. For instance, when looking at the second tower sequence, some students determine the number of toothpicks in the nth tower by counting the base lines and number of toothpicks in the tower separately. This leads to the expression $n + 3n + n$. Other students count the left base, followed by the vertical segments on the left of the tower, the horizontal segments, the vertical segments on the right of the tower, and finally the right base. Doing so leads to the expression $n + n + n + n + n$. As students compare their rules with one another, they can come to a better understanding of what it means for algebraic expressions to be equivalent. Since the different ways of counting yield the same result, students have a contextual justification for why $n + 3n + n$ is equivalent to $n + n + n + n + n$. Students who come to this understanding in advance of others can begin to explore the meaning of slope by comparing the explicit rule $5n$ (second sequence) to the explicit rule $3n + 2$ (first sequence), noting what the coefficient of n says about rate of increase from one tower to the next. Comparing the two explicit rules and their accompanying visual patterns also provides justification for not conjoining the terms in $3n + 2$ to produce $5n$.

Implementing the Common Core

See Homework Task 8 to consider student work on a visual pattern task that demonstrates understanding of how to "interpret the structure of expressions" and "write expressions in equivalent forms to solve problems" (Content Standard A-SSE).

Understanding Slope

The concept of slope is so central to beginning algebra that the logistics of teaching and learning it have received a substantial amount of research attention. One of the main messages from research is that instruction should not be limited to the procedural aspects of learning symbols used to represent slope and computational formulas. Conceptual issues should also be emphasized, including the study of **physical** and **functional contexts for slope** (Stump, 2001a). Physical contexts involve determining the steepness of objects such as ramps and hills. Functional situations involve interpreting slope as a measure of change in situations such as distance versus time or cost per hour. Prompting students to explore both kinds of situations can foster better understanding of the meaning of slope and its applications.

Both functional and physical applications for slope can pose difficulties for students. Stump (2001b) asked precalculus students to interpret the meaning of slope in both settings. In one problem involving a physical situation, students were shown three ramps and three decimal numbers, .83, 1.3, and 1.75, representing slopes to be matched with the ramps. Only half of the students were able to find correct matches. Some interpreted slope as the height of a ramp. In another physical situation, students were asked to interpret the meaning of "7% grade" on a road sign. Very few included the ratio of rise to run in their interpretations. In a functional situation, students were asked to examine graphs of wheel revolutions versus pedal revolutions for various gears on a bicycle. Only about half of the students mentioned that it would represent the ratio of the number of wheel turns to the number of pedal revolutions. Similar difficulties occurred when students were asked to interpret the slope of a graph showing profit from a dance as a function of ticket sales. These findings indicate that teachers need to provide opportunities for students to examine both functional and physical situations, and to help them understand slope both as a measure of physical steepness and as a measure of rate of change. In some cases, teachers need to supplement their curriculum materials to include both functional and physical problems.

Some of students' difficulty in thinking of slope as a ratio can be traced back to specific teaching episodes. Lobato, Ellis, and Muñoz (2003) described classroom-based factors that lead students to think of slope as a difference rather than a ratio. One of these was the tendency to use the phrase "goes up by" when discussing slope. Students in Lobato and colleagues' study interpreted the phrase "goes up by" to refer to four different things: the difference in y-values, the difference in x-values, the scale of the y-axis, and the scale of the x-axis. The teacher, on the other hand, interpreted "goes up by" to mean the amount of increase in y for every unit increase in x. These differing interpretations meant that the teacher and students were unknowingly speaking different languages on many occasions, as in the following exchange:

Ms. R: . . . and so if you, if I ask you what's the rate of change of this graph, what do they really want to know, do you think?

Salvador: A rate?

Ruben: How much it's going up by.

Ms. R: How much it's going up by. That's really good, Ruben . . . (p. 16)

In an earlier interview, Ruben interpreted the phrase "goes up by" to mean the scale of the *x*-axis. The teacher was not aware of his interpretation, assumed that Ruben was using the phrase in the same sense she was, and hence did not probe for further clarification. Students' tendency to interpret slope as the difference in y-values was reinforced by the types of tables students were asked to examine in class. They began an instructional unit with a table for a linear equation where the change in *x* from cell to cell was 1. The teacher asked students to focus on the change in *y*-values to determine slope. Some students continued to focus exclusively on the change in *y* in tables where the jump between *x*-values was no longer 1. These classroom episodes are illustrative of how students may form narrow procedural understanding of slope and not understand it conceptually.

> ### Implementing the Common Core
>
> See Clinical Task 13 to assess a student's understanding of slope and y-intercept within a given context (Content Standards 8.EE, 8.SP, and S-ID.7).

TECHNOLOGY CONNECTION

The Slope Game

One way to develop students' conceptual understanding of slope is to have them play "the slope game" (Kunkel, Chanan, & Steketee, 2009). To play, one student begins by constructing several lines using Geometer's Sketchpad and measuring their slopes with the software. The slopes are then dissociated from the lines, as shown in Figure 8.15. Another student then attempts to match the correct slope to each line. After matches have been attempted, the software can relabel the lines with their defining points to allow students to check the answers. As they play the game, students should be encouraged to produce lines with positive and negative slopes, as well as some that are horizontal and vertical. Including all of these types of lines helps build students' understanding of the range of values slope may take on. To extend the slope game and to address functional interpretations of slope, students can be asked to write slope descriptions to match various graphs. For example, a student may write, "We will make $5.16 for every dance ticket sold," and then use Geometer's Sketchpad to produce a line with a slope of 5.16. After producing several more descriptions and corresponding graphs, the student can scramble the descriptions and graphs and challenge another student to match them. Discussing the reasons for students' successes and failures in the game can prompt them toward deeper conceptual understanding of slope.

(continued)

(Continued)

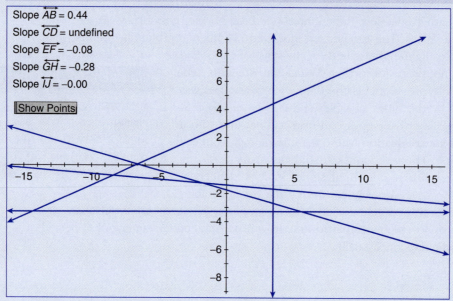

Figure 8.15   Screenshot from the slope game with Geometer's Sketchpad.

Follow-up questions:

1. See Figure 8.15. Match the slopes on the left side of the figure to the appropriate lines.

2. Would you use the slope game activity before or after students have learned a formula for computing slope? Explain.

TECHNOLOGY CONNECTION

Data Collection Devices

Another technology-based strategy for developing conceptual understanding of slope involves the use of a data collection device called a calculator-based ranger (CBR). A CBR can gather many types of data from real-world situations involving functions and send the data to a graphing calculator to be plotted. In one CBR activity (Herman & Laumakis, 2008), students are shown various graphs of distance versus time for a person walking along a path (Figure 8.16). They are asked to walk in front of the CBR to reproduce the path. As they gather data, students are asked how they could make the slope of a given graph or portion of a graph steeper. They are also asked to give an interpretation of slope that fits the context of the situation. Along with providing experiences interpreting slope conceptually, the activity prompts students to interpret the y-intercept and to think about equations to fit curves. The activity also addresses a common misconception about distance-versus-time graphs: the idea that such graphs show a picture of the path traveled (Janvier, Girardon, & Morand, 1993). Students may begin the activity with this sort of misconception, but then reject the idea when they find it to be inadequate for producing the desired graphs with the CBR.

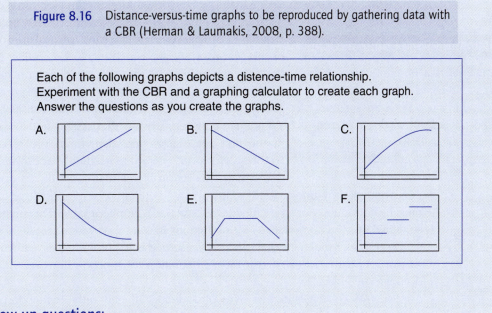

Figure 8.16   Distance-versus-time graphs to be reproduced by gathering data with a CBR (Herman & Laumakis, 2008, p. 388).

Each of the following graphs depicts a distence-time relationship. Experiment with the CBR and a graphing calculator to create each graph. Answer the questions as you create the graphs.

A.   B.   C.

D.   E.   F.

Follow-up questions:

1. Interpret the slope and y-intercept of each graph shown in Figure 8.16. What do they represent in terms of the real-world situation being discussed?

2. How might the CBR activity be used to teach students about the concepts of velocity and acceleration? What types of assessment questions could you ask to bring these ideas to the forefront?

Building a Concept Image for Function

A theme you may see running through this chapter, and others, is that students' notions about mathematical concepts are often quite idiosyncratic. It has been illustrated that students' interpretations of distance-versus-time graphs, for instance, may differ from those intended by teachers. Tall and Vinner (1981) used the phrases **concept image** and **concept definition** to capture the difference between students' ideas about concepts and accepted formal theory. Concept images consist of the mental pictures, properties, and processes students associate with a given concept. Concept definitions consist of the words used by the discipline to specify a concept. Students' concept images are built up through both in-school and out-of-school experiences. Often, students' concept images do not match formal concept definitions. Take, for example, young students' experiences with multiplication. After multiplying whole numbers in school for an extended time, students may draw the reasonable conclusion that multiplication is an operation that makes things bigger. Everyday language supports this idea about multiplication (e.g., "I multiplied my cash by putting it in the bank"; "The rabbits multiplied quickly"). With an entrenched concept image like this, it is difficult to understand how and why multiplication can actually decrease the size of an initial quantity (e.g., $5 \times \frac{1}{3}$).

Function is one area in which students' concept images often conflict with concept definitions. Simply memorizing a "correct" definition for function often does not resolve the conflict. Vinner and Dreyfus (1989) described a striking **compartmentalization phenomenon** that occurs among students, which involves a conflict between the concept image and the concept definition held by a student. Some of the students in Vinner and Dreyfus's study provided an acceptable definition of function as a "correspondence between two sets that assigns to every element in the first set exactly one element in the second set" (p. 360). However, the students then made claims when doing mathematics problems that contradicted this definition. Some of the students in the study operated under the assumption that functions must be continuous, or that a function is a formula, graph, or table. This suggests that students' previous experiences with functions in school consisted largely of working with continuous functions that can be represented with graphs, tables, and equations. Knowing how to state a correct formal definition for function does not necessarily dislodge such ideas.

> **Implementing the Common Core**
>
> See Clinical Task 14 to investigate the relationship between the treatment of functions in a text and a student's personal definition for the concept of function (Content Standards 8.F and F-IF.1).

Even (1993) suggested that some of students' difficulty with the function concept can be traced back to teachers' understanding of it. When prospective secondary teachers Even studied were asked questions designed to assess their understanding of functions, several problematic ideas surfaced. Some of them defined a function as something involving "operations on numbers," even though the modern definition of function does not specify that the sets involved need to be numerical. Others defined a function as an equation, expression, or formula, even though the modern definition does not require the mapping between domain and range to be specifiable by an equation or formula. Still others thought that all functions should be continuous and smooth, a thought that actually resonates with earlier historical notions of function. When asked how they would define the concept of function for students, many referred to the vertical line test rather than providing a definition. This type of response suggests that students may be asked to work only with functions that can be represented by means of a conventional coordinate graph. If students work with such a narrow range of functions, the concept images they develop for function will be quite limited.

> **Implementing the Common Core**
>
> See Homework Task 9 to think about how to "relate the domain of a function to its graph, and where applicable, to the quantitative relationship it describes" (Content Standard F-IF.5).

> **STOP TO REFLECT**
>
> Not all functions can be represented by tables, graphs, and equations. For example, students in a school may be assigned to either algebra class or study hall during fifth period. The students can be thought of as constituting the domain of the function, and the assignment options the range. A functional relationship exists—each element of the domain is assigned to an element of the range, but no domain element can be assigned to more than one range element. Provide several more examples of such functions. Explain how your examples conform to an accepted formal definition of function.

CONCLUSION

Algebra is a gateway to the study of more advanced mathematics, so developing students' algebraic thinking is one of the most important tasks mathematics teachers carry out. Algebraic thinking is multi-faceted and consists of such elements as doing-undoing, building rules, and abstracting from computation. These and other elements of algebraic thinking come into play as students interpret symbols and conventions, carry out procedures, and work with functions. We know of several difficulties students may encounter within these broad categories of algebraic activity. A few of the examples described in this chapter were the reversal error, quasi-rules for simplifying expressions, concatenation, operational understanding of the equal sign, and compartmentalization in thinking about functions. To address these frequently encountered challenges to learning, teachers should carefully consider the possible benefits and drawbacks of using such tactics as the balance model for equations, multiple representations, computer algebra systems, graphing calculators, and algebra tiles. The success or failure of any given teaching strategy should ultimately be judged by its ability to enhance students' algebraic thinking. Making such judgments requires that teachers carry out continuous assessment and adjust instruction according to students' observed learning needs. To help get you started in thinking about how to design and adjust instruction to match students' learning needs in algebra, sample four-column lessons are shown at the end of the chapter.

VOCABULARY LIST

After reading this chapter, you should be able to offer reasonable definitions for the following ideas (listed in their order of first occurrence) and describe their relevance to teaching mathematics:

HOMEWORK TASKS

1. Write an algebra word problem that can be solved by using literal symbols. Also write a solution to the problem. The solution should involve using literal symbols in at least two different ways (as variables, parameters, constants, unknowns, etc.). In your written solution, identify the different types of literal symbols you used.

2. Consider the following situation: "At Mindy's restaurant, for every four people who ordered cheese-cake, there were five who ordered strudel" (J. Clement et al., 1981, p. 287). Give a possible student error in translating the situation to an equation, along with at least two possible explanations for the cause of the error. Then show how the problem could be solved using each of the following representations: a table, a graph, and an equation.

3. Consider the following problem: "A primary school with 345 pupils has a sports day. The pupils can choose between inline skating, swimming, and a bicycle ride. Twice as many students choose inline skating as bicycling, and there are 30 fewer pupils who choose swimming than in-line skating. 120 pupils want to go swimming. How many choose in-line skating and bicycling?" (Van Dooren et al., 2003, p. 33). Solve the problem in two different ways: by using an algebraic equation and by using arithmetic. Which way of solving the problem is more efficient? Justify your position.

4. Write a rational inequality for which the solution using algebraic symbol manipulation requires division by a negative at some point. Show how students might attack the problem using a number line, function graphs, and algebraic symbols. Give examples of correct as well as incorrect thinking students may use while employing the three strategies. Explain how you would try to help students overcome their incorrect thinking patterns.

5. Do an online search for the phrase "computer algebra systems." Find the manufacturer websites for two different computer algebra systems (CASs) currently on the market. Build a table that will enable you to compare salient features of each CAS. Then identify one built-in feature on each CAS that could be used in a second-year algebra class. Describe how you would have students use the identified features to explore issues that would be difficult to explore in a conventional paper-and-pencil environment.

6. Consider the following two questions: Why are the second differences constant in $y = x^2$? Are the second differences constant for any given quadratic? Give well-developed answers to each question from two different perspectives: (1) the perspective of a first-year high school student who has just successfully completed an introductory algebra course, and (2) the perspective of a senior in high school who has just successfully completed an introductory calculus course.

7. Joe puts tiles around square pools for a living. He uses the same general design each time, and an example of how he tiles around a 3 × 3 pool is shown below:

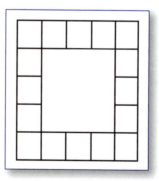

How many tiles would Joe need for a 4 × 4 pool? How many for a 10 × 10 pool? How many for an $n \times n$ pool? Explain your reasoning and show all your work.

8. Explain the thinking patterns that may lead students to derive each of the following explicit rules *just* from examining the diagram in Homework Task 7: $4n + 4$, $4(n + 1)$, $2n + 2(n + 2)$, $(n + 2)^2 - n^2$. Assume the students have not learned, or at least do not make use of, algebraic rules for factoring, multiplying, or any kind of simplifying, but instead reason strictly from the diagram.

9. If students were to make graphical representations of the function rules they discovered in Homework Task 8, would it make sense for them to connect the points they plotted for the graphs? Why or why not?

CLINICAL TASKS

1. Interview a teacher at your clinical site to learn about the school's philosophy regarding providing students access to algebra (e.g., do all students take algebra? If so, at what point? If not, why not? Do all students who take algebra experience a rigorous curriculum?). Then summarize the similarities and differences between the main aspects of the school's philosophy and the NCTM Equity Principle.

2. Examine an algebra textbook from your clinical setting. Describe the role that functions play in the textbook. Is function a central or peripheral concept? Provide several examples of problems and exercises from the book to support your answer. Also provide bibliographic information (title, author[s], date, and publisher) for the text.

3. Show the five equations in Figure 8.2 to a student who has completed at least one year of high school algebra. Ask him or her to explain the similarities and differences among the five equations. Also ask the student to explain the purposes for which each type of equation might be used. Summarize the student's responses. Describe any surprising or unexpected patterns you see in the student's responses.

4. Observe an algebra lesson that incorporates the use of literal symbols. Give several examples of how literal symbols are used by the teacher, the students, and the curriculum materials. Categorize the examples into groups. Form and justify conjectures about which categories of literal symbol use may have been most difficult for students to understand.

5. Write three equations in which literal symbols are used in different ways. Interview at least one student and ask him or her to write a word problem that can be solved by using each equation. Ask the student to think aloud while doing this. Record the student responses and describe any factors that appear to cause difficulty with the task.

6. Ask a class of students to complete the student-professor problem (J. Clement et al., 1981) individually, in writing. Tally how many times the reversal error occurs. Then, interview at least two students who made the reversal error. Ask them to solve the problem while thinking aloud, and then write an explanation for the causes of their error.

7. Interview at least three students. Describe each student's grade level and the mathematics course he or she is currently taking. During the interview, point to the "=" symbol in the equation $2 + 4 = 6$ and ask the student to tell you the name of the symbol and what it means. Repeat for the equation $y + 5 = x + 6$. Use examples of student responses from your interviews to support your answers to the following questions: (1) Which students held an operational view of the equal sign? A relational view? (2) Did any students give a relational response for one of the equations and an operational response for the other?

8. Interview at least three students who have studied algebraic expressions. Ask each of them if it is possible to further simplify the expression $3a + 2$. If students cojoin the terms, ask them to explain why they did so. If students do not cojoin terms, ask them to explain why the expression can be simplified no further. Describe the patterns you see in the students' thinking. If you were teaching these three students algebra, how would you build on the thinking the students exhibited during the interviews?

9. Find a page of exercises from an algebra textbook requiring students to simplify algebraic expressions. Interview at least three students, asking them to think aloud as they solve the exercises. Describe the strategies each student used. Then try to classify the strategies within one of Demby's (1997) categories. Explain why you placed each student response within a given category.

10. Ask an algebra student to judge whether or not each of the following sets of expressions are equivalent.

    1. $\frac{6x^2+1}{x^2}$ and $6x + \frac{1}{x^2}$

    2. $\sqrt{x} + \sqrt{y}$ and $\sqrt{x+y}$

    3. $\cos(3x)$ and $\cos3(x)$

    4. $(x+y)^2$ and $x^2 + y^2 + 2xy$

    5. $2x$ and x^2

    Ask the student to think aloud as he or she reasons through each set. Describe the justifications the student gives for each set of expressions and explain what the justifications reveal about areas of strength and weakness in the student's algebraic thinking.

11. Ask a student who has taken algebra to determine the value of the "?" in the expression $?x + 3y = -6$. Show the student the graph of the function corresponding to $?x + 3y = -6$ (choose a value for "?" and produce a correct, clearly labeled graph on grid paper). Ask the student to show all work and to talk aloud while doing the problem. Describe how the student went about doing the problem. Based on the student's response, do you think he or she understands the Cartesian connection? Justify your answer.

12. Ask a student to solve the beam problem shown next (from Lannin, Barker, & Townsend, 2006, p. 13). Describe the strategy the student uses to solve the problem and classify it as recursive, explicit, or somewhere in between.

Beams are designed as a support for various bridges. The beams are constructed using rods. The number of rods used to construct the bottom of the beam determines the length of the beam. Below is a beam of length 4.

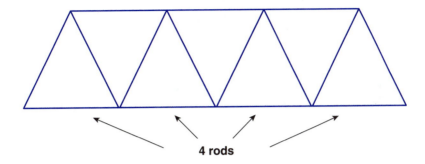

4 rods

How many rods are needed to make a beam of length 5? Of length 10? Of length 20? Of length 34? Of length 76?

How many rods are needed for a beam of length 223?

Write a rule or a formula for how you could find the number of rods needed to make a beam of any length. Explain your rule or formula.

13. Ask a student who has completed an algebra course the following questions:

   a. If you were driving along a mountain road and saw a road sign that said, "7% grade," how would you interpret it?

   b. If asked to measure the steepness of a wheelchair ramp outside the school building, how would you do so?

   c. Examine the graph below showing the amount of water in a tank versus the time elapsed since a hose started to fill it. What does the slope of the line tell you? The *y*-intercept?

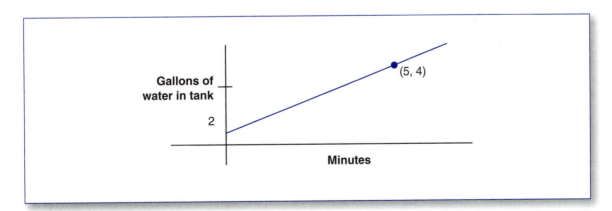

Describe how the student responded to tasks a, b, and c. Based on the responses, draw conclusions about the strength of the student's understanding of slope in functional and physical situations.

14. Examine an algebra textbook to see if and how it defines *function*. Then examine the examples of functions given throughout the text. Drawing upon the examples, make a hypothesis about the types of concept images students studying from the text may form about the concept of function. Then ask a student who has studied from the text to give you a definition for function and three examples. Describe the student's response and comment on the adequacy of his or her concept image for function.

VIGNETTE ANALYSIS ACTIVITY

Focus on Looking for and Expressing Regularity in Repeated Reasoning (CCSS Standard for Mathematical Practice 8)

Items to Consider Before Reading the Vignette

Read CCSS Standard for Mathematical Practice 8 in Appendix A. Then respond to the following items:

1. Drawing on your own experiences as a mathematics student, provide specific examples of teaching practices that can help students look for and express regularity in repeated reasoning.

2. Drawing on your own experiences as a mathematics student, provide specific examples of teaching practices that might prevent students from looking for and expressing regularity in repeated reasoning.

3. Factor $x^2 + 6x + 8$, $x^2 - 5x + 6$, and $x^2 - x - 6$ with algebra tiles.

4. Factor $2x^2 + 5x + 3$ and $x^2 + 3x + 2$ with algebra tiles.

5. Algebra tiles are often called an "area model." Why is this title appropriate?

6. Should students learn how to multiply and factor polynomials with algebra tiles only after learning conventional algorithms for these tasks? Explain your position.

Scenario

Mrs. Neal had just returned from the NCTM annual national conference. During the conference, she had attended several sessions providing ideas for teaching introductory algebra. She was now going through the difficult process of translating the ideas she learned to her own classroom context. She decided to try to master just one major idea at a time rather than overhauling all of her algebra teaching approaches at once. Mrs. Neal was especially intrigued by the algebra tile approach to teaching polynomial multiplication and factoring (described earlier in this chapter). The tiles offered a way to represent work with polynomials visually, so Mrs. Neal thought that some of her more reluctant learners would find the approach appealing. The first day she planned to use algebra tiles, she entered her classroom with a great deal of optimism.

The Lesson

Mrs. Neal had a specific sequence of events in mind for the lesson. At the outset, she wanted to demonstrate how to factor $x^2 + 6x + 8$, $x^2 - 5x + 6$, and $x^2 - x - 6$ with algebra tiles. Students would then factor

$2x^2 + 5x + 3$ and $x^2 + 3x + 2$ on their own for practice. After factoring each one, they would compare results with a partner and share their results with the large group. The homework assignment would be to factor eight more polynomials by drawing diagrams of algebra tiles on grid paper. Some of the homework problems were similar to those done in class (e.g., $x^2 + 6x + 9$). Other homework problems would require much larger pictorial representations (e.g., $x^2 + 15x + 36$). Mrs. Neal did not want to force students to use the algebra tile representation, so those who did not want to draw pictorial representations in class or for the homework would be allowed to use a different method.

To begin the lesson, Mrs. Neal told students the name of each piece in the algebra tile set. She said that factoring a polynomial was done by arranging the pieces representing the polynomial into a rectangle. During her demonstrations at the beginning of the class, students got the idea that factoring was a task something like putting together a jigsaw puzzle. After students had arranged the pieces for a given polynomial into a rectangle, Mrs. Neal would ask, "How many columns are there?" and "How many rows are there?" to lead students to the correct answer. Students were not at any point asked to think of factoring as the task of determining the dimensions of a rectangle with an area specified by a given polynomial. As Mrs. Neal did the initial problems on factoring in front of the classroom, she did not immediately show students how to form the rectangles needed, but instead took suggestions from the class. During class work time, she asked some students to demonstrate their strategies for forming rectangles for different polynomials. When students finished their demonstrations, Mrs. Neal allowed them to return to their seats while she explained the strategies the students used in arranging the tiles.

As students began to do their homework problems independently, Mrs. Neal asked them to check their work by using procedures for polynomial multiplication they had learned earlier in the year. Most students complied, and they judged the correctness of their answers by how well the algebra tile solution matched the solution from the known procedure. Several students began to use the previously learned procedure to solve their homework problems and then use the algebra tiles only after figuring out the answers by hand. The homework exercises were all very similar. In each one, students were given a polynomial and asked to sketch an algebra tile representation and provide its factors. Mrs. Neal required the students to draw their sketches for homework exercises on graph paper, as shown below for factoring $x^2 + 2x + 1$.

As the class session neared the end of its 45-minute time allotment, Mrs. Neal stated, "I hope that today has helped you to visualize factoring so that those of you who have struggled with it in the past feel a little bit better about it." During the lesson, she had, in fact, seen signs that students who had been reluctant to try factoring polynomials in the past were actually engaging in the task. Since these students were able to solve many of the exercises given during the lesson, Mrs. Neal decided that algebra tiles provided a good alternative to other approaches to factoring. Upon completing the lesson, she was anxious to share the algebra tile approach with her colleagues.

Questions for Reflection and Discussion

1. Which aspects of the CCSS Standard for Mathematical Practice 8 did Mrs. Neal's students seem to attain? Explain.

2. Which aspects of the CCSS Standard for Mathematical Practice 8 did Mrs. Neal's students not seem to attain? Explain.

3. What necessary background knowledge must students have in order to understand algebra tile representations of polynomials?

4. Is factoring $x^2 + 15x + 36$ with algebra tiles a good exercise to assign students? Why or why not?

5. What kinds of strategies can be used to help students transition from using algebra tiles to using more abstract strategies for polynomial factoring and multiplication?

6. How could Mrs. Neal have made a better connection between the concept of area and the algebra tile representations for polynomials?

7. Was it a good idea for Mrs. Neal to have her students check their solutions to the homework problems by using a previously learned procedure? Why or why not?

8. What are the advantages and disadvantages of having students draw algebra tile diagrams on graph paper?

RESOURCES TO EXPLORE

Books

Lloyd, G., Beckmann, S., Zbiek, R., & Cooney, T. (2010). *Developing essential understanding of functions for teaching mathematics in Grades 9–12*. Reston, VA: NCTM.

Description: This book is designed to help teachers develop in-depth conceptual understanding of functions. It addresses the function concept, covariation and rate of change, families of functions and modeling, and combining and transforming functions.

Graham, K., Cuoco, A., & Zimmerman, G. (2010). *Focus in high school mathematics: Reasoning and sense-making in algebra*. Reston, VA: NCTM.

Description: This volume supplements NCTM's *Focus in High School Mathematics* by giving detailed attention to reasoning and sense-making in the context of algebra. It provides a discussion of links between algebra and geometry, building equations and functions, and teaching the notation of formal algebra.

Friel, S., Rachlin, S., & Doyle, D. (2001). *Navigating through algebra in Grades 6–8*. Reston, VA: NCTM.

Description: This book provides activities for teaching algebra in an inquiry-oriented manner to middle school students. The activities are designed to help students understand patterns, relations, and functions; analyze change in multiple contexts; explore linear relationships; and use algebraic symbols.

Burke, M., Erickson, D., Lott, J.W., & Obert, M. (2001). *Navigating through algebra in Grades 9–12*. Reston, VA: NCTM.

Description: This book provides activities for teaching algebra in an inquiry-oriented manner to high school students. The activities are designed to help students understand the language of algebra, the notion of variable, function representations, algebraic equivalence, and change.

Greenes, C. E. (Ed.). (2008). *Algebra and algebraic thinking in school mathematics* (Seventieth yearbook of the National Council of Teachers of Mathematics). Reston, VA: NCTM.

Description: The chapters in this yearbook provide a broad perspective on the teaching of algebra, from the early grades to secondary school to teacher education efforts. Chapter authors discuss historical perspectives on algebra in the curriculum, the nature of algebraic thinking, research on the learning of algebra, and algebra classroom activities.

Chazan, D. (2000). *Beyond formulas in mathematics and teaching*. New York, NY: Teachers College Press.

Description: The author presents an autobiographical account of his efforts to teach algebra to diverse students. By reading Chazan's story, the reader has the opportunity to develop a deeper understanding of algebra, students, and classroom dynamics.

Driscoll, M. (1999). *Fostering algebraic thinking: A guide for teachers Grades 6–10*. Portsmouth, NH: Heinemann.

Description: The text describes habits of mind that are characteristic of algebraic thinking. The book is designed to help teachers understand how to help students make the transition to the study of formal algebra.

Website

Insights Into Algebra 1: Teaching for Learning: **http://www.learner.org/resources/series196.html**

Description: This is an eight-part video series designed to help teachers improve the manner in which they teach concepts typically found in middle and high school algebra. Programs in the series deal with linear functions and inequalities, systems of equations and inequalities, quadratic functions, exponential functions, and modeling.

FOUR-COLUMN LESSON PLANS TO HELP DEVELOP STUDENTS' ALGEBRAIC THINKING

Lesson Plan 1

Based on the following NCTM resource: Kukla, D. (2007). Graphing families of curves using transformations of reference graphs. *Mathematics Teacher, 100,* 503–509.

Primary objective: Students will understand transformations of absolute value functions

Materials needed: Page-size pieces of graph paper, poster-size pieces of graph paper, graphing calculators

Steps of the lesson	Expected student responses	Teacher's responses to students	Assessment strategies and goals
1. Ask students to produce a table of values for the function rule $y = |x|$.	Some students may not recall the meaning of the absolute value symbol.	Ask various students to explain the meaning of $|x|$. Try to elicit the idea that it is the distance from 0 to x on a number line.	Provide specific values for x that are positive, negative, and 0. Then ask students who have not spoken yet to compute $|x|$.
2. Ask students to predict the shape of the graph of $y = |x|$ on the page-size pieces of graph paper.	Students may try to predict the shape of $y = |x|$ by thinking first about the graph of $y = x$.	Ask how the graphs of $y = |x|$ and $y = x$ are likely to be similar and different. Prompt students to produce a table of values for $y = x$ if necessary.	Examine students' tables and graphs for $y = |x|$ and $y = x$ to see how they deal with negative values for x.
3. Break the class up into three groups. Each group receives a large piece of graph paper. The first group will be responsible for graphing $y = |x| + 3$, $y = |x| - 3$, and $y = |x| + 6$ on the same sheet of graph paper. The second group will graph $y = |x| + 3$, $y = |x-3|$, and $y = |x+6|$. The third group will graph $y = 3|x|$, $y = -3|x|$, and $y = 6|x|$.	Some groups will notice that their graphs are transformations of the parent function, $y = |x|$. Some students will be able to explain how the constants influence the graphs. Some groups may obtain straight lines rather than V-shaped graphs based on the tables of values they produce.	Students who recognize the transformed graphs and the influence of constants will be asked to write explanations for why the graphs are transformed as they are. Students who produce partial graphs will be prompted to expand their tables of values so they see the graphs' V shapes.	Examine students' written explanations for why the graphs change as they do to assess their understanding of the connections among graphical, symbolic, and tabular representations of absolute value functions.
4. Have each group of students present the graphs they produced to the rest of the class. During each presentation, they should explain how the graphs they made compare to $y = |x|$. They should also explain the reasons for similarities to and differences from $y = |x|$.	Students may begin to make generalizations such as the following: $y = |x| + c$ shifts $y = |x|$ either up or down, depending on the sign of c; $y = |x + c|$ shifts $y = |x|$ left or right, and $y = c|x|$ makes $y = |x|$ either narrower or wider.	Prompt students to put their generalizations in writing. Have each group write down at least one generalization, and emphasize that potential generalizations will be explored further.	Assess the verbal and written generalizations students produce. If any incorrect generalizations are given, let them stand for the time being. In the next step of the lesson, students will further test the generalizations.

5. Distribute graphing calculators. Ask students to produce several of their own examples of functions of the form $y = |x| + c$, $y = |x + c|$, and $y = c|x|$. After graphing several examples with the calculator, they should explain why they think the generalizations given by each group are correct or incorrect.	Some students will choose values of c that make salient features of graphs fall outside the standard viewing window. Some will refer to aspects of symbolic and tabular representations of the functions to justify their generalizations.	Students who have trouble setting the graphing window should be encouraged to look back at the previous examples to choose a useful window. Students should be prompted to go beyond citing specific examples to explain why the generalizations are true.	Collect students' written work from this step as a summative assessment. They may finish during class or at home. Assess whether or not students can explain how and why the change caused by c influences all of the representations of a function.

How This Lesson Meets Quality Control Criteria

- *Addressing students' preconceptions:* This lesson builds on everyday phrases such as stretching, shrinking, and shifting; draws on previous study of function graphs; and uses the metaphor of distance to explain absolute value.
- *Conceptual and procedural knowledge development:* Students are encouraged to discuss the meanings of absolute value symbols and understand connections among various representations.
- *Metacognition:* Students are prompted to look for patterns in the output of their graphing calculators; students are asked to talk and write about their thinking.

Lesson Plan 2

Based on the following NCTM resource: Pace, C. L. (2005). You read me a story, I will read you a pattern. *Mathematics Teaching in the Middle School, 10,* 424–429.

Primary objective: Students will construct explicit rules to describe linear algebraic patterns

Materials needed: Pattern block manipulatives

Steps of the lesson	Expected student responses	Teacher's responses to students	Assessment strategies and goals
1. Show students the following pattern, constructed using equilateral triangles in a set of pattern blocks: Ask them to describe the pattern verbally and predict the perimeter of a row of five triangles.	Some students will count the number of triangle sides rather than the number of segments that produce the perimeter. This will lead some of them to say that the perimeter of the first figure is 3, the perimeter of the second is 6, the perimeter of the third is 9, and so forth.	Direct students' attention toward the outside edge of each figure. Measure out the perimeter of the chalkboard with a meter stick to emphasize the idea that the perimeter of an object consists of iterations of a segment length.	Ask students to explain why the perimeter of the fifth object is 7 rather than 15. This may be a writing exercise or just a quick verbal question directed toward one student, depending on how many students are having difficulty.

(Continued)

(Continued)

Steps of the lesson	Expected student responses	Teacher's responses to students	Assessment strategies and goals
2. Ask students to predict the perimeter of a row of 100 triangles. Have them work with a partner to solve the problem.	Some students will want to try a recursive strategy all the way up to the 100th pattern. Other students will see that the perimeter of a given row equals the number of triangles in it plus two.	Acknowledge students' recursive strategies as valid approaches to the problem, but ask students using recursive strategies to look for "shortcuts," or more efficient ways to solve the problem.	After students work with a partner, ask them to verbally report on their strategies and conclusions to the rest of the class.
3. Ask students to state the perimeter of a row of n triangles.	Some students will arrive at the rule $p = n + 1$ because the perimeter increases by one for each successive object. Some students will arrive at the explicit rule $p = n + 2$ by noticing that the perimeter of a row of triangles equals the number of triangles plus two.	Engage students in discussion of the meanings of the variables n and p, since it is anticipated that students will use n in the two different ways described. Discussion should aim to bring out the different usages.	Encourage students to write the meanings for n and p in their own words and discuss and compare different interpretations. The goal is to move students toward the meanings inherent in the explicit rule.
4. Ask students to use the square pieces in the pattern block set to form the following pattern: Ask them to determine the perimeter of the nth figure if one additional square is added each time.	Students will arrive at the explicit rule $p = 2n + 2$ or $p = 2(n + 1)$. The former comes from doubling the length of the horizontal pieces of variable length and adding the two constant vertical side lengths. The latter comes from looking at the base and height of each figure put together for a total length of $n + 1$ and then doubling it to get the entire perimeter.	As students work on their explicit rules and explanations, select a variety of students to go to the board to present their work. Make sure to select students who can represent each of the various rules that have been formulated.	Ask students to write about how the explicit rules they have constructed compare to the others that have been presented. The goal is to see if they understand that the same explicit rule has a number of equivalent representations.
5. Ask students to compare the explicit rule for the triangle pattern to the explicit rule for the square pattern and write about similarities and differences between the two.	Some students will compare tables of values for the two rules; some will use graphs; some will focus on the symbolic form. Some may focus on two or three of the representations.	As students write about similarities and differences, select a variety of students to go to the board to present their work. Choose students on the basis of the various algebraic representations they have used to compare and contrast the explicit rules.	After various student approaches to the problem have been presented, ask students to describe how the tables, graphs, and equations for the three rules are similar and different. Then ask them to try to formulate a pattern that will yield a nonconstant rate of change.

How This Lesson Meets Quality Control Criteria

- *Addressing students' preconceptions:* This lesson connects to the experiential and school-based task of measuring the perimeter of objects, entails using familiar shapes to build patterns, and prompts students to move beyond recursive reasoning.
- *Conceptual and procedural knowledge development:* This lesson develops the concept of equivalence and foreshadows the concept of nonlinear functions. Students also become familiar with variable notation.
- *Metacognition:* Students compare their strategies to those of others and examine the efficiency of the rules they formulate.

Lesson Plan 3

Based on the following NCTM resource: Hyde, A., George, K., Mynard, S., Hull, C., Watson, S., & Watson, P. (2006). Creating multiple representations in algebra: All chocolate, no change. *Mathematics Teaching in the Middle School, 11,* 262–268.

Primary objective: Students will understand connections among graphical, tabular, and symbolic representations of a function.

Materials needed: Graph paper

Steps of the lesson	Expected student responses	Teacher's responses to students	Assessment strategies and goals
1. Pose the following problem to students and encourage them to ask questions about its meaning: "If you have $10 to spend on $2 Hershey Bars and $1 Tootsie Rolls, how many ways can you spend all your money without receiving change? All chocolate, no change" (Hyde et al., 2006, p. 262).	Some questions about the context and meaning of the problem will arise, such as: (1) How much tax do you have to pay on the candy? (2) Is it OK to buy just one kind of candy, or do you have to buy the same amount of each? (3) Is it possible to buy half of a Hershey Bar or Tootsie Roll?	In response to question 1, clarify that you are working with a somewhat idealized situation involving no sales tax. In response to question 2, redirect students to the question so they can see that you could actually spend all the money on one type of candy if desired. In response to question 3, leave the possibility of working with fractional parts of candy open, but encourage them to work with whole pieces first.	As you respond to questions about the context of the problem, check for understanding. For example, you might ask, "Given that there is no sales tax, how much would it cost to buy three Hershey Bars and two Tootsie Rolls? Does buying three Hershey Bars and two Tootsie Rolls spend all of your money? If you were to buy just Hershey Bars, how many would you get if you spent all of your money? How much would you expect a Tootsie Roll and a half to cost?"
2. Ask students to give possible solutions to the problem.	Students will give a variety of responses, such as, "Four Hershey Bars and two Tootsie Rolls," "Five Hershey Bars and zero Tootsie Rolls," and "Ten Tootsie Rolls and zero Hershey Bars." Responses will not necessarily be given in a systematic, organized manner.	Suggest systematically organizing the responses students contribute into a table that accounts for all possible combinations.	As class discussion takes place, ask students if all possible combinations have been accounted for. Continue asking for additional combinations until all have been identified in the class-constructed table.

(Continued)

(Continued)

Steps of the lesson	Expected student responses	Teacher's responses to students	Assessment strategies and goals
3. Ask students to look for patterns in the table representing the problem (*H* represents number of Hershey Bars, *T* represents number of Tootsie rolls): 	*H*	5	4	3	2	1	0
T	0	2	4	6	8	10		Some will notice that as *H* decreases by 1, *T* increases by 2. Some will notice that there is a point at which *H* is 0 and another where *T* is 0.	As students identify patterns in the table, ask them why those patterns exist. For example, ask students who notice that as *H* decreases by 1, *T* increases by 2 to explain why by referring back to the context of the problem.	Evaluate students' responses to your probing questions for connections back to the context of the problem. Students should be able to say, for example, that, reading from left to right, each jump on the table is analogous to exchanging a Hershey bar for two Tootsie rolls.
4. Ask students to graph the relationship of *H* against *T* in the coordinate plane, and to explain what the graph shows about the relationship between *H* and *T*.	There will be questions about what should go on the horizontal axis and what should go on the vertical. There will be questions about whether or not the points should be connected once they are graphed.	Recommend that the class agree on which axes should be assigned *H* and *T*, emphasizing that this is just for the sake of uniformity. Show a connected graph and ask if noninteger solutions fit the context of the problem.	During class discussion, assess whether or not students understand that noninteger solutions do not fit the context of the original problem. Ask students how the tabular pattern of *H* decreasing by 1 and *T* increasing by 2 is shown in the graph.
5. Ask students to write a number sentence or equation that describes the relationship between *H* and *T*.	Students will decide that a general form for the number sentence would be (money spent on Tootsie Rolls) + (money spent on Hershey Bars) = (total budget of $10)	Ask students to express their general form in terms of *H* and *T*. They should use the relationship shown in the table and graph to obtain $2H + T = 10$.	As a homework assignment, ask students to write a word problem about a budget and two items to buy. They are then to use tables, graphs, and equations to describe it.

How This Lesson Meets Quality Control Criteria

- *Addressing students' preconceptions:* Connections to experiential knowledge are made by connecting mathematics to the context of buying items. The context of the problem is thoroughly discussed at the outset to avoid having the context interfere with mathematical problem solving. The problem can initially be approached with multiple strategies learned in previous schooling.
- *Conceptual and procedural knowledge development:* Making connections among function representations contributes to conceptual understanding. Producing tables and graphs requires students to draw on procedural, computational knowledge.
- *Metacognition:* Students are asked to examine the tables they produce to see if all possibilities have been taken into account.

Lesson Plan 4

Based on the following NCTM resource: King, S. L. (2002). Sharing teaching ideas: Function notation. *Mathematics Teacher, 95,* 636–639.

Primary objective: Students will understand the meaning of function notation and the concept of function.

Materials needed: Handheld computer algebra system for each student

Steps of the lesson	Expected student responses	Teacher's responses to students	Assessment strategies and goals
1. Begin the lesson by writing the word *function* on the board and then ask students to write a personal definition of the word. After they write for a few minutes, ask them to share their definitions with the class.	Students will write and share a variety of definitions for the term, some from previous school experiences and others from out-of-school experiences. Some, for example, may think of bodily functions; others may recall the term from previous classes.	Emphasize the mathematical sense of the word *function* by using the metaphor of a machine that takes inputs, performs the same operations on each one, and then produces an output. Each input produces at most one output.	Ask students to write a sentence to explain the difference between everyday definitions of the word *function* and the mathematical definition. When they share these in class, see if they distinguish between the two.
2. Distribute handheld CAS units to students. Demonstrate how to enter $f(x) = 3x - 5$; have students enter the same on their calculators. Before going any further with the CAS, ask students to predict the result when the CAS is used to produce the output of $f(0)$, $f(5)$, and $f(10)$.	Many students will interpret $f(0)$, $f(5)$, and $f(10)$ as representing "f times 0," "f times 5," and "f times 10" because of their previous experiences with literal symbols, parentheses, and adjacent symbols in algebra.	After students write their predictions down and share them with the rest of the class, ask them to go ahead and have the CAS compute $f(0)$, $f(5)$, and $f(10)$. Then have them write a sentence to explain how their predicted results compared to the CAS output.	Circulate about the room to examine students' written comparison of the initial conjecture to the results of the CAS output to see which students, if any, still hold on to the conception that $f(x)$ represents "f times x" in this context, and to see which students write in terms of input and output of a function.
3. Relate students' work on the CAS back to the function machine metaphor introduced earlier in class, comparing x to the input and $f(x)$ to the output. Then ask students to predict CAS outputs for $x = 0$, 5, and −3 for the function $g(x) = 4x^2 - 1$.	Some students will not immediately recognize that g can be used in the same manner as f when writing an equation for a function. Other students will immediately transfer their knowledge from the previous function situation. Some students will make procedural mistakes with order of operations and squaring a negative number.	Remind students that the choice of literal symbols in algebra is often arbitrary, and that there is nothing to prevent g from serving the same role as f in this situation. Let students know it is often convenient to label functions with different letters in situations where two or more functions are to be compared or discussed. After they have made their predictions, ask students to have the CAS compute $g(0)$, $g(5)$, and $g(−3)$.	Circulate about the room to find students who obtain results from the CAS that do not align with their original predictions. Determine the source of the discrepancy between the prediction and the CAS output. When possible, have a student who received the predicted output discuss the problem with a student who did not receive the predicted output.

(Continued)

(Continued)

4. Ask students to predict CAS outputs for $x = 0, 5,$ and -3 for the function $h(x) = |x| - 2$.	Some students will have difficulty determining the absolute value of -3 or interpreting the absolute value symbol.	After students have made their predictions, ask them to have the CAS compute $h(0)$, $h(5)$, and $h(-3)$.	Use the same assessment strategy/goal as in step 3 of this lesson.
5. Ask students to predict CAS outputs for $x = 0, 5,$ and -3 for the function $k(x) = 4$.	Most students will expect to get various outputs, just as they did with the previous functions used in the lesson.	After students have made their predictions, ask them to have the CAS compute $k(0)$, $k(5)$, and $k(-3)$.	Use the same assessment strategy/goal as in step 3 of this lesson. For homework, have students invent five of their own functions that can be used to challenge a classmate the next day.

How This Lesson Meets Quality Control Criteria

- *Addressing students' preconceptions:* Students' out-of-school experiences with the word *function* are compared to the mathematical meaning at the beginning of the lesson. Interference from previous school experiences with multiplication notation is addressed.
- *Conceptual and procedural knowledge development:* The meaning of function is developed alongside notational conventions for representing functions.
- *Metacognition:* Students are asked to compare their predictions for inputs against CAS output and reconcile differences.

Chapter 9

DEVELOPING STUDENTS' STATISTICAL AND PROBABILISTIC THINKING

Statistics and probability are quickly rising to more prominent roles in school curricula. Once reserved primarily for college, they have now become integral parts of students' studies in Grades K through 12 (Friel, 2008). Their rise to prominence began in earnest when *Curriculum and Evaluation Standards for School Mathematics* (National Council of Teachers of Mathematics [NCTM], 1989) introduced a statistics and probability standard. The NCTM (2000) *Principles and Standards for School Mathematics* supplemented the *Curriculum and Evaluation Standards* with a data analysis and probability standard outlining suitable content for Grades pre-K through 2, 3 through 5, 6 through 8, and 9 through 12. Following the release of the NCTM standards, individual states built data analysis and probability content into their own curriculum standards (Lott & Nishimura, 2005). To help supplement, extend, and flesh out NCTM's vision of developing students' understanding of statistics and probability, the American Statistical Association produced *Guidelines for Assessment and Instruction in Statistics Education* (GAISE; Franklin et al., 2007). The *Common Core State Standards* (National Governor's Association for Best Practices & Council of Chief State School Officers, 2010) also put high priority on the study of statistics and probability in Grades 6 through 12. This increase in attention to curriculum standards for statistics and probability underscores the idea that those subjects are no longer to be pushed to the curricular background.

WHAT ARE STATISTICAL AND PROBABILISTIC THINKING?

Amid all of the curricular activity surrounding statistics and probability, it can be surprisingly easy to lose sight of some fundamental questions: What are statistical and probabilistic thinking? How are they related to one another? Surprisingly, statisticians and educators are really just in the beginning stages of grappling with these questions. Since one's answers to these questions will influence instructional goals, it is important to consider current thinking about the meanings of the phrases *statistical thinking* and *probabilistic thinking*.

Statistical Thinking Elements

Statistical thinking is really just one of three terms often used to describe important aspects of statistical cognition. Two other aspects of statistical cognition are **statistical literacy** and **statistical reasoning**. Ben-Zvi and Garfield (2004) noted a lack of uniformity in the use of the three terms. They are often used interchangeably by statisticians and educators. Adding further complexity to the issue, those who do agree on definitions for the terms acknowledge that they are not distinct categories. There is a degree of overlap among them. Despite such difficulties in definition, Ben-Zvi and Garfield argued that it is important to discuss the meanings of the three terms. In considering the meaning of each term, important learning goals for students are brought to light.

Statistical Literacy

Ben-Zvi and Garfield (2004) described statistical literacy as including "basic and important skills that may be used in understanding statistical information or research results" (p. 7). Their characterization of statistical literacy emphasizes being a critical consumer of the statistics one is bombarded with in the media and other sources each day. J. M. Watson (2000) described three **tiers of statistical literacy** that are relevant to interpreting statistics critically:

1. A basic understanding of statistical terminology

2. An understanding of statistical language and concepts when they are embedded in the context of wider societal discussion

3. A questioning attitude that can apply more sophisticated concepts to contradict claims made without proper statistical foundation (p. 54)

Each tier builds on the one before it. That is, a grasp of tier 1 is necessary to function at tier 2, and a grasp of tier 2 is necessary to function effectively at tier 3. Tiers 1 and 2 are generally targeted by school curricula. Tier 3, however, is sometimes not addressed as explicitly. Teachers need to critically examine their own curricula to ensure that tier 3 is addressed.

J. M. Watson (2000) illustrated the three tiers of statistical literacy by describing students' interpretations of a newspaper article. The article stated that a cause-and-effect relationship existed between the use of motor vehicles and heart disease deaths. The headline claimed, "The family car is killing us" (p. 55). Students were asked to respond to several questions about the article, each representing competencies from one of the three tiers of statistical literacy. Tier 1 competencies included the ability to name and produce graphs illustrating cause-and-effect claims such as the one made in the newspaper. Students functioning at tier 2 could explain how graphical representations applied to comparison of the number of heart disease deaths and the number of motor vehicles on the road. At tier 3, students could go beyond representing the situation mathematically and question the claim that automobiles cause heart disease. These students identified factors aside from cars that contribute to the breakdown of the environment and hence may cause heart disease. They also identified societal trends that had occurred alongside the increase in automobiles, such as worse dieting habits and increased stress levels. These tier 3 ideas run counter to the common misconception, often presented in society, that correlation implies causation.

Statistical Thinking

Ben-Zvi and Garfield (2004) described statistical thinking as "an understanding of how and why statistical investigations are conducted and the 'big ideas' that underlie statistical investigations" (p. 7). The GAISE report endorses the idea that the best way to develop statistical thinking is to be immersed in the **process of statistical investigation**. The components of the statistical investigation process outlined in GAISE are (1) formulating questions, (2) collecting data, (3) analyzing data, and (4) interpreting results. Although it may be natural in some investigations to proceed in order from component 1 to component 4, there are many instances in which this does not occur. For instance, after collecting data on the effectiveness of a drug (component 2), researchers may notice during analysis (component 3) that the drug impacts males and females differently. This may motivate formulation of a new question (representing a move back to component 1) about whether or not there are gender differences relevant to the drug's effectiveness. The dynamic, iterative nature of the process of statistical investigation is best understood when students engage in formulating some of their own questions for investigation and collect data themselves in order to answer them.

> **Implementing the Common Core**
>
> See Clinical Task 1 to assess students' abilities to "evaluate reports based on data" (Content Standard S-IC.6) when reading an article from the news media.

> **Implementing the Common Core**
>
> See Homework Task 2 to examine how much attention is given to the process of statistical investigation in the *Common Core.*

> In your past experiences as a student of statistics, have your teachers encouraged you to engage in the process of statistical investigation? How is teaching through the process of statistical investigation different from other approaches to teaching statistics?
>
> **STOP TO REFLECT**

Examining how statisticians conduct statistical investigations can shed additional light on important elements of statistical thinking. By interviewing practicing statisticians, Wild and Pfannkuch (1999) identified competencies that are important in the process of statistical investigation. The specific types of thinking exhibited by statisticians include recognizing the need for data to answer quantitative questions, explaining and dealing with variation that occurs when collecting data, using statistical models, and making sense of statistics within different contexts for data collection. Statisticians must also be able to imagine possibilities for different plans of attack in generating study designs, seek information and ideas to address gaps in the knowledge they bring to an investigation, and decide what to believe and discard when judging the results of investigations. Effective statisticians also possess character traits such as skepticism, curiosity, openness to ideas that challenge their preconceptions, and perseverance.

Statistical Reasoning

Ben-Zvi and Garfield (2004) defined statistical reasoning as "the way people reason with statistical ideas and make sense of statistical information" (p. 7). This can include interpreting data, statistical displays, and summary statistics. It can also involve making connections among concepts, recognizing how data and probability relate to one another, and explaining statistical processes. An instrument called the Statistical Reasoning Assessment (SRA; Garfield, 2003) contains several examples of items that require statistical reasoning, including those shown in Figure 9.1.

| **Figure 9.1** | Excerpt from the Statistical Reasoning Assessment (SRA; Garfield, 2003, p. 36). |

9. Which of the following sequences is most likely to result from flipping a fair coin 5 times?
   a. HHHTT
   b. THHTH
   c. THTTT
   d. HTHTH
   e. All four sequences are equally likely

10. Select one or more explanations for the answer you gave for the item above.
   a. Since the coin is fair, you ought to get roughly equal numbers of heads and tails.
   b. Since coin flipping is random, the coin ought to alternate frequently between landing heads and tails.
   c. Any of the sequences could occur.
   d. If you repeatedly flipped a coin five times, each of these sequences would occur about as often as any other sequence.
   e. If you get a couple of heads in a row, the probability of tails on the next flip increases.
   f. Every sequence of five flips has exactly the same probability of occurring.

11. Listed below are the same sequences of Hs and Ts that were listed in item 9. Which of the sequences is *least* likely to result from flipping a fair coin 5 times?
   a. HHHTT
   b. THHTH
   c. THTTT
   d. HTHTH
   e. All four sequences are equally likely

Implementing the Common Core

See Clinical Task 2 to investigate students' correct and incorrect reasoning in tasks that require them to "understand and evaluate random processes underlying statistical experiments" (Content Standard S-IC).

Implementing the Common Core

See Homework Task 3 to compare and contrast statistical and mathematical reasoning in terms of what it takes to "make sense of problems and persevere in solving them" (Standard for Mathematical Practice 8).

The sample SRA items in Figure 9.1 require making connections between data and probability. Given a probabilistic situation, students are to predict the likelihood of generating different sets of data. Although all four sequences of coin flips are equally likely, students may hold misconceptions about the situation presented in the sample items. Common reasoning patterns that interfere with coming to the correct conclusion are represented by choices a, b, c, and e in item 10. Throughout this chapter, we continue to explore prevalent cognitive patterns that help and impede the learning of statistics and probability.

Before moving on, it is worth noting that statistical reasoning is not precisely the same as mathematical reasoning. Use of data and context differ in mathematics and statistics (delMas, 2004). In mathematics, data and context are sometimes used to help illustrate or bring out a mathematics concept. However, in mathematics problems we generally seek to strip away contextual details to see mathematical patterns. On the other hand, a frequent goal in statistical data analysis is to understand the interaction between data and context. This may involve a degree of shuttling back and forth between data and context (Wild & Pfannkuch, 1999). Analyzing data grounded in specific contexts is necessary for this sort of reasoning to occur.

Is statistics a branch of mathematics, or is statistics a discipline in its own right? How might a teacher's statistics instruction be influenced by his or her beliefs about the relationship between statistics and mathematics?

STOP TO REFLECT

Elements of Probabilistic Thinking

In many cases, it is difficult to separate probabilistic thinking from statistical literacy, thinking, and reasoning. For example, knowledge of probability is needed to solve the statistical reasoning tasks shown in Figure 9.1. Knowledge of probability helps students determine if a given data set is particularly unusual or not. Probability models form the basis of formal statistical inference (which will be discussed in the Advanced Placement statistics section of Chapter 11). Therefore, rather than trying to draw a sharp distinction between statistical and probabilistic cognitions, this section will explain three conceptions of probability that were not explicitly mentioned in the discussion of statistical cognition: frequentist, classical, and subjective. These conceptions are implicit in curricular recommendations for teaching probability in several countries (G. A. Jones, Langrall, & Mooney, 2007).

Frequentist Probability

Frequentist probability is sometimes referred to as experimental or a posteriori probability. Under this approach, probability can be defined as "the ratio of the number of trials favorable to the event to the total number of trials" (G. A. Jones et al., 2007, p. 912). The frequentist approach has been influential in school curricula in the recent past. Although not explicitly mentioned in most curriculum documents and textbooks, its influence can be seen in the increased use of classroom probability simulations. In such simulations, students may be given a scenario like the one shown in Figure 9.2.

In approaching the task shown in Figure 9.2, students may make initial estimates of how many boxes they would expect to buy to obtain all three cards. Teachers can encourage students to share their initial conjectures and compare them with one another. Then the class can systematically collect data to estimate the probability of obtaining the cards in different numbers of cereal box purchases. For instance, some students may estimate that they need to buy just three boxes of cereal to obtain all of the cards. They can reflect on the accuracy of their estimation by carrying out a probability simulation with a six-sided die. Rolling a one, two, or three could represent

Figure 9.2 Superhero card probability simulation task.

A cereal company is giving away superhero cards in each box of cereal. There are three superhero cards: Spiderman, Batman, and Superman. The probability of getting a Spiderman card in any given box is $\frac{1}{2}$, the probability of getting a Batman card in any given box is $\frac{1}{6}$, and the probability of getting a Superman card in any given box is $\frac{1}{3}$. How many boxes do you think you would have to buy to obtain all three superhero cards?

Implementing the Common Core

See Clinical Task 3 to assess students' ability to "design and use a simulation to generate frequencies for compound events" (Content Standard 7.SP.8.C) when working the problem shown in Figure 9.2.

obtaining a box with a Spiderman card, since there is a 50% chance of getting such a card in any given box that is purchased. Similarly, rolling a four could represent obtaining a Batman card, and rolling a five or six could represent Superman. As the dice are rolled, students can keep track of their results in a tabular or graphical display. The class might then pool their results to get a better picture of how accurate their initial estimates were. Such an activity reflects a frequentist approach to probability because it asks students to use data generated during simulation to estimate the likelihood of different outcomes. The more trials conducted, the more accurate the estimates are likely to be.

Classical Probability

Classical probability is sometimes referred to as theoretical or a priori probability. Under this approach, probability is defined as "the ratio of the number of favorable outcomes to the total number of outcomes, where outcomes are assumed to be equally likely" (G. A. Jones et al., 2007, p. 912). In some cases, like the task in Figure 9.2, theoretical probability can be challenging to determine. In other cases, it is impossible to do so. An example of a problem that can be approached theoretically, however, is shown in Figure 9.3.

Figure 9.3 Coin flipping task.

Suppose you are playing a carnival game that involves flipping two balanced coins simultaneously. To win the game, you must obtain "heads" on both coins. What is your probability of winning the game?

Implementing the Common Core

See Clinical Task 4 to assess students' ability to "find probabilities of compound events using organized lists, tables, tree diagrams, and simulation" (Content Standard 7.SP.8) when working the problem shown in Figure 9.3.

The theoretical probability for the task shown in Figure 9.3 can be computed using Figure 9.4. It shows that the total number of outcomes is four, and that there is just one favorable outcome. Hence, the theoretical probability of winning the game is one-fourth. Although determining the theoretical probability in this case may seem straightforward, students often have considerable difficulty doing so. Approximately 50% of 12th-grade students believe there is a 50% chance of winning the type of game described in Figure 9.3 (Zawojewski & Shaughnessy, 2000). To convince students that such ideas need to be revised, a frequentist approach can be useful. As students simulate many trials of the game and see that they win approximately 25% of the time, they may become convinced that there really is not a 50% chance of winning. This can motivate a more careful study of the theoretical chances of winning. There are many similar situations where frequentist and classical approaches to measuring probability can interact productively with one another. From a curricular point of view, the two approaches can be considered complementary rather than in opposition.

Figure 9.4   Display for solving the coin flipping task.

	Heads	Tails
Heads	HH (you win)	TH (you lose)
Tails	HT (you lose)	TT (you lose)

Provide your own example of a situation where using both frequentist and classical approaches to probability can lead to deeper student understanding.

STOP TO REFLECT

Subjective Probability

While frequentist and classical approaches to probability are well entrenched in school curricula, the subjective approach is not yet as influential. **Subjective probability** "describes probability as a degree of belief, based on personal judgment and information about an outcome" (G. A. Jones et al., 2007, p. 913). Subjective probability is important because there are many situations in which frequentist and classical approaches are not helpful.

Consider, for example, estimating the probability that a given team will win a football game. In practice, the frequentist approach is not sufficient because a game is a onetime event that is not repeated over and over again. The theoretical approach is also insufficient, since the dynamics of a football game are not as cleanly defined as they are in a situation like the coin flipping game shown in Figure 9.3. Accurate estimates of a team's chances of winning a game usually rest on well-informed subjective judgment based on an individual's knowledge of football and the teams and players involved. Such estimates can vary widely, and not all estimates are of equal value.

At times, flawed probabilistic thinking processes can be revealed when students give their rationale for subjective estimates. For example, when asked about a team's chances of winning a football game, some students estimate it to be one-third based on the fact that there are three possible outcomes: a win, a loss, or a tie (Albert, 2006). This underscores the importance of eliciting subjective probability estimates from students and probing for justification. As errant probabilistic thinking processes are detected, they can be addressed in subsequent instruction.

The increased attention to statistics and probability in curriculum documents has helped to motivate a great deal of research on students' learning of the subjects. Having a sense of what the research says about the nature of students' statistical and probabilistic learning can help teachers design instruction that is responsive to students' needs. Therefore, the bulk of this chapter examines salient research findings about students' statistical literacy, thinking, and reasoning as well as their probabilistic thinking. Discussion of the research findings is organized under three main topics: statistical study design, understanding and using descriptive tools in statistics, and making probabilistic inferences and judgments.

Statistical Study Design

The GAISE document emphasized the importance of immersing students in all phases of statistical investigation, from question formulation to interpretation of results. Unfortunately, issues of statistical study design are often underemphasized in school curricula. Shaughnessy (2007) cautioned against this widespread underemphasis, stating,

> If students are given only prepackaged statistics problems, in which the tough decisions of problem formulation, design, and data production have already been made for them, they will encounter an impoverished, three-phase investigative cycle and will be ill-equipped to deal with statistics problems in their early formulation stages. (p. 963)

Hence, to help students develop their knowledge of statistics to the fullest extent possible, it is important to pose problems that require engagement with all phases of statistical investigation.

Formulating Questions

Formulating a question for statistical investigation may seem, on the surface, to be a fairly straightforward enterprise, but research shows its many intricacies. Heaton and Mickelson (2002) identified some of the intricacies in their research on a group of prospective teachers doing statistical investigations. The prospective teachers were to design a statistical investigation related to the teaching practices in their practicum schools. The assignment was open ended in that they were to choose a classroom aspect of interest and then formulate a question that could be addressed by using quantitative measures. Various topics were chosen, such as looking at the amount of wait time teachers allowed before and after answering questions, tracking how often teachers used the chalkboard, and investigating how much time teachers spent with students. Heaton and Mickelson observed that one weakness in the questions was that they were relatively uninteresting "How much" and "How many" questions that would not yield much insight about teaching. Posing an interesting and fruitful question thus emerged as one of the challenges individuals may face in formulating statistical questions. A second challenge that emerged involved posing quantifiable questions. Some of the questions formulated by the prospective teachers could not be addressed with statistical measures. Questions falling into this category included, "What is the classroom like?" and "Why are particular activities chosen to do with children?" Heaton and Mickelson's findings emphasize students' need for assistance in formulating interesting, quantifiable questions when given the task of designing their own statistical investigations.

In forming interesting, quantifiable questions, students also need to understand that the choice of question heavily influences the rest of the phases of the investigation. Heaton and Mickelson (2002) found that even when prospective teachers formed promising questions, the nature of the question made data collection difficult. For example, one prospective teacher in their study chose to focus on how many students raised their hands in response to different types of teacher questions. However, upon entering the classroom in an attempt to collect data, it was discovered that students raised and lowered their hands much too quickly for one to obtain an accurate count. Groth and Powell (2004) described how the questions posed by high school students complicated the data collection and analysis phases of the investigative cycle. One student was interested in comparing the distances walked each day by students in different grade levels. He quickly discovered several ways that the device he planned to use to record steps per day could report inaccurate data. A pair of students in the same class wanted to take random samples of students in the school to compare the blood pressure of athletes to those of nonathletes. They learned about the difficulty of obtaining truly random samples in practice, since some teachers would not let students out of class to have their blood pressures taken. From one perspective, these might be considered failed projects, since students did not anticipate how difficult it would be to gather data to answer the questions they posed. However, these are more productively viewed as rich learning opportunities, since students gained firsthand experience of the messiness and complexity inherent in carrying out statistical investigations from beginning to end.

Drawing Samples

Many statistical questions require drawing samples. Because of this, a fair amount of research has focused on students' abilities to draw representative samples from populations of interest. Students have sound intuitions about some sampling situations. Nearly two-thirds of students responding to a question on the National Assessment of Educational Progress (NAEP) were able to identify the location within a school building most suitable for drawing a representative sample (Zawojewski & Shaughnessy, 2000). They identified the school cafeteria as a more suitable place for sampling than other locations such as the faculty room, an algebra class, or the guidance office. Additionally, many middle school students are able to find problems with the quality of samples and detect potential bias when given school-based sampling situations (V. R. Jacobs, 1999). Therefore, research suggests that sampling is an accessible topic of study for middle and high school students.

Although sound intuitions about sampling exist among middle and high school students, problematic beliefs and notions exist as well. Some students believe participant self-selection is a good way to obtain a representative sample because "everyone has a chance to participate" (V. R. Jacobs, 1999). Students may also place too much or too little confidence in random samples (Rubin, Bruce, & Tenney, 1991). Students placing too much confidence in samples tend to believe that even small random samples should be perfectly representative of the populations from which they are drawn. Students placing too little confidence in samples may want to survey the entire population, even when it would not be feasible to do so. Teachers need to help students navigate a middle ground between the belief that samples tell us nothing about a population and the belief that samples tell us everything about it.

To help teachers anticipate the types of ideas about sampling students may bring to the classroom, J. M. Watson and Moritz (2000) described six prevalent categories of thinking. Categories 1 through 3 included students who tended to trust small samples, categories 4 and 5 comprised students who trusted large samples, and category 6 consisted of students who expressed no consistent preference about sample size. Category 1 students were called "small samplers without selection." Like other students in categories 1 through 3, students in category 1 thought of samples as being a

Implementing the Common Core

See Clinical Task 5 to use the J. M. Watson and Moritz (2000) categories to assess a student's ability to "use random sampling to draw inferences about a population" (Content Standard 7.SP) and to "make inferences and justify conclusions from sample surveys" (Content Standard S-IC).

small bit of something, like a sample in a grocery store. Students in categories 1 through 3 were content with sample sizes of less than 15 when larger samples would have been appropriate. Students in the first category, however, did not suggest methods of sampling from a population when asked to do so, or else they gave idiosyncratic methods. Category 2 students, "small samplers with primitive random selection," differed from category 1 students in that they felt it was important to get a random sample from a population, though they gave no methods for doing so. Category 3 students, "small samplers with preselection of results," suggested selecting study participants by looking at their personal characteristics, not recognizing the bias that would bring to the sampling process.

Categories 4 and 5 in the J. M. Watson and Moritz (2000) study described the thinking of students who were "large samplers." Large samplers preferred samples of size 20 or greater, or samples composing a given portion of the population. Category 4 students were "large samplers with random or distributed selection." Category 5 students, "large samplers sensitive to bias," gave examples of a broad range of situations in which samples would be needed, including surveys. Category 4 students sometimes used the word *average* in describing samples, while category 5 students used both *average* and *representative* in reference to samples. Students in both categories 4 and 5 suggested drawing samples using a random process or geographic distribution of participants. Category 5 students, however, expressed concerns about bias in the selection of samples, and they were able to identify biased samples in newspaper articles reporting the results of surveys.

Category 6 students in the J. M. Watson and Moritz (2000) study were called "equivocal samplers." They, like students in other categories, were able to provide examples and descriptions of samples. However, unlike other students, they were sometimes indifferent about sample size. When they depended on small samples, they were often able to use appropriate selection methods or show some sensitivity to bias. In cases where they relied on large samples, they sometimes used inappropriate selection methods. Category 6 illustrates that even though sample size is an important characteristic for many students in judging the validity of a sample, it is not a determining factor for all students.

Given the diversity of categories of thinking that students exhibit, it is natural to wonder how teachers might help students develop sound ideas about sampling. Schwartz, Goldman, Vye, and Barron (1998) proposed having students solve sampling problems for situations set in various contexts. They noted that context heavily influences students' solutions. For example, students are often comfortable with random sampling when drawing marbles, but not when conducting an opinion survey. Students may worry that when drawing a random sample to take a survey, not everyone is given a chance to participate. They might also express preference for a voting or census model rather than random selection to determine who responds to a survey. As ideas like these arise, teachers can direct students' attention toward similarities between the contexts of drawing marbles and conducting surveys. This can be done in large-group discussions about students' survey sampling methods or through written feedback on individual student papers. When students see similarities in random sampling between drawing marbles and conducting surveys, they can begin to view random sampling in research situations appropriately.

Random sampling is, of course, just one weapon in the arsenal of a well-armed statistician. Since it is one of the most common statistical methods in middle and high school curricula, it has been given the most attention in this chapter. Looking ahead, additional methods and considerations for empirical study design are discussed in Chapter 11, which includes an overview of Advanced Placement (AP) statistics. Various additional data gathering strategies, such as experimental design, are integral to the AP course.

UNDERSTANDING AND USING DESCRIPTIVE STATISTICAL TOOLS

While statistical study design is underemphasized in many school curricula, there is usually no lack of emphasis on descriptive statistical tools. In this chapter, the phrase *descriptive statistical tools* refers to data displays and summary statistics. Students' thinking about data displays will be discussed first, and then attention will shift to their thinking about averages and other summary statistics. Although the study of descriptive statistical tools is pervasive, they are not always taught effectively. Research illuminates characteristics of students' thinking that may prompt reconsideration of the manner in which descriptive statistical tools are ordinarily taught.

Data Displays

Friel, Curcio, and Bright (2001) identified four factors that influence students' graph comprehension: purposes for using graphs, task characteristics, disciplinary characteristics, and student characteristics. Each factor will be explored below to help characterize students' thinking about data displays.

Factor 1 in Friel and colleagues' (2001) framework relates to the reasons graphs are used. Graphs can be used, for example, as exploratory devices early in data analysis, or they may communicate the results of a completed investigation. In school settings, however, graph construction is often seen as an end in itself. That is, students are often asked to construct a display not for the sake of making sense of a set of data or communicating findings, but simply to complete the task of constructing the display. Roth and McGinn (1997) warned against producing graphs just for the sake of producing graphs. They recommended that classroom discussions consist of having students use graphs to convey arguments about data they have collected. During such discussions, students should be encouraged to use a variety of data representations to convince others that they have drawn sound conclusions. This may require tasks such as rescaling axes and constructing different representations for the same data set. Discussions like these have two distinct benefits: Students gain better understanding of the characteristics of data displays, and they come to see graphing as a useful tool in statistical discourse. By avoiding tasks that require students to produce graphs as ends in themselves and encouraging more meaningful activities, teachers can help build students' graph comprehension.

Factor 2 in Friel and colleagues' (2001) framework also relates to the characteristics of tasks involving graphs. Two such characteristics are visual decoding and making judgments from data. Students' responses to **visual decoding tasks** (i.e., directly reading data from a graph or table) are generally easier to assess than **judgment tasks** (i.e., drawing plausible conclusions about a situation through interpretation of associated data). Visual decoding tasks can be marked "right" or "wrong" unambiguously, but grading students' judgments can involve gray areas. Students generally do much better on the literal reading of values on a graph than on tasks that require making inferences (Zawojewski & Shaughnessy, 2000). Given the importance of statistical literacy for all students, judgment tasks should not be avoided just because they are more difficult for students to perform and for teachers to assess. Most of the data displays encountered outside of school require some form of judgment, so helping students make sound judgments is an important instructional goal.

STOP TO REFLECT	Write an example of a visual decoding task you could give to students. Also write an example of a judgment task you could give to students. What are the important differences between the two items you wrote? Which task is more likely to help build students' conceptual understanding? Why?

Gal (1998) offered helpful advice for posing and assessing judgment tasks. He recommended not giving specific hints about where to look in data displays on tasks requiring interpretation of tables and graphs. Questions should also be worded to make it clear that an opinion or judgment is called for rather than a specific number. For instance, instead of asking, "What is the mean salary?" for a set of data shown in a graph, one could ask, "What is the typical salary?" and ask students to explain their choices. Judgment tasks also elicit the most information from students when they are set in familiar contexts and involve questions students want or need to answer. Using the case of typical salary to illustrate this point, one might give students data displays showing salaries for summer jobs at several different businesses. Their task could then be to decide which job to take, based on their characterizations of the typical salary at each business. Setting judgment problems in familiar, motivating contexts helps ensure that students use the full power of their reasoning.

Once students' opinions and judgments have been elicited, assessment can take several factors into account. Some things to look for in students' opinions include the following:

The quality of the reasoning on which they are based

The reasonableness of any arguments presented

The nature of and the relevance of the evidence/data used in their creation or justification

The adequacy of methods employed to generate, process, or analyze the evidence/data (Gal, 1998, p. 278)

Gal (1998) recommended constructing rubrics for opinion tasks to facilitate the assessment process. Highest scores on the rubric may be awarded for responses that

are reasonable in light of the given data, refer to relevant and sufficient evidence, make correct use of technical terms, appropriately reference statistical indices . . . , consider (where relevant) issues of variation and the reliability of data, [and] make sensible assumptions about the sources of the data and about the context in which the problem is embedded and in which data originated. (Gal, 1998, pp. 290–291)

Of course, additional criteria may be appropriate for different types of judgment tasks.

Factor 3 in Friel and colleagues' (2001) framework relates to disciplinary characteristics. The discipline of statistics involves measuring variation, displaying different types of data, and using graphs of varying degrees of complexity. Data displays that show each individual data value, such as stem and leaf plots, line graphs, and tables, are generally easier for students to read than those that condense the data, such as boxplots (Zawojewski & Shaughnessy, 2000).

Implementing the Common Core

See Homework Task 4 to use Gal's (1998) criteria to construct a rubric to assess students' responses to a task for which they must "make inferences and justify conclusions from sample surveys" (Content Standard S-IC).

Figure 9.5 illustrates the difference between data displays that show all data values and those that condense the data. It shows several different displays for the same set of test score data. The table in Figure 9.5 simply shows each of the individual data values. The dotplot next to the table also shows each individual value, but in graphical form. The top histogram is a bit more complicated to read, because it shows how many scores fall into each interval rather than each individual value. The bottom histogram uses a bin width of 10, providing a much different picture. For example, it is not entirely clear what the maximum and minimum values in the data set are in the histogram that uses a bin width of 10. The boxplot clearly displays the maximum and minimum values, but it does not show the individual values within each quartile. Each display fills a distinct role and presents different levels of reading difficulty.

Factor 4 shifts attention from the characteristics of data displays and the discipline of statistics to the characteristics of graph readers. The graph reader's mathematical background and familiarity with contexts in which graphs are set both influence graph comprehension. Mooney (2002) observed that some students exhibit **statistical intuition** when required to perform graph construction and interpretation tasks. Statistically intuitive students are "able to construct representations by balancing quantitative and contextual aspects of the data and by determining the reasonableness

Figure 9.5   Data displays presenting different levels of reading difficulty.

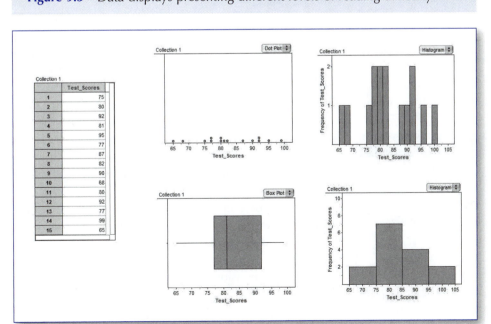

of inferences" (Mooney, 2002, p. 49). For example, one student in Mooney's study used both the data at hand and the context of the situation to extend a partial line graph showing the temperature in a given city over a period of several hours. In the process of data analysis, students with statistical intuition might also consider multiple representations and discuss their effectiveness for displaying data for a particular purpose. Hence, developing students' statistical intuition requires prompting them to consider both mathematical and contextual aspects of situations requiring data analysis.

Although knowledge of context can play a helpful role in data interpretation, it can also be problematic. Batenero, Estepa, Godino, and Green (1996) illustrated this point with tasks that required students to interpret two-way tables. One of the two-way tables they asked students to interpret is shown in Figure 9.6.

Figure 9.6 Bronchial disease and smoking (Batanero, Estepa, Godino, & Green, 1996, p. 168).

In a medical center 250 people have been observed in order to determine whether the habit of smoking has some relationship with bronchial disease. The following results have been obtained:

	Bronchial disease	No bronchial disease	Total
Smoke	90	60	150
Not smoke	60	40	100
Total	150	100	250

Using the information contained in this table, would you think that, for this sample of people, bronchial disease depends on smoking? Explain your answer.

When presented with the bronchial disease task and data, 17- to 18-year-old students gave several different types of responses. Some completely disregarded the data and opted instead to rely on previous theories about the dependence of illnesses on smoking. This sort of response reflects a somewhat extreme focus on context. Other students based their judgment on the fact that there were more smokers than nonsmokers with bronchial disease in the sample. Although this is true, this response overlooks the fact that the percentage of smokers in the sample with bronchial disease is the same as the percentage of nonsmokers with the illness. Still others based their judgment on examining just one cell of the table (e.g., observing a high number of smokers with the disease). The different responses to the task illustrate how contextual knowledge and proportional reasoning can interact to influence interpretations of data.

From the preceding discussion of factors that influence graph comprehension, several implications for teaching can be drawn. The first implication pertains to sequencing.

Implementing the Common Core

See Clinical Task 6 to construct a rubric suitable for grading students' responses to the problem shown in Figure 9.6, which requires one to "understand independence and conditional probability and use them to interpret data" (Content Standard S-CP).

As seen in the contingency table task in Figure 9.6, proportional reasoning plays an important role in drawing inferences from certain types of data displays. Given this phenomenon, Friel and colleagues (2001) recommended delaying the introduction of some data displays until students develop proportional reasoning. Proportional reasoning generally develops in middle and high school, although there are many students who still reason additively in those grade levels (recall the discussion in Chapter 7). There are several displays, in addition to contingency tables, that require proportional reasoning. Two of the most notable are boxplots and histograms. Since these displays condense the data, they can be unintelligible to students who reason strictly in terms of raw frequencies. To move students toward understanding condensed displays, it can be helpful to compare and contrast them with noncondensed displays.

TECHNOLOGY CONNECTION

Dynamic Statistics Software

Dynamic statistics software can facilitate comparison of condensed and noncondensed displays, because it allows students to efficiently produce and manipulate multiple representations of data. Figure 9.5, for example, was produced with the software program Fathom. After students have produced the displays shown in Figure 9.5, they can highlight different values in the table to see their positions in each data display, as shown in Figure 9.7. Red areas on the screen represent the highlighted data value in the table. Students can click through the entire table, and corresponding portions of each data display will

Figure 9.7   Using Fathom to compare different representations of the same set of data.

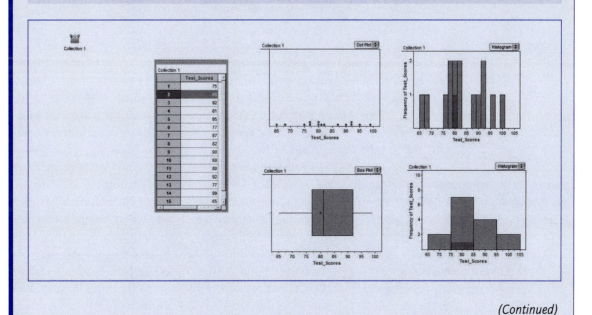

(Continued)

(Continued)

appear in red. Students can make displays like this very quickly in Fathom, literally within a minute or two, because it is a drag-and-drop environment. This allows students more time to focus on the characteristics of graphs and the different ways they display data without becoming bogged down in tedious calculations and point plotting.

Follow-up questions:

1. Should students be required to know how to produce boxplots and histograms by hand before using dynamic statistics software? Why or why not?

2. The two histograms shown in Figure 9.7 display the same set of data but look dramatically different. Why?

Transnumeration

Implementing the Common Core

See Homework Tasks 5 and 7 to begin to think about how to help students "use appropriate tools strategically" (Standard for Mathematical Practice 5) when dynamic statistics software is available.

Producing multiple representations of data, as shown in Figure 9.7, is an essential part of a process that has been called **transnumeration**. Shaughnessy and Pfannkuch (2002) identified three aspects that characterize transnumeration-type thinking:

- capturing measures of the real system that are relevant,
- constructing multiple statistical representations of the real system, and
- communicating to others what the statistical system suggests about the real system. (p. 256)

The first and third bullet points are important to heed when having students work in dynamic data environments. Recall the arguments earlier in this chapter about avoiding tasks that involve producing data displays just for the sake of producing them. When students make decisions about what to measure in a given situation and then communicate their results to others, there is an inherent purpose for the graphs they produce. It is important not to prescribe what sort of graph should be drawn to facilitate statistical representation. Students should have opportunities to construct and compare various displays and determine what each reveals about the data at hand. When Shaughnessy and Pfannkuch (2002) allowed students to construct their own graphs to determine the typical wait time between eruptions of the "Old Faithful" geyser, students discovered that plotting consecutive wait times revealed an oscillating pattern that was not shown in histogram and boxplot representations. The usefulness of the oscillating graph came to light as students communicated their results with one another.

Implementing the Common Core

See Homework Task 6 for a problem that requires the reader to "summarize numerical data sets in relation to their context" (Content Standard 6.SP) and to "summarize, represent, and interpret data on a single count or measurement variable" (Content Standard S-ID).

TECHNOLOGY CONNECTION

Informal Representations With Dynamic Statistics Software

When students engage in transnumeration activities, they should not be limited to producing conventional data displays. Sometimes nonconventional displays make data analysis more accessible to students. For example, scatterplots are conventional representations for displaying bivariate data. However, they can be difficult for students to understand initially. Konold (2002) gave examples of alternative representations students may use. One alternative way to represent bivariate data is to produce ordered case value bars, which depict ordered pairs from bivariate data sets as bar lengths placed in order according to the first value of each ordered pair (Figure 9.8). This type of representation emerged as a natural way for middle school

Figure 9.8   Ordered case value bars representation for bivariate data.

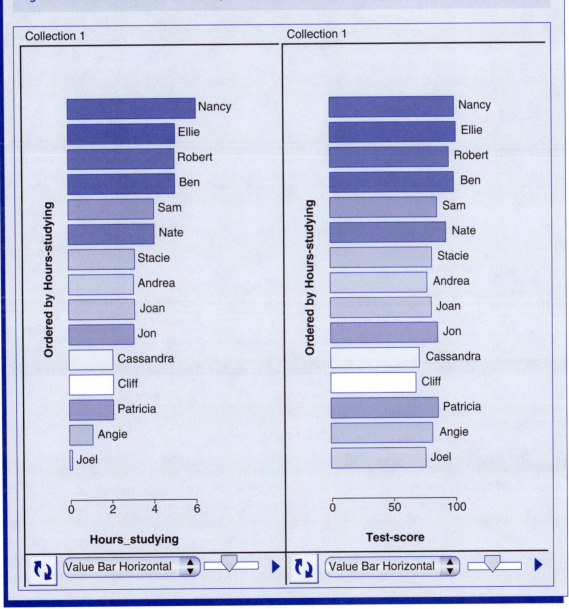

students to look for relationships between variables in a classroom where they were encouraged to construct their own representations for a data set and report the results (Cobb, 1999). Ordered case value bars can be constructed using the informal dynamic statistics software package TinkerPlots.

An alternative representation for bivariate data is a **sliced scatterplot**. As its name implies, this representation takes a conventional scatterplot and partitions it into sections. The mean *y*-value of the data points within each vertical slice may be computed to provide a better picture of the overall trend of the data. Noss, Pozzi, and Hoyles (1999) found that the sliced scatterplot representation helped nurses see trends in data showing the relationship between blood pressure and age. Part of the reason for the sliced scatterplot's effectiveness in revealing trends in data is that it allows students to focus on one section of the graph at a time rather than leading them to try to take in the entire data display immediately. A sample sliced scatterplot is shown in Figure 9.9. This nonconventional data representation can also be produced with TinkerPlots.

Figure 9.9   Sliced scatterplot representation for bivariate data.

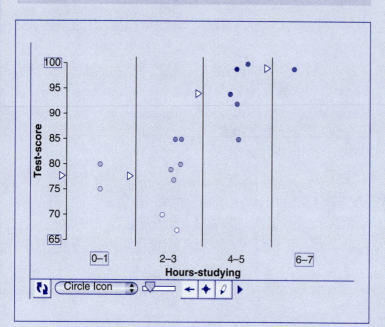

Follow-up questions:

1. How are ordered case value bars similar to scatterplots? How are they different?

2. Would it be a good use of instructional time to have students produce the data displays shown in Figures 9.8 and 9.9 by hand? Why or why not?

Averages

Implementing the Common Core

See Clinical Task 7 to observe how case value bars, sliced scatterplots, and scatterplots can help students "investigate patterns of association in bivariate data" (Content Standard 8.SP).

As with research on the teaching and learning of data displays, one of the primary messages from research on averages and other descriptive statistics is that students should use descriptive statistics with understanding. Research shows that students have little difficulty computing descriptive statistics (Tarr & Shaughnessy, 2007) but have considerable difficulty understanding what they mean (Mokros & Russell, 1995).

An implication for teaching is to avoid situations where students are asked to compute the value of a descriptive statistic merely for the sake of computing it. Instead, students should learn about descriptive statistics in contexts where they are naturally useful.

Because averages (in the broad sense—e.g., mean, median, and mode) are some of the most widely used descriptive statistics, students' thinking regarding them has received a great deal of research attention. Konold and Pollatsek (2002) described four interpretations of average that appear in mathematics curricula: average as a data reducer, as a fair share, as a typical value, and as a signal in noise. The four interpretations frame the discussion that follows.

Average as a data reducer comes into play in situations where it helps condense a set of numbers down to a single value, as in the following problem:

> For a class party, Ruth bought 5 pieces of candy, Yael bought 10 pieces, Nadav bought 20, and Ami bought 25. Can you tell me in one number how many pieces of candy each child bought? How did you decide on that number? (Strauss & Bichler, 1988, p. 70)

Reducing the key element of the situation to a single number removes the burden of recalling all of the individual values in the data set. At the same time, there is a danger of oversimplifying the situation. Although averages are often convenient summary statistics, in many situations it is also desirable to look at other statistics, such as the range, maximum value, minimum value, and interquartile range, to obtain a more complete portrait of the data. In data sets with extreme values, one might also consider using the median as the single value that best characterizes the data set rather than the mean, since the median is resistant to outliers.

Another interpretation of the concept of average is that of **average as a fair share**. This interpretation applies specifically to the arithmetic mean. When thought of as a fair share, the arithmetic mean describes the amount of a given quantity shared among a given number of individuals. The fair share interpretation has a unique etymological connection with the word *average* itself. Rubenstein and Schwartz (2000) explained that the word comes from the Arabic term *awariyah*. *Awariyah* was the word used to describe the losses to be shared by investors when shipments were damaged or lost at sea. For instance, if 1,000 units of currency were lost from a shipment paid for by 10 investors, each investor's fair share of the loss would be 100 units. The word became part of the Italian language as *avaria* and later became *average* in English. Hence, the fair share interpretation is one of the earliest contexts for the concept of average.

The fair share interpretation of average also lends itself well to visual representation. The left side of Figure 9.10 shows a graph representing the number of cookies that five different people have. The right side shows a graph representing the result of distributing the cookies evenly among the five people. It shows that the fair share is four cookies per person.

The mean number, or fair share, can be determined simply by moving the squares in the graph on the left side of Figure 9.10 around until every person has the same number of cookies. Such activities form the basis for understanding the add-and-divide algorithm for the mean. Determining the total number of cookies is analogous to adding up all of the data values, and distributing them evenly is analogous to dividing by the number of values in the set. In the situation shown in Figure 9.10, the fair share is a whole number. Connections to rational numbers can be made by changing the

Figure 9.10   Representation of cookie sharing scenario.

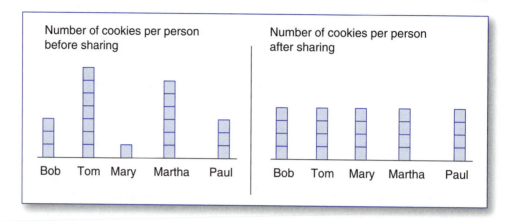

total number of cookies slightly. For example, one could start with 21 cookies and then ask students how they would split the leftover cookie among the five individuals. Grounding the add-and-divide algorithm in a meaningful context such as this is important because many students, including those in secondary school who have seen the algorithm many times, view it as just another procedure to be carried out and have little understanding of it (Mokros & Russell, 1995).

Another interpretation of the mean commonly found in curriculum materials is **average as a typical value**. Typical value problems involve selecting statistical measures to concisely describe the characteristic values of data sets. Figure 9.11 shows a typical value problem like those often given on the NAEP.

Figure 9.11   Typical value problem of the kind often given on the NAEP.

The table below shows the daily attendance at two bowling alleys for 5 days and the mean (average) and median attendance.

	Alley 1	Alley 2
Day 1	99	70
Day 2	89	99
Day 3	92	68
Day 4	13	73
Day 5	88	101
Mean (average)	76.2	82.2
Median	89	73

a) Which statistic, the mean or the median, would you use to describe the typical daily attendance for the 5 days at Alley 1? Justify your answer.

b) Which statistic, the mean or the median, would you use to describe the typical daily attendance for the 5 days at Alley 2? Justify your answer.

On one administration of the NAEP, when given a problem like the one shown in Figure 9.11, only 4% of 12th graders provided responses that were scored at the highest possible level on the NAEP rubric (Zawojewski & Shaughnessy, 2000). Scoring at the highest possible level entailed identifying an outlier and making a statement about the impact of the outlier on the mean. Students tended to prefer the mean for both sets of data, regardless of the distribution. Zawojewski and Shaughnessy (2000) conjectured that one of the reasons for preferring the mean is that it is emphasized more heavily in traditional curricula. In this case, the characteristics of the task itself may also have influenced students' responses. The mean was given the label "average," implying that the median is not a form of average. In addition, some students may have chosen the same statistic for each bowling alley to facilitate comparison between the two, even though no comparison is explicitly requested in the task.

Although typical value, fair share, and data reduction interpretations of the mean are fairly prevalent in curriculum materials, viewing **average as a signal in noise** is not (Konold & Pollatsek, 2002). The signal-in-noise interpretation characterizes average as a way to make sense of data generated by a noisy process that has a distinct signal to be identified. Repeated measurement problems help illustrate some of the characteristics of the signal-in-noise interpretation. Such problems might involve weighing an object several times and obtaining a slightly different weight each time. The average then provides a way to cut through variation due to measurement error to obtain an accurate estimate of the actual weight of the object.

> **Implementing the Common Core**
>
> See Clinical Task 8 to understand how students "draw informal comparative inferences about two populations" (Content Standard 7.SP) when working the problem shown in Figure 9.11.

> **Implementing the Common Core**
>
> See Clinical Tasks 9 and 10 to investigate how a textbook addresses various conceptual and procedural characteristics of averages (Content Standards 6.SP, 7.SP, and S-ID).

IDEA FOR DIFFERENTIATING INSTRUCTION   Repeated Measures Task

Encouraging students to invent their own strategies for repeated measures tasks can reveal many different levels of thinking. One such task, accessible to students with various statistical backgrounds, involves asking for an opinion about the true weight of a fish weighed several times on the same scale. Seven different scale outputs (in pounds) were obtained: 19.2, 21.5, 10.1, 23.1, 22.0, 20.3, and 21.8 (Groth, 2005). The task is meant to prompt students to consider how the measurement of 10.1 might have occurred. When left to invent their own strategies, some students disregard the 10.1 as an extreme error and take the average of the rest. Others use the informal strategy of spot checking the data to obtain a reasonable estimate for the weight of the fish. Still others simply take the average of all the measurements, including 10.1, producing an unreasonably low estimate. As teachers encourage students to share these types of strategies in class, it can be emphasized that informal strategies for data analysis are sometimes more powerful than formal ones. When students do use formal strategies for data analysis, they should be asked to check their estimates against the data and their context. Doing so can help them avoid blind application of algorithms and unreasonable conclusions.

Implementing the Common Core

See Homework Task 8 to analyze students' responses to the fish weighing task, which requires them to demonstrate "understanding of statistical variability" (Content Standard 6.SP) and to account for "possible effects of extreme data points" (Content Standard S-ID).

Descriptive Statistics Beyond Averages

Although averages receive a great deal of curricular attention, it should not be concluded that they are the only types of statistics students should use to analyze and compare distributions of data. It is also important for students to attend to the shape and spread of data. Each of these characteristics can be of special interest, depending on the situation. For example, if one is interested in how well a test distinguishes among different individuals' ability, it may be helpful to look at the spread of the data. If scores in various groups are clustered tightly around a single value, one might judge the test to be less effective than if scores are spread over a wide range. Hence, using averages to analyze data and compare groups is not always desirable. An average should be seen as just one characteristic of a distribution alongside other characteristics such as shape and spread (Biehler, 1994).

In examining shape, center, and spread of distributions, informal strategies can be quite effective. For example, if data sets being compared have the same number of values, then comparing the totals for each set can be just as effective as computing a conventional measure of center (J. M. Watson & Moritz, 1999). Visual comparisons of central clumps or clusters in the data can also be viable in some cases (Cobb, 1999). Visually comparing data sets before selecting formal tools can help foster the good habit of examining data to see if it is appropriate to apply formal inferential techniques such as t-tests (J. M. Watson & Moritz, 1999). (Formal inferential techniques are discussed more extensively in the AP Statistics section of Chapter 11). Encouraging students to examine the different characteristics of a distribution has the added benefit of helping them see the data as an aggregate rather than just a collection of individual points. The aggregate view does not develop automatically and can require a substantial number of experiences with exploratory data analysis to emerge (Cobb, 1999). In short, formal statistics such as the mean, the median, and the mode are quite useful when it comes to comparing data sets, but students should also be encouraged to use other summary statistics and informal strategies in examining data and comparing groups.

Procedural Knowledge Issues

Up to this point in the chapter, the conceptual side of understanding summary statistics has been emphasized, but not the procedural. Part of the reason for this focus is that conceptual aspects are generally more difficult for students than computational aspects (Zawojewski & Shaughnessy, 2000). Nonetheless, research has documented some procedural aspects that cause difficulty for students as well. These include executing the algorithm for the mean in reverse, computing weighted averages, and deciding what to do with zero values in a data set when computing an average. Each is discussed in the following.

Some situations require calculation of the mean in reverse. For instance, a student with one test left to take in the semester may be trying to obtain a certain final average semester grade. In such a problem, the mean score is known and one of the data points is unknown. Cai (2000) gave another example of a problem where this type of use of the mean was required. The problem stated that a student was selling hats for a club at school over a four-week period. In the first week, nine hats sold, in the second week three hats sold, and in the third week six hats sold. The problem was to determine how many hats the student had to sell during the fourth week so that the average number of hats sold per week would be seven. Cai gave the problem to middle school students in the United States and in China. Most of the U.S. students did the problem incorrectly. One prevalent incorrect strategy was to average the first three weeks of hat sales and give the result as the answer. Another prevalent incorrect strategy was to add 3 to the total number of hats sold during the first three weeks, so that when the total number of hats was divided by 3, the result would be 7. Chinese students fared better on the task, primarily because they were more likely to approach the situation algebraically, setting up and solving the equation $\frac{9+3+6+x}{4} = 7$ for x. As a result of these findings, Cai recommended that curricula be structured to give students opportunities to work with algebraic representations for the mean in the middle grades. Doing so can help students apply the algorithm more flexibly when necessary.

Another problematic procedural issue involving the mean is computation of weighted averages. Pollatsek, Lima, and Well (1981) interviewed college students to obtain responses to the task shown in Figure 9.12.

When initially approaching the task, most of the college students interviewed simply added 3.22 and 3.78 and divided by 2, even though twice as much time was spent at college B. When students gave this sort of response, Pollatsek and colleagues (1981) probed further by asking, "Now, suppose the student had spent one semester at college A and seven semesters at college B. Then what would his GPA be for all his college work?" (p. 196). Most students did not change their responses even after being presented with the more extreme situation. Some students recognized the need to revise their estimate, but felt they would need GPA information from each semester to do so.

One suggestion Pollatsek and colleagues (1981) gave for helping students realize that a simple mean is not adequate for situations like the GPA problem is to use a **mean as balance point** metaphor. With such a metaphor, the mean is analogous to a scale fulcrum that is closer to some of the data values than others. Figure 9.13 shows a balance scale diagram for the GPA problem.

Figure 9.12 Weighted average problem (Pollatsek, Lima, & Well, 1981, p. 195).

A student attended college A for two semesters and earned a 3.22 GPA. The same student attended college B for four semesters and earned a 3.78 GPA. What is the student's GPA for all his college work?

Figure 9.13 Balance scale diagram for GPA problem.

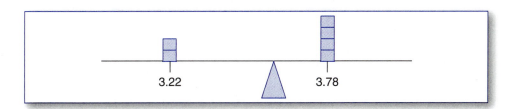

Although the balance scale diagram shown in Figure 9.13 is not sufficient in and of itself to determine the weighted average, it does provide an opportunity to highlight some essential features of the problem. Each small rectangle in the diagram represents one semester's work (assuming the semesters contained equal loads of coursework). The fulcrum, indicated by the triangle, represents the mean. If one thinks of the mean in terms of a point of balance, it follows that it must be closer to 3.78 than to 3.22 to balance the scale, since there is more weight on 3.78. Pollatsek and colleagues (1981) conjectured that the balance scale metaphor for the mean could help prevent students from making gross errors when computing weighted averages, since the diagram provides a way to reflect on the reasonableness of answers.

STOP TO REFLECT

Explain why it makes sense, mathematically, to think of the mean as the balance point for a set of data.

Another procedural issue to consider in regard to the mean involves students' thinking with data sets that include values of zero. Strauss and Bichler (1988) found that students may not believe that zero should be included in the calculation of the mean if it occurs in a data set. Some students tried to justify their exclusion of zero computationally, providing rationales such as "You don't have to take the zero into account because when you add or subtract, you don't add or subtract the zero" (p. 77). Chazan (2000) provided further insight about why students may think it appropriate to exclude zero values when computing the mean. He gave high school students the task of determining the average bonus from a data set that included values of zero. Students did not think that the zero values would be appropriate to include in the calculation, because a zero bonus indicated that no bonus was received at all. Given this caveat, they felt it would be more accurate to take into account only the bonus amounts greater than zero. This instance illustrates the impact that the context of a statistics problem can have on students' solution strategies.

One possible cause of students' procedural difficulties with descriptive statistics is the language teachers use to describe them. Groth and Bergner (2006) found imprecision in the manner that some prospective teachers talked about the median and the mode. The median was frequently referred to as "the middle number," without any discussion of ordering the data values from greatest to least. Even when there was discussion of ordering the data, the use of the phrase "the middle number" provided no information about the course of action to take for a data set with an even number of values. The vast majority of prospective teachers in the study referred to the mode as "the most frequent number." One substantial problem with such a definition is that the mode is often not a number at all. In fact, the mode is most useful when describing categorical data. For instance, if one took a survey on the favorite colors of students in a class, the mode would not be a number, but rather a category, such as "red." Hence, one way for teachers to eliminate some of students' procedural difficulties with descriptive statistics is to be mindful of language use.

MAKING PROBABILISTIC INFERENCES

Although some statistical tasks students encounter can be approached strictly with descriptive tools such as graphs and summary statistics, others require probabilistic thinking. Such tasks often present substantive challenges, since probability can be one of the most difficult subjects to understand. Researchers have documented many common reasoning heuristics and misconceptions that lead students astray. A sound understanding of probability is needed for tasks such as making inferences from a sample back to a population. Understanding random variation that occurs as samples are drawn is necessary for drawing meaningful inferences. This section discusses probability reasoning heuristics, misconceptions about probability, and aspects of students' understanding of variation.

Misconceptions About Conditional Probability

Situations involving conditional probabilities often elicit student misconceptions. One such reasoning pattern is the **Falk phenomenon**, named after the researcher who documented it (Falk, 1986). It occurs when students believe it does not make sense to compute the probability of an event when the outcome of a second, related follow-up event is given (see Figure 9.14, part b).

Students generally have little problem figuring out that the probability requested in part *a* of the problem in Figure 9.14 is $\frac{1}{3}$. Since it is given that a green marble has been drawn, one green marble remains in the urn along with two red ones. However, when asked to respond to part b, students usually react quite differently. Some believe that it does not even make sense to talk about the situation, thinking that the outcome of the first draw must be known in order to make a statement about the second draw. Others believe that the answer to part b is $\frac{1}{2}$, thinking it appropriate to look only at the numbers of marbles in the jar at the outset of the experiment, since the result of the first draw is unknown. Whereas part a is straightforward because it is easy to visualize the events occurring in sequence, part b requires a sort of backward inference.

A demonstration with physical materials can help students overcome the Falk phenomenon. Give students an opaque jar and tell them that it contains two green and two red marbles. Have them take a marble out of the jar at random and place it in their pocket without looking at it. Next, students should draw another marble out of the jar at random and look at its color. Suppose the marble is green. At this point, students can be asked to determine the probability that the first marble drawn (the marble in

Figure 9.14   Problem that elicits the Falk phenomenon.

A jar contains two red marbles and two green marbles. Two marbles are drawn at random, one after the other, without being placed back into the jar. What is the probability that

a. the second marble drawn is green, given that the first marble drawn was green?
b. the first marble drawn is green, given that the second marble drawn was green?

the pocket) was a green one. Since the second draw removed one green marble from the collection, there is a probability of $\frac{1}{3}$ that the marble in the pocket is green. This is equivalent to saying that the probability of obtaining a green marble on the first draw, given the information that a green marble was drawn on the second, is $\frac{1}{3}$.

Another common misconception about conditionals is called **confusion of the inverse** (Falk, 1986). This is essentially the belief that, in all cases, $P(A|B) = P(B|A)$. Confusion of the inverse occurs frequently in the context of interpreting medical tests. It is often believed that the probability of an illness given a positive test result is the same as the probability of a positive result given the presence of the illness. In reality, these two probabilities are often quite different. Consider a test that catches 99% of all cases of an illness. Suppose that 10% of all patients who receive the test actually have the illness. Further, suppose we know that when the illness is not present, there is still a 1% chance of a positive test result (i.e., a "false positive"). Figure 9.15 shows the data one would expect to generate after giving the test to 1,000 patients.

For the hypothetical medical test under consideration, the probability of having the illness given a positive test result is 99/108, or about 92%. The presence of false positives makes this probability less than 99%. On the other hand, the probability of a positive test result given that the illness is present is 99/100. If the probability of a false positive were higher, the differences between the two probabilities would diverge even more.

Figure 9.15 Results of a hypothetical medical test.

	Positive test result	Negative test result	TOTAL
Illness present	99	1	100
No illness present	9	891	900
TOTAL	108	892	1000

IDEA FOR DIFFERENTIATING INSTRUCTION   Parallel Probability Tasks

Attending carefully to the phrasing of student tasks can help counteract common misconceptions. J. M. Watson and Moritz (2002), for instance, found that middle school students were more successful with conditional probability tasks when asked for frequencies rather than probabilities. They compared students' performance on the two tasks shown in Figure 9.16.

When answering Q3 in Figure 9.16, students were less likely to say that both events had the same probability (i.e., they were less likely to exhibit confusion of the inverse). This suggests that teachers should design lessons where students reason from data to make sense of probability statements (Shaughnessy,

Figure 9.16 Conditional probability tasks in different formats (J. M. Watson & Moritz, 2002, p. 66).

Q3. Please estimate:

(a) Out of 100 men, how many are left-handed.
(b) Out of 100 left-handed adults, how many are men.

Q4. Please estimate:

(a) The probability that a woman is a school teacher.
(b) The probability that a school teacher is a woman.

2003). Such data-based approaches to probability hold promise for making the subject more understandable. The two tasks shown in Figure 9.16 could be used as parallel tasks in classrooms where students differ in their ability to work with probability statements. The task framed in terms of frequencies deals with the same content as the other task, yet provides a more accessible entry point to the content for some individuals. Teachers can help students see connections between the two tasks by leading discussion between groups of students who are assigned each one.

Additional Common Reasoning Heuristics and Misconceptions

Like situations involving conditional probabilities, situations involving compound probabilities can be challenging for students. One particularly salient difficulty is associated with the **conjunction fallacy** (Kahneman & Tversky, 1983). Students exhibiting the conjunction fallacy believe that conjunctive probabilities are greater than individual probabilities when they are not (i.e., thinking that $P(A \cap B) < P(A)$ when $P(A \cap B) < P(A)$). As an illustration of the conjunction fallacy, consider the personality description presented in Figure 9.17.

Figure 9.17   Profile eliciting the conjunction fallacy (Kahneman & Tversky, 1983, p. 297).

Linda is 31 years old, single, outspoken, and very bright. She majored in philosophy. As a student, she was deeply concerned with issues of discrimination and social justice, and also participated in anti-nuclear demonstrations.

When given Linda's profile (Figure 9.17), 85% of college students in a sample studied by Kahneman and Tversky believed that the statement "Linda is a bank teller and is active in the feminist movement" was more probable than "Linda is a bank teller."

Given the scenario shown in Figure 9.17, explain why the probability that "Linda is a bank teller and is active in the feminist movement" is actually less likely than the statement "Linda is a bank teller." Use a table or diagram (e.g., a Venn diagram) to support your explanation. Repeat for the scenarios and statements shown in Figure 9.18.

STOP TO REFLECT

To help students overcome the conjunction fallacy and other probability misconceptions, Fast (1997) recommended the use of analogies. He administered a written test designed to elicit common probability misconceptions. He then administered a parallel test, containing analogous items, to help students detect errors in reasoning they may have made on the first test. To illustrate this approach for the conjunction fallacy, consider the two items shown in Figure 9.18. Item 1 appeared on the first version of the test, and item 2 on the second. Item 1 in Figure 9.18 is similar to the "Linda" item in Figure 9.17. Since a vivid picture is painted of Tom's personality,

Figure 9.18 Analogous items for the conjunction fallacy (Fast, 1997, pp. 341, 342).

Item 1: Tom lives in the Yukon. He drives long distances, often on snow-covered roads and rough or muddy trails. Sometimes he must cross shallow rivers and climb steep hills with his vehicle. Which of the two statements below has greater chance of being true?

   a) Tom drives a truck
   b) Tom drives a 4-wheel drive truck

Item 2: You are driving down a highway when you are passed by a car. Which of the two statements below has a greater chance of being true?

   a) The car is red
   b) The car is a red convertible

students are likely to say that event b is more likely than event a. Item 2 is mathematically parallel, but psychologically different. The item does not paint a portrait of a particular individual, but asks students to make a judgment about a situation they may approach more objectively. Comparing the mathematical principles of item 2 to those of item 1 can help students overcome the conjunction fallacy.

Closely related to the conjunction fallacy is the **availability heuristic** (Tversky & Kahneman, 1974). Individuals using the availability heuristic to reason about a situation overestimate the likelihood of a given event because of their personal experiences. For example, someone with several friends who have been in car accidents may overestimate his or her own chance of being in an accident. The psychological pull of the immediacy of the experience with car accidents trumps data from the larger population that would place a lower probability on a given individual's being in an accident. It has been noted that although the availability heuristic leads to inaccurate mathematical judgments, it is not necessarily bad to use when reasoning about a situation (Shaughnessy & Bergman, 1993). Someone who overestimates his or her likelihood of being in a car accident, for example, may be more cautious when driving. Though an inaccurate mathematical judgment may be made, one could argue that it is actually beneficial to employ the availability heuristic in such cases.

STOP TO REFLECT

Give at least three examples of situations where employing the availability heuristic may lead to an inaccurate mathematical judgment, yet may be beneficial to the individual employing it. Be sure to explain why the availability heuristic applies to each situation.

Another error in probabilistic judgment has been called the **outcome approach** (Konold, 1995). Students exhibiting the outcome approach interpret a probability task in terms of whether or not something will happen rather than as a statement about its likelihood. For example, Konold reported on college students' interpretation of the statement that there was a 70% chance of rain on a given day. Over 40% of the students felt the statement would be very accurate if it rained between 95% and 100% of the time the statement was made. Students giving such a response seemed to believe that a 70% chance of rain should be interpreted to mean it is almost certain to rain that day. The "70%" appeared to convey no useful quantitative information about the situation to them. Only 32% of students surveyed selected the interval 65% to 74%, which contained the 70% in the original statement. Students using the outcome approach seem to believe that in probabilistic situations, it is one's task to say whether or not a given event will occur rather than to quantify its likelihood of occurrence.

Implementing the Common Core

See Clinical tasks 11 and 12 to investigate students' ability to understand "conditional probability and the rules of probability" (Content Standard S-CP) and design teaching interventions to address misconceptions.

When judging the probability of various events, some students believe that all random events are equally likely. This belief has been called the **equiprobability bias** (LeCoutre, 1992). Many events that involve chance evoke this bias. For example, Groth (2008) studied the conversation that occurred among prospective teachers regarding a problem requiring judgment about who should be chosen to take the last shot in a basketball game. One player had a shooting percentage of 50%, and the other had a shooting percentage of 70%. An additional caveat was that the 50% shooter had made more of his recent shots than the 70% shooter. Some of the prospective teachers participating in the discussion expressed the opinion that it did not matter who took the last shot, since the selected player would either make or miss it. Such a statement assigns an equal probability to making or missing the shot. Equiprobability bias resembles the outcome approach in that students disregard pertinent quantitative information.

Some probability heuristics relate to what students expect to obtain when drawing a sample from a population. One of these is the **representativeness heuristic** (Tversky & Kahneman, 1974). Students exhibiting the representativeness heuristic may expect that even very small samples will represent the population from which they were drawn. An example of a problem likely to elicit the representativeness heuristic is shown in Figure 9.19.

Figure 9.19   A version of the hospital problem (Merseth, 2003, p. 69).

A town has two hospitals. On the average, there are 45 babies delivered each day in the larger hospital. The smaller hospital has about 15 births each day. Fifty percent of all babies born in the town are boys. In one year each hospital recorded those days in which the number of boys born was 60% or more of the total deliveries for that day in that hospital. Do you think it's more likely that the larger hospital recorded more such days than the smaller hospital or that the two recorded roughly the same number of such days?

When Groth (2006a) asked prospective teachers to solve the hospital problem shown in Figure 9.19, the vast majority of them used the representativeness heuristic. A typical response was

I believe that the number of days the number of boys was 60% or greater is equal in the larger and smaller hospital because if you take 60% for each hospital, they each come out to be equal, percentage-wise. 9 to 15 is 27 to 45. (p. 59)

Implementing the Common Core

See Clinical Task 13 to design a rubric for assessing students' responses to the hospital problem (Figure 9.19), which requires "making inferences and justifying conclusions" (Content Standard S-IC).

Such a response shows a grasp of proportional reasoning but also reflects a belief in the **law of small numbers** (Tversky & Kahneman, 1974). The law of small numbers, in contrast to the law of large numbers, is the belief that sample size is not likely to influence the accuracy with which a sample statistic reflects a given characteristic of the population. Although proportional reasoning is an important part of understanding probability and statistics, it is not sufficient. It must be accompanied by knowledge of variation, a topic to be taken up next.

Understanding Random Variation

Variation is a core statistical idea that occurs in many different contexts. Examples from earlier in this chapter include repeated measurement problems and the hospital problem. Recently, an increasing amount of research has been done to explore students' understanding of variation. Shaughnessy (2007) described a series of tasks involving sampling candy from a bowl that yield insight about students' understanding of variation. Though the candy task can take on a number of different forms, the basic premise is that one is taking a handful of candies, at random, from a bowl. The distribution of colors in the bowl is known. For example, in one version of the task, students are told that a bowl has 100 wrapped candies in it. Given that 20 are yellow, 50 are red, and 30 are blue, students are asked to predict the number of red candies when several random samples are drawn (with replacement). When given this task, some students respond that the number of reds drawn should always be very close to 50%, reflecting a belief that samples are always representative of the population. Some even express the belief that every sample drawn should have exactly 50% reds, an idea reminiscent of the representativeness heuristic. On the other end of the spectrum, some believe that the number of reds will vary widely from sample to sample, since the process is random.

Shaughnessy, Ciancetta, and Canada (2004) provided a general framework describing how middle and high school students tend to respond to the candy task. The framework consists of three types of reasoning: additive, proportional, and distributional. Additive reasoners relied on the absolute number, or frequencies, in the mixture, expecting more reds just because there were more reds in the population. Proportional reasoners expected the sample to have about 60% reds because the population did. Distributional reasoners expected about 60% reds but also expected some random variation to occur as candies were drawn. Surprisingly, 25% of the high school students and 15% of the middle school students studied actually expected to obtain the same number of red candies on each draw, even when explicitly asked if there might be variation from sample to sample. The fact that high school students performed worse than middle school students on predicting variation is worth noting. It suggests that as proportional reasoning develops more fully, intuitions about statistical variation may actually regress. Therefore, as proportional reasoning develops in secondary school, teachers must be conscious of providing opportunities for students to retain and build intuitions about the nature of random behavior.

Another study demonstrating the need for high school students to develop more powerful intuitions about random phenomena was conducted by Batanero and

Serrano (1999). They asked 14- to 17-year-old students to judge whether each of the coin flipping sequences shown in Figure 9.20 was made up or actually produced by flipping a coin 40 times. Students were told that some of the sequences were produced by flipping a coin 40 times and others were just made up. ("H" indicates obtaining heads, and "T" indicates obtaining tails.)

Students adopted a variety of strategies in judging whether the coin flipping sequences were made up or from actual data: looking for patterns, looking at how frequently a given result occurred, looking for runs in the data, and at times simply attributing results to luck. When students argued that a given result was random, they generally justified their argument by referring to the presence of unpredictability and irregular patterns. The presence of regular patterns and long runs of heads or tails were associated with lack of randomness, even though runs frequently occur in random strings.

> **Implementing the Common Core**
>
> See Homework Task 9 to "use random sampling to draw inferences about a population" (Content Standard 7.SP) when working a version of the candy task and to "develop a probability distribution for a random variable defined for a sample space in which theoretical probabilities can be calculated; find the expected value" (Content Standard S-MD.3).

Figure 9.20   "Real" and "made-up" coin flipping sequences (Batanero & Serrano, 1999, p. 560).

Coin flipping sequence 1:	T T T H T H H T T T H T H H H H H T H T H T T H T H H T T T T H H H T H H H T H H
Coin flipping sequence 2:	H T H T T H H T H T H H T T H T T H H T T H T T H H T T H T H T H T H T H T T H T
Coin flipping sequence 3:	H T T T H T T H H H T H T T T T T H T H T H H T H T T H H H H T T T H T T H H H
Coin flipping sequence 4:	H T T T H T T H T H T T T T H T T T T T H H T T T H T T H T T H T T T T T H T T T H T

Which sequences in Figure 9.20 do you believe to be real? Which do you believe to be made up? Explain.

STOP TO REFLECT

One way to help students develop sound intuition about random behavior is to engage them in probability simulations. NCTM (2000), for example, recommended that high school students use simulations to estimate answers to probability questions like the one shown in Figure 9.21.

Figure 9.21   Problem for probability simulation (NCTM, 2000, p. 333).

How likely is it that at most 25 of the 50 people receiving a promotion are women when all the people in the applicant pool from which the promotions are made are well qualified and 65% of the applicant pool is female?

Implementing the Common Core

See Clinical Task 14 to help a student understand how to use simulation for the problem shown in Figure 9.21 to "decide if a specified model is consistent with results from a given data-generating process" (Content Standard S-IC.2).

Students can simulate the problem with a random number generator on a calculator, a table of random digits, or another device for generating random numbers. The numbers 1 through 65 could represent females, and 66 through 100 could represent males. Then, 50 random numbers could be produced to simulate the situation where qualified individuals are chosen for the position at random, with no inherent bias. The proportion of females in the simulated sample could then be compared to the proportion of males to see how close each is to one-half. To get an even better estimate, several samples of 50 could be drawn. The process of drawing many samples can be facilitated with dynamic statistics software or online applets (see, for example, www.rossmanchance.com).

Although probability simulations have the potential to help students see the link between probability and statistics and to build sound intuitions about random behavior, they, like other teaching methods, are not a panacea. Pfannkuch (2005) pointed out that researchers are just beginning to study how simulations help students understand the link between theoretical and experimental probability. Research that has been done on high school students' work with probability simulations describes both productive and unproductive reasoning patterns that can emerge. Zimmerman and Jones (2002), for example, found that high school students were generally successful in constructing valid probability generators. Some of the students also believed that experimental probability should approach theoretical probability as the number of trials increased. Along with these encouraging results, however, some problematic beliefs were found. Some students felt that simulation should not be used to model real-world problems. Some also exhibited the representativeness heuristic while engaged in simulation, believing that a sample should always be reflective of the parent population from which it is drawn. These findings highlight the importance of monitoring students' work while they engage in simulation. Teachers can study beliefs students have about simulations as a first step in designing instruction that builds on helpful beliefs and challenges unhelpful ones.

CONCLUSION

Current curriculum documents make clear that the study of statistics and probability is no longer to be reserved just for the college level. Secondary students should develop statistical literacy, thinking, and reasoning. Curricula should provide a firm grounding in frequentist, classical, and subjective interpretations of probability. Some of the challenges that may emerge for students as they study statistics include difficulties in forming fruitful statistical questions and understanding what graphs and summary statistics say about data situated in various contexts. There are also several prevalent probability misconceptions, such as the conjunction fallacy, the availability heuristic, and the law of small numbers. Students' difficulties can be addressed within the context of conducting statistical investigations, using dynamic data software to facilitate transnumeration, and doing probability simulations. As students engage in such activities, they can appreciate the entire process of statistical investigation and develop sounder intuitions about random behavior and variation. Sample four-column lessons are provided at the end of this chapter to spark your thinking about how the ideas presented in this chapter can be translated into classroom instruction that helps develop students' thinking.

VOCABULARY LIST

After reading this chapter, you should be able to offer reasonable definitions for the following ideas (listed in their order of first occurrence) and describe their relevance to teaching statistics and probability:

Statistical thinking   254

Statistical literacy   254

Statistical reasoning   254

Tiers of statistical literacy   254

Process of statistical investigation   255

Frequentist probability   257

Classical probability   258

Subjective probability   259

Visual decoding tasks   263

Judgment tasks   263

Statistical intuition   265

Dynamic statistics software   267

Transnumeration   268

Ordered case value bars   269

Sliced scatterplot   269

Average as a data reducer   271

Average as a fair share   271

Average as a typical value   272

Average as a signal in noise   273

Mean as a balance point   275

Falk phenomenon   277

Confusion of the inverse   278

Conjunction fallacy   279

Availability heuristic   280

Outcome approach   280

Equiprobability bias   281

Representativeness heuristic   281

Law of small numbers   282

HOMEWORK TASKS

1. Compare the NCTM (2000) *Principles and Standards for School Mathematics* (http://standards
.nctm.org) with the *Guidelines for Assessment and Instruction in Statistics Education* Pre-K–12
report (www.amstat.org/education/gaise/). In the NCTM document, focus specifically on the recommendations for teaching statistics in Grades 6 through 8 and 9 through 12. In the GAISE document, focus specifically on levels B and C. Identify two content items that receive more emphasis in the NCTM document than in the GAISE document. Then identify two content items that receive more emphasis in the GAISE document than in the NCTM document. Give quotes from the NCTM and GAISE documents to support your responses. Then make an argument for giving either increased or decreased curricular attention to one of the content items you identified.

2. Study the high school and middle school curriculum documents adopted by the state in which you plan to teach. Do the documents explicitly identify the process of statistical investigation, as outlined in GAISE, as an important object of study? If so, provide quotes from the curriculum documents to show that the process is identified as an explicit object of study. If not, provide recommendations regarding how the curriculum documents might be rewritten to better incorporate the process of statistical investigation.

3. George Polya (1945) outlined a four-step mathematical problem solving process: (1) Understand the problem, (2) devise a plan, (3) carry out the plan, and (4) look back. (For more details on this process, see Polya's original work and/or do an Internet search on "Polya's problem solving process"). Compare and contrast Polya's mathematical problem solving process to the investigative process for

statistical problem solving given in the Pre-K–12 GAISE report. Describe at least two similarities and two differences between the two problem solving processes. Explain how the differences between the two processes might influence the manner in which you teach statistics.

4. Groth (2003) posed the following judgment task to a group of high school students:

Joe went to a shopping mall in Florida from 8:00 A.M. until 9:00 A.M. on February 20 and stopped people for interviews. He asked each person he stopped, "May I please interview you for a study I am doing about people who live in Florida?" He asked for the following information from all 20 people who agreed to do interviews:

   a. Age

   b. Annual income

   c. Favorite color

   d. Political party affiliation

   e. Whether or not they considered themselves computer literate

   f. How many hours per week they slept

Comment on Joe's method of collecting data. How useful do you think the information he obtained is? Why?

Here are some of the responses students gave:

Sample Response 1

Ok, um to a certain extent that might be useful to him, but since it was kind of random, um, he might have talked to more people of a certain age or political party, and I don't know, but it might have been affected also by him being in a mall. For example, real elderly people might not necessarily be going to a mall, or you just might not have the same, different types of people there, so you might have different results. So, maybe he should be more careful in making sure that he's going to different types of areas that have different people so that he's correctly talking to the people and finding out the amounts of different kinds of people.

Sample Response 2

Well, it depends on what he wanted to know about the people who live in Florida. Twenty people really doesn't do anything; you can't sum up the state of Florida with 20 people. And he's only limiting his interviews to people that go to the mall, because there's a lot of people who don't go to the mall that day, or people who usually go to the mall are either . . . are like, in the middle age group. People that are really old tend not to go to the mall that much. I don't know, some of these questions don't seem to be of that much importance, but I don't know what he's going to do with it. Um . . . yeah, I don't . . . to get an accurate study, he'd have to ask more than 20 people . . . from different places, like they would have to be random, and they couldn't be one right after another, because, I don't know, it would only take 15 minutes, and 15 minutes out of like a whole day doesn't really include that many people, or that big of a variety.

Sample Response 3

Um, well I think it's kind of biased, because, I don't know if Feb. 20 was, like a weekday, but, um, he's only collecting data from people who don't have a morning job, or a full-time job. Or, um, he's not taking into consideration students who might be in school at that time. And he's missing people that may not use the mall as their main shopping center. And he might be targeting people who have a higher annual income, because

---

some people might not be able to afford either just the . . . to go to the mall, or transportation. Favorite color wouldn't really be affected, so it would probably be affecting age and annual income, mostly, and how many hours per week each person sleeps, because it's kind of early.

Sample Response 4

Um, his method of collecting data is pretty biased, because it's all based on, like, voluntary responses. So, it's not like he's going to get too broad of a spectrum . . . well, I think he could, but it's all, like, voluntary, and it's all people who are going to be at the mall. So there could be, I guess, people who don't go to the mall that he's missing, and it's in the morning. And for age, what if he talks to a girl who doesn't want to tell her real age, and she could lie. She could go, "I'm 21" when she's actually 30. Annual income people could lie about, too, if they're not too comfortable with how much money they make. Favorite color, people will probably tell the truth; political party, they will probably tell the truth. Computer literate, that they'll probably tell the truth. How many hours of sleep . . . he could do that . . . I don't see what it has to do with Florida. Or favorite color.

Sample Response 5

Well, it's, like, a good starting point, but he needs to come back at a later time. Because people who shop at that time of day are going to be different than people who shop in the evening or afternoon or something.

Sample Response 6

Well, I'm not really sure what he was trying to find out, but it seems like if he's trying to figure out the type of people who live in Florida—I think the age one is good, because a lot of people tend to think that more older senior citizens live in Florida. So, that could show if it really was. And annual income could show that, too. I don't really know about favorite color, because I don't think that living in Florida would have anything to do with what your favorite color was. Political party affiliation, that could also show how people think in Florida. Whether or not they consider themselves computer literate—I think you could be computer literate if you lived in Florida or if you lived in Alaska, I don't think that would really matter. And how many hours per week each person sleeps, that could be relevant, too, because they think of Florida as, like, a retirement community. So, maybe those people tend to sleep more than the average person who works all day. So, I think that could be useful, too.

Sample Response 7

I think that the questions are good, but, like, the time that he did it is not good, because the time he did it, like, 8 A.M. to 9 A.M., a lot of older people are awake, but not a lot of younger people. Also, he asked people in a mall, and that's not really that good, because people at malls, they usually have more income, because they're there for a purpose—to buy something. And I think that if he did more of a random sample of people, that he would get a better idea for each.

Sample Response 8

When I think of Florida, I think a lot of people go there for tourism, but I guess he said, "Do you live in Florida," so it would be OK. Then, the other thing I would think is during that time that not a lot of working class people, besides the people that work there, are going to be at the mall. So usually those old walkers, that walk around the mall, he would get a lot of their surveys. Basically, that's all he would get, because the kids would all be at school, since it was a weekday and everything.

Sample Response 9

I really don't think there's that many people that go shopping from 8 to 9 in the morning. And the people that are there are just the old people who are walking around, so . . . and not everybody goes to the mall.

Rank the nine sample responses in order from least developed to most developed. Draw on Gal's (1998) criteria for ranking the responses as appropriate. Add additional criteria for ranking the responses as necessary. Then produce a rubric that would be suitable for grading students' responses to the task.

5. Download the Instructor's Evaluation Edition of Fathom from www.keypress.com. Then browse through the sample activities available on the website. Select one sample activity likely to address a difficulty students often have in learning statistics. Write a summary of the components of the activity and explain why you would expect it to be helpful for statistics students. Be sure to include the name of the curriculum module and the title of the activity in your report.

6. Some of the data Shaughnessy and Pfannkuch (2002) provided about the minutes between Old Faithful eruptions are shown below. Suppose that someone is asking you how long she will need to wait between eruptions of Old Faithful. Produce at least three graphs, using at least three days' worth of data for each one. Explain what each graph helps reveal about Old Faithful's eruption times, and then decide which one you would show the person who wants to know. Justify your decision.

Old Faithful Daily Data Sets—Minutes Between Eruptions																		
1)	86	71	57	80	75	77	60	86	77	56	81	50	89	54	90	73	60	83
2)	65	82	84	54	85	58	79	57	88	68	76	78	74	85	75	65	76	58
3)	91	50	87	48	93	54	86	53	78	52	83	60	87	49	80	60	92	43
4)	89	60	84	69	74	71	108	50	77	57	80	61	82	48	81	73	62	79
5)	54	80	73	81	62	81	71	79	81	74	59	81	66	87	53	80	50	87
6)	51	82	58	81	49	92	50	88	62	93	56	89	51	79	58	82	52	88
7)	52	78	69	75	77	53	80	55	87	53	85	61	93	54	76	80	81	59

7. Download the Instructor's Evaluation Edition of TinkerPlots from www.keypress.com. Then browse through the sample activities available on the website from the curriculum modules. Select one sample activity likely to address a difficulty students often have in learning statistics. Write a summary of the components of the activity and explain why you would expect it to be helpful for statistics students. Be sure to include the name of the curriculum module and the title of the activity in your report.

8. Groth (2003) interviewed high school students to obtain responses to the following task:

A fisherman weighed the first fish he caught seven different times on the same scale. Here are the measurements (in pounds) that he came up with: 19.2, 21.5, 10.1, 23.1, 22.0, 20.3, 21.8. What do you think the true weight of the fish was? Why?

Students gave these responses:

Student Response 1

[Adds up weights on calculator and divides by 7.] It's probably around 19.7 pounds. I just added up all the weights he came up with and divided by 7.

Student Response 2

[Uses calculator.] 21.316. I took the average of the greatest six, and just threw out the 10.1, because I don't think that's right, anyway.

Student Response 3

I think I'd pretty much disclude 10, because it seems like that's an obvious error. [Uses calculator to find the average of the rest.] I would say, like, 21.3 pounds. Well, I got rid of 10 because that seems like an obvious error. Then I added up all the other numbers and then divided by it, minus, like—since I took out 10, there were six numbers, so then I divided it by 6.

Student Response 4

Twenty-one point three. I just discarded the 10 because it was so extreme, and I took the average of the six remaining measurements.

Student Response 5

I would throw the weights into lists. [Puts data into the TI-83.] Um, the true weight would be around 23.1 pounds. I would throw the 10.1 out because it would throw off the average weight. I took the mean of the measurements, but I didn't use the 10.1 pounds—it's a low value and it would throw off the more realistic weight.

Student Response 6

I would probably just add up the rest of them, not including the 10.1, and just take the average of that . . . [uses calculator] . . . so that came out to be 21.3, which seems about right, just about in the middle or below, so I would just leave out that 10.1, because I think he screwed up.

Student Response 7

About 22. It looks like about the average. I ignored 10.1, because like I said earlier, some outliers you just have to ignore, and realize there was something wrong with the data or something. So, I took the average of the other six, and it's about 21.5, or 22, somewhere around there.

Student Response 8

Since 10 is the only one pretty much out of whack there, I'd say it's . . . mmm . . . about 21 pounds. Kind of throwing out the 10.1, because he must have been holding the fish wrong or something when he was holding that one. So, kind of looking at the other numbers in my head, I decided on an average around 21.

Student Response 9

Twenty point nine. Yeah, I just put those weights in numerical order, and then . . . uh, oh, wait, no, I counted wrong. 21.5. I thought that the middle one was between 20.3 and 21.5, but the middle number is 21.5. Or you could add them all up and divide them by . . . [picks up calculator] . . . that's quite a bit of a difference. OK, I'm changing my answer again to 20.6. Because if you just added all those weights together and then divided by 7, you got 19.7 as the average. If you just looked in the middle, you got 21.5. So, then I just add 21.5 to 19.7 and divided the two, and got 20.6. So, I just did the average of both averages.

Student Response 10

Well, I'd have to say the true weight was probably 20, just taking the average of all the weights that they came up with.

Student Response 11

Um, probably between like . . . 21.6 and 21.7, because there are two that are . . . the 21.8 and 21.5 are so close together, and those two are so close together that it was probably around there.

(Continued)

(Continued)

Student Response 12

I would probably just find the average of the highest one and the lowest one, but I think he should just get a new scale. [Uses calculator.] If you find the average of the highest and the lowest, it's 16.05. [Interviewer: So you would say the true weight was 16.05?] Probably around there, but he should check it on another scale. That's probably just the closest he could get with those measurements.

Student Response 13

Looking at it, I would say the fish was about 21 pounds. I just kind of spot checked it. It seemed like there was a couple of 21s and a low 22, so right around there.

Student Response 14

I actually think that the 10.1 was the exact amount of the fish. I would say this just because if you think about it, if the fish was going around on the scale, he probably put down his hands to control the fish, and that probably added a good 10 pounds. It just matters how hard he pushed down on the fish. I think when the fish was still, he got the 10.1, I think that's how he got the right answer. A 23-pound fish is a pretty big fish. There's a lot of 10.1 fish caught.

Student Response 15

Well, definitely not 10.1, but somewhere around 21 pounds or 22 pounds. Well, I took out the 10.1, and I just kind of averaged in my head. There was two 21, 22 . . . about 22-pound ones, and I just made an assumption after just looking at it, kind of.

Rank the 15 sample responses in order from least developed to most developed. Draw on Gal's (1998) criteria for ranking the responses as appropriate. Add additional criteria for ranking the responses as necessary. Then produce a rubric that would be suitable for grading students' responses to the task.

9. Design a probability simulation that uses a table of random numbers to model the situation of drawing candies, with replacement, from a bowl of 1,000. We know that 200 are yellow, 500 are red, and 300 are blue. Draw 30 samples of size 5 and construct a graph to show how many reds you obtain in each sample. Then repeat, drawing 30 samples of size 20. Compare the shape, center, and spread of the two data displays you produced, and explain any differences you observe. Finally, provide and justify a reasonable estimate of the number of red candies you might obtain if you drew out a random sample of 10. Your estimate need not be a single number.

CLINICAL TASKS

1. Find an article in a newspaper or online source that makes a statistical claim. Give the article to a group of middle or high school students and ask them to evaluate the claim. Have the students individually write the reasons why they agree or disagree with it. Then collect their work and analyze it. Which, if any, of the three tiers of statistical literacy are represented in the students' work? Include excerpts from the work samples to support your response.

2. Ask a group of students to write responses to the questions shown in Figure 9.1. Analyze the responses for evidence of correct and incorrect reasoning. Which misconceptions, if any, do the students exhibit? How many students exhibit each misconception? How many exhibit correct reasoning? Describe a teaching strategy you would use to help students overcome the misconceptions associated with incorrect responses.

3. Ask a student to think aloud as he or she solves the task shown in Figure 9.2. Ask him or her to give an initial estimate of the likelihood of obtaining all three cards after purchasing only three boxes of cereal. Then ask him or her to devise and carry out a plan for systematically testing the accuracy of the initial estimate. If the student says that he or she cannot think of a way to test the estimate, provide hints to get him or her started. Keep track of and report how the student approaches the task and the assistance that you provide. Based on your observations, describe strengths and weaknesses in the student's understanding of the simulation problem.

4. Ask a group of students to write a solution to the task shown in Figure 9.3. Then write a report that summarizes the students' responses. Do their responses reflect Zawojewski and Shaughnessy's (2000) finding that many students think there is a 50% chance of winning this sort of game? Give examples of student work to support your explanation.

5. Interview a student about his or her ideas about sampling, using the interview questions shown below, adapted from J. M. Watson and Moritz (2000):

---

1. Have you ever heard the word *sample*? If so, where? What does *sample* mean?

2. Suppose a newspaper stated, "Researchers studied the weight of ninth-grade children in Wisconsin. They used a sample of ninth-grade children from the state." What does the word *sample* mean in this newspaper article?

3. Why do you think the researchers studied a sample of ninth graders rather than all ninth graders in Wisconsin?

4. Do you think the researchers used a sample of about 10 ninth graders? Why or why not? How many students should they have used for their sample? Why?

5. How should the researchers have chosen students for their sample? Why?

6. The researchers went to two schools: one city school and one rural school. Each school had about 50% boys and 50% girls. The researchers took a sample of 50 students from the city school and 20 students from the rural school. One of these samples was unusual in that more than 80% were boys. Is the unusual sample more likely to have come from the sample of 50 or from the sample of 20, or are both samples equally likely to have been the unusual sample? Please explain your answer.

Based on the student's responses to the questions, in which of the six categories about conceptions of sampling would you place him or her? Provide excerpts from the interview to support your categorization.

6. Have a class of students give written responses to the contingency table task shown in Figure 9.6. Sort the written responses into piles according to the strategies and justifications the students provide. Rank the responses in order from least developed to most developed. Then produce a rubric that would be suitable for grading students' responses to the task.

7. Obtain a copy of a textbook that includes a section on scatterplots. Use a data set from that section of the textbook to produce a set of ordered case value bars, a scatterplot, and a sliced scatterplot.

Then show three students all three data representations. Each student should see the representations in a different order. One student should see the scatterplot first, another should see it second, and the final student should see it third. Ask each student to describe what each representation reveals about the data. Describe the students' responses and explain similarities and differences you see across responses.

8. Have a class of students complete the typical value task shown in Figure 9.11. Sort the written responses into piles according to the strategies and justifications provided. Rank the responses in order from least developed to most developed. Then interview at least two students to better understand why they answered the task as they did. One of the students you interview should have given a more fully developed written response than the other. Write a report on the similarities and differences in the strategies used and justifications given by the two students you selected for interviews.

9. Examine a textbook that presents the concept of average. Determine which of the following interpretations of average are present in the textbook: average as a data reducer, as a fair share, as a typical value, and as a signal in noise. Give specific examples of problems from the textbook that fit each category and justify your categorizations. In cases where there are no problems in the textbook fitting a given interpretation, write problems you would use to introduce students to the missing interpretation. Explain how the problems you wrote fit each interpretation.

10. Examine a textbook that presents the concept of average. Determine which of the following procedural applications of average are present in the textbook: computing the mean in reverse, computing a weighted average, and calculating the mean of a data set that includes one or more zero values. Give specific examples of problems from the textbook that fit each category and justify your categorizations. In cases where there are no problems in the textbook fitting a given category, write problems you would use to introduce students to the missing procedural application. Explain how the problems you wrote fit each category.

11. Design a conditional probability problem of your own that may elicit confusion of the inverse. Also design a brief teaching intervention that could help students overcome confusion of the inverse. Interview at least two students and ask them to think aloud as they solve the problem you designed. If they exhibit confusion of the inverse, use the teaching intervention you designed to help them overcome it. Write a report in which you describe the problem you designed, the teaching intervention you designed, the students' responses to the problem, and the students' responses to your teaching intervention (if applicable).

12. Design a conditional probability problem of your own that may elicit the outcome approach. Also design a brief teaching intervention that could help students reject the outcome approach. Interview at least two students and ask them to think aloud as they solve the problem you designed. If they exhibit the outcome approach, use the teaching intervention you designed to help them reject it. Write a report in which you describe the problem you designed, the teaching intervention you designed, the students' responses to the problem, and the students' responses to your teaching intervention (if applicable).

13. Have a class of students give written responses to the hospital problem shown in Figure 9.19. Be sure to ask the students to give written explanations for their answers. Sort the written responses into piles according to the strategies and justifications the students provide. Rank the responses in order from least developed to most developed. Then produce a rubric that would be suitable for grading students' responses to the task.

14. Interview a student and ask him or her to make a conjecture about the answer to the task shown in Figure 9.21. Then ask him or her to construct a probability simulation to estimate an answer. If the student has not yet studied probability simulations, demonstrate how the situation can be modeled with a random number generator of your choice. Also show the student how technology like Fathom or an online applet (such as one shown at www.rossmanchance.com) can facilitate drawing many samples. After helping the student with the simulation task, ask him or her to write an original simulation problem and solve it. Describe the student's problem and solution strategy, and make a conjecture about which aspects of probability simulation the student understands and which aspects the student does not understand.

| VIGNETTE ANALYSIS ACTIVITY | **Focus on Looking for and Making Use of Structure (CCSS Standard for Mathematical Practice 7)** |

Items to Consider Before Reading the Vignette

Read CCSS Standard for Mathematical Practice 7 in Appendix A. Then respond to the following items:

1. Drawing on your own experiences as a mathematics student, provide specific examples of teaching practices that can help students look for and make use of structure.

2. Drawing on your own experiences as a mathematics student, provide specific examples of teaching practices that might prevent students from looking for and making use of structure.

3. You have a bag with three green marbles and two blue marbles.

   a. Suppose you draw a marble at random and it is blue. You replace the marble and draw again. What is the probability of getting a green marble on the next draw?

   b. Suppose you draw a marble at random and it is green. You do not replace the marble, and you draw again. What is the probability of getting a green marble on the next draw?

   c. Suppose you drew two marbles at random, without replacement. The second marble you drew was green. What is the probability that the first marble you drew was blue?

4. Write your own example of a conditional probability problem that involves a real-world context other than drawing objects from a container.

Scenario

Mrs. Baker was teaching the concepts of conditional probability and independence to her students for the first time. In college, she had had little chance to study these ideas. Her first degree was in accounting, which required her to take an introductory statistics course. After a few years as an accountant, she had decided to become a teacher. Now she was surprised to see conditional probability and independence in the high school curriculum; she had thought students ordinarily first encountered these ideas in college. The courses she had taken to become certified to teach had not discussed these ideas, either. Mrs. Baker found herself in the position of trying to stay at least one section ahead of her students in the class textbook. She did her best to make sense of conditional probability and independence by studying the teacher's edition of her students' textbook. Despite her efforts, her lessons sometimes took unexpected twists and turns.

The Lesson

Mrs. Baker always began her lessons with "math brush-up" exercises. She and her teaching colleagues wrote sets of brush-up exercises together at the beginning of each month. They believed that the exercises would help students retain information from previous lessons for the end-of-year state test. The brush-up activity sheet for this month had 30 exercises. Students were to choose four of them to complete at the beginning of the lesson. In some exercises, students were given information in a diagram and asked to determine the area of a geometric figure. In others, they were asked to do computations such as multiplying fractions, subtracting decimals, and dividing integers. In still others, they were asked to solve algebraic equations and inequalities. Students spent the first 20 minutes of the 75-minute class period working the exercises and having Mrs. Baker check their solutions.

After the brush-up exercises, Mrs. Baker reviewed different ways to write probabilities. The following dialogue occurred:

Mrs. Baker: What were the three different ways you could write probability? The probability of something happening?

Jenny: With a colon, and . . .

Mrs. Baker: With a colon is one.

Nate: As a fraction.

Mrs. Baker: OK, one other one.

Keesha: Couldn't you do it like 5 to 2, something like that?

Mrs. Baker: That's a fraction. Oh, you mean with words . . .

Keesha: With words, yeah . . .

Mrs. Baker: Oh, you mean with the word *to*. Sorry about that, yeah.

Satisfied with the students' responses, Mrs. Baker assigned several exercises involving simple probability situations and asked students to write the answer for each one in three different ways.

While working the assigned exercises, some students asked if they should simplify the fractions they gave as answers. Mrs. Baker told them that it did not matter if they simplified or not. When it looked as if most students had finished, Mrs. Baker asked the class to compare answers with others sitting at their tables. She directed them to make sure that each person had written the answer in three different ways. As students worked, she noticed that none of them were writing probabilities in percent form. This concerned her, since it was one of the representations she wanted students to know. She gathered the entire class back together for a large-group discussion, during which she stated that probabilities can be written as percentages. She also reviewed the procedure for converting a fraction to a decimal and a percentage. Students then worked on converting all of their fractional answers to percentages. When they had finished, Mrs. Baker called on students to say their answers for the rest of the class. When students did not give correct answers, she called on others until the correct answer was given.

During the final 25 minutes of the lesson, Mrs. Baker introduced an activity designed to help students understand the distinction between dependent and independent events. Students received a sheet with six problems. Some were intended to illustrate the idea of independent events, others to illustrate dependent events. One of the problems, for example, was to determine the probability of rolling a one on a six-sided die and landing on green on a given spinner. Another problem was to determine the probability of drawing a given marble out of a bag, given that a marble of a certain color had been drawn and not replaced. Mrs. Baker told students to work in groups, sort the problems into categories, and explain their reasoning. She hoped they would sort "without replacement" situations into a category so she could introduce the idea of conditional probability.

As students worked, they found a number of ways to sort the problems into categories. Some separated "high probability" situations from "low probability" situations. Others sorted problems with odd numbers in the denominators of their answers from those with even numbers. A few separated "realistic" situations from "fake" ones. The number of different categorization schemes that emerged was much larger than Mrs. Baker expected. Additionally, none of the groups of students separated "with replacement" situations from "without replacement" ones. Since Mrs. Baker considered this activity to compose the "inquiry-oriented" portion of the lesson, she was hesitant to tell the students how she had actually hoped they would sort the problems.

As the class period wound down, Mrs. Baker decided to gather students back together as a large group one final time. She announced, "What we'll do tomorrow is we'll come back to this with a little bit different information and a little bit different problems. I have another way that we can do this to see if you can find what you need. You know what I'm saying?" At this point, some students became restless, sensing that they were not doing the task correctly. Peter spoke up and said, "Can't you just tell us how to do it now?" Mrs. Baker replied, "No, I'm not just going to tell you, I'm not going to just do it. Think about the two—there are two things happening—see if you can write it down so you can remember where we left off. You're doing two things." The class period came to a close with Mrs. Baker and her students feeling uneasy about the ending point for the day.

After the lesson, Mrs. Baker reflected on its effectiveness. During a recent professional development session, the presenters had said that inquiry-oriented lessons facilitate students' learning to a greater extent than do teacher-directed ones, but she had her doubts. The teacher-directed portion of the lesson that day had gone much more smoothly than the inquiry-oriented portion. She wondered, "Is inquiry-oriented instruction just a fad that will soon be on the scrap heap of education reform?" With that thought in mind, she began to plan the next day's lesson.

Questions for Reflection and Discussion

1. Which aspects of the CCSS Standard for Mathematical Practice 7 did Mrs. Baker's students seem to attain? Explain.

2. Which aspects of the CCSS Standard for Mathematical Practice 7 did Mrs. Baker's students not seem to attain? Explain.

3. What are some possible drawbacks to using the "math brush-up" activity described at the beginning of the lesson?

4. As the classroom teacher, what remarks would you make to students during the dialogue about different ways to represent probability? What kinds of questions would you ask them about the representations?

5. Do you agree with Mrs. Baker's comment that it does not matter if one simplifies a probability expressed as a fraction? Under what circumstances might it be best not to simplify?

6. At one point in the lesson, Mrs. Baker called on students to give the answers to exercises. When they were incorrect, she called on other students to get the correct answers. Was this a good instructional decision? Why or why not?

7. Comment on the time Mrs. Baker spent on each portion of the lesson. Describe adjustments you would make to allow more class time for directly addressing the ideas of conditional probability and independence.

8. Comment on the effectiveness of the problem sorting activity. What modifications could be carried out to make the activity more effective?

9. Would you describe Mrs. Baker's problem sorting activity as "inquiry oriented?" Explain.

10. If you were in Mrs. Baker's position, what would you do during the next class session? Why?

RESOURCES TO EXPLORE

Books

Bright, G. W., Brewer, W., McClain, K., & Mooney, E. S. (2003). *Navigating through data analysis in Grades 6–8.* Reston, VA: NCTM.

Description: The authors provide activities for teaching data analysis in an inquiry-oriented manner to middle school students. Activities address doing exploratory data analysis, comparing sets with equal and unequal numbers of elements, and exploring bivariate data.

Bright, G. W., Frierson, D., Tarr, J. E., & Thomas, C. (2003). *Navigating through probability in Grades 6–8.* Reston, VA: NCTM.

Description: The authors provide activities for teaching probability in an inquiry-oriented manner to middle school students. Activities address the concept of probability, probability distributions, the law of large numbers, and connections between probability and statistics.

Burrill, G. F. (Ed.). (2006). *Thinking and reasoning with data and chance* (Sixty-eighth yearbook of the National Council of Teachers of Mathematics). Reston, VA: NCTM.

Description: This yearbook provides a broad perspective on the teaching of statistics and probability throughout Grades K through 12. Chapter authors discuss methods for engaging students in authentic data analysis, the nature of students' statistical thinking, and differences between mathematics and statistics.

Burrill, G., Franklin, C. A., Gobold, L., & Young, L. (2003). *Navigating through data analysis in Grades 9–12.* Reston, VA: NCTM.

Description: The authors provide activities for teaching data analysis in an inquiry-oriented manner to high school students. Activities address decision making, variability, sampling, categorical data, numerical data, and designing studies.

Franklin, C., Kader, G., Mewborn, D., Moreno, J., Peck, R., Perry, M., & Scheaffer, R. (2007). *Guidelines for assessment and instruction in statistics education (GAISE) report: A pre-K–12 curriculum framework.* Alexandria, VA: American Statistical Association.

Description: This American Statistical Association–endorsed document provides guidance on teaching statistics in Grades pre-K through 12 in a manner aligned with the NCTM standards. It provides suggestions about how to engage students in the process of statistical investigation during classroom activities.

Huff, D. (1954). *How to lie with statistics.* New York, NY: W. W. Norton.

Description: This book is a classic and lively exposition on some common misuses of statistics. It is rich with examples teachers can use to illustrate how averages, graphs, and probability calculations are often used in deceptive ways.

Shaughnessy, J. M., Barrett, G., Billstein, R., Kranendonk, H. A., & Peck, R. (2004). *Navigating through probability in Grades 9–12.* Reston, VA: NCTM.

Description: The authors provide activities for teaching probability in an inquiry-oriented manner to high school students. Activities address probability as long-run relative frequency, sample spaces, conditional probability, and expected value.

Shaughnessy, J. M., & Chance, B. (2005). *Statistical questions from the classroom.* Reston, VA: NCTM.

Description: In teaching secondary school statistics, a number of conceptually oriented questions often arise. The authors provide answers to common questions such as, "What is a *p*-value?" and "What are degrees of freedom and how do we find them?"

Shaughnessy, J. M., Chance, B., & Kranendonk, H. (2009). *Focus in high school mathematics: Reasoning and sense making: Statistics and probability.* Reston, VA: NCTM.

Description: The authors of the text illustrate how teachers can carry out the vision of reasoning and sense-making set forth in *Focus in High School Mathematics* in the context of statistics and probability. Activities include engaging students in analyzing data from a census, the Old Faithful geyser, and the Olympics.

Website

American Statistical Association Education Website: **http://www.amstat.org/education/**

Description: The American Statistical Association maintains a page with a variety of links of interest to teachers. Resources on the page include information on teacher workshops and student competitions as well as a listing of useful websites for teachers.

FOUR-COLUMN LESSON PLANS TO HELP DEVELOP STUDENTS' STATISTICAL AND PROBABILISTIC THINKING

Lesson Plan 1

Based on the following NCTM resource: Groth, R. E. (2007). Reflections on a research-inspired lesson about the fairness of dice. *Mathematics Teaching in the Middle School, 13*, 237–243.

Primary objective: Engage students in making and testing conjectures about the fairness of random number generators

Materials needed: Poster board and markers for students, several packs of Post-it notes, three types of six-sided dice (some sold as being "fair" [called yellow dice in this lesson plan], some that are heavily weighted to favor certain rolls [purple; available from toy store websites], some that are slightly shaved to favor certain rolls but look similar to the fair ones [red; available from casino supply websites])

Steps of the lesson	Expected student responses	Teacher's responses to students	Assessment strategies and goals
1. Introduce a game involving two players. The first player scores a point every time a one, two, or three is rolled. The second player scores a point every time a four, five, or six is rolled. The first player to score 10 points wins. Ask students if they consider the game to be fair or not.	Some students will believe the game is fair because each player has the same number of positive outcomes. Other students will consider some numbers to be "luckier" than others. Most students will not mention that the dice themselves need to be fair to make it a fair game.	Ask students to share their written responses about the fairness of the game. Try to elicit responses from students who think the game is fair as well as from those who feel some numbers are naturally luckier than others.	Note the students who seem to think that some numbers are naturally luckier than others. Throughout the lesson, look for opportunities to challenge this belief.
2. Pair students and have them play a round of the game described at the beginning of the lesson. Give them the purple dice to play with, but don't tell them that they are weighted.	Most students will quickly notice that the dice they have been given to play the game are not fairly weighted. Some will begin to physically examine the dice. Others will hold onto the belief that one number is just luckier than the others.	Pose the following question for group brainstorming: "How can we determine if a given die is fair?" Record students' responses to the brainstorming question on the board.	Track how many students propose physical examination of the dice and how many propose rolling them. Ask those who propose rolling them how many rolls it would take to decide whether the dice are fair or not.
3. Give each pair of students a yellow die and a red die. Tell them that one is sold as a fair die and the other is not. State that their job will be to make a conjecture about which one is fair and then test it by	Some students will make initial conjectures through physical examination strategies like trying to balance each die on one of its corners. Others will begin to roll the dice and gather data on the outcomes. Students may produce a variety of	If students try only physical examination strategies, ask them if they believe that gathering data from rolls of the dice would be helpful. If students roll the dice only a few times, ask them if they think that gathering data from	Examine the types of graphs students use to display their data. If some displays are more effective than others, choose a variety to be presented during large-group discussion. In examining the displays, decide if it would be

Steps of the lesson	Expected student responses	Teacher's responses to students	Assessment strategies and goals
gathering data. Each pair of students should make a poster to display their findings and support their final conclusion regarding which dice are fair.	graphs and tables to support their conclusions.	more rolls would better support their conclusion.	fruitful to introduce additional representations during large-group discussion.
4. Ask students to present their posters to the rest of the class.	Some pairs of students will have come to the conclusion that the yellow dice are fair, some will conclude that the red ones are fair, and some will conclude that both are fair. Most will want the teacher to tell them which conclusion is correct.	Do not tell students which conclusion is correct at this point. Ask them to evaluate the arguments that have been presented to determine if the dice are fair or not. In particular, try to elicit the idea that larger samples of data can help support conclusions.	Ask students whether or not larger sample sizes provide stronger support for conclusions. Have those who answer "Yes" engage in discussion with those who answer "No."
5. Have each pair of students roll a yellow die 10 times and a red die 10 times. The outcome of each roll should be recorded on a Post-it note. After completing the rolls, students should place the Post-it notes on the board in an organized display. Once the display is posted, ask if the new data displays support any of the conjectures that have been made about the dice.	Students will look for salient features in the data displays, particularly peaks and valleys, to try to determine which dice are fair. There may still be doubt about which dice are fair, causing some students to revert back to physically examining the dice.	Ask students what the distribution for a completely fair die would look like. As ideas are generated, choose students to sketch a hypothetical graph of a fair distribution on the board.	Examine the graphs that students generate to show a hypothetical distribution for a die that is completely fair. If a distribution that is "flat" (i.e., uniform) is not suggested, introduce such a graph into the conversation and ask students to interpret its meaning.
6. Once the key ideas of (1) gathering more data to reach a sound conclusion and (2) using a uniform distribution as a model for a "fair dice" situation have been introduced, let students know that the yellow dice were sold as fair dice and the red dice were not.	Students may wonder why the data gathered during the lesson do not decisively support one conclusion or the other regarding the dice.	Take the opportunity to emphasize the point that there is variation in data, and that in statistics, the solutions to problems are not always as clear-cut as in mathematics. Emphasize the idea that due to variation in the manufacturing process, even dice that are sold as fair may have imperfections that affect the data.	Ask students to complete a writing exercise about a situation parallel to what they encountered in this lesson: "How would you go about testing the fairness of a random number generator on a calculator?"

How This Lesson Meets Quality Control Criteria

- *Addressing students' preconceptions*: Experiential knowledge with fair games and dice is elicited at the beginning of the lesson. School-based knowledge of graphical displays of data is drawn on and assessed throughout the lesson.
- *Conceptual and procedural knowledge development*: The lesson builds intuitive understanding of the law of large numbers and the concept of uniform distribution. Producing tables and graphs requires students to draw on procedural, computational knowledge.
- *Metacognition*: Students are asked to make conjectures about the fairness of dice and then test them by gathering and examining data. Students' conjectures and conclusions are examined and reexamined in interactions with small-group members, the entire class, and the teacher as he or she monitors the small-group work.

Lesson Plan 2

Based on the following NCTM resource: Morita, J. G. (1999). Capture and recapture your students' interest in statistics. *Mathematics Teaching in the Middle School, 4,* 412–418.

Primary objective: To understand a method for estimating a population parameter from sample statistics

Materials needed: Large clear plastic bin, large bag of cheddar-flavored Goldfish crackers, small bag of pretzel Goldfish crackers, Post-it notes, paper cups, plastic sandwich bags

Steps of the lesson	Expected student responses	Teacher's responses to students	Assessment strategies and goals
1. Let students know that the purpose of the day's lesson is to develop methods to estimate the number of fish in a lake. Dump a large container of cheddar Goldfish crackers into a clear plastic bin to simulate fish in a lake. Ask students to estimate how many Goldfish are in the bin. Do not, at any point, tell them precisely how many cheddar Goldfish are in the bin.	Some students will estimate the number of fish along the length, width, and height of the bin and multiply the three together to estimate the number of fish. Others will suggest that it is necessary to count all of the fish in the lake. Some will provide estimates that are wild guesses.	When students provide estimates that are wild guesses, encourage them to listen to other students' more systematic estimation strategies. When students suggest counting all of the fish in the lake, draw them back into the real-world context of the problem: It is usually impossible to count all of the fish in any given lake.	Look for evidence of number sense in the estimation strategies students propose. Ask students how the bin with Goldfish differs from an actual lake. If the idea that fish are not as nicely stacked together in the real-life situation does not arise, introduce it into the conversation. This will help motivate the search for statistical estimation strategies.
2. Dump about 450 pretzel Goldfish into the bin with the cheddar Goldfish. Tell students the number of pretzel Goldfish. Say that the pretzel Goldfish represent fish that have been caught and tagged. Mix the fish together to represent the tagged fish mingling with the others. Then have a student draw a sample of fish out of the	Students may notice that the ratio of tagged fish to the total number of fish varies from sample to sample. As students see the ratios for each sample being written on the board, they may begin to ask how the data they are gathering might be useful.	As students notice how the proportions vary from sample to sample, ask them why this variation might occur. Also ask them to think of other kinds of real-world sampling situations that involve random variation. As students begin to ask how the data might be of use, remind them of the	After all students have had a chance to draw a sample, ask them to map this activity back to the real-world context. Ask how a biologist might go about tagging fish and then resampling them. Assess whether or not students can map the Goldfish cracker sampling simulation

Steps of the lesson	Expected student responses	Teacher's responses to students	Assessment strategies and goals
bin at random using a sandwich bag as a glove and a paper cup as a net. Have the student count and report the number of tagged and untagged fish in the sample and write on the board the ratio of tagged fish to the total number in the sample. Have the student replace and remix the fish, and then pass the bin to the rest of the students so they can draw samples.		original question. The goal will be to estimate the total number of fish in the lake. They should be prompted to start thinking about how the ratios and method they are observing could be used to make an approximation in a real-world setting involving fish in a lake.	back to sampling from a lake. Ask for similarities and differences between the classroom simulation and the actual practice of tagging and resampling.
3. Ask students to examine their own individual sample data. Pose the question of how the data could be used to estimate the total number of fish in the lake. Give students a few minutes to answer the question on their own, share their answer with a partner, and then share during a whole-class discussion.	Some students will set up a proportion in the following manner: $$\frac{sample\ number\ tagged}{total\ number\ in\ sample} = \frac{population\ number\ tagged}{total\ number\ in\ population}.$$ Some students will multiply 450 by the fraction $\frac{total\ number\ in\ sample}{sample\ number\ tagged}$. Some students will have difficulty with the assumption that the sample proportion tagged is approximately equal to the population proportion tagged.	Ask students to make judgments about which estimation strategies are most likely to be effective. For students who have difficulty with the assumption that the sample proportion tagged is approximately equal to the proportion tagged, clarify some of the assumptions in the problem. We are assuming that no fish are more likely than others to be caught in a net. Also, the assumption is that the population remains stable throughout the sampling process—there is no gain or loss of fish.	Ask students to explain why setting up and solving a proportion makes sense in the sampling/resampling context. Check for genuine proportional reasoning rather than mere execution of an algorithm. Ask students if data from a single sample can be trusted, given the variation in the observations. Check to see if they understand the need to deal with the variability from sample to sample.
4. Have students get into groups of four. Ask each group to discuss how they could combine their individual estimates into a single group estimate. Give a writing assignment to explain their individual and group estimation strategies and why they make sense.	In forming a single group estimate, some students will: - Choose estimates near the middle - Compute average values - Use intuition in place of examining the data	Emphasize the need for examining data to answer the question of interest. Do not rush students to use formal measures such as averages, since visual estimates may be just as effective. Compare the results of using formal measures with the results of using visual estimates.	Gather the students' written work to determine the types of strategies groups believed to be reasonable. On the basis of the answers they provide, decide whether or not it is necessary to pose a task that is statistically parallel to this one in a future lesson.

(Continued)

(Continued)

Steps of the lesson	Expected student responses	Teacher's responses to students	Assessment strategies and goals
5. Gather students back together for a whole-class discussion and have them compare their estimates and estimation strategies with one another.	Students will give estimates that may be fairly far apart from one another. Groups using similar estimation strategies may have estimates that are closer together.	As students give estimates, construct a Post-it note histogram to show how much variation occurred from sample to sample.	Ask students to examine the Post-it note histogram and write their best estimate of the number of fish in the lake. Ask them to justify their estimates. In their written responses, look for evidence of reasoning about shape, center, and spread.

How This Lesson Meets Quality Control Criteria

- *Addressing students' preconceptions*: Students' experiential knowledge of nature is tapped to motivate the lesson at the outset. School-based knowledge of proportional reasoning is drawn on and built throughout the lesson.
- *Conceptual and procedural knowledge development*: Formulas for volumes and procedures for solving proportions arise as different estimates of the number of fish are given. The statistical concepts of sampling and variation are featured throughout the lesson.
- *Metacognition*: Students compare their initial estimates of the number of fish in the lake to the estimates they obtain from statistical sampling. Students write about why their individual and group sampling methods make sense.

Lesson Plan 3

Based on the following NCTM resource: Groth, R. E., & Powell, N. N. (2012). Using research projects to help develop high school students' statistical thinking. In J. Newton & S. Kasten (Eds.), *Reasoning and sense-making activities for high school mathematics: Selections from* Mathematics Teacher (pp. 275–279). Reston, VA: NCTM.

Primary objective: To engage students in the study of all phases of the investigative cycle in statistics (over a period of approximately two weeks)

Materials needed: Dynamic statistics software, several sheets of poster board

Steps of the lesson	Expected student responses	Teacher's responses to students	Assessment strategies and goals
1. Let students know that they will be designing their own statistical research projects from start to finish. Use student poster competition guidelines to help get them started, such as those on www.amstat.org or	Some students will formulate questions that are interesting but not easily quantifiable (e.g., "What is it like to be a ninth grader in our school?") Some pairs of students will find it difficult to agree on a research question of	When students formulate questions that are not easily quantifiable, talk through the rest of the phases of statistical investigation with them to raise doubts about data collection and analysis related to the question. Encourage students to	Assess students' questions for two main elements: (1) Are the research questions of interest easily quantifiable? (2) Are the research questions interesting enough to keep students engaged for two weeks? To assess students' progress toward these

Steps of the lesson	Expected student responses	Teacher's responses to students	Assessment strategies and goals
www.causeweb.org. Pair students up and allow each pair one to two days to formulate a statistical question of interest to investigate.	mutual interest. Some students will formulate easily quantifiable but relatively uninteresting questions.	examine the existing data files available on dynamic statistics software to spark ideas if they find it difficult to agree on a mutual question of interest. Remind students who choose uninteresting questions that they will have to live with their question for about two weeks.	assessment goals, circulate among the pairs to see what kinds of questions they are formulating and to offer advice. Also have pairs of students present their research questions to the entire class near the end of the first session or the beginning of the second. Once they have been presented, provide feedback and have the rest of the students do the same.
2. Have students write a plan describing how their study will be carried out. The plan should include the main research question, data gathering methods, data analysis methods, and a conjecture about what kinds of conclusions they will be able to draw from the study once it is completed.	Students may propose a variety of different studies, including research on the association between two variables (e.g., the relationship between test scores and students' distance from the front of the room during testing); comparative studies (e.g., blood pressure of athletes vs. blood pressure of nonathletes); and descriptive studies (e.g., determination of the favorite type of music among students in the school).	As plans for studies are formulated, encourage students to pilot data collection procedures on one another to the fullest extent possible to help reveal flaws in study design. For example, a group that plans to give a survey to a large portion of the school should try it out with a few students in class to see if any questions are unclear, leading, or ambiguous.	Check to see that the instruments and study design procedures are reasonable before allowing students to gather a great deal of data. Having students present their written plans to the class before carrying out data collection can be used as an assessment strategy. This opens the studies up once again for critique by the teacher and by peers.
3. Have the students use the instruments they designed for their studies to gather data.	At this point, to gather data, students may need to leave the room and go to data gathering locations such as the gym and other classrooms. As students gather data, they may begin to discover additional limitations in the design of their data collection instruments or overall plans.	Arrange the logistics for student supervision while they are out of the room as far in advance as possible. Enlist the help of teacher assistants, other teachers, hall monitors, and other adults in the school building. Consider the flaws and imperfections students discover in their data collection plans and instruments to be a natural part of learning the intricacies of statistical investigation.	When the data collection portion of the final student project is assessed, concentrate on fidelity to the original study design rather than attainment of a "perfect" set of data. Concentrate on students' ability to reflect on problems that arose while conducting their investigations and their descriptions of how they would improve the study the next time around.

(Continued)

(Continued)

Steps of the lesson	Expected student responses	Teacher's responses to students	Assessment strategies and goals
4. Have students analyze the data they gathered and draw conclusions about the answer to their initial research question.	Students may use a variety of data displays they have encountered in past classes, such as scatterplots, boxplots, histograms, and line graphs. Students may also use a variety of summary statistics, such as mean, median, mode, range, standard deviation, and interquartile range.	Challenge students on their choice of data displays and summary statistics. For example, in cases where students use the mean, ask why the median is not a better choice. Encourage the use of dynamic statistics software to analyze data to allow the main focus to be on interpretation rather than computation.	As students work, check for a good fit between the tools they use to analyze the data and their research questions. Also check to see that they are discussing the output carefully when using the statistical software and not simply playing with a variety of graphs and statistics without reflecting on their meaning.
5. As a culminating activity, have each group of students present their findings to the rest of the class. At the end of each presentation, allow time for students to ask questions about each study.	Students may initially be hesitant to ask questions about studies done by their peers or to criticize them. In communicating statistical results, students may use language imprecisely (e.g., stating that a regression line allows one to "determine" values rather than to "estimate" them; using the term *outlier* when it may not be reasonable to do so).	Allow a great deal of wait time after each presentation has concluded for students to begin to ask questions of the presenters. If no questions are forthcoming, ask students to break into small groups to come up with questions to ask. If students do not question imprecise use of statistical language, question it yourself after students have finished asking their own questions about the presentations.	Use an evaluation form (such as the one on page 109 of the article this lesson plan is based on) for written feedback. Teachers and students can use the form to give written feedback on each presentation. The form assesses successful completion of all phases of statistical investigation as well as quality of presentation.

How This Lesson Meets Quality Control Criteria

- *Addressing students' preconceptions*: Students engage with out-of-school knowledge in the process of formulating a research question of interest. Students engage school-based knowledge in choosing summary statistics and graphs for the data they gather.
- *Conceptual and procedural knowledge development*: The multifaceted concept of statistical investigation is the central focus of study for this two-week lesson. Procedures related to dynamic statistics software use help support the statistical investigation process.
- *Metacognition*: Students are asked to make initial conjectures about the conclusions they expect to draw from the studies they design, and then to revisit the reasonableness of those conjectures throughout the statistical investigation process.

Lesson Plan 4

Based on the following NCTM resource: Canada, D. L. (2008). The known mix: A taste of variation. *Mathematics Teacher, 102,* 286–291.

Primary objective: To build students' intuition about random variation in a sampling context
Materials needed: Several sheets of poster board, several jars of plastic chips (each containing 30 green and 70 yellow chips), Fathom software

Steps of the lesson	Expected student responses	Teacher's responses to students	Assessment strategies and goals
1. Pose the following problem to students: A committee of 10 members of a school band is to be chosen from a set of 30 males and 70 females. How many females might be on such a committee? (Canada, 2008, p. 287.) Ask students to share initial expectations and estimates.	Some students will focus strictly on proportions and predict that precisely seven females will be on the committee. Some students will acknowledge that the number of females may vary slightly from the expected value of seven. Some may specify wide or narrow ranges of variability.	Try to help students understand that some random variation is bound to occur in such a situation by drawing an analogy to what happens when one draws chips from a hat. Emphasize the importance of drawing several random samples to make reasonable estimates.	Record students' estimates on the board as they share them. This will provide a picture of how many students reason additively, proportionally, and distributionally about the situation. The initial estimates can be revisited later in the lesson to see if any shifts in reasoning have occurred.
2. Put students into groups of three to four. Have each group construct a graph to display the results they would expect from drawing 10 names out of a hat (with replacement) 30 times. Have them make a dotplot on a piece of poster board to display the number of females they would expect to obtain in each sample. When the posters are completed, display them in the front of the classroom and hold a discussion about them.	Students will tend to do the following with their data displays: - Make 7 the modal number of females - Show a gradual increase up to the mode and then a gradual decrease - Include a few cases where 10 females were selected for the committee but few or no cases where zero were selected - Include a few cases where there are three or fewer females selected for the committee	Probe students' thinking as they examine and comment on the posters by asking questions such as - Would it be possible to have a situation where the mode is not 7? How likely is such a situation? - Would it be possible to have a situation where there are more ones than threes? How likely is such a situation? - Do all of the possible values have to occur in a sample of 30?	Ask students to revisit their initial conjectures about the answer to the committee problem. Specifically ask if anyone would change his or her initial estimate after seeing posters from other groups in the class. In cases where students do change their initial estimates, ask them to explain the reasons why they changed. This may provide an opportunity for other students to rethink their initial estimates as well.
3. Have each group do a simulation to refine their initial estimate. Give each group a jar with 30 green and 70 yellow chips and ask them to draw 30 samples of size 10. They should record the number of yellow chips drawn in each sample and produce a new dotplot titled "Actual Results" to display alongside the previously constructed posters.	Students will use a variety of strategies for remixing the chips when they replace them, including stirring, shaking vigorously, and shaking weakly. Some may forget to remix. Students will exhibit varying degrees of care in drawing random samples. Some may look into the jar as they draw out samples.	As students work, emphasize the importance of thoroughly remixing the chips before drawing out a new random sample. As students work, emphasize the importance of drawing a truly random sample and not looking at the chips in the jar as they draw from it.	As students work, ask them why it is important to remix. Ask specifically if a jar that was not remixed could reasonably be expected to yield a random sample. If necessary, draw an analogy to reshuffling a deck of cards. As students work, ask them why sampling blindly is important.

(Continued)

(Continued)

Steps of the lesson	Expected student responses	Teacher's responses to students	Assessment strategies and goals
4. Line up the groups' actual results posters beneath their initial posters. Ask them to look across the top row as well as between rows, comparing initial results to actual results. During whole-class discussion, ask them to share what they notice and what seems to be surprising.	Observations about the posters will likely include the following: - Not as many of the actual results posters have modes of 7 as the expected results posters. - Some actual results posters do not include values of 10. - Few groups actually got values below 3 or 4. - The actual results posters are more "staggered" or "uneven" than the initial posters.	As students share their observations about the graphs, record them on the board. Try to elicit responses from each of the groups. If some of the bulleted items in the column immediately to the left do not emerge during class discussion, look for opportunities to bring them up during whole-class discussion.	Monitor the informal and formal language students use to describe the posters. Look for evidence that they see clusters and groups in the data. When formal terms are used to describe the data displays, probe students to explain what the formal language means and why it is (or is not) appropriate for the given situation.
5. Have students simulate drawing 30 samples using Fathom, and have them produce graphs of the results with the software. They should repeat this several times and then make comparisons to the class posters.	Students will see the work with Fathom as a natural extension of the work they did with the chips. They will also see it as a much quicker way to gather sample data than drawing all of the samples by hand.	As students work, encourage them to continue drawing 30 samples several times, and to compare the resultant graphs to one another. Ask them to identify any patterns they notice among the Fathom graphs.	As students work, ask them to write about how the data displays generated via computer simulation compare to those generated in class. Collect students' written explanations to assess their thinking.
6. Have students simulate drawing several more than 30 samples of size 10 by using Fathom.	By this point in the lesson, students may see it as desirable to dramatically increase the number of samples drawn, since the computer software is powerful enough to draw many more samples than one would have time for within the bounds of a class period.	Encourage students to experiment with sample sizes to make a conjecture about when the shapes and centers of the graphs produced with the software begin to more closely resemble the shapes and centers of the graphs on the initial posters students produced during the lesson.	Ask students to explain, writing in a document created using word processing software, how the shape, center, and spread of the distribution change as the number of samples increases. Have students copy and paste graphs from Fathom into their written explanations to help provide support.

How This Lesson Meets Quality Control Criteria

- *Addressing students' preconceptions*: Students' experiential knowledge about variation is drawn on by making an analogy between random variation in a chip drawing situation and the initial problem. School-based knowledge is drawn on in constructing graphs.
- *Conceptual and procedural knowledge development*: The concept of variation, which is often under-emphasized in statistics, is at the core of the lesson. Procedures for constructing data displays by hand are reviewed or learned as posters are constructed. Procedures for using computer software to conduct simulation are reviewed or learned as the software is used to extend the initial exploration.
- *Metacognition*: Students compare estimates about the initial problem situation to the results they obtain through gathering data. Students write about the patterns that they see in the data they gather via computer simulation.

Chapter 10

DEVELOPING STUDENTS' GEOMETRIC THINKING

[handwritten margin notes: deductive system of thought in math geom. — practical · theoretical]

The teaching of geometry has a long, rich history. One of the most important works in the discipline, Euclid's *Elements*, dates back to approximately 300 B.C. In it, Euclid compiled and extended the work of his predecessors. Boyer and Merzbach (1989) characterized the *Elements* as the most influential textbook in history, and speculated that it may be second only to the Bible in terms of number of editions published. Euclid's *Elements* reflects the Greek commitment to establishing and teaching a deductive system of thought in mathematics. Studying such a system can be contrasted with learning geometry for more practical purposes, such as determining measurements. Even though the Greek word *geometria* is rooted in *geo*, meaning "earth," and "metron," meaning measurement (Rubenstein & Schwartz, 2000), the *Elements* does not emphasize the practical measurement aspect. Between the time *Elements* was written and the establishment of modern schools, the question of whether to emphasize the theoretical or practical aspects of geometry has persisted (Stamper, 1906).

The first attempts to bring geometry to schools in the United States focused on the study of deductive reasoning. Geometry made its way into the high school curriculum in the 1840s as colleges added it to their admissions requirements (Herbst, 2002). During this period, students were expected to memorize proofs from textbooks with expositions of Euclidean geometry. It was not until near the beginning of the 20th century that having students produce their own original geometric proofs became a widespread practice. The move toward having students produce original proofs was largely catalyzed by Bull Wentworth's textbook series, which became known for its inclusion of "originals" and came to dominate the textbook market (Donoghue, 2003).

At the outset of the 21st century, new questions about middle and high school geometry exist. Two major questions that have emerged in the past few decades are (1) how does geometric thinking develop? and (2) what role should dynamic software packages play in teaching and learning geometry? Researchers have devoted a great deal of attention to each of these questions. Their findings have important implications for teachers and, accordingly, will be discussed in detail in this chapter.

[handwritten margin notes: Geometric Thinking — dynamic software]

WHAT IS GEOMETRIC THINKING?

[margin handwriting: Two perspectives — normative geometric thinking vs. thinking as students learn geometry — "thinking as they do the Math"]

To understand how to support the development of students' geometric thinking, it can be helpful to begin by considering its nature. Geometric thinking can be examined from two different perspectives: (1) the thinking of mathematicians as they are engaged in doing geometry (hereafter referred to as **normative geometric thinking**), and (2) the thinking exhibited by students as they learn geometry. Since the ultimate goal of instruction should be to help students engage in normative geometric thinking, it is important to understand the maturation process leading to its attainment. To provide perspective on normative thinking, Cuoco, Goldenberg, and Mark's (1996) **geometric habits of mind** are considered below. Then, work inspired by the research of Dina and Pierre van Hiele (van Heile, 1986) is discussed as a means of understanding students' thinking while learning geometry.

Normative Geometric Thinking: Habits of Mind

*[margin handwriting: *Habits of mind]*

Geometers exhibit a variety of habits of mind in carrying out geometric investigations. These include using proportional reasoning, using several languages at once, using a single language for everything, reasoning about systems, studying change and invariance, and analyzing shapes (Cuoco et al., 1996). As a starting point for understanding normative geometric thinking, each habit of mind is considered and illustrated below.

Proportional reasoning is a vital element in reasoning about things such as vectors, fractals, and theorems about planar objects. As an illustration of the central role of proportional reasoning in geometry, consider the diagram shown in Figure 10.1.

Suppose we know that circle C in Figure 10.1 has a radius of 5, $\overline{AB}$ is a diameter, $\overline{BD}$ is a tangent, and $m\overline{BD} = 7$. From that information, it is possible to determine $m\overline{AE}$. First, note that $m\overline{AB} = 10$ because it is twice the length of the radius. In addition, $m\angle ABD = 90°$ because the intersection of a tangent line to a circle and its diameter

Figure 10.1 Diagram for a geometric proportional reasoning problem.

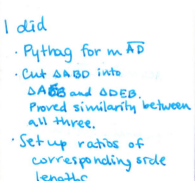
*[margin handwriting: I did
· Pythag for m $\overline{AD}$
· Cut ΔABD into ΔAEB and ΔDEB. Proved similarity between all three.
· Set up ratios of corresponding side lengths.]*

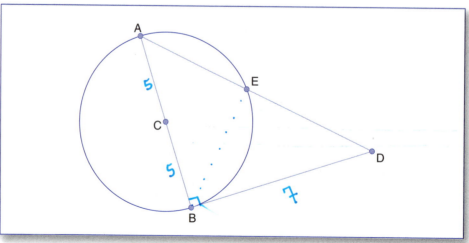

forms a right angle. This means that ABD is a right triangle, so the Pythagorean theorem can be applied to determine $m\overline{AD}$. Proportional reasoning then becomes the key to solving the problem if one looks for a triangle similar to triangle ABD that contains $\overline{AE}$. This sort of proportional reasoning problem often appears on teacher certification exams (Educational Testing Service, 2009).

Determine m $\overline{AE}$ in Figure 10.1. Show all of your work and explain how you reached your conclusion.	**STOP TO REFLECT**

In addition to using proportional reasoning, Cuoco and colleagues (1996) noted that there are times when geometers use several languages at once and times when they use a single language for everything. It is often the case that multiple techniques can be used to solve a problem. Coordinate geometry and vectors are among the languages that can be brought to bear in solving a single geometric problem. While these languages often come to the aid of geometers, geometry itself assists many other branches of mathematics. Euclidean geometry, for example, provides language to talk about algebraic objects such as the coordinate $(3, 4)$ (a point), $y = 3x + 1$ (a line), and $x + 2y + 3z = 1$ (a plane). Other branches of mathematics, such as number theory, can also be simplified by using the geometric concept of point. Number theorists call the points in the Cartesian plane with integer coordinates lattice points. Lattice points can be considered fundamental objects of study in number theory. Essentially, the relationship between the language of geometry and that of other branches of mathematics is reciprocal. Geometry often comes to the aid of other branches of mathematics, just as tools from other branches can be useful for geometric problems.

language

Another hallmark of geometric reasoning is the richness of the systems that geometers construct and work within. Euclid's geometric system was based on five postulates. Controversies surrounding the fifth postulate, often called the *parallel postulate*, provoked the creation of several alternative, non-Euclidean systems. Referring to Figure 10.2, the parallel postulate essentially states that if $m\angle CAB + m\angle DBA < 180°$, then $\overrightarrow{CA}$ and $\overrightarrow{DB}$ must eventually intersect on the side of $\overrightarrow{AB}$ where $\angle CAB$ and $\angle DBA$ are situated. Many efforts were made to prove that Euclid's parallel postulate was actually just a consequence of the previous four. One individual who attempted such a proof was Nikolai Lobachevsky. In the early 19th century, he came to believe that no such proof was possible, and his attention shifted to designing a valid geometric system based in part on an axiom directly contradicting the parallel postulate (Boyer & Merzbach, 1989). Upon the publication and widespread acceptance of his results, normative modes of geometric reasoning were permanently shaken. Euclidean geometry was no longer considered the sole arbiter of absolute truth, since Lobachevsky's geometry, and several others, proved to be logically consistent as well. Geometers henceforth did not restrict their investigations to the Euclidean plane, and they began to study new ideas such as spherical and hyperbolic geometry.

Rich systems as work environment

Figure 10.2   Illustrating Euclid's parallel postulate.

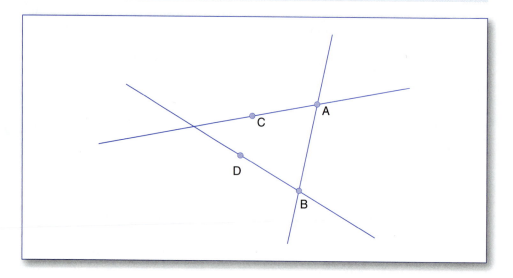

As geometers work within systems, they often explore characteristics of geometric objects that change as well as those that remain invariant. To illustrate the exploration of change, consider the three diagrams shown in Figure 10.3.

Figure 10.3   Changing the location of the intersection of two segments.

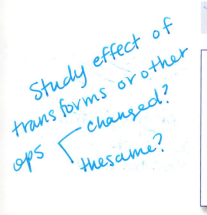

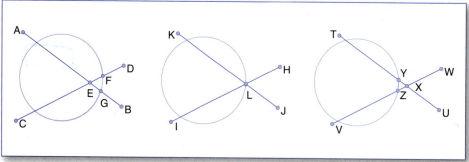

Geometers are interested in exploring such matters as how the arc lengths and angles in Figure 10.3 relate to one another. In the circle farthest to the left, the point of intersection between two chords lies in the interior of the circle. In the second diagram, the intersection point lies on the circle. Finally, in the third diagram, it lies on the exterior. Arc and angle relationships change as the point is moved. Along with change, invariance is interesting to geometers. Cuoco and colleagues (1996) noted that looking for invariants under geometric transformation is a particular point of interest. For example, identifying points that do not move under transformation can help one find the location of the center of a rotation.

Given the preceding examples of normative geometric thinking, it probably goes without saying that studying shapes is another fundamental geometric habit of mind. Three categories pertinent to the study of shape are classification, analysis, and representation (Cuoco et al., 1996). Shapes can be classified into categories based on attributes such as congruence, similarity, symmetry, self-similarity, and topology (Senechal, 1990). Analysis in geometry regularly includes looking for lines of symmetry, using lattices, and dissecting shapes (Senechal, 1990). Representations can come in the form of physical models of geometric objects, maps, shadows and lenses, drawings, and computer graphics (Senechal, 1990). These forms of representation support visual thinking needed for advancements in mathematics and science.

> **Implementing the Common Core**
>
> See Homework Task 1 to use dynamic geometry software to explore the diagrams shown in Figure 10.3, which lead students to "understand and apply theorems about circles" and "find arc lengths" (Content Standard G-C).

Students' Geometric Thinking: van Hiele Levels

It is important for teachers to know that acquisition of normative geometric habits of mind is generally not a rapid process. In their doctoral dissertation work, Dina and Pierre van Hiele identified several levels of development through which students tend to pass in learning geometry (van Hiele, 1986). These are commonly referred to as the **van Hiele levels**. The van Hiele levels have been used extensively to guide investigations of students' geometric thinking. Researchers have invested a great deal of time in testing the levels against empirical classroom data. Though some have questioned the descriptive power of the levels and have proposed refinements to them, the van Hiele levels remain one of the most influential lenses for studying the development of students' geometric thinking.

Battista (2009) provided a summary of current thought on the van Hiele model and the characteristics of each level. Level 1 is referred to as **visual-holistic reasoning**. At this level, students can name shapes when they are shown to them. However, the names are based on the general appearance of the shapes rather than on careful analysis of their properties. A student reasoning at level 1, for example, may see rectangles and squares as completely different kinds of shapes simply because they perceive rectangles to be "longer" than squares. At this point, students characterize shapes by general appearance rather than by carefully comparing components such as sides and angles. Many level 1 students are also affected by the orientation of a shape. For instance, if a square is rotated from its conventional position, they may consider it to be a "diamond" rather than a square (see Figure 10.4).

Battista (2009) called van Hiele level 2 **descriptive-analytic reasoning**. At this level, students begin to differentiate among shapes by analyzing their component parts. Hence, rather than considering an object to be a rectangle because it is long and skinny, students begin to focus on properties such as angles at intersections of segments and how segments are oriented relative to one another. Students' descriptions of these properties may consist of informal language until the formal terms *point*, *segment*, and *angle* are learned in school. Although level 2 students begin to describe shapes in terms of their component parts, they do not make connections

Figure 10.4   The same square in two different orientations.

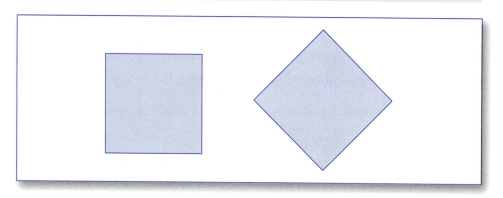

among the descriptions they give. For example, they may give definitions for *square* and *rectangle* that capture many of the pertinent characteristics of each shape, yet not understand how a square can be considered a special type of rectangle.

Students do begin to see relationships among definitions for geometric shapes at van Hiele level 3, **relational-inferential reasoning** (Battista, 2009). Inferences about the characteristics of shapes are generally made from observing many examples. For instance, after constructing and measuring many parallelograms, students may conclude that their opposite sides are always congruent because the property holds in all of the examples they consider. Constructing definitions for shapes by drawing on multiple examples prepares students to reason hierarchically about the definitions. Since squares and rectangles can both be described as quadrilaterals with four right angles, for instance, students can begin to see the logic of categorizing a square as a special type of rectangle. Even those who see this sort of logic, however, sometimes initially resist imposing a hierarchy on definitions.

Attainment of van Hiele level 4, **formal deductive proof** (Battista, 2009), is the goal of most high school geometry courses. At level 4, students understand the importance of undefined terms, definitions, axioms, and theorems in deductive reasoning. They can construct proofs by drawing on given information and using previous results to build a deductive argument. Common tasks at the high school level that require level 4 thinking include proving that two triangles in a diagram are similar or congruent by using theorems such as side-angle-side and angle-side-angle. High school geometry textbooks are often replete with such exercises.

Battista (2009) called the final van Hiele level **rigor**. At this level, students are able to reason about alternative axiomatic systems. They can understand that more than one logically consistent system of geometry exists. Although the study of non-Euclidean geometries is usually left to university-level mathematics courses, some mathematics educators advocate them as enrichments to conventional high school courses. House (2005), for example, provided ideas for teaching **taxicab geometry** at the secondary school level. Taxicab geometry redefines the conventional concept of distance in plane geometry. Superimposing a grid on the plane helps illustrate the difference in definitions (see Figure 10.5).

The left side of Figure 10.5 shows how distance is measured in conventional plane geometry. The distance between (−4, 3) and (5, −1) is measured "as the crow flies" and can be determined using the distance formula ordinarily taught in algebra.

Handwritten margin notes:

Level 3: relational-inferential reasoning

use examples (many) to see how shape characteristics relate

Level 4: Formal Deductive Proof

construct proof from combo of givens with previous results

Level 5: Rigor

can reason about alternative axiomatic systems

Figure 10.5   Two different ways to measure the distance between (–4, 3) and (5, –1).

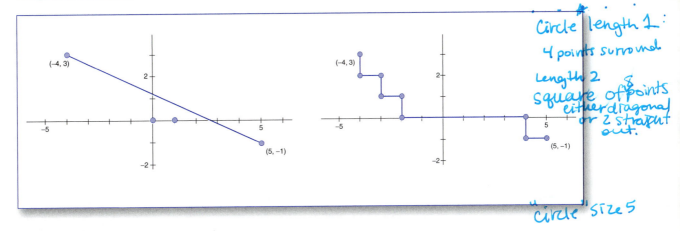

(handwritten margin notes)

Circle length 1:
4 points surround
Length 2
square of points
either diagonal
or 2 straight
out.

circle size 5

The right side of Figure 10.5 shows the measurement of distance in taxicab geometry. The coordinate grid can be thought of in terms of city streets, and distance is measured by considering the shortest path a taxi could drive in commuting between two points on the grid. Counting the number of unit grid lengths traced out by the shortest path gives the distance. Redefining the concept of distance in this manner can be used as a springboard for a task such as asking students to determine the conditions under which Euclidean and taxicab distances are the same and when they are different. Students may also be asked to describe what a circle would look like in taxicab geometry, given that a circle is defined as the set of all points equidistant from a given point (House, 2005).

(handwritten margin note)

$d = x + y$

STOP TO REFLECT

Sketch several examples of taxicab geometry circles. Explain how taxicab geometry circles are similar to and different from Euclidean circles. Also write a distance formula that can be applied to find the distance between any two points in taxicab geometry, and compare and contrast this formula with the one for finding the Euclidean distance between two points in the Cartesian plane.

Some of those who have challenged the descriptive power of the van Hiele model question whether students progress through the levels in a linear fashion. Gutiérrez, Jaime, and Fortuny (1991) argued that it is more accurate to speak of students' thinking in terms of degree of acquisition of van Hiele levels rather than as a progression of discrete jumps from one level to the next. Their data suggested that students may function at several different levels simultaneously. Despite this potential limitation, Gutiérrez (1992) found van Hiele levels to be useful for characterizing students' reasoning with three-dimensional objects. In addition, Jaime and Gutiérrez (1995) used van Hiele levels to describe students' reasoning about geometric transformations.

Perhaps the most vivid insights yielded by research on students' acquisition of van Hiele levels are the characterizations of students' thinking after completion of high school geometry courses. Recall that the goal of most high school geometry courses is to help students attain van Hiele level 4, deductive reasoning. In a study of 2,700 students from five different states, Usiskin (1982) found that most students did not progress beyond the first two van Hiele levels, even after completing high school geometry. Senk (1985) underscored Usiskin's results by finding that only 30% of students from proof-oriented geometry courses attained 75% mastery in proving. Even those who demonstrated mastery often did not know the purpose of constructing proofs. Without such understanding, proof becomes a meaningless, mechanistic ritual to perform rather than a means of building knowledge in mathematics. Clearly, conventional high school geometry courses have fallen well short of the goal of helping all students attain van Hiele level 4.

The preceding discussion of the van Hiele levels suggests aspects of students' geometric thinking that need teachers' attention. Students' reasoning about fundamental geometric objects and their definitions is emphasized in the first three van Hiele levels. Students' ability to construct and understand proofs compose the core of the fourth level. Reasoning about geometric measurement is inherent in understanding both shapes and proof. A particular form of geometric measurement emphasized in school curricula is trigonometry. Some school geometry curricula are also beginning to emphasize more contemporary topics such as transformations, tessellations, chaos, and fractals. Therefore, students' patterns of thinking in regard to all of the preceding curricular areas will be explored next.

> **Implementing the Common Core**
>
> See Clinical Task 2 to administer a comprehensive assessment of students' ability to construct geometric proofs (Content Standards G-CO, G-SRT, and G-GPE) and use the assessment data to drive instructional decisions.

Understanding Fundamental Shapes and Their Definitions

Students build their own personal definitions for shapes from examples they see in school and in everyday life. Through their experiences, students build **prototypes** that can become quite strong and influential in their thinking. Prototypes can be described as mental images that exemplify categories (Lakoff, 1987). Those that capture many of the relevant aspects of a category can be helpful in learning mathematics, and those that are more limited can be detrimental (Presmeg, 1992). Unfortunately, students often build limited prototypes for geometric objects, such as (1) an altitude always lies inside a triangle, (2) diagonals always lie inside a polygon, (3) right triangles have their right angles oriented toward the bottom in a diagram, and (4) the base of an isosceles triangle is positioned at the bottom of a diagram (Hershkowitz, Bruckheimer, & Vinner, 1987). In regard to numbers 3 and 4, individuals often find it more difficult to recognize isosceles and right triangles when presented with nonprototypical images. The influence that prototypes exert on individuals' thinking can make it difficult for them to understand and accept formal definitions that conflict with their entrenched prototypes. When possible, teachers should select unusual examples to share with students to challenge and uproot limiting, entrenched prototypes.

> **Implementing the Common Core**
>
> See Clinical Tasks 3 and 5 to explore students' attention to precision (Standard for Mathematical Practice 6) when writing geometric definitions.

Another issue to deal with in teaching formal definitions for geometric objects is that there exist both **hierarchical** and **partitional definitions** (de Villiers, 1994). Hierarchical definitions, as the term suggests, establish a system in which a hierarchy of concepts can be formed. In the case of quadrilaterals, many textbooks adopt a hierarchical classification scheme in which a square is a special type of rectangle, a rectangle is a special type of parallelogram, and a parallelogram is a special type of quadrilateral. A primary reason for adopting hierarchical definition schemes is that they simplify the process of deductive proof. For instance, under the hierarchical system suggested above, if one can prove that a property is true for all parallelograms, then the property automatically applies to all squares as well. Unfortunately, students often resist adopting hierarchical definition schemes because of entrenched prototypes that lead them to favor partitional ones. Partitional definitions, as the name suggests, partition concepts into separate, mutually exclusive bins. Many students have prototypical images of squares and rectangles that suggest one is not a subset of the other. This leads them to favor distinct definitions for the two concepts rather than overlapping ones.

IDEA FOR DIFFERENTIATING INSTRUCTION   Discussing Prototypes

Encouraging students to discuss the their personal prototypes for geometric objects is essential to helping them develop richer prototypes and accept more efficient hierarchical definitions. De Villiers, Govender, and Patterson (2009) suggested that it is inefficient for teachers to simply present definitions to students and expect them to commit them to memory. Instead, students should be encouraged to trace out some of the reasoning that went into the formation of the definition. One way to start engaging students in this reasoning process is to ask them to draw as many examples of a shape as possible. As students do this, teachers can gauge the range of personal prototypes students hold for the shape. Individual prototypes can then be shared and discussed publicly. In one instance (Groth, 2006b), this process helped a class negotiate a shared definition for *trapezoid*. Two definitions emerged as students shared drawings of their personal prototypes: (1) a quadrilateral with at least one pair of parallel sides and (2) a quadrilateral with exactly one pair of parallel sides. The class discussed the consequences of each definition in terms of which types of shapes would be considered trapezoids and which would not. Such considerations led some to favor one definition over the other. In formulating and debating the two definitions, the class was able to explore some of the considerations that go into creating geometric definitions, rather than just seeing the definitions in their finished form. Students at many different levels of understanding were able to participate in the process because gaining access to the task required reflection on personal prototypes rather than complete knowledge of formal definitions.

TECHNOLOGY CONNECTION

Dynamic Geometry Software, Drawings, and Constructions

Dynamic geometry software environments (DGEs) can also be used to help students form more powerful prototypes for concepts and understand formal definitions. DGEs such as Geometer's Sketchpad, Cabri, and GeoGebra can be used to prompt students to consider the fundamental characteristics of objects they are asked to construct. Consider the task of constructing a square in Geometer's Sketchpad. When asked to do

(Continued)

(Continued)

so, many students produce a drawing of a square rather than a construction (Hollebrands & Smith, 2009). That is, they often use the segment tool to construct four segments that appear to be of equal length and perpendicular at the appropriate intersections. A **DGE drawing** ceases to be a square when dragged, but a **DGE construction** remains a square when dragged. See Figure 10.6 for

Implementing the Common Core

See Homework Task 2 to get a start on helping students "use appropriate tools strategically" (Standard for Mathematical Practice 5) when dynamic geometry software is available.

an illustration of how a drawing loses its "squareness" when dragged, and see Figure 10.7 for a construction of a square that remains a square when dragged. In general, objects that are drawn lose their properties when dragged, but those that are constructed maintain them. Prompting students to make constructions rather than drawings can help them carefully examine the properties of the given shape.

Figure 10.6   A DGE drawing of a square before and after dragging vertex C.

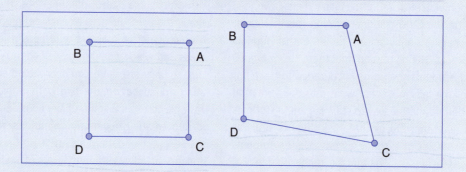

Figure 10.7   A DGE construction of a square before and after dragging vertex C.

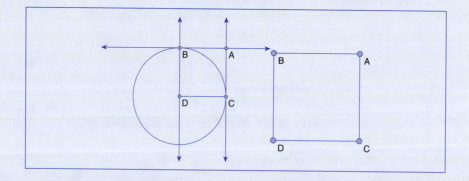

The drawing in Figure 10.6 was produced simply by using the segment tool to construct four connected segments and arranging them into a square-looking object. When producing such a drawing, students may go so far as to use the angle and side measurement capabilities of the DGE to ensure that they have four right angles and four congruent sides. However, as soon as one of the sides or vertices is dragged, the drawing loses its squareness. This problem can be overcome by producing a construction as shown in Figure 10.7. To produce the square figure shown, the segment tool was used to produce DC. Then the perpendicular line command was used to construct BD. The intersection point of circle D was then constructed with the appropriate software command, and another perpendicular line was constructed through point B. One last perpendicular line was constructed through point C, and its intersection with BA was constructed to form the fourth vertex of the square. Since the object of interest was the square, all extraneous por-

Implementing the Common Core

See Clinical Task 4 to assess a student's ability to construct a square with dynamic geometry software (Content Standard G-CO.12).

tions of the construction were taken away by using the appropriate Hide Objects command. Dragging sides and vertices in the construction produces new squares rather than destroying its squareness.

K. Jones (2000) found that asking students to produce constructions, rather than drawings, can help improve their understanding of formal definitions and relationships among geometric objects. He began instruction by asking students to produce objects in a DGE that could not be "messed up" by dragging. This request prompted students to move beyond simply producing drawings of the objects. In one case, students were to construct a rhombus and explain why it was a rhombus. In another case, they were asked to produce a rectangle that could be dragged to make a square, and then to explain why all squares can be considered rectangles. Later on, they were asked to construct a kite that could be dragged to produce a rhombus and explain why rhombi can be considered kites. By the point in the instructional sequence when the kite task was given, students had become increasingly formal in their use of mathematical statements. Instead of using informal language or the language of the software in their explanations, they had transitioned to using formal geometric language. This transition was demonstrated by the increasing sophistication of their statements about squares and rectangles. Initially, a square was considered to be a type of rectangle simply because it "looked like one." Using the DGE led students to refine their justification by stating that a rectangle can be "dragged into a square." Finally, at the end of the unit, students discussed the fact that both are "quadrilaterals with four right angles." Producing constructions of objects within the DGE led to this gradual adoption of normative geometric language and modes of thinking.

Implementing the Common Core

See Homework Task 3 to try your hand at making geometric constructions with dynamic geometry software (Content Standard G-CO.12).

Follow-up questions:

1. What other types of quadrilaterals could be constructed in a DGE? Explain how you would construct at least one other quadrilateral.

2. Why is it important for teachers to know the distinction between "drawings" and "constructions" in DGEs?

TECHNOLOGY CONNECTION

Analyzing Premade Dynamic Constructions

Another DGE-based strategy that can facilitate understanding of definitions of shapes and relationships among them is having students work with premade constructions rather than producing their own. The Shape Makers environment (Battista, 2003) supports such an approach. Shape Makers comes packaged

(Continued)

(Continued)

with premade constructions for shapes such as squares, trapezoids, parallelograms, and rectangles. Yu, Barrett, and Presmeg (2009) described two types of tasks they asked students to do when using Shape Makers. In one type of task, students experimented with different shape makers to produce given pictures made of shapes (e.g., a person made of squares, rectangles, circles, and other shapes). Carrying out these tasks prompted students to examine the range of capabilities for each shape maker. In another type of task, students experimented with shape makers to see which ones would produce different shapes (a square, a rhombus, a trapezoid, etc.). As they worked through the tasks, students were encouraged to write down their thoughts and conjectures and share them with the instructor and with each other. Exploring the capabilities of each shape maker and making their thinking processes explicit helped students refine personal prototypes for the geometric shapes under consideration.

Follow-up questions:

1. In what types of situations would you want your students to work with premade constructions? In which situations would you want them to construct their own?

2. What kinds of conjectures about relationships between shapes would you expect students to make when interacting with Shape Makers? Which important geometry concepts could they learn in the process?

UNDERSTANDING AND CONSTRUCTING GEOMETRIC PROOFS

Developing normative, formal definitions for shapes lays a foundation for understanding geometric proofs. Such a developmental progression is suggested by the van Hiele levels, since level 4 thinking involves the capability to deal with proof. Unfortunately, as noted earlier, far too few students are successful in understanding proof upon completing their high school geometry courses (Senk, 1985; Usiskin, 1982). The following discussion describes some of the reasons for the widespread failure to understand proof and some steps that can be taken to remedy the situation.

The Intellectual Need for Deductive Proof

A major cause of the difficulty with proof in schools appears to be that students often feel no intellectual need to reason deductively (Hershkowitz et al., 2002). **Deductive reasoning** involves incorporating accepted statements such as theorems, postulates, and definitions into a logical argument. For example, if one wishes to prove that all triangles contain 180 degrees, a deductive argument that involves theorems about alternate interior angles could be used as a key aspect. Simply measuring the angles in several different triangles would not be sufficient, since it is not possible to measure every triangle that could possibly be constructed. However, students are often satisfied of the truth of a conjecture after seeing only a few specific cases where it holds up. Most of the geometry students Koedinger (1998) interviewed were satisfied that the diagonals of kites are perpendicular after seeing a few specific examples. Students with this tendency have been called **naive empiricists** (Balacheff, 1988) because they rely entirely on the empirical evidence produced by a finite number of cases.

Naive empiricism, though a prevalent cause of difficult with proof, is not the only cause. Some students do not even go so far as to gather empirical evidence to support a conjecture. Instead, they exhibit external proof schemes, believing that truth is established by appealing to outside authority (Harel & Sowder, 1998). In addition, even those who know the mechanics of deductive proof may not see an intellectual need for it. Some of the students interviewed by Koedinger (1998) were actually able to write a deductive proof showing that diagonals of a kite are perpendicular. However, these students usually only did so when the interviewer used the prompt "Do a proof like you do in school." Furthermore, some students who can follow a deductive proof may not be convinced that it covers all cases. Chazan (1993) found that some students wanted additional empirical examples to support a deductive proof after it had been written. These results support the idea that lack of intellectual need, perhaps even more than lack of knowledge of the mechanics of proof, is a key roadblock.

> ### Implementing the Common Core
>
> See Clinical Task 6 to probe a student's intellectual need to prove a theorem about the sum of the measures of the interior angles of a triangle (Content Standard G-CO.10).

TECHNOLOGY CONNECTION

Using DGEs for Proof Activities

DGEs present both challenges and opportunities when it comes to helping students see the need for deductive proof. DGEs allow users to produce many examples very quickly. For instance, a student may quickly become convinced that the interior angles of all triangles sum to 180 degrees while dragging vertices and observing that the sum remains constant (Figure 10.8).

Figure 10.8   Empirical examples of triangle measures produced in a DGE.

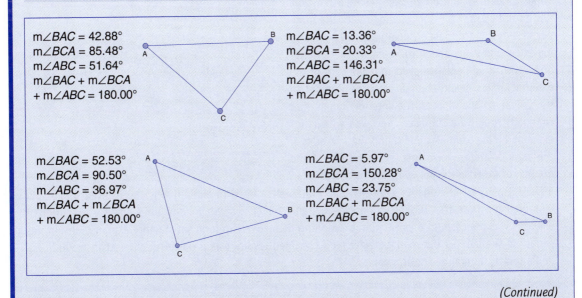

(Continued)

(Continued)

Since empirical examples are so easy to come by in a DGE, the potential exists for students to see even less need for deductive proof (Mariotti, 2001). However, it is also true that mathematicians often examine many empirical examples in order to become convinced that a proposition is worth attempting to prove (de Villiers, 1998). Although students, particularly naive empiricists, may become too confident in conclusions gained from analyzing empirical examples, analyzing examples intelligently is part of normative geometric thinking. Hence, it does not seem reasonable to prohibit students from using DGEs for proof-oriented activities. Instead of asking *if* students should be allowed to use DGEs, a more productive question is *how* teachers can help students use them appropriately.

One way to help students engage in productive work in a DGE is to carefully choose the questions they address with the technology. De Villiers (1998) recommended having students make conjectures about patterns they see when exploring within a DGE, and then asking them to explain why the conjectures are true. In one activity, for example, students were asked to construct a triangle and the midpoints on each of its sides. They were then to connect the midpoints to their opposite vertices to form the medians of the triangle (Figure 10.9) and state a conjecture about the medians. Students tested their conjectures by dragging the triangle and observing what happened to the medians as the triangle become obtuse, scalene, and right. After drawing tentative conclusions about the situation, students shared with one another. They were then asked to explain why the conjecture was true by explaining it "in terms of other well-known geometric results" (p. 392). After constructing deductive explanations, students again shared them with one another to identify areas of agreement and disagreement as well as the explanations that seemed most satisfactory. Such an approach stands in contrast to conventional proof instruction, where students are generally given statements and then asked to prove them. In de Villier's activities, students had roles in formulating the conjectures to be proven, which helped create an intellectual need to explain why the conjectures were true.

Hadas, Hershkowitz, and Schwarz (2000) used a different approach to establishing an intellectual need for deductive proof. They set up a situation where conjectures students formed while examining empirical examples proved to be incorrect. In an introductory activity, students were asked to determine the sum of the interior angles in a polygon, and to notice that the sum changed with the number of sides. From their observations, they were to make a conjecture about the sum of the exterior angles. Most students believed that the sum of the exterior angles would change with the number of sides, just as the sum of the interior angles had changed. They were surprised, however, when they examined more examples and found that the sum of the exterior angles was constant, regardless of the number of sides on the polygons they constructed in a DGE. This sparked students' curiosity, and many felt a need to explain why the exterior angle sum remained constant. The students did not believe that the DGE constructions provided an explanation of why this was the case, so they set about reasoning deductively to form satisfactory explanations. In this situation, the conflict between expected and obtained results was a powerful catalyst in moving students toward formal geometric proof.

Figure 10.9   Medians of a triangle.

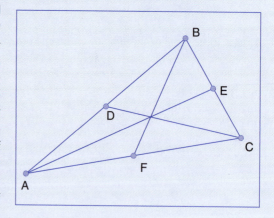

Hadas and Hershkowitz (1998) also used the element of surprise to help prompt students toward forming deductive proofs. They asked students to make a conjecture about a diagram where one of the angles of a triangle was trisected with two segments (Figure 10.10). Most students believed that the side opposite the trisected angle would be split into three congruent parts. When students tried to support this conjecture by measuring empirical examples, they were not able to produce any examples to support the conjecture. Students drew on geometric properties they had learned in the past to begin to explain why the conjecture actually was not true. As in the Hadas, Hershkowitz, and Schwarz (2000) study, the elements of uncertainty and contradiction of intuitive beliefs provoked an intellectual need for deduction.

Since naive empiricism is deeply ingrained in the thinking of many students as they study geometry, teachers should not expect rapid mastery of deductive proof. Moving from empirical observations to deductive reasoning is a process that takes time. DGE use is likely to be most effective in moving students toward proof when adequate instructional time is invested (Marrades & Gutiérrez, 2000). Though it may seem inefficient at first to allow extensive amounts of time for examining examples, making conjectures, and explaining why the conjectures are true, such a sequence of activities is optimal for helping students begin to engage in normative geometric thinking patterns.

> **Implementing the Common Core**
>
> See Homework Task 4 to prove a theorem about the medians of triangles (Content Standard G-CO.10).

Figure 10.10   A triangle with one angle trisected by two segments.

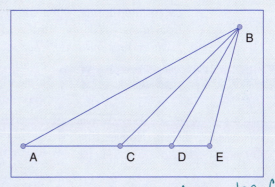

Follow-up questions:

1. Why is the sum of the exterior angles for any polygon 360°? Provide a deductive explanation.

2. Provide your own example of a counterintuitive geometric property that students could investigate in a DGE. How would students' DGE findings for your example conflict with their intuition?

Establishing a Classroom Culture of Proof

A DGE is just one possible element of a classroom environment that promotes a culture of proving. Martin, McCrone, Bower, and Dindyal (2005) studied a geometry classroom where students successfully made conjectures, provided justifications, and built chains of reasoning. To encourage these behaviors, the teacher posed open-ended tasks, placed responsibility for reasoning on the students, and analyzed their reasoning to determine when further coaching was necessary. In one classroom episode, the teacher began by giving students the open-ended task of writing down everything they knew about a pair of congruent pentagons. The task led one student to conjecture that the distances between nonadjacent corresponding vertices in each pentagon were equal. The teacher asked students to investigate the conjecture through paper folding, affirmed the truth of the statement, and then asked students to explain why it was true. In response, students attempted to construct deductive explanations. Some of the attempts at deductive explanation were initially unsuccessful. The teacher introduced counterexamples highlighting the portions of the students' arguments that needed to be rethought. Students persisted in their attempts to construct deductive explanations because the teacher's feedback and interaction indicated that their attempts were valued, even if they were not initially correct.

Many teachers believe that providing a specific proof writing format is another way to support students' attempts to construct proofs. The two-column proof format, in which statements are written in one column and corresponding reasons in the other, is deeply ingrained in the culture of teaching geometric proof in the United States (Herbst, 2002). A typical two-column proof from high school geometry is shown in the top portion of Figure 10.11. In recent decades, the two-column format has been criticized. Schoenfeld (1988) remarked that constructing geometric proofs can become a ritualistic and mechanistic enterprise when the form of a proof is emphasized more than its substance. Moreover, mathematicians do not hold themselves to using the two-column format when constructing proofs. In an attempt to shift students' focus from the two-column format to the actual substance of proofs, *Curriculum and Evaluation Standards for School Mathematics* (National Council of Teachers of Mathematics [NCTM], 1989) identified two-column proof as a topic that should receive less attention. Proof itself was still to be an important part of the curriculum, but NCTM recommended de-emphasizing the two-column format in an attempt to shift attention toward the quality of students' deductive reasoning and away from their ability to adhere to a specific format. Two other possible formats for writing proofs, flowcharts (McMurray, 1978) and paragraphs (Brandell, 1994), are shown in Figure 10.11.

> ### Implementing the Common Core
>
> See Clinical Task 7 to inquire about a teacher's strategies for using different formats to help students construct proofs (Content Standards G-CO, G-SRT, and G-GPE).

Any given format for writing a proof has potential weaknesses and strengths. A potential weakness of any form is that students may begin to focus more on form than on substance, as noted earlier. The primary strength of paragraph proofs is that they closely resemble the types of proofs constructed by mathematicians. Therefore, students who can read and construct paragraph proofs may be in better position to succeed in college mathematics. Nonetheless, other formats may be useful for scaffolding students' thinking so that they can ultimately master paragraph proof. Even the two-column format, which has been somewhat demonized in the recent past, can serve a useful scaffolding role when used appropriately. Weiss, Herbst, and Chen (2009) noted that the two-column format can be a useful tool for outlining the general structure of a proof. In one classroom they observed, a student sketched a general structure for a proof using two columns, initially skipping some of the reasons in the second column but returning to fill them out later. When used in this way, the two-column format can help students organize their thinking. On the other hand, when teachers insist that two-column proofs be filled out in a linear fashion, with each step justified before another step may be written, students' thinking is constrained. In general, any proof format is valuable when used to help facilitate thinking rather than impede it.

Under any form of proof, care must be taken to ensure that students interpret the accompanying diagrams correctly. Battista (2007) described a variety of ways that students misinterpret the intended meanings of geometric diagrams. One common misinterpretation is to believe that a deductive proof only covers the specific diagram

Figure 10.11   The same proof written in three different formats.

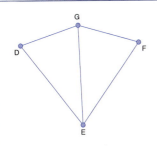

Suppose we know that $\overline{EG}$ bisects ∠DGF and that ∠FEG ≅ ∠DEG. Prove that triangles DGE and FGE are congruent.

Two-column proof:

Statements	Reasons
1. $\overline{EG}$ bisects ∠DGF	1. Given
2. ∠DGE ≅ ∠FGE	2. The two angles formed by an angle bisector are congruent.
3. $\overline{GE}$ ≅ $\overline{EG}$	3. Reflexive property
4. ∠FEG ≅ ∠DEG	4. Given
5. Triangles DGE and FGE are congruent.	5. Angle-side-angle postulate

Flowchart proof:

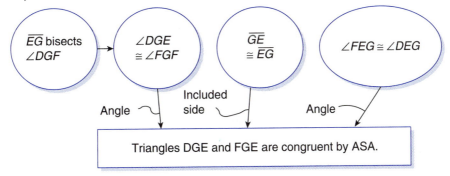

Paragraph proof:

The two triangles can be shown congruent by using the angle-side-angle congruence postulate. First, note that ∠DGE ≅ ∠FGE because $\overline{EG}$ bisects ∠DGF, and the bisector of an angle splits it into two congruent angles. Next, we know that $\overline{GE}$ ≅ $\overline{EG}$ by the reflexive property. Finally, it was given that ∠FEG ≅ ∠DEG. This shows that two corresponding angles in each triangle and their included corresponding sides are congruent. Hence, triangles *DGE* and *FGE* are congruent by the angle-side-angle (ASA) congruence postulate. Q.E.D.

accompanying the proof. In reference to Figure 10.11, for example, some students may believe that if $\overline{EG}$ were lengthened, a new proof would be required, even though the essential structure of the situation would remain unchanged. Another common misinterpretation is to believe that features of a diagram can disprove a theorem established deductively. Suppose, for example, a teacher drew a diagram of a circle and a tangent line, and then drew a radius out to the tangent line that appeared to intersect it at an acute angle. Such a diagram will lead some students to believe that the intersection of a radius and a tangent line does not always form a right angle. To help students avoid misinterpretations like these, it is important to explicitly discuss the meanings of geometric diagrams. Students need to understand that diagrams simply serve as (sometimes imperfect) visual props for working toward a deductive proof.

Geometric Measurement

Along with the study of shapes and proofs, measurement undergirds most geometry courses taught in middle and high schools. As in other areas of the mathematics curriculum, there are many student thinking patterns that should be taken into account in planning instruction. Students' thinking about measurement of length, area, volume, and angles is discussed below.

Measuring Length

An item from the National Assessment of Educational Progress (NAEP), commonly called the "broken ruler problem," provides a good starting point for discussing students' understanding of length measurement. The version of the problem given to students in 2003 is shown in Figure 10.12.

Figure 10.12   Broken ruler problem from 2003 National Assessment of Educational Progress.

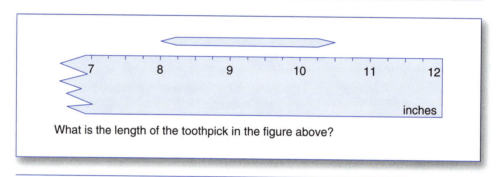

What is the length of the toothpick in the figure above?

Source: http://nces.ed.gov/nationsreportcard/naepdata/.

Approximately 42% of eighth graders (Blume, Galindo, & Walcott, 2007) and 20% of high school seniors (Struchens, Martin, & Kenney, 2003) answer the broken ruler problem incorrectly. There are several ways students can go wrong. Some have difficulty reading fractional lengths. Even those who can do so may say that the toothpick is 10½ inches long because they are accustomed to reading the location of the endpoint on the right side of the ruler to determine a measurement. Those who do attempt to determine the distance between the two endpoints of the toothpick may do so incorrectly by counting each mark above a whole number as one unit of length. That is, since the whole numbers 8, 9, and 10 fall within the length of the toothpick, students may incorrectly conclude that the toothpick is 3½ inches long.

> **Implementing the Common Core**
>
> See Clinical Task 8 to investigate students' attention to precision (Standard for Mathematical Practice 6) when measuring length with a broken ruler.

Miscounting the units in the length of an object like the toothpick in the broken ruler problem suggests a fundamental misunderstanding of length measurement. Students need to understand that the distance between two consecutive whole numbers, rather than a mark above a number, represents a unit of length. One strategy for helping students understand this important characteristic of measurement involves setting aside formal units of measure and rulers. Van de Walle (2001) recommended that students measure length by iterating an informal unit along an object. For example, students might use their hand spans to measure the length of a table. Doing so reinforces the idea that one essentially lays the same unit end to end over and over again to measure length. When repeatedly iterating the informal unit becomes tedious, students can abbreviate the process by making their own measuring sticks by taping together several copies of the unit. This can help them understand how conventional rulers abbreviate the iteration process. Finally, when students obtain different measurements for the same table because their hands are different lengths, the concept of formal units can be introduced and appreciated for its ability to facilitate discussions about the length of an object.

Understanding how the iteration of units composes the foundation of measurement can also be useful when students attempt to estimate. Another problem that causes difficulty on the NAEP involves estimating the length of one object with another. On the 2003 NAEP, students were asked to estimate the length of the 882-foot-long cruise ship *Titanic*. They were to choose the most accurate estimate from among several options: 2 moving van lengths, 50 car lengths, 100 skateboard lengths, 500 school bus lengths, or 1,000 bicycle lengths. Only 39% of eighth graders answered correctly (Blume et al., 2007). The low rate of success on the item suggests a lack of opportunity to think about measuring an object through the iteration of units. Although such opportunities would ideally occur in the lower grades, middle and high school teachers may need to address gaps in students' understanding by providing opportunities for measuring via iteration of units in the later grades.

Another way to build students' understanding of length measurement involves using a **geoboard**. A geoboard is essentially a pegboard on which rubber bands can be strung to create geometric objects. Dot paper can be used in place of a physical geoboard. Some online applets also replicate physical geoboards (Figure 10.13). Ellis and Pagni (2008) described an instructional sequence for using the geoboard to help students understand lengths not represented by whole numbers. They asked students to determine the lengths of the sides of simple objects, such as squares and rectangles, formed on the geoboard. Students could do so by counting the number of units (rather than the number of pegs) along each side. They then asked students to determine the lengths of diagonals on the geoboard. This prompted students to use the Pythagorean theorem. If a length of 1 is assigned to the distance between neighboring pegs on the geoboard, then the shortest diagonal is $\sqrt{1^2 + 1^2} = \sqrt{2}$. Once that length has been established, students can take on more complicated problems, such as determining how many segments of $\sqrt{10}$ can be found on the geoboard.

> **Implementing the Common Core**
>
> See Homework Task 5 to try your hand at a geoboard task that requires students to "understand and apply the Pythagorean Theorem" (Content Standard 8.G).

Figure 10.13 A geoboard from the National Library of Virtual Manipulatives (http://nlvm.usu.edu/en/nav/vLibrary.html).

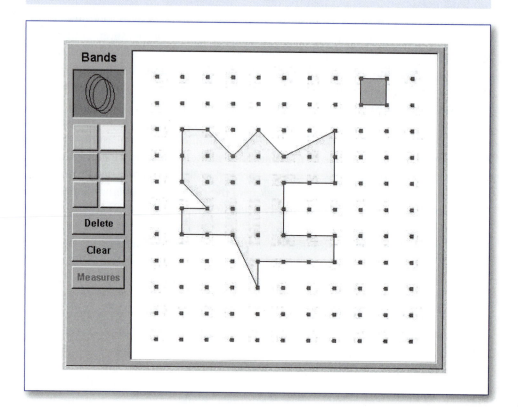

Measuring Area

Students' difficulties in measuring length feed difficulties in measuring area. NAEP data indicate that only about 24% of eighth graders can determine the surface area of a given rectangular prism (Blume et al., 2007). In addition, teachers are sometimes uncertain about relationships between perimeter and area of geometric objects. Ma (1999) asked teachers how they would respond to a student who claimed that the area of a closed figure always increases as its perimeter increases. Many of the teachers from the United States responded that they would ask the student to produce several examples to verify the claim. Few responded that they would guide students toward counterexamples that would actually refute the claim, like the one shown in Figure 10.14.

> ### Implementing the Common Core
>
> See Clinical Task 9 to explore how well students "construct viable arguments" (Standard for Mathematical Practice 3) when discussing the relationship between area and perimeter of a shape.

One of the prominent portions of the curriculum where measurement of length and area interact is the Pythagorean theorem. Although the Pythagorean theorem is often used simply to determine the length of a side of a right triangle when only the other two side lengths are provided, it also expresses an interesting relationship

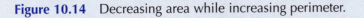

Figure 10.14   Decreasing area while increasing perimeter.

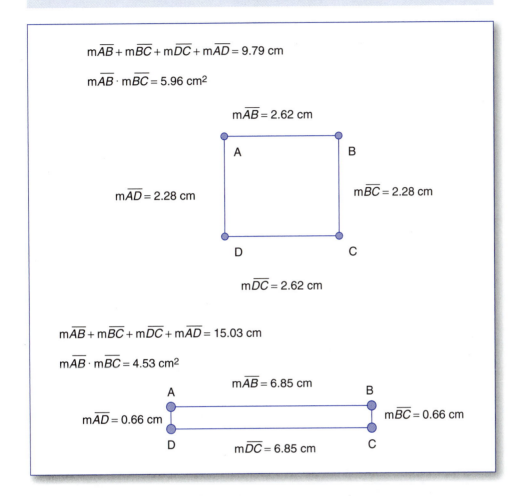

among areas of squares. Essentially, it states that the sum of the areas of the squares constructed on the legs of a right triangle will equal the area of the square constructed on the hypotenuse (Figure 10.15). There are many ways to help students understand the Pythagorean theorem in terms of area. Yun and Flores (2008), for example, suggested using jelly beans as informal units of area measure-

Implementing the Common Core

See Homework Task 6 for an opportunity to "explain a proof of the Pythagorean Theorem" (Content Standard 8.G.6).

ment. They constructed several cardboard containers in the shape of the diagram shown in Figure 10.15. They then asked students to fill each square portion with a layer of jelly beans and describe the relationships among the areas of the squares. Yun and Flores recommended extending the activity by having students examine containers with circles and other shapes constructed on the sides of the right triangle to conjecture whether or not the relationship between the areas still holds. As in length measurement, work with informal units can be followed by work with formal units represented by unit squares.

Figure 10.15 Expressing the Pythagorean theorem in terms of area.

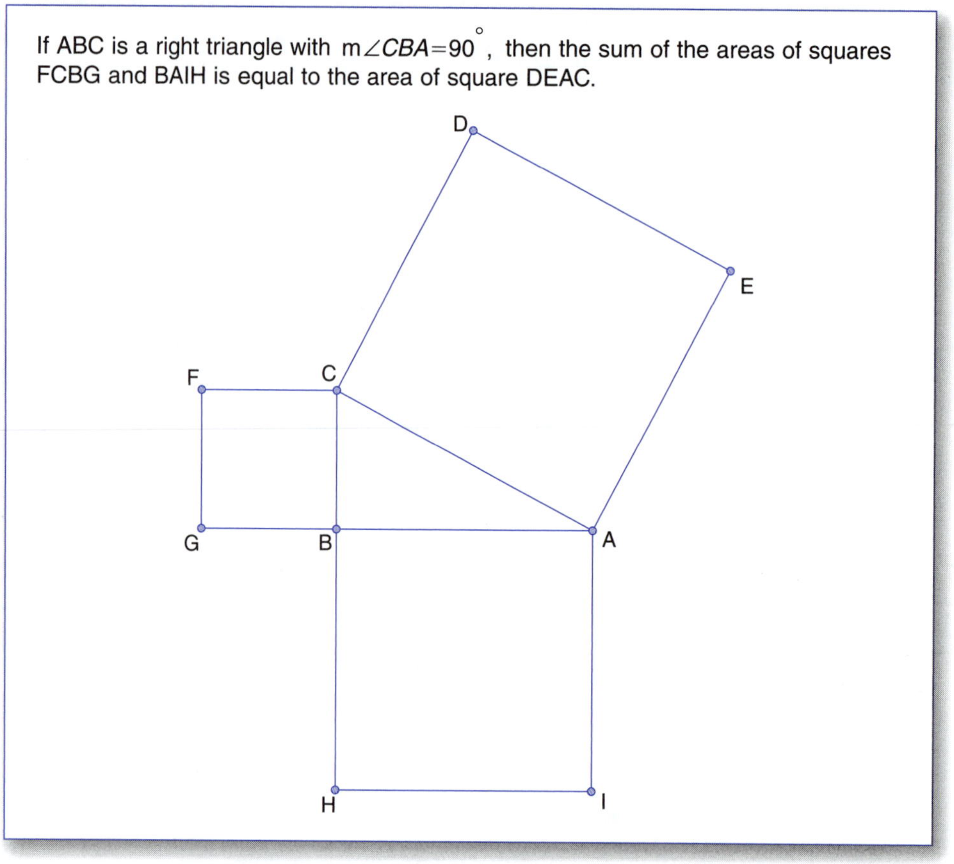

If ABC is a right triangle with $m\angle CBA = 90^\circ$, then the sum of the areas of squares FCBG and BAIH is equal to the area of square DEAC.

IDEA FOR DIFFERENTIATING INSTRUCTION Measuring Nonrectangular Areas

Encouraging students to measure nonrectangular areas with unit squares sets the stage for diverse, rich thinking strategies to emerge. Hodgson, Simonsen, Lubek, and Anderson (2003) described an activity that required students to measure the area of the state of Montana. When presented with the task, students suggested superimposing a grid of unit squares on a map. They then proposed methods to obtain more accurate measurements, such as making the squares progressively smaller to minimize empty spaces within the grid. Utley and Wolfe (2004) suggested using the unit squares on geoboards as a means of measuring the areas of different shapes. Since geoboards have unit squares built into their structure, they can be used to visualize area for a variety of shapes, such as the trapezoid shown in Figure 10.16. One might form a rectangle around the

Implementing the Common Core

See Homework Task 7 to explore how a geoboard can be used for problems that require students to solve mathematical problems involving areas of unusual shapes (Content Standards 6.G and 7.G).

trapezoid and then cut out two triangles, or cut the shape itself into a rectangle and two triangles. Students may also devise alternative strategies for determining its area. It is important to note that knowing the formula for the area of a trapezoid is not necessary to perform the task. Determining the area is actually a good precursor to deriving a formula, as students can be prompted to find shortcuts for calculating area after using more time-consuming visual methods repeatedly. The formula can then be understood and appreciated as an abbreviation of the visual methods rather than simply as a teacher-invented recipe for determining area. Students who discover a formula for the area of a trapezoid in advance of others can be encouraged to devise additional area formulas for more complex geometric shapes.

Figure 10.16   A geoboard representation of a trapezoid.

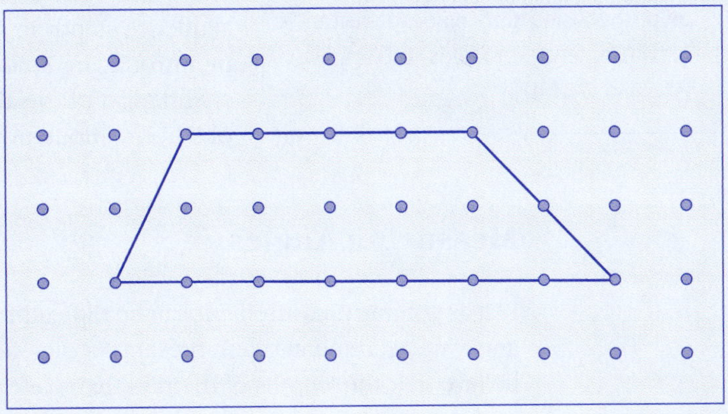

Describe at least four different ways to determine the area of the trapezoid shown in Figure 10.16. Explain your reasoning completely.

STOP TO REFLECT

Measuring Volume

As with area measurement, it is important for students to develop conceptual knowledge of volume before dealing with procedural formulas. Students who are very good at memorizing and using formulas to determine volume often have little understanding of what the formulas mean. NAEP results show that approximately 45% of 12th-grade students do not know that 48 cubic inches represents a measure of volume (Battista, 2007). Without knowledge of the fundamental unit that composes volume, students have little chance to develop conceptual understanding of volume measurement.

Battista (1999) described an instructional sequence for helping students develop conceptual understanding of volume. Instruction began by asking students to determine the number of cubes it would take to fill different boxes represented by two-dimensional drawings (see, for example, the diagrams in Figure 10.17). Students made conjectures, compared them with one another, and then tested the conjectures using actual cubes and boxes. Discrepancies between predicted and actual results prompted students to go back and revise their thinking. In some cases, students found their predictions to be incorrect because they double-counted cubes or omitted cubes in the middle of a box in forming their predictions. Some students began the task of counting the number of cubes in each box by using skip counting, but gradually moved to multiplication as a more efficient strategy. After working with

Implementing the Common Core

See Clinical Task 10 to investigate a student's ability to decompose a trapezoid into simpler shapes (Content Standard 6.G.1) to determine its area.

Implementing the Common Core

See Clinical Task 11 to assess a student's ability to determine the volume of prisms using nets made of rectangles (Content Standards 6.G.4, 7.G, 8.G, and G-GMD).

several boxes (i.e., rectangular prisms), students moved on to explore other solids, such as pyramids. At the end of the instructional sequence, students were able to successfully enumerate cubes in 3D arrays. Battista's instructional sequence stands in stark contrast to typical units that begin by introducing students to formulas for volume and then spend most of the allotted time having students repeatedly practice computation with those formulas.

Measuring Angles

Just as volume measurement can be difficult for students to understand conceptually, angle measurement often presents a significant cognitive hurdle. Some students believe that the lengths of the rays that make up an angle influence its measure (i.e., as the rays become longer, the angle measure increases even though the rays remain in the same orientation to one another; Struchens et al., 2003). Students also have a difficult time conceiving of angle measurement as an amount of turn (Mitchelmore & White, 2000). This particular student difficulty can be partially explained by the language teachers use to describe angles. Browning and Garza-Kling (2009) found that prospective teachers tend to describe angles as "corners" or "something you measure in degrees." Such descriptions of angles fail to emphasize their usefulness for measuring the amount of turn from one position to the next.

Given students' difficulty in conceiving of angle measurement as an amount of turn, geometry instruction should explicitly address this characteristic. Browning and Garza-Kling (2009) fostered understanding in this area by asking students to imagine that the degree, a standard unit of angle measurement, had not yet been invented. Students were to devise their own strategies for measuring angles. They settled on forming wedges and iterating them until they had completely measured out an angle. Since each group of students began with a different-size wedge, the importance of

Figure 10.17　A net and a representation of its three-dimensional structure when folded.

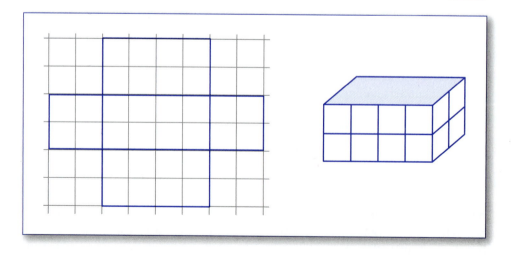

having a standard wedge to measure all angles became apparent so that each group would assign the same measure to a given angle. This led to understanding and appreciation of a degree as a standard "wedge size" used for angle measurement. Later on, Browning and Garza-Kling asked students to work with a graphing calculator applet showing angles being swept out by turning segments. Such experiences highlighted important concepts that compose the foundation of angle measurement.

> **Implementing the Common Core**
>
> See Homework Task 8 for an opportunity to design a manipulative to help students understand the concept of angle measurement (Content Standard 8.G).

TRIGONOMETRY

Many of the geometric concepts discussed so far, particularly the idea of angle, form the core of secondary school trigonometry. Despite the importance of the subject, research on the teaching and learning of trigonometry is in its infancy. It will be important for teachers to track developments in this field of research in the coming years as it continues to develop. Insights from existing research are discussed next.

In recent years, researchers have highlighted potential conceptual difficulties with common approaches to teaching trigonometry. P. W. Thompson (2008) claimed that the trigonometry of right angles and the trigonometry of periodic functions are often treated in isolation. Weber (2008) agreed, noting that the calculation of ratios in static triangles is often overemphasized in comparison to time spent building functional understanding. Thompson went on to note that trigonometry is often taught procedurally, with little emphasis on the conceptual underpinnings of angle measurement. Because of this, students may be able to transition between radians and degrees without understanding that the two are essentially just different units for measuring angles. In trigonometry, as in many other areas of the mathematics curriculum, fundamental problems with teaching and learning appear to be often rooted in an imbalance between emphases on procedural and conceptual knowledge.

Weber (2008) described a teaching sequence that departs from conventional approaches to trigonometry. He started instruction by asking students to work with a circle with a radius of 1 and centered on the point (0, 0) (i.e., a **unit circle**) on graph paper. On the unit circle, students drew angles and approximated the values of different trigonometric functions by measuring the coordinates of the intersection of the terminal segment and the unit circle. In Figure 10.18, for instance, students could estimate sine by estimating the y-coordinate of the intersection and cosine by estimating the x-coordinate. After associating sine with y-coordinates and cosine with x-coordinates, Weber had students estimate sine and cosine for several different examples. Along the way, students were asked to determine the exact values for sine and cosine of $0°, 90°, 180°, 270°$ and $360°$ without measuring. With this background, students were prepared to approach conceptual trigonometric tasks, such as determining whether or not it is possible to have a situation where $\sin(x) = 2$, deciding on the sign of $\cos(300°)$, and determining if $\sin(23°)$ is larger or smaller than $\sin(37°)$. Weber (2005) reported that this general approach helped students develop a deep understanding of trigonometric functions.

Figure 10.18 Unit circle diagram.

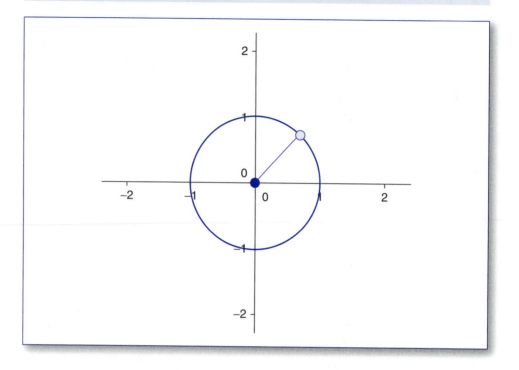

Cavey and Berenson (2005) described how teachers can improve their understanding of right triangle trigonometry by engaging in a modified version of Japanese lesson study. They traced the learning of one preservice teacher, Molly, during the process of collaborative planning and modification of lessons. Initially, when asked to teach a unit on right triangle trigonometry, all Molly could recall about the topic was the mnemonic "SOHCAHTOA." The acronym provided a means for calculating sine, cosine, and tangent, but little else (i.e., the acronym states that sine is "opposite [O] over hypotenuse [H]," cosine is "adjacent [A] over hypotenuse [H]," and tangent is "opposite [O] over adjacent [A]"). Through collaborative planning, Molly was able to take her teaching of trigonometry beyond this simple mnemonic. One idea she gained, for example, was using a **clinometer** (Figure 10.19), a tool consisting of a protractor, straw, washer, and string that can be used to measure angles in real-world situations involving right triangle trigonometry. In revising and extending her lessons, she also enriched her understanding of the mathematical concepts of ratio and similarity. Her improved content knowledge helped enhance the lessons she taught. Thinking about the mathematics within the context of her own practice proved to be a crucial element in Molly's development. Sharing her plans with others was a key mechanism in helping her identify and address gaps in her mathematical and pedagogical knowledge.

Implementing the Common Core

See Clinical Task 12 to assess students' ability to "solve problems involving right triangles" (Content Standard G-SRT) and "extend the domain of trigonometric functions using the unit circle" (Content Standard F-TF).

Implementing the Common Core

See Clinical Task 13 to assess students' ability to use trigonometric ratios to solve applied problems (Content Standard G-SRT.8) with a clinometer.

Figure 10.19   Components of a clinometer.

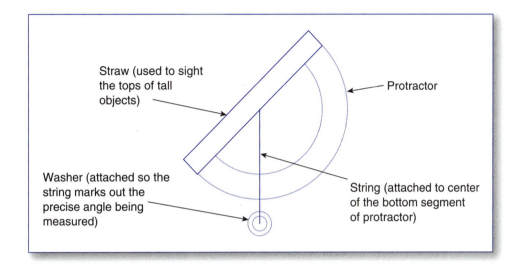

Straw (used to sight the tops of tall objects)

Protractor

Washer (attached so the string marks out the precise angle being measured)

String (attached to center of the bottom segment of protractor)

Describe how students could use a clinometer to measure the heights of very tall objects, such as the school building or trees surrounding it. Write at least three exercises you would ask students to do outside using clinometers. Include a diagram showing proper use of the clinometer to measure a tall object.

STOP TO REFLECT

CONTEMPORARY TOPICS IN GEOMETRY

As geometry continues to develop as a discipline, the school curriculum should respond accordingly. There has, of course, been much advancement in the discipline of geometry since the time of Euclid, yet the study of plane geometry and traditional measurement continue to dominate the curriculum. Contemporary topics that can be included in secondary curricula are transformation geometry, tessellations, fractals, and chaos (National Governor's Association for Best Practices & Council of Chief State School Officers, 2010; NCTM, 2000). As these ideas are relatively new to school curricula, we are just beginning to investigate optimal ways of teaching and learning them. Nonetheless, existing mathematics education research does provide some useful insights.

Transformation Geometry

Isometries can be defined as transformations in the plane that preserve the distance between points (Jaime & Gutiérrez, 1995). Isometries typically included in the high school curriculum include translations, rotations, and reflections. Isometries can be produced by a variety of methods, including paper folding, using dynamic geometry software,

Implementing the Common Core

See Homework Task 9 to explore a variety of transformation geometry problems (Content Standards 8.G, G-CO, and G-SRT) that can be approached using a Mira manipulative.

using online applets such as those available on the National Library of Virtual Manipulatives, and employing a Mira tool. Figure 10.20 shows isometries involving a hexagon.

Figure 10.20 Isometries involving a hexagon.

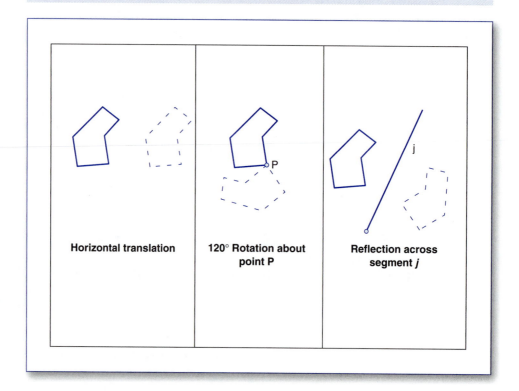

Horizontal translation

120° Rotation about point P

Reflection across segment *j*

Jaime and Gutiérrez (1995) suggested using the van Hiele levels as a structure for determining goals for the study of plane isometries. They identified van Hiele level 3 as a suitable target for secondary school activities involving isometries, since most courses at this level aim to help students construct chains of deductive reasoning. Aiming for van Hiele level 3 means going beyond having students produce transformations. Jaime and Gutiérrez recommended tasks such as asking students to explain why the product of rotations is equivalent to a translation when the sum of the rotation angles is a multiple of 360°. Another recommended task was to explain why the product of two rotations is either a rotation or translation. Writing explanations for why transformations behave as they do can help students build the deductive reasoning skills characteristic of the higher van Hiele levels.

Tessellations

Students who have studied some transformation geometry can appreciate **tessellations**, which are tilings of a plane that do not contain any gaps or overlaps. Three regular polygons will tile the plane in this manner: triangles, squares, and hexagons. These three tile the plane because the measures of their interior angles are divisors of 360°. Figure 10.21 shows tilings done using an applet on the National Library of

Virtual Manipulatives website. Using applets, software programs, or just paper and scissors, students can make and test conjectures about the kinds of shapes that will tile the plane.

Figure 10.21   Tessellations using regular polygons.

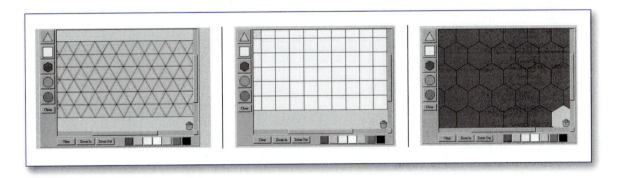

Perhaps the best-known tessellations occur in the artwork of M. C. Escher. Examples of how Escher incorporated tessellations in his paintings can be seen on www .mcescher.com.

Escher went beyond merely tiling the plane with shapes. He performed transformations on the shapes to form unique figures and then used them to tile the plane. Shockey and Snyder (2007) described an approach to helping students produce Escher-like tessellations. They asked students to take a square, cut a design along one edge of it (corner to corner), and then translate the design to the opposite side of the square (Figure 10.22). When the design is used to tile the plane, an Escher-like picture is formed. One could perform the same procedure with the other pair of opposite sides of the square to produce a different portrait. Students can also experiment to find other shapes and transformations that tile the plane.

Fractals and Chaos

Concepts from fractal geometry have begun to make their way into middle and high school curricula. **Fractals** can be described in the following terms:

> Roughly speaking, fractals are complex geometric shapes with fine structure at arbitrarily small scales. Usually they have some degree of self-similarity. In other words, if we magnify a tiny part of a fractal, we will see features reminiscent of the whole. Sometimes the similarity is exact; more often it is only approximate or statistical. (Strogatz, 1994, p. 398)

Sierpinski's triangle (Figure 10.23) is one well-known fractal. Notice the self-similarity that exists within it. Strogatz noted that other fractals resemble naturally occurring objects such as clouds, coastlines, and blood vessel networks. Their ability to capture characteristics of natural objects, along with the fact that many discoveries in fractal geometry have been made in the very recent past (Devaney, 1998), make fractals a potentially exciting addition to school curricula.

Figure 10.22   Creating a simple Escher-like picture.

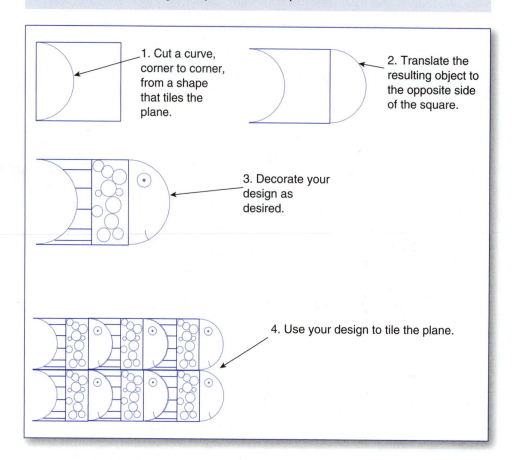

1. Cut a curve, corner to corner, from a shape that tiles the plane.

2. Translate the resulting object to the opposite side of the square.

3. Decorate your design as desired.

4. Use your design to tile the plane.

Figure 10.23   Sierpinski's triangle generated with an online applet (http://curvebank.calstatela.edu/sierpinski/sierpinski.htm).

Devaney (1998) described a game that can be used to introduce students to the mathematics underlying the construction of Sierpinski's triangle. He described the rules of the *chaos game* (Barnsley, 1989) in the following terms:

> First pick three points—the vertices of a triangle (any triangle works—right, equilateral, isosceles, whatever). Name one of the vertices 1,2, the second 3,4, and the third 5,6. The reason for these strange names is that we will use the roll of a die to determine the moves in the game. To begin the game, choose any point in the triangle. This point is the seed for the game. (Actually, the seed can be anywhere in the plane, even miles away from the triangle). Then roll a die. Move the seed halfway toward the named vertex: If 1 or 2 comes up, move the point half the distance to the vertex named 1,2. Now erase the original point and repeat this procedure, using the result of the previous roll as the seed for the next: Roll the die again to move the new point half the distance to the named vertex, and then erase the previous point. (p. 92)

After following the instructions above for a small number of rolls, students should connect the points generated. After many trials, regardless of the initial seed used, Sierpinski's triangle will begin to emerge from the pattern. The chaos game can be extended by asking students to experiment with changing the rules and observing the pattern generated, or by starting with a picture of a fractal and attempting to discover the rules that generated it.

CONCLUSION

Geometry presents many interesting ideas to study, dating from the historical era before Euclid to the present day. The fundamental goals of school geometry include understanding shapes and their definitions and constructing proofs. Currently, we know that far too few students completing high school geometry are able to write and understand proofs. Therefore, teachers must look for opportunities to help students move past naive empiricism and develop deductive reasoning. Dynamic geometry software can help progression toward this goal when used appropriately. As students move toward deductive proof, they should also develop an understanding of measurement as the process of iterating a given unit. The iteration process applies to the measurement of length, area, volume, and angles. Geoboards, rectangular grids, cubes, and wedges are among the physical tools helpful for developing students' conceptual understanding of measurement. Trigonometry takes angle measurement as one of its central objects of study, and at the same time presents an opportunity to further develop students' understanding of functions. Relatively recent developments in geometry related to transformations, tessellations, and chaos help add vibrancy to the subject. Sample four-column lessons are provided at the end of the chapter to further spark your thinking about teaching geometry.

VOCABULARY LIST

After reading this chapter, you should be able to offer reasonable definitions for the following ideas (listed in their order of first occurrence) and describe their relevance to teaching mathematics:

Euclid's *Elements*   307

Normative geometric thinking   308

Geometric habits of mind   308

van Hiele levels   311

HOMEWORK TASKS

1. Describe relationships that exist between the angles and arcs shown in Figure 10.3. Use dynamic geometry software and geometry textbooks as resources as necessary. Show all your work and justify your reasoning.

2. Suppose your school is searching for a DGE software program to adopt for its high school geometry courses. Examine the websites of at least three DGEs (e.g., Cabri, Geometer's Sketchpad, GeoGebra; do an Internet search to find the official websites). Then write a letter to the high school mathematics department chair recommending one of the software programs for adoption. Justify your adoption recommendation with details about the unique aspects of the selected DGE.

3. Use DGE software to construct a rhombus, a rectangle, and a kite that cannot be "messed up" (i.e., they retain their defining characteristics when dragged). Show and describe all steps in the construction. Then hide all portions of the construction extraneous to each shape, as done for the square on the right-hand side of Figure 10.7.

4. State a conjecture about the medians of a triangle (see Figure 10.9). Provide evidence that your conjecture may be true by giving paper-and-pencil examples that are carefully constructed or producing examples within a DGE. Then prove the conjecture deductively, drawing on other well-known results from geometry.

5. Determine how many segments of $\sqrt{10}$ can be found on a 10×10 geoboard. Also describe how to determine the number of segment lengths of $\sqrt{10}$ you can find on an $n \times n$ geoboard. Show all of your work and justify your answer.

6. Do an Internet search on "proofs of the Pythagorean theorem." Choose two proofs you would share with a high school geometry class you might teach in the future. Explain the essential similarities and differences between the logic of each of the proofs you choose.

7. Form a nonrectangular shape on a geoboard and describe how to determine its area. Your shape should be concave and have at least five sides. Then form another concave, rectangular shape on the geoboard and ask a classmate to find its area. The shape you give to your classmate should present a significant challenge.

8. Draw a diagram of a manipulative that would help students understand the idea of angle measurement in terms of the iteration of wedges. Write three progressively more difficult tasks that can be

solved by using the manipulative. Explain how your tasks could help students begin to understand and appreciate standard units of angle measure (e.g., degrees) and standard instruments for angle measurement (e.g., protractors).

9. Do an Internet search on "Mira geometry tool." Describe four different types of exercises that can be done using a Mira. Be sure to cite the website from which each exercise idea was generated.

10. Describe how students could use a clinometer to measure the heights of very tall objects, such as the school building or trees surrounding it. Write at least three exercises you would ask students to do outside the classroom using clinometers. Include a diagram showing proper use of the clinometer to measure a tall object.

CLINICAL TASKS

1. Download the University of Chicago's van Hiele geometry test (http://ucsmp.uchicago.edu/van_Hiele.html) and administer it to a class of high school geometry students (be sure to request permission to do so, as noted on the project website). Score the test using one of the scoring schemes provided. Construct a data display that could be given to the teacher of the class to summarize the van Hiele levels of the students. Describe how the information in the data display could be used to guide decisions about instruction.

2. Download the University of Chicago's geometry proof test (http://ucsmp.uchicago.edu/van_Hiele.html) and administer it to a class of high school geometry students (be sure to request permission to do so, as noted on the project website). Score the test using one of the scoring schemes provided. Construct a data display that could be given to the teacher of the class to summarize the proof construction abilities of the students. Describe how the information in the data display could be used to guide decisions about instruction.

3. Interview at least three students. Ask each one to draw as many examples of trapezoids as he or she can. After the students have drawn several examples, ask them to write a definition for the word *trapezoid*. Describe what the students' drawings and definitions reveal about their personal prototypes for the concept.

4. Ask a student to use a DGE to construct a square that cannot be "messed up" when dragged (i.e., it remains a square even when portions of it are dragged). Describe how the student responds to the task. In your description, be sure to note whether a drawing or a construction is produced.

5. Ask a student to work with premade constructions in a DGE for a rhombus, a rectangle, and a kite. Ask the student to drag each shape and write about how it changes and stays the same under drag. After experimenting with each shape, have the student write a definition for each shape. Write a report that critiques the student's reasoning process and final definition. In your report, be sure to note whether the final definitions for each shape are hierarchical or partitional.

6. Interview a student who has completed high school geometry and provide him or her access to a ruler and protractor. Ask the student if it is true that all triangles have interior angles whose measures sum to 180 degrees. Then ask the student to justify his or her position. If the student does not use a deductive proof, ask whether he or she can produce a proof like those done in high school geometry classes. Write a report describing what you learned about the student's thinking. In your report, be sure to address whether or not the student exhibited an intellectual need to produce a deductive proof.

7. Interview a geometry teacher to determine which forms of proof he or she encourages students to use. In particular, ask if two-column, paragraph, or flowchart proofs are used. Try to determine why

the teacher uses the form(s) of proofs mentioned. Ask for specific examples of proof exercises students are required to complete. Write a report describing the forms of proof the teacher uses and the types of exercises students are expected to complete. Also give a personal reaction to the rationale he or she provides for choosing forms of proof for use in class.

8. Have a class of students complete the "broken ruler" problem shown in Figure 10.12. In looking through students' work on the item, note how many correctly answered the item and how many did not. Then analyze the papers of students who provided incorrect responses. Describe the type of mistake each student made. Then select two students who provided incorrect responses for interviews. Question the students to understand why they answered incorrectly. Write a report of your findings, along with a general strategy you would use to help students overcome misconceptions associated with this type of problem.

9. Interview three students. Ask each of them to evaluate the truth of the following statement: "As the perimeter of a closed shape increases, its area increases as well." Ask the students to justify their reasoning. Write a report on how the students evaluated the statement, including any diagrams or work they produced in the process of doing so. Then, based on what you learned about the students' reasoning, write a follow-up task you would ask them to solve. The follow-up task should be designed to help extend or correct the thinking patterns you observed.

10. Ask a student to determine the area of the trapezoid shown in Figure 10.16 without using a formula learned in school. Write a report that describes the strategies he or she used to determine its area. To prepare for the possibility that the student is not able to determine the area, write a set of hints you would use to prompt him or her in the right direction.

11. Ask a student to predict how many cubes it would take to fill the box shown in Figure 10.17. Also create a pattern picture of your own without an accompanying box picture and ask the student to predict how many cubes it would take to fill it if the pattern picture were folded into a box. Have the student check each prediction by using physical materials. Write a report that describes the student's initial predictions for each task and any revisions the student made to his or her conjectures after working with the physical materials.

12. Ask at least three students who have studied trigonometry if it is possible to have a situation where $\sin(x) = 2$. Also ask each student to determine the sign of $\cos(300°)$, (i.e., positive or negative). Finally, ask if $\sin(23°)$ is larger or smaller than $\sin(37°)$. Have each student explain his or her reasoning for each task. If students use a calculator, ask them if they can also solve each task without a calculator. Describe how each student responded to the tasks and then identify the conceptual and procedural elements implicit in their thinking.

13. Ask a class to do a task of your own design that requires a clinometer. Have them show all work involved in solving the task and then hand it in. Referring to the students' work, the level of challenge the tasks presented, and logistical issues, suggest ways to improve the activity the next time you use it.

14. Play the chaos game with a class of students. Split students into several small groups and have them record their results on clear overhead transparencies. Each transparency should have the three initial vertices for the game in the same location. Consider giving each group a movable dot so they do not have to erase points while playing the game. When the students are finished playing in small groups (let them do approximately 10 trials per group), overlay the transparencies on one another and note how closely they resemble a Sierpinski's triangle. Drawing on your experiences of playing the chaos game in class, take a position on the appropriateness of fractal geometry for middle and/ or high school students. Defend your position by drawing on the observations you made while playing the chaos game with your students.

VIGNETTE ANALYSIS ACTIVITY	Focus on CCSS Standards for Mathematical Practice 1–8

Items to Consider Before Reading the Vignette

1. Reread each of the CCSS Standards for Mathematical Practice in Appendix A. Which of these standards have you seen most often in classes you have observed? Which have you seen least often?

2. Provide a statement of the triangle inequality. Describe a strategy you could use to help students understand the inequality.

3. Suppose that in triangles ABC and DEF, side *AB* is congruent with side *DE* and side *AC* is congruent with side *DF*. We also know that the measure of angle A is greater than the measure of angle D. What can we conclude about the lengths of sides *BC* and *EF*? Why?

4. Item 3 above suggests a geometric theorem commonly called the "hinge theorem." Why do you think it has this name? How would you help students understand the connection between the name of the theorem and its content?

Scenario

Mr. Martz was just beginning his student teaching semester. He had serious misgivings about embarking on a career in teaching. During his classroom observations, he had noticed that many students did not have what he considered "basic skills," such as the ability to solve simple equations and to factor and multiply polynomials. They also seemed indifferent about studying mathematics, whereas he loved the subject. Now he would be responsible for teaching the students he observed. Nervous about how he would be effective as their classroom teacher, he had frequent conversations with his mentor teacher and university supervisor about strategies he could employ. Mr. Martz was not completely convinced that the strategies they suggested would work, but nonetheless did try to take some of their advice into account. In one of the first geometry lessons he was responsible for teaching, students were to learn the geometric "hinge theorem" (see items 2 and 3 in the previous section). In teaching the lesson, Mr. Martz relied on a combination of his own intuition about students and the advice he received from his mentor and university supervisor.

The Lesson

The lesson began with a warm-up activity intended to review the triangle inequality, an idea taught the previous day. Mr. Martz put three sets of segment measurements on the board and asked students to determine whether or not they would form a triangle. The sets were the following:

a. 1, 5, 7

b. 3, 6, 8

c. 4, 3, 7

In part a, students were to notice that $5 + 1 < 7$, so the sides could not form a triangle. In part b, they were to see that any combination of two side lengths added together would be greater than the remaining side length. Therefore, the side lengths in part b would form a triangle. In part c, students needed to reason that since $4 + 3 = 7$, the segments could not form a triangle. Mr. Martz felt these three cases would be adequate for reviewing the main aspects of the triangle inequality.

After about five minutes of socializing with one another, students settled in and began to write the warm-up activity in their notebooks. As Mr. Martz circulated about the room, he was surprised to see that many did not know how to start. As he answered students' questions, he felt as if he were reteaching the previous day's lesson on an individual basis to each of the 25 students in the classroom. He was especially disheartened because the previous day's lesson had involved using a concrete manipulative, popsicle sticks, to teach the triangle inequality. This strategy was recommended by both his university supervisor and mentor teacher. Sensing Mr. Martz's frustration, his mentor teacher asked the class to stop working on the warm-up problems and direct their attention to her. The mentor teacher reminded students of the previous day's popsicle stick activity and asked students to think of the segment lengths in terms of the popsicle sticks. Questions about how to start the activity then subsided as students seemed to connect the work done with the popsicle sticks to the task at hand. Some students requested rulers so they could draw the popsicle sticks to scale to solve the review exercises. Within 10 minutes, students were ready to move on.

Next, Mr. Martz started what he considered to be the main part of the lesson: teaching the hinge theorem. He asked students to pair up. After milling about for a couple of minutes, each student seemed to have found a partner. The pairs of students were directed to use protractors to obtain two angles. One pair member was to produce a 50° angle, and the other was to produce a 30° angle. Popsicle sticks were to be used to indicate the side lengths in each angle. After the angles had been formed with the popsicle sticks, students were to measure the distance from the tip of one stick to the other in each angle. Mr. Martz had originally planned to have students do these measuring and constructing activities on their own, but now decided to show students how to do each step in the process at the document camera in front of the room, fearing that students left to their own devices might do some steps incorrectly. When Mr. Martz finished demonstrating the steps in the activity, he told students to notice that the distance between popsicle stick tips was greater for the 50° angle than for the 30° angle.

After doing the popsicle stick demonstration for the class, Mr. Martz asked students to break up from their pairs and return to individual work. Students noisily gathered their protractors, popsicle sticks, and pencils to return to their original seats. When most students had settled in once again, Mr. Martz distributed a set of class work exercises. In each exercise, the students were given a pair of triangles. In each triangle, the length of two sides and the measure of their included angle was given. From this information, students were to determine whether the length of the nonincluded side was greater in the first triangle or the second. Mr. Martz showed students how to do the first exercise and then directed them to finish the remaining exercises on their own.

Within 10 minutes, all students had either finished or stopped doing the class work problems and begun to socialize. A few had their heads down and were sleeping. Mr. Martz had not anticipated that his students would finish this portion of the lesson so quickly. He considered starting the next day's lesson, but there were only seven minutes left in the class period. Instead of starting something new, he opted to give them their homework assignment. Mr. Martz loudly announced the page number and exercise numbers for the homework. Some students wrote the information down while others continued to talk. A few opened their books to begin the assignment, but most kept talking. Mr. Martz had a sinking feeling that his lesson had ended with a thud.

Questions for Reflection and Discussion

1. Which CCSS Standards for Mathematical Practice are most evident in the vignette? Which are least evident? In regard to those that are least evident, how could the lesson be improved to help students better work toward the standards?

2. Comment on the numbers Mr. Martz chose for the warm-up exercises. Is his number choice helpful for bringing out the main aspects of the triangle inequality? What improvements could be made?

3. How could the review of the triangle inequality have gotten off to a smoother start? Suggest specific steps to be taken.

4. Comment on Mr. Martz's use and organization of group/pair work. Were there times he should have used it and did not? How could the process of forming groups be made more effective and efficient? What kinds of tasks would be meaningful for pairs or groups to do in the context of learning the hinge theorem?

5. How could Mr. Martz get students more involved in discovering the hinge theorem?

6. Did the structure of the lesson give Mr. Martz good opportunities to assess students' geometric understanding? Why or why not?

7. Would dynamic geometry software be helpful for enhancing any portions of this lesson? Why or why not?

RESOURCES TO EXPLORE

Books

Albrecht, M. R., Burke, M. J., Ellis, W., Kennedy, D., & Maletsky, E. (2005). *Navigating through measurement in Grades 9–12.* Reston, VA: NCTM.

Description: The authors provide a collection of activities for teaching measurement in an inquiry-oriented manner to high school students. Activities address the process of measurement, using formulas to measure complex shapes, discovering and creating measurement formulas, and measuring with technology.

Battista, M. T. (2003). *Shape Makers.* Emeryville, CA: Key Curriculum Press.

Description: This book supports students' work with Shape Makers in the Geometer's Sketchpad dynamic geometry environment. As students manipulate preconstructed shapes, they develop deeper understanding of a hierarchy of quadrilaterals.

Clements, D. H. (Ed.). (2003). *Learning and teaching measurement* (Sixty-fifth yearbook of the National Council of Teachers of Mathematics). Reston, VA: NCTM.

Description: This yearbook contains a number of articles relevant to teaching and learning measurement in secondary school. Articles for secondary school teachers focus on estimating areas of irregular shapes, exploring measurement through literature, and using geoboards to teach measurement.

Craine, T. V. (Ed.). (2009). *Understanding geometry for a changing world* (Seventy-first yearbook of the National Council of Teachers of Mathematics). Reston, VA: NCTM.

Description: This yearbook consists of a collection of articles useful for teaching various aspects of geometry. Articles address topics such as teaching geometry for conceptual understanding, using interactive geometry software, having students discover geometric theorems, and exploring fractals in nature.

Day, R., Kelley, P., Krussel, L., Lott, J .W., & Hirstein, J. (2002). *Navigating through geometry in Grades 9–12.* Reston, VA: NCTM.

Description: The authors provide activities that can be used to teach geometry in an inquiry-oriented manner to high school students. Activities address geometric transformations, similarities, and fractals.

Pugalee, D. K. Frykholm, J., Johnson, A., Slovin, H., Malloy, C., & Preston, R. (2002). *Navigating through geometry in Grades 6–8*. Reston, VA: NCTM.

Description: The authors provide activities that can be used to teach geometry in an inquiry-oriented manner to middle school students. Activities address characteristics of shapes, coordinate geometry, transformations, and visualization.

Websites

Learning Math: Geometry: **http://www.learner.org/resources/series167.html#program_descriptions**

Description: This website contains videos relevant to teaching geometry in the middle school. Programs of interest to middle school teachers deal with proof, the Pythagorean theorem, similarity, and solids.

van Hiele Levels and Achievement in Secondary School Geometry: **http://ucsmp.uchicago.edu/resources/van-hiele/**

Description: This website describes the work of the Cognitive Development and Achievement in Secondary School geometry project. It provides insight on a large sample of secondary students' understanding of geometry, and it also contains tests that can be used to assess students' van Hiele levels and proof writing abilities.

FOUR-COLUMN LESSON PLANS TO HELP DEVELOP STUDENTS' GEOMETRIC THINKING

Lesson Plan 1

Based on the following NCTM resource: Groth, R. E. (2006). Expanding teachers' understanding of geometric definition: The case of the trapezoid. *Teaching Children Mathematics, 12,* 376–380.

Primary objective: To help students understand two commonly accepted definitions of *trapezoid* and the consequences of each for quadrilateral classification schemes.

Materials needed: Paper, pencil, chalkboard, computers with Internet access

Steps of the lesson	Expected student responses	Teacher's responses to students	Assessment strategies and goals
1. Put the following writing prompt on a screen or chalkboard at the beginning of the lesson: "Give an example of a word that has more than one definition. The word does not have to be from mathematics. Write at least two different definitions for it."	Students may list words whose different meanings can be determined from the context in which they are used (e.g., *hack* can mean to physically strike something or to break into a computer). Some may list words that can be either nouns or verbs (e.g., *storm*).	Emphasize that words can often be assigned a variety of definitions. Mention that even some words in mathematics, such as *trapezoid*, can take on different meanings in different definitional systems.	After students have written for a few minutes, have some of them share their responses with the rest of the class. Ask students for examples of mathematical words that can be defined in different ways. This will help assess whether or not they believe that only one definition can be "right" in mathematics.
2. Have students work in small groups to produce as many examples of trapezoids as possible. After working for a few minutes, each group should send a representative to the chalkboard to post the examples they generated.	Some groups will restrict their examples to quadrilaterals with *exactly* one pair of parallel sides. Other groups will include examples of quadrilaterals with *at least* one pair of parallel sides. Some groups will produce shapes that are not quadrilaterals.	At this point in the lesson, do not censor the shapes on the chalkboard. The purpose of this portion of the lesson is to engage students in brainstorming, not to formalize their thinking. The shapes on the board will serve as a catalyst for discussions throughout the rest of the lesson.	Ask students which shapes on the board should be considered trapezoids and which should not. This should help elicit students' current ideas about how trapezoids are to be defined. It will also get students thinking about how they might formally define a trapezoid so that the definition includes shapes they believe to be trapezoids.

(Continued)

(Continued)

Steps of the lesson	Expected student responses	Teacher's responses to students	Assessment strategies and goals
3. Have students return to small-group work. Give them the task of writing a formal definition for the word *trapezoid* based on the discussion and on examples that have been given up to this point.	Some groups will write definitions with more detail than necessary. Others will not include enough detail to guide the reader to produce an example of something they would consider a trapezoid.	Draw attention to weaknesses in students' definitions. For those that include too much detail, use just the necessary components of the definition to produce examples. For those that do not include enough detail, produce examples that are based on the definition to highlight the inadequacy.	Ask students to write second drafts of their group definitions based on the class discussion that occurred after the first drafts were shared. The second drafts will become their working definitions for the next step in the lesson.
4. In groups, students should do Internet searches on "define trapezoid." Have each group use a different search engine (e.g., Google, Yahoo!, Bing) so they obtain slightly different results. They should be prepared to share their results with the rest of the class.	Students will likely find three types of definitions for *trapezoid*: a quadrilateral with exactly one pair of parallel sides, a quadrilateral with at least one pair of parallel sides, and nonmathematical definitions.	Ask each group to report the definitions they found. After they have reported, point out the three types of definitions. Emphasize the two mathematical types of definitions as being commonly used in mathematics curriculum materials.	As students report on the definitions they have found, assess whether or not they found the two types of mathematical definitions. If they did not find both types, be sure to introduce the missing type of definition into the class conversation.
5. Based on the discussion of definitions found online, have students once again produce as many examples of trapezoids as possible.	Some will still resist the idea that more than one legitimate definition can exist, and others will use both definitions to produce examples.	Emphasize the importance of consistency within a system rather than one absolutely "correct" definition for every possible system.	Give a writing exercise: "Can squares, rhombi, and parallelograms ever be considered trapezoids? Why or why not?"

How This Lesson Meets Quality Control Criteria

- *Addressing students' preconceptions:* This lesson connects to school-based and experiential knowledge of defining concepts and expands on students' previous knowledge of geometry by introducing the idea of the existence of more than one legitimate definition.
- *Conceptual and procedural knowledge development:* The lesson addresses the concept that different consequences follow different definitions of a shape and introduces two different definitions that essentially specify procedures to produce trapezoids.
- *Metacognition:* Students are encouraged to compare their thinking with that of classmates at several different points in the lesson, particularly during group discussion. Students are encouraged to compare their thinking to information found online.

Lesson Plan 2

Based on the following NCTM resource: Britton, B. J., & Stump, S. L. (2001). Unexpected riches from a geoboard quadrilateral activity. *Mathematics Teaching in the Middle School, 6,* 490–493.

Primary objective: To help students develop and use strategies for sorting quadrilaterals into families

Materials needed: One geoboard for each student, one piece of dot paper for each student

Steps of the lesson	Expected student responses	Teacher's responses to students	Assessment strategies and goals
1. Have students work in groups and attempt to find all possible quadrilaterals that can be formed on a 3 × 3 section of a geoboard. They should keep track of the quadrilaterals they find by sketching them on a piece of dot paper.	Students will discuss what it means to have different quadrilaterals. Some will interpret "different" to mean noncongruent. Others will think that two shapes that are the same except for size are not different.	While working with groups, encourage them to think of "different" as meaning "noncongruent." Congruence and noncongruence can be determined by laying one shape on top of another and trying to line the two up.	Ask students if a large square and a smaller square can be considered "different." If they do understand that the two squares are considered different in the context of this activity, they are ready to engage fully in the lesson.
2. Have groups share the results of their geoboard explorations. Begin by having each group share one quadrilateral they formed. Continue having groups share one quadrilateral at a time until they have no new different ones to share.	In some cases, groups will present shapes that are congruent to a shape already presented. Groups may not consider the possibility of having concave quadrilaterals.	When groups present shapes congruent to those already presented, ask if a transformation could be performed on the shape to make it the same as another. If concave quadrilaterals are not presented, show an example of one and have students look for the rest of the possible concave quadrilaterals.	Keep track of all examples presented to determine if students have identified all 16 possible quadrilaterals. In classes where some of the quadrilaterals are not identified, provide hints that will guide students to discover them (as with the concave quadrilaterals mentioned in the cell to the left).
3. Take all 16 possible geoboard quadrilaterals and sort them into two groups. In one group, place all shapes that have an obtuse interior angle. Place the rest of the shapes in a second group. Tell students that all of the quadrilaterals in one group have an attribute that is shared by none of the quadrilaterals in the other group. Ask them to identify the attribute.	Some students may immediately notice that obtuse angles set one group apart from the other. Some may have difficulty expressing their thoughts in formal geometric language. Some may identify attributes that are actually shared by both groups.	If a student immediately identifies obtuse angles as the relevant attribute, do not immediately comment on the correctness of the answer. Instead, encourage the class to look for additional possible differentiating attributes. When students use informal language (e.g., "pointy" rather than "acute"), introduce the corresponding formal term.	When students have difficulty expressing their thinking in terms of formal geometric language, look for opportunities during the discussion to assess their acquisition of the language. For example, students who originally do not use the formal term *acute* should be asked to provide descriptions of shapes that contain acute angles at various points in the class discussion.

(Continued)

(Continued)

Steps of the lesson	Expected student responses	Teacher's responses to students	Assessment strategies and goals
4. Give students a worksheet that shows all 16 possible quadrilaterals on dotpaper. Have students cut the shapes apart and classify them in at least two different ways. They should be prepared to present their classification schemes to the rest of the class.	Students may use various different properties to form different classifications, including: number of sets of parallel sides, number of right angles, and type of symmetry. For example, some may put all quadrilaterals with one right angle into one group and the rest of them into another.	Encourage students to go beyond the stated requirements of the task by finding as many ways as possible to sort the quadrilaterals. Encourage and support their use of formal language in describing their categorization ideas to one another.	Assess whether or not students use classification schemes different from those discussed during class. In cases where students do not come up with their own original categorization schemes, prompt them to devise some.
5. As an extension to the main activity for the day, encourage students who finish early to determine the areas of the 16 different shapes introduced during the lesson.	Some students will be eager to take on the task, having exhausted interesting ways to categorize the shapes. Others will need time to continue to categorize the shapes.	Encourage students who work on the area task to share their strategies with one another. Time permitting, choose a few students to present strategies to the entire class.	As students present their area measurement strategies, look for evidence of original thought rather than mere use of previously learned formulas for determining area.

How This Lesson Meets Quality Control Criteria

- *Addressing students' preconceptions:* The informal language and sorting strategies students have learned outside of school are connected to formal language and sorting strategies commonly used in geometry.
- *Conceptual and procedural knowledge development:* Procedurally, students encounter definitions for different geometric shapes; conceptually, they come to understand and appreciate the thinking processes involved in sorting and defining.
- *Metacognition:* Students are asked to examine their thinking to determine whether or not they have produced all possible shapes on a 3 × 3 section of a geoboard; they are asked to examine their use of the word different in reference to shapes.

Lesson Plan 3

Based on the following NCTM resource: Kaufmann, M. L., Bomer, M. A., & Powell, N. N. (2009). Want to play geometry? *Mathematics Teacher, 103,* 190–195.

Primary objective: To help students understand and appreciate the axiomatic structure of geometry

Materials needed: A sheet of poster board for each group of students, markers, household objects to be used for games (see Step 1)

Steps of the lesson	Expected student responses	Teacher's responses to students	Assessment strategies and goals
1. Students should be divided into groups of two to four each. Give each group a miscellaneous set of household objects (e.g., buttons, pins, egg cartons, dice, balls, marbles). Tell each group to devise a game that uses the objects and consists of at least five rules.	Students will be able to draw on their out-of-school knowledge of games to construct rules. However, for some games, it is likely that some rules may contradict each other. It may also be that some rules are not complete. Some rules may also be repetitive.	Allow students to work freely in groups at this point in the lesson. Do not intervene to correct them at this point, since the next step of the lesson involves peer review. One of the objectives of peer review is to develop skill at noticing possible mistakes.	Observe and listen to students as they work and take note of which games have rules that are contradictory, incomplete, or repetitive. Also note the specific flawed rules. This information will be needed later in the lesson.
2. Tell students that they will be reviewing one another's games before they will be marketed. Lead a whole-class discussion about how to determine if the rules for a game are reasonable.	Some students will give ideas that correspond to the three main categories of interest: contradictory, incomplete, and repetitive rules. Others will give ideas not related to these categories.	List all student ideas on the board or a screen as they are given. Near the end of the discussion, highlight the student suggestions that correspond to the three main categories of interest.	Monitor student contributions for evidence of suggestions aligning with the three categories of interest. Encourage further discussion until all three categories have arisen.
3. Have each group construct a poster to display the title of their game and its rules. Posters will then be sent out for review by classmates.	Since the previous portion of the lesson dealt with identifying contradictory, incomplete, and repetitive rules, examples of such rules should be identified by students.	Draw upon the assessment information gained in Step 1 in the lesson to try to ensure that students are not missing important flaws in the games they are reviewing. Draw attention to flaws that students do not identify if they incorrectly believe they have spotted them all.	Assess the students' critiques against the observations you made in assessing Step 1. If students missed a substantial number of flaws while doing peer reviews, choose sample games to critique together during whole-class discussion.
4. Build the analogy between the games the students constructed and axiomatic systems. To do this, ask questions such	Students with some previous knowledge of geometric proof may connect the rules of a game to axioms, and	Scaffold students' learning by asking for specific examples of if-then statements, conjunctions, and the	By listening to students' responses, assess whether or not they have difficulty understanding the role of if–then statements or the

(Continued)

(Continued)

Steps of the lesson	Expected student responses	Teacher's responses to students	Assessment strategies and goals
as the following: "What role do the rules play?" "What role do if–then statements play?" "What happens if you change conjunctions like *and* or *or*?" "What happens if you remove one of the rules?" "What elements of the game correspond to geometric theorems and undefined terms?"	understand that removing one rule changes the system substantially. These students may also speak of objects used to play the game as undefined terms, and of plays that occur during the game as theorems. Even those without much proof experience may recognize the role of if-then statements and how a rule is often changed substantially when a conjunction is changed.	effect of removing a rule from the game or changing a conjunction. Have them draw examples from their own games or from the games they reviewed. Introduce formal geometric language as necessary if it does not arise in the conversation.	impact of changing a conjunction. If so, have them play the game again, this time with one of the rules containing an if–then statement or conjunction changed. They can then report back on their observations.
5. Ask students to compare the U.S. government to an axiomatic system. Lead a brainstorming session about how the U.S. government is axiomatically different from other forms of government.	Students will offer various ideas. They may compare the U.S. Constitution to a set of postulates, compare legislation to theorems, and give reasons for adhering to different axiomatic systems.	Record student ideas on the board as they are offered. Encourage students to construct analogies about postulates, theorems, and axiomatic systems if they do not arise naturally.	Assess the strengths and weaknesses of the analogies students offer. In cases where analogies are greatly stretched, ask if a different analogy that uses the same formal terminology can be constructed.

How This Lesson Meets Quality Control Criteria

- *Addressing students' preconceptions:* Students' out-of-school experiences with rules of games are drawn on to build the idea of an axiomatic system. Students' school-based knowledge of government provides another analogous situation for study.
- *Conceptual and procedural knowledge development:* The activities develop the overall concept of an axiomatic system. Students work with the set of permissible procedures within the systems they investigate.
- *Metacognition:* After receiving classmates' critiques, students are prompted to reexamine the rules they established for the games at the beginning of class. Students reexamine the rules they established by comparing them to geometric axioms.

Lesson Plan 4

Based on the following NCTM resource: Buhl, D., Oursland, M., & Finco, K. (2003). The legend of Paul Bunyan: An exploration in measurement. *Mathematics Teaching in the Middle School, 8,* 441–448.

Primary objective: To model length, area, and volume measurement and connect them to scale factor and proportion

Materials needed: Cardboard replica of an ax; dimensions of length, breadth, average depth, maximum depth, and volume for a local lake; modeling clay for student use; a piece of grid paper for each student; a set of cubes for each student

Steps of the lesson	Expected student responses	Teacher's responses to students	Assessment strategies and goals
1. Read a version of the legend of Paul Bunyan to the class (see www .paulbunyantrail.com/). Be sure to emphasize (1) the size of Paul Bunyan's blue ox, Babe; (2) the dimensions of Paul Bunyan's skillet; and (3) the volume of the lake in which Paul Bunyan worked. After reading the story, ask students to identify connections to geometric measurement.	Students may identify length, area, and volume as important elements in the story. These elements pertain to Babe's height, Paul's skillet, and Paul's length, respectively. Students may identify additional elements in the story that may be measured.	List all student responses as they are offered. To summarize their responses and lead into the next portion of the lesson, emphasize ideas that include measurement of length, area, and volume. Acknowledge the validity of geometric measurement ideas that fall outside these three aspects as well.	As students offer ideas, start asking which units are commonly used to measure length, area, and volume. Also ask them to explain why these measures are used. Students' responses will begin to provide a sense of whether they understand geometric measurement conceptually.
2. Show students a cardboard replica of an ax. Mention that the distance between Babe's horns was 42 ax handles. Then ask students to estimate the following lengths in terms of ax handles and also in terms of the distance between Babe's horns: (1) the length of the school building and (2) the distance from school to home.	NAEP data show that students often have difficulty measuring one distance in terms of another. Students may want to use a standard unit of measurement instead of trying to estimate lengths using nonstandard units of measure.	Remind students that length measurement, whether using standard or nonstandard units, consists of iterating the units end to end so there are no gaps or overlaps. If they struggle to begin the tasks, have them take the cardboard ax and measure out 42 ax handles to understand the size of the unit of measure being used.	Assess students' work for reasonable estimates at this point. It is not necessary that they know the exact length of the school building or the exact distance from school to home. However, before moving on to the next few tasks, make sure their estimates reflect reasonable approximations.
3. Tell students that Paul Bunyan's skillet covered an acre of land (43,560 ft²). Ask them to determine its radius.	Setting up the proportion incorrectly can produce an answer of 50,824 feet for the radius.	If students obtain 50,824 feet, ask them to draw a diagram of a circle with such a radius and determine its area.	Ask students who obtain 50,824 feet for the radius why this cannot be a correct answer.
4. Tell students that Paul Bunyan was supposed to have created lakes by stomping through muddy land. The students' task will be to examine data regarding the dimensions of	Students may take several factors into account, including the possible dimensions of Paul's foot, the amount one would expect a	Encourage students to think about the dimensions of a foot in terms of area. If necessary, have students trace out a foot on grid paper and then think	A broad range of answers are possible, but the key thing to assess is whether or not the answers are justified by mathematics and the context of the story.

(Continued)

(Continued)

Steps of the lesson	Expected student responses	Teacher's responses to students	Assessment strategies and goals
a lake to argue whether or not this is a reasonable extension to the legend. Choose a local lake and provide data on its length, breadth, average depth, maximum depth, and volume. For dimensions of the Great Lakes, do an Internet search on "Great Lakes dimensions."	foot to sink in mud, and whether or not the dimensions of Paul's foot would line up with the dimensions of a lake. They might also consider whether or not several footprints end to end would measure out the given dimensions of a lake.	about how many pieces of grid paper would be needed for Paul's foot. If it is necessary to help students think about the reasonableness of volume measurement, have cubes available that they can use to create a scale model of the chosen lake.	Students may come to different conclusions about whether or not it is a reasonable extension of the legend to say that Paul created lakes with his footprints, but their justifications must be reasonable.
5. To conclude the activity, have students use modeling clay to construct scale models of several of the objects of the lesson, including Paul, Babe, and Paul's frying pan.	Students will draw on measurements done earlier in the lesson to construct Paul and Babe. The frying pan may be constructed as a square and then molded into a circle. Some students will finish early.	Ask students who finish the task before others to construct a scale model of the lake explored earlier. Those still working can check their models by comparing them with one another.	At the end of the lesson, have students share their scale models with the rest of the class. As they present, check to see whether the dimensions they have chosen are reasonable.

How This Lesson Meets Quality Control Criteria

- *Addressing students' preconceptions:* Students' out-of-school experiences with legends and tall tales are engaged as they evaluate the sizes of objects from one particular tale. Students begin to make connections among the previously learned concepts of distance, area, volume, ratio, and proportion.
- *Conceptual and procedural knowledge development:* Students are prompted to explore the concept of measurement in terms of the foundational principle of iteration of a unit. Students draw on procedural knowledge in solving proportions.
- *Metacognition:* Students are prompted to evaluate the reasonableness of their measurements by checking their results against the description of events provided in the legend. The teacher plays a role in prompting students to rethink their estimations when they are not reasonable.

Chapter 11

DEVELOPING STUDENTS' THINKING IN ADVANCED PLACEMENT COURSES

Advanced Placement (AP*) courses in statistics and calculus have become increasingly poplar in high schools. Table 11.1 gives a snapshot of the amount of growth that AP Statistics and AP Calculus have experienced since 1997 (the first year the AP Statistics examination was offered). AP Statistics is designed to be the equivalent of an introductory, non-calculus-based semester of college statistics. AP Calculus AB addresses roughly the equivalent of first-semester college calculus, and AP Calculus BC addresses roughly the first two semesters of college calculus. AP courses have become very attractive to students, parents, teachers, and school administrators because students can gain college credit before leaving high school if they earn acceptable scores on AP end-of-course examinations.** The rapid growth of AP courses makes it likely that high school teachers now entering the profession will have a chance to teach or launch an AP course sometime during their careers.

Teaching an AP course is a tremendous amount of work, but there are many professional development opportunities designed specifically to help with the task

Table 11.1 Growth of the AP program since 1997 (College Board, 2010).

	Number of schools offering the course (1997)	Number of students taking the examination (1997)	Number of schools offering the course (2009)	Number of students taking the examination (2009)
AP Statistics	752	7,667	5,707	116,876
AP Calculus AB	8,793	111,834	12,423	230,588
AP Calculus BC	2,284	22,668	5,083	72,965

*AP is a registered trademark of the College Board, which was not involved in the production of and does not endorse this product.

**Acceptable scores are determined on a college-by-college basis. AP exams are graded on a 5-point scale. Some universities award credit for a score of 3 or higher. Others require a score of 4 or higher. A handful of universities do not accept AP credit, but this is the exception rather than the rule.

(College Board, 2010). The College Board holds workshops and institutes throughout the United States every year. At these workshops and institutes, presenters share information about the general content of AP exams, common patterns of student thinking, and classroom-tested pedagogical strategies. The same type of information is provided at the national AP conference each year. Workshops, institutes, and conferences provide opportunities to network with other AP teachers. Networking opportunities are also available online, as the College Board hosts a listserv for each of the AP courses. Teachers regularly pose and respond to questions about teaching AP Calculus and AP Statistics via email on the listserv.

Another unique networking and professional development opportunity occurs each summer, after the AP examinations have been administered, when open-ended items need to be graded. During the first two weeks in June, a large team of high school teachers and university professors is assembled to handle the task. Grading open-ended items provides an opportunity to learn about students' thinking in regard to the course content and to compare one's professional judgment about assessing students' work to that of other teachers. Teachers can apply to be included on a grading team if they teach an AP Statistics or Calculus course.

The number of professional development opportunities associated with AP illustrates that teaching such a course is a craft one must continually refine. Hence, this chapter does not attempt to provide you with everything you need to teach the AP courses, but it does provide a brief overview of AP Statistics and Calculus, some of the salient student thinking patterns you can expect to encounter, and pedagogical strategies to help develop students' thinking.

What Is Advanced Mathematical Thinking?

AP courses contain college-level content. The type of mathematical thinking required at the college level is often labeled **advanced mathematical thinking** (AMT; Mamona-Downs & Downs, 2002). Although there has been substantial debate about the essential components of AMT (Artigue, Batanero, & Kent, 2007), some features do stand out as being particularly important. In general, there is a fair amount of agreement that it is important to engage students in the thinking processes involved in producing mathematics, and not just in studying the theoretical products that have been handed down by mathematicians (Mamona-Downs & Downs, 2002). Fostering AMT can be conceptualized as inducting students into the same sort of thinking processes that mathematicians use in their profession (Harel, Selden, & Selden, 2006). It should be noted that under this conceptualization, AMT is not strictly the province of AP courses. Teachers should help students engage in the thinking processes involved in producing mathematics at all grade levels, as discussed in the previous chapters.

Mathematicians go beyond executing routine procedures and algorithms in conducting their work. Defining is a central activity in AMT, because it creates a structure that relates concepts to one another (Harel et al., 2006). Such activity involves not merely memorizing definitions, as is frequently done in school mathematics, but actually writing definitions that build on one another to form a conceptual structure for a field of study. The definitions encountered in developing AMT can be less intuitive than those in precollege experiences, because they are often abstracted from mathematical structure rather than from sense experience

(Mamona-Downs & Downs, 2002). To the extent feasible, students need to reconstruct these conceptual structures for themselves if they are to develop AMT. Not all of the reconstruction of a field such as calculus will occur within the first semester or two of study, but the beginning semesters do need to lay the foundation for students' later work.

Another caveat to consider in the discussion of AMT is that statistical thinking is actually quite different from mathematical thinking. If one conceives of advanced statistical thinking as the type of thinking done by statisticians (Pfannkuch & Wild, 2004), a number of differences emerge. Rossman, Chance, and Medina (2006) discussed some of the most important ones:

- In solving a mathematical problem, the goal is often to strip away the context in order to see an underlying mathematical pattern. In statistics, it is necessary to take the context into account throughout the problem solving process because the context helps provide a meaningful interpretation of data.

- In statistics, there is a need to skillfully define measures of abstract ideas. Mathematical problems often involve measuring relatively concrete quantities such as a person's height, but statistical problems often involve measuring abstract quantities such as intelligence. There can be considerable debate about how best to define measures to quantify such abstract ideas.

- Statistical thinking involves analyzing study design, and not just the numerical data produced during a study. This means that statistical thinking includes many non-mathematical aspects, such as critiquing the quality of questions on a survey and the method in which it was administered to a sample of a population.

- Statistics does not produce deductive proofs in the manner of mathematics. Instead, statistical conclusions are offered with degrees of qualification. Individuals analyzing the same data set may draw different defensible conclusions from it depending on the measures and methods selected for use.

To be sure, statistics draws on mathematics and cannot survive without it, but at the same time the central modes of inquiry employed by statisticians and mathematicians are not identical.

Despite the differences between advanced mathematical and statistical thinking, a central guiding principle can be taken forward into the following discussion of fostering students' thinking in AP courses: AP Calculus and AP Statistics should not be taught in an exclusively procedural manner. AP Calculus students need to be able to do more than just compute derivatives and integrals, and those taking AP Statistics should not merely be taught to carry out statistical tests in a mechanistic fashion. Approaching an AP course in a procedural fashion does little to help lay the foundation for postsecondary study. To the extent possible, AP Calculus students should have opportunities to reason as mathematicians do, and AP Statistics students should have opportunities to reason like statisticians. Since AP programs are rapidly growing in size, the calculus and statistics courses offered in high schools will increasingly bear the responsibility for providing future mathematicians and statisticians with the first in-depth look at their professional fields. The courses also provide the only glimpse many students will have of postsecondary mathematical study, since those going into nonscientific fields often fulfill college mathematics requirements by earning AP credit.

FOSTERING STUDENTS' THINKING IN AP CALCULUS

The Calculus Reform Movement

The current AP Calculus AB and BC courses reflect the influence of the **calculus reform movement** that began among university mathematicians in the 1980s and continues to the present. In the 1980s, widespread dissatisfaction arose with traditional calculus courses because they served as filters to block students from advanced mathematical and scientific study rather than pumps to propel students into such fields (Steen, 1988). No single approach was universally endorsed as a "silver bullet" solution to the problems of traditional calculus courses, but curricular ideas from Harvard University influenced the direction that many calculus courses would take, including AP Calculus. **Harvard calculus** became a shorthand phrase for a curriculum developed by the Calculus Consortium at Harvard University (Hughes-Hallet, Gleason, et al., 1994) with funding from the National Science Foundation.

The guiding principles of Harvard calculus have been summarized in the following manner by the Calculus Consortium:

- Rule of Four: Where appropriate, topics should be presented geometrically, numerically, analytically, and verbally.

- Problem Driven: Formal definitions and procedures evolve from the investigation of practical problems. Whenever possible, the authors start with a practical problem and derive the general results from it. These practical problems are usually, but not always, real world applications.

- Open-Ended Real World Problems: The real world problems are open-ended, meaning that there may be more than one solution depending on a student's analysis. Many times, solving a problem relies on common sense ideas that are not stated in the problem but which students will know from everyday life.

- Plain English: These books present the main ideas of calculus in plain English to encourage the students to read it in detail, rather than just reading the worked out examples. (John Wiley & Sons, 2010, n.p.)

Although AP Calculus should not be thought of as strictly congruent to Harvard calculus, the influence of the four guiding principles above is evident as one examines course descriptions and sample items from past examinations. The **rule of four,** for example, explicitly appears in the AP Calculus course description (College Board, 2009) in much the same form as it does in the Harvard materials.

The calculus reform movement among university mathematicians helped motivate in-depth study of the manner in which students think about concepts in introductory courses. As calculus curricula were redesigned, researchers sought to explore inroads and obstacles to students' learning of the subject. Some of the mathematics education research findings most pertinent to teaching AP Calculus are discussed below. For convenience, the research is discussed under the categories of limits, derivatives, and integrals, though there is conceptual overlap among the three categories. Examining research in these three areas can help provide guidance for the tasks of designing and implementing curricula and instruction that support students' attainment of conceptual understanding in calculus.

Students' Understanding of Limits

A formal definition for the limit of a function is shown in Figure 11.1. Limits are essential to the disciplinary foundation of calculus, so they should have a role in introductory courses. However, the definition is so notoriously difficult for students to understand (S. R. Williams, 1991) that it has become a central dilemma in teaching introductory calculus. Some of the primary difficulties students have in coming to understand limits are discussed below. Additionally, some potentially promising methods for teaching limits to beginning calculus students are described.

Szydlik (2000) investigated calculus students' understanding of limits. Some of the students interviewed for the study believed that a function can never reach its limit, which conflicts with the formal definition. These students argued that a function can come "really, really close" to the limit, but never reach it. When asked to analyze limits associated with $f(x) = 7$, they began to make exceptions to their intuitive rules for limit, considering them to be special exceptions to the general rule. A related erroneous belief some students held was that $.\overline{9} \neq 1$ because the process of adding 9s to the decimal representation never ends. Students may, however, revise their beliefs about the value of $.\overline{9}$ when asked to write $.\overline{3}$ as a fraction (Tall & Vinner, 1981). In general, it appears that students are more likely to regard the limit as unreachable if they have external sources of conviction rather than internal ones (Szydlik, 2000). Students with **external sources of conviction** look to authorities such as the teacher, the answers in the back of a text, a calculator, or some other external source to determine mathematical correctness. In Szydlik's (2000) study, students with **internal sources of conviction**, who look to their own reasoning to establish mathematical correctness, were more likely to hold limit definitions aligned with mathematical convention.

Along with believing limits to be unreachable, some calculus students think about limits in terms of the metaphors of **collapse** or **approximation**. For example, when Oehrtman (2009) investigated students' thinking about the derivative as a limit, some described the gradual collapse of the rise and run triangle showing the slope of a secant line to a graph. Some also used the collapse metaphor when discussing integrals, imagining the rectangles used to approximate the area under a curve collapsing into line segments. The idea of approximation was often associated with the collapse metaphor. For instance, some students described using a series of secant lines to approximate the tangent line to a curve. These students sometimes thought of the secant as collapsing into the tangent. Additionally, an approximation metaphor was

Figure 11.1   Formal definition of limit of a function (Swokowski, 1991, p. 53).

Let a function f be defined on an open interval containing a, except possibly at a itself, and let L be a real number. The statement

$$\lim_{x \to a} f(x) = L$$

means that for every $\varepsilon > 0$, there is a $\delta > 0$ such that

if $0 < |x - a| < \delta$, then $|f(x) - L| < \varepsilon$

prevalent when students were asked about repeating decimals and Taylor series. When asked to comment on these ideas, "students described the limit (infinite series) as the value being approximated, partial sums as approximations, and the difference between the two (remainder) as the error" (Oehrtman, 2009, p. 414).

Students' approximation metaphors for limits may be drawn on to help them develop understanding of a conventional formal definition for limit. Oehrtman (2009) recommended asking the following series of questions in concert with approximation metaphors to help students transition to understanding the formal definition of limit:

> (a) What is being approximated? (b) What are the approximations? (c) What are the errors? (d) Given an approximation, how can you find the bound on the error? (e) Given a desired bound on error, how can you generate an approximation with that level of accuracy? (p. 421)

Asking the five questions consistently can help students develop a rigorous, formal understanding of the definition of limit (Oehrtman, 2008).

In general, research suggests that the most promising approaches to helping students understand limits involve starting with their existing intuitions. Beginning with a formal definition for limit is likely to fail. Accordingly, the course description for AP Calculus AB and BC calls for helping students develop "an intuitive understanding of the limiting process," "calculat[e] limits using algebra," and "estimat[e] limits from graphs or tables of data" (College Board, 2009, p. 7). To develop an intuitive understanding of the limiting process, students may use tables of values on graphing calculators or computer software packages to analyze the behavior of a function numerically around a given point. Examining numerical behavior can help students deal with limits when strictly algebraic representations are provided, as shown in Figure 11.2. Additionally, students should determine limits in situations where a graph is provided without a table or algebraic formula. The sample question shown in Figure 11.3 exemplifies such a situation.

Figure 11.2 Limit sample item 1 from AP Calculus examination (College Board, 2009, p. 19).

11. What is $\lim_{x \to \infty} \dfrac{x^2 - 4}{2 + x - 4x^2}$?

(A) -2

(B) $-\dfrac{1}{4}$

(C) $\dfrac{1}{2}$

(D) 1

(E) The limit does not exist.

Source: Copyright © 2009 The College Board. Reproduced with permission. http://apcentral.collegeboard.com

In sum, in the AP course, students should work with limits of functions using a variety of representations. Functions may be represented graphically, algebraically, numerically, or with any combination of representations. As students work with these representations, they can begin to form an intuitive foundation for conventional, formal definitions of limit. As with many concepts in mathematics, limits are best approached by scaffolding students' intuition toward formal disciplinary conventions rather than attempting to discount or circumvent intuition. Whereas starting instruction on limits by having students memorize a formal definition has a high likelihood of failure, engaging students in examining multiple representations and then monitoring and building on the intuitive ideas they form can help lay the foundation needed to develop sound understandings of limits.

Figure 11.3   Limit sample item 2 from AP Calculus examination (College Board, 2009, p. 23).

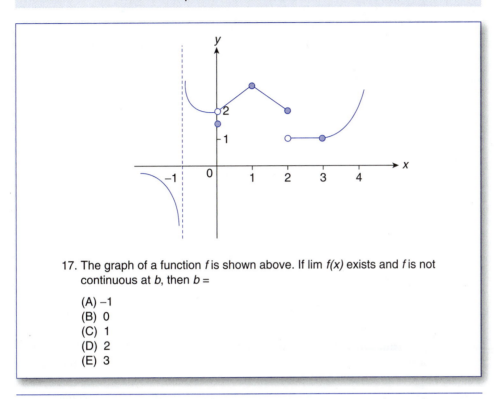

17. The graph of a function *f* is shown above. If lim *f(x)* exists and *f* is not continuous at *b*, then *b* =

(A) –1
(B) 0
(C) 1
(D) 2
(E) 3

Students' Understanding of Derivatives

Since derivatives are defined as limits, the two ideas are conceptually linked to one another. In AP Calculus courses, students are to understand "derivative defined as the limit of the difference quotient" (College Board, 2009, p. 7). A sample question to test students' understanding of this idea is shown in Figure 11.4. In the AP courses, students are to become proficient in working with the concept of derivative, the derivative at a point, derivative as a function, second derivatives, application of derivatives, and computation of derivatives.

A key to understanding the derivative at a point is conceptualizing it as the limit of average rate of change. The notion of average rate of change itself can pose considerable difficulty for calculus students. They may have difficulty distinguishing between average rate of change and rate of change at a given point on a

Figure 11.4   Item eliciting use of the limit definition of derivative (College Board, 2009, p. 17).

11. What is $\lim\limits_{h \to \infty} \dfrac{\cos\left(\dfrac{3\pi}{2} + h\right) - \cos\left(\dfrac{3\pi}{2}\right)}{h}$?

(A) 1

(B) $\dfrac{\sqrt{2}}{2}$

(C) 0

(D) –1

(E) The limit does not exist.

curve (Orton, 1983a). Bezuidenhout (1998) found that calculus students sometimes use idiosyncratic methods for computing average rate of change. Rather than using the formula $\frac{f(b) - f(a)}{b - a}$ to determine the average rate of change of f over the interval $[a, b]$, students sometimes compute the derivative of f at several points along the interval $[a, b]$ and then divide by the number of points they used. In such cases, students draw on their previous understanding of the concept of average, but use it in a nonconventional way to compute the average rate of change of a function. Without a solid understanding of the idea of average rate of change of a function, it is not possible to understand the derivative as the limit of average rate of change.

TECHNOLOGY CONNECTION

Examining Local Linearity

To understand the relationship between average and instantaneous rate of a function, students should distinguish between secant and tangent lines to a graph. Ferrini-Mundy and Lauten (1993) recommended that teachers use the zooming capabilities built into graphing calculators to have students examine secant lines to graphs of functions where the x-values for the points of intersection between the line and the graph are very close to one another. Many software packages, including Fathom, Mathematica, and Maple, also have zooming capabilities. As the x-values come close to one another, the secant line begins to closely approximate a tangent line. Students can be encouraged to use tangent line approximations to make conjectures about the value of a derivative by examining several examples. For instance, determining the slope of the line going through (x, x^2) and $(x + c, (x + c)^2)$ can lead students to conjecture that the derivative of $y = x^2$ is $y' = 2x$ if they experiment with several values of c and x. This sort of zooming and conjecturing can help students examine the idea of local linearity— that is, the idea that repeated zooming eventually "straightens out" a graph of a function that is differentiable at a given point (Figure 11.5).

Figure 11.5 Zooming in on the graph of $y = x^2$ to investigate local behavior.

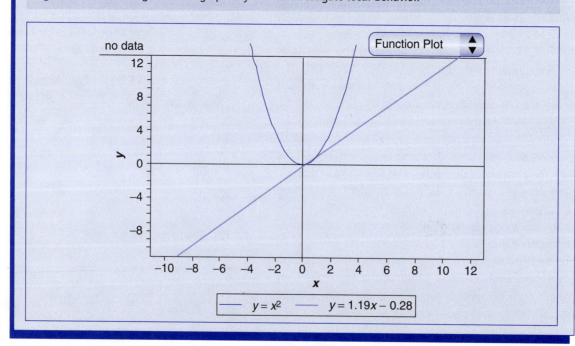

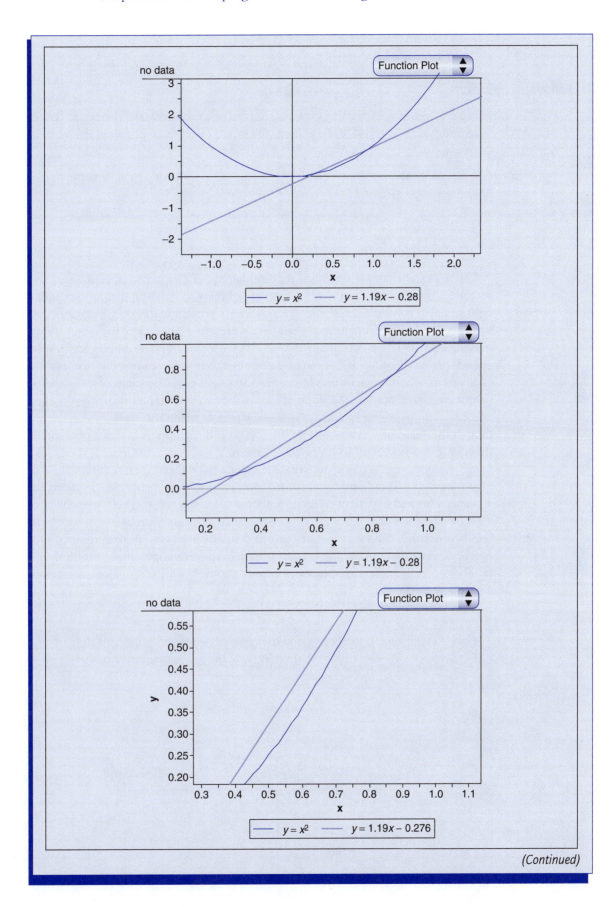

(Continued)

(Continued)

Follow-up questions:

1. How can examining the slope of the line going through (x, x^2) and $(x + c, (x + c)^2)$ lead students to conjecture that the derivative of $y = x^2$ is $y' = 2x$? Experiment with several values of c and x, show your work, and explain your reasoning.

2. Will examining local linearity always help students form conjectures about the value of a derivative at a given point? Why or why not?

Developing Flexible Thinking

As students examine graphical representations of functions, tangent lines, and secants, it is important for them to develop flexibility in thinking that facilitates efficient problem solving. Such thinking is not always present among beginning calculus students. For instance, in one study, students were asked to examine the graph shown in Figure 11.6 and then determine $f(5)$ and $f'(5)$ (Asiala, Cottrill, Dubinsky, & Schwingendorf, 1997). The value of $f(5)$ can be determined simply by reading the y-coordinate of the point where tangent line L touches the graph of f. The value of $f'(5)$ can be determined by computing the slope of L using the two labeled points. However, when asked to determine the value of $f(5)$, some students wrote the equation for L, even though the value of $f(5)$ was plainly displayed on the graph. A number of students thought it was necessary to know the equation $y = f(x)$ to determine $f(5)$ or $f'(5)$. Some students went so far as to write the equation for L and then differentiate it to determine $f'(5)$. The desire for algebraic formulas was particularly overpowering for students who had experienced a conventional calculus course. Their inflexibility of thought suggests that teachers of introductory calculus should ask students to regularly perform tasks that require reasoning strictly from graphical representations. Algebraic representations are powerful and have advantages that graphical and tabular representations do not, but not every problem solving situation requires an algebraic formula.

Figure 11.6 Graph to accompany the tasks of determining $f(5)$ and $f'(5)$ (Asiala, Cottrill, Dubinsky, & Schwingendorf, 1997, p. 404).

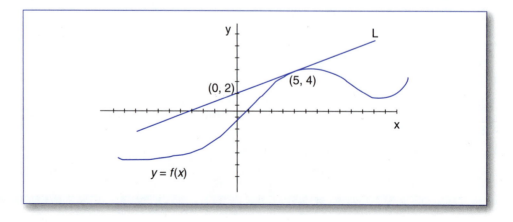

Another instance of difficulty with graphical reasoning was documented when students were given the task shown in Figure 11.7 (Baker, Cooley, & Trigueros, 2000). To successfully complete the task, students had to coordinate information across intervals and understand the meanings of the calculus concepts embedded in it. Coordinating the information across intervals and understanding the calculus concepts both posed difficulties. Some students were not able to envision a function that was both decreasing and concave up. Some thought that the first derivative provided information about concavity and the second derivative was to be used only to determine inflection points. A few believed that if the derivative of a function is zero at a given point, its graph must cross the x-axis at that point. It was also difficult for some students to think of a graph as having a cusp at $x = -4$; the ideas of continuity, cusp, and differentiability were not always clearly understood. The vertical tangent present at $x = 0$ was another source of difficulty. In general, however, students did well in graphing features related to the first derivative. This led Baker and colleagues to conjecture that the first derivative may be overemphasized, at the expense of other concepts, in conventional introductory calculus courses. Teachers need to balance attention to the first derivative with problems that involve the other concepts embedded in the task shown in Figure 11.7.

Figure 11.7   Sketching the graph of a function, given information about derivatives and limits (Baker, Cooley, & Trigueros, 2000, p. 563).

Sketch a graph of a function h that satisfies the following conditions:

h is continuous;

$h(0) = 2$, $h'(-2) = h'(3) = 0$, and $\lim_{x \to 0} h'(x) = \infty$;

$h'(x) > 0$ when $-4 < x < -2$ and when $-2 < x < 3$;

$h'(x) < 0$ when $x < -4$ and when $x > 3$;

$h''(x) < 0$ when $x < -4$, when $-4 < x < -2$, and when $0 < x < 5$;

$h''(x) > 0$ when $-2 < x < 0$ and when $x > 5$;

$\lim_{x \to -\infty} h(x) = \infty$; and $\lim_{x \to \infty} h(x) = -2$

Sketch a graph of a function that satisfies all of the conditions specified in Figure 11.7.

STOP TO REFLECT

Since graphical representations of functions are more prevalent in reform-oriented calculus courses like AP, it is important for teachers to probe and understand the nature of students' visual thinking. Aspinwall, Shaw, and Presmeg (1997) described the thinking of a student, Tim, who had just completed a year of introductory calculus. For one task, Tim was shown a parabola and asked to graph its derivative. He graphed the derivative function for the parabola with some success without translating the graph to a formula. When questioned about what would happen as the parabola expanded to the left and to the right, he stated that the parabola would eventually run up against vertical asymptotes. Tim could not shake this visual image. When he thought about the parabola's algebraic representation, $y = x^2$, he began to question why the derivative would be $y' = 2x$ (a straight line). Tim also had difficulty reconciling his visual image of a parabola having asymptotes with the domain of the parabola being all real numbers. Given the images that interfered with Tim's thinking, Aspinwall and colleagues cautioned that in courses emphasizing visual representations, some students may be influenced by uncontrollable mental images. Teachers need to become aware of these images by asking students to describe their visual thinking as they work on tasks that emphasize graphical representations. As teachers become aware of the images, they can be addressed and challenged during instruction.

STOP TO REFLECT

Explain why the parabola $y = x^2$ does not have vertical asymptotes.

Although visual representations of the derivative are emphasized in reform-oriented calculus courses, it is important to help students recognize connections among algebraic representations as well. Clark and colleagues (1997) found that students may know rules for differentiating power functions, trigonometric functions, implicitly defined functions, and composite functions but not see them as being linked via the chain rule. Students can often successfully differentiate functions falling into each category without understanding that they are actually employing the chain rule. Some, for instance, can state the derivatives of $\sin x^3$ and various composite functions simply because they have memorized the derivatives by rote. Although rote memorization produces correct answers in some cases, it leads to misconceptions about important elements of the disciplinary structure of calculus. To help students understand this structural aspect of calculus, Clark and colleagues recommended that "more emphasis should be given to recognizing a function as the composition of other functions as well as to the relationships among various problem situations which embody the chain rule concept" (p. 360).

Students' Understanding of Integrals

In AP Calculus, students are to study antidifferentiation as well as differentiation. The study of antiderivatives poses a variety of procedural challenges. Computational errors occur when students try to determine the antiderivatives of power functions that have negative or fractional exponents. For instance, when asked to compute antiderivatives of expressions with negative exponents, it is not uncommon for students to write that $-(n + 1)$ is one greater than $-n$ (Orton, 1983b). Students may also have difficulty determining the antiderivative of a function expressed as $y = k$, where k represents a constant. Those who do not understand the difference between variables and constants are apt to believe that the antiderivative of such a function is $\frac{k^2}{2}$. The general procedure of "adding one to the exponent and dividing by the new exponent" appears to be a very resilient part of the concept image students form for antiderivatives, even if they apply it incorrectly. When asked to explain what an anti-derivative is, many students who have taken calculus reference this "reverse power rule" as the essence of antidifferentiation (Gonzalez-Martin & Camacho, 2004; P. W. Thompson, 1994).

Of course, there is a great deal more to know about antidifferentiation than the so-called reverse power rule. Students in the AP course are required to use multiple representations, including tables and graphs, when reasoning about antiderivatives as well as any other idea in calculus. For example, given Figure 11.8 and the information that $f(0) = 0$, students should be able to produce the graph of $f(x)$.

Examining two students' responses to the item shown in Figure 11.8 helps illus-trate the role of visual reasoning in such tasks. The two students, Jack and Bob, could both do routine symbolic computation of integrals, but Bob was more successful than Jack on the graphical task (Haciomeroglu, Spinwell, & Presmeg, 2009). Jack relied on visual thinking alone. Doing so allowed him to sketch a roughly accurate shape for the antiderivative function, but not to determine its minimum value. Bob also drew on visual thinking to sketch an accurately shaped antiderivative graph. Like Jack, he recognized that the antiderivative function reached a minimum at $x = 1$, was decreas-ing up to that point, and was increasing after. Unlike Jack, Bob was able to determine that the minimum value of the antiderivative function was $-\frac{1}{2}$ because he calculated the area of the triangle formed by the x-axis, the y-axis, and the graph of $f'(x)$. Bob also drew on his knowledge of antidifferentiation involving symbolic representa-tions to reason that $f(x)$ must be quadratic with $f'(x)$ being linear. The cases of Bob and Jack illustrate that when students are given graphical tasks, they should be encouraged to coordinate visual thinking with other bits of knowledge that can be mined from the situation. Blending visual and analytic thinking when necessary to analyze a situation has been called **harmonic thinking** (Haciomeroglu et al., 2009).

Figure 11.8   Graph of $f'(x)$ to prompt students to produce the graph of $f(x)$ given that $f(0) = 0$ (Haciomeroglu, Aspinwall, & Presmeg, 2009, p. 141).

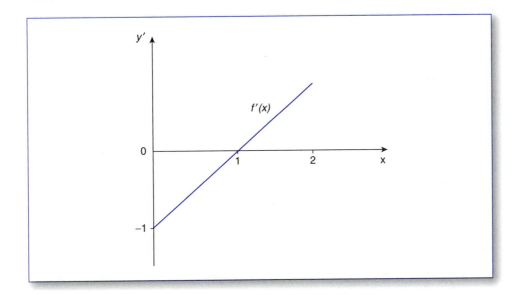

Teachers can help students mine knowledge from antidifferentiation problems involving graphs by asking questions that prompt examination of important features (Haciomeroglu et al., 2009):

- What are the powers of the derivative and antiderivative graphs?
- How do slopes of tangent lines to the antiderivative graph change at key points?
- What are the local maxima and minima of the antiderivative graph?
- Where are the inflection points of the antiderivative graph? Where is it concave up? Concave down?
- Where is the antiderivative graph increasing? Decreasing?

Asking students to respond to questions like these can help scaffold their thinking. As such questions are posed in connection with a variety of tasks, students can internalize the questions and ultimately ask them spontaneously when necessary.

As students begin working with definite integrals, additional challenges arise. The AP course description calls for students to "understand the meaning of the definite integral both as a limit of Riemann sums and as the net accumulation of change" (College Board, 2009, p. 6). Procedural aspects of Riemann sums that are problematic for students include interpreting conventional summation notation and determining the heights of individual rectangles used to approximate the area beneath a curve (Bezuidenhout & Olivier, 2000; Orton, 1983b). A key conceptual challenge in dealing with definite integrals is interpreting the sign of the result obtained from evaluating an integral. For instance, some take a negative sign to indicate that the curve itself must have a negative slope, and others disregard the sign altogether (Orton, 1983b). These procedural and conceptual challenges help define critical aspects of the task of teaching about definite integrals.

Antidifferentiation and definite integration are both pertinent to the fundamental theorem of calculus. It can be stated in two parts, as shown in Figure 11.9. To understand part I of the fundamental theorem, one must be able to work with functions defined as integrals. Some items on the AP examinations test students' ability to do so, as shown in Figure 11.10. Functions defined as integrals pose many conceptual hurdles. When asked to identify local maxima for a function such as g (Figure 11.10), some students instead identify local maxima for f. Others, when asked to identify local

Figure 11.9 Fundamental theorem of calculus (Swokowski, 1991, p. 283).

Suppose f is continuous on a closed interval $[a, b]$.

Part I If G is the function defined by

$$G(x) = \int_a^x f(t)\, dt$$

for every x in $[a, b]$, then G is an antiderivative of f on $[a, b]$

Part II If F is any antiderivative of f on $[a, b]$, then

$$\int_a^b f(x)\, dx = F(b) - F(a)$$

Figure 11.10   An AP item requiring analysis of a function defined as an integral (College Board, 2009, p. 34).

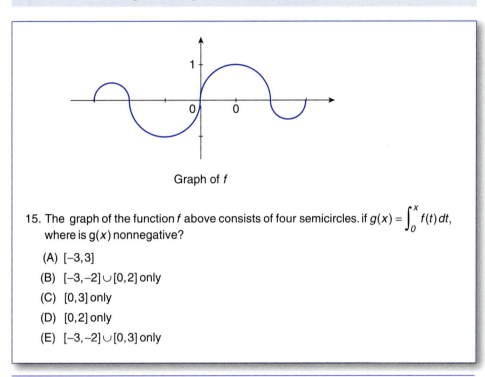

Graph of f

15. The graph of the function f above consists of four semicircles. if $g(x) = \int_0^x f(t)\,dt$, where is $g(x)$ nonnegative?

(A) $[-3, 3]$

(B) $[-3, -2] \cup [0, 2]$ only

(C) $[0, 3]$ only

(D) $[0, 2]$ only

(E) $[-3, -2] \cup [0, 3]$ only

maxima for *g*, start with the correct strategy of analyzing the accumulated area between *f* and the *x*-axis. However, this strategy may be applied incompletely when students add all areas together rather than subtracting areas below the *x*-axis from those above it. This causes students to conclude that *g* increases over the entire interval because area between the graph of *f* and the *x*-axis continues to accumulate across the entire interval (Bezuidenhout & Olivier, 2000). Part II of the fundamental theorem involves reasoning with the definite integral, and key procedural and conceptual difficulties in doing so were described earlier.

Examine the distracters for the multiple-choice question shown in Figure 11.10. Describe incorrect patterns of thinking that would lead students to select each of the distracters. Describe how you could help students choosing each distracter revise their thinking.

STOP TO REFLECT

Given the importance of the fundamental theorem to introductory calculus, it is crucial to focus on how students' conceptual and procedural difficulties associated with understanding it can be addressed. One approach that has been successful involves having students examine definite integrals that can be computed by using geometric formulas, such as $\int_0^x 2t\,dt$ (H. L. Johnson, 2010). The value of the integral is equivalent to the area of the right triangle in the first quadrant with a base length of x and a height of $f(t)$ (see Figure 11.11). Students can try several values for x and observe how the value of the integral changes. Results for several values of x can be tracked in a table (see Table 11.2). Tracking results in a table helps students understand what it means to have a function defined as an integral—a key to understanding part I of the fundamental theorem of calculus—because the table helps show accumulated area as a function of x (and not of t). Points from the table can then be plotted on a graph (see Figure 11.12), and students can attempt to fit a curve through them. Seeing that the points trace out a parabola can lead students to suspect that $\int_0^x t\, dt = x^2$ To follow up, and to lay further groundwork for student understanding of part II of the fundamental theorem of calculus, students can be asked to change the lower bound of the integral to different values. For example, repeating the process described above with a different lower bound can lead students to conjecture that $\int_2^x 2t\, dt = x^2 - 2^2$

After working with definite integrals that can be computed precisely with simple geometric formulas, students can work with tasks that involve approximating the area beneath a curve with rectangles. A good starting point for investigation is to approximate $\int_0^x t^2\, dt$ for various values of x (Gordon & Gordon, 2007). Students can fill in a table of values, as in Table 11.2, by calculating the areas of rectangles that approximate the area between the curve and the horizontal axis. Plotting the points from the table of values can lead students to conjecture that $\int_0^x t^2\, dt = \frac{x^3}{3}$. Changing the lower boundary

Figure 11.11   Graph of $y = 2t$ used for integral exploration activity.

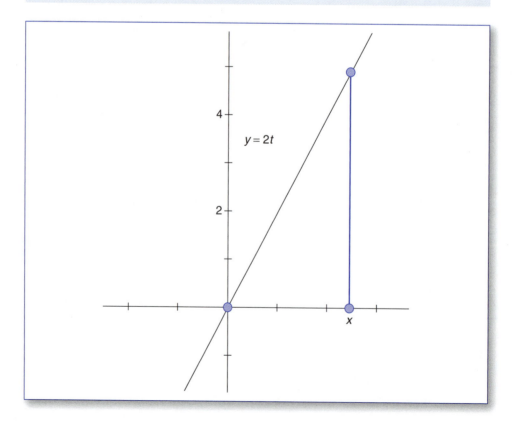

Table 11.12   Table used to track values of a definite integral for several different upper boundaries.

x	$\int_0^x 2t\,dt$
0	0
$\dfrac{1}{2}$	$\dfrac{1}{4}$
1	1
$1\dfrac{1}{2}$	$\dfrac{9}{4}$
2	4
$2\dfrac{1}{2}$	$\dfrac{25}{4}$
3	9
$3\dfrac{1}{2}$	$\dfrac{49}{4}$
4	16

and repeating the activity can lead to the broader conjecture that $\int_a^x t^2\,dt = \frac{x^3}{3} - \frac{a^3}{3}$. After students have used rectangles to approximate integrals associated with a variety of types of functions, it becomes natural to conjecture that part II of the fundamental theorem (Figure 11.9) holds. Additionally, if students are asked to make the approximating rectangles narrower and narrower as they work with different types of functions, the idea of the definite integral as a limit of Riemann sums can be grasped intuitively. With an intuitive understanding and some firm conjectures about the fundamental theorem in hand, students have the background to appreciate formal proofs of it (see the first four-column lesson plan at the end of this chapter for more details on this approach).

Summary

Historically, introductory calculus has served as a filter to disqualify many students from pursuing mathematics-related careers. Recognizing this problem, mathematicians and mathematics educators have sought to reform the teaching of introductory calculus. The AP Calculus courses reflect major trends in the reform movement by endorsing an intuitive approach to introductory calculus. Sound intuitions about limits, derivatives, and integrals can help counteract the development of many

Figure 11.12   Graph of $\int_0^x 2t\,dt$ versus x.

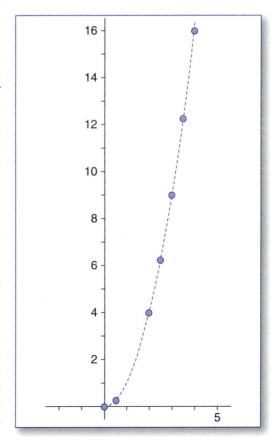

of the student misconceptions documented in the research literature. Intuitive arguments and investigations involving inductive reasoning, as described previously, help set the stage for rigorous deductive reasoning and proof in later courses.

FOSTERING STUDENTS' THINKING IN AP STATISTICS

The Statistics Reform Movement

AP Statistics, like AP Calculus, reflects the influence of a reform movement initiated at the university level. The primary tenets of the statistics reform movement can be found in the *Guidelines for Assessment and Instruction in Statistics Education* (GAISE) college report endorsed by the American Statistical Association (Aliaga et al., 2010). It offered six recommendations for the reform of introductory statistics courses:

1. "Emphasize statistical literacy and develop statistical thinking" (p. 14). Classrooms reflecting this recommendation provide opportunities for students to choose research questions and techniques, not just carry out teacher-prescribed procedures.

2. "Use real data" (p. 16). One source of real data is that generated by the class. Examining classmates' responses to survey items can lead to new questions and the design of new survey instruments to investigate them.

3. "Stress conceptual understanding, rather than mere knowledge of procedures" (p. 17). As in mathematics, procedural understanding without conceptual understanding is very limited. Learning underlying concepts facilitates the learning and retention of important procedures.

4. "Foster active learning in the classroom" (p. 18). Students should be encouraged to discuss and analyze data rather than simply following lists of step-by-step procedures.

5. "Use technology for developing concepts and analyzing data" (p. 19). Technology is essential to the practice of statistics. It also supports the ability to focus on conceptual understanding by automating calculation and graph production, allowing many trials of simulations, and revealing links among representations of data.

6. "Use assessments to improve and evaluate student learning" (p. 21). Formative and summative assessments should take a variety of forms, such as critiques of articles from the media, student projects, and open-ended investigative tasks.

The six pedagogical recommendations from the **GAISE college report** can be applied to teaching a broad range of content in the introductory course.

While GAISE provides recommended pedagogies for introductory statistics, the AP Statistics course description outlines the main content themes to be addressed. Four broad themes define the scope of the AP course (College Board, 2008):

1. Exploring data: Describing patterns and departures from patterns

2. Sampling and experimentation: Planning and conducting a study

3. Anticipating patterns: Exploring random phenomena using probability and simulation

4. Statistical inference: Estimating population parameters and testing hypotheses (p. 5)

These four themes will be used as an organizing framework for the following discussion of students' thinking regarding college-level statistics and strategies for helping it develop.

Theme 1: Exploring Data

AP Statistics students are to engage in exploratory data analysis with both univariate and bivariate data sets. Graphs and descriptive statistics are important tools in exploratory analysis. Constructing graphs and computing descriptive statistics are not to be ends in themselves, but such tools should be put to work in the context of posing and answering questions about data sets. Chapter 9 dealt extensively with developing students' understanding of exploratory data analysis and its accompanying descriptive tools, so this chapter will not revisit the topic in great detail. However, some time will be spent on examining the teaching and learning of bivariate data analysis, since it is a key theme of the AP course and was not discussed at length in Chapter 9. Two core ideas in the study of bivariate data are considered in the following: (1) reasoning about correlation and causation and (2) understanding correlation coefficients. The role of technology in teaching these ideas is also examined.

Reasoning About Correlation and Causation

The GAISE college report endorsed the use of real data for exploratory analysis. A pedagogical advantage to doing so is that real data sets are likely to capture students' interest. Using real data sets also helps students understand that there can be multiple plausible interpretations of a given situation. However, with these pedagogical advantages come challenges. Students' knowledge and beliefs about contexts in which real data are generated can interfere with their ability to draw sound statistical conclusions. In the case of analyzing bivariate data, students may at times wrestle with illusory and spurious correlations.

Illusory correlation exists when a student believes that two uncorrelated variables are actually correlated (Garfield & Ben-Zvi, 2008). For example, one may have a strong belief that scores on the SAT examination and college GPAs have a strong positive relationship. If the belief is strong enough, contradictory empirical data may not even change an individual's perception of the relationship. Rather than being persuaded by the data, those holding an illusory correlation may instead view cases that confirm the existing belief about SAT scores versus college GPA as more credible than cases that do not. Essentially, the individual's beliefs serve as an overpowering filter through which the data are processed. Students are susceptible to forming illusory correlations within any context likely to elicit strong opinions. Teachers should consider contexts of problems in advance to anticipate student beliefs that may need to be addressed during instruction. Problematic contexts should not necessarily be avoided, but some time should be spent engaging students in dialogue about why individual anecdotes and experiences should not supplant clear patterns detected in data.

Another barrier to the intelligent analysis of bivariate data is **spurious correlation**. Spurious correlations are those that exist between variables that are not causally related. For instance, the number of firefighters responding to a fire and the amount of damage caused to a home are likely to have a strong positive relationship. Care must be taken, however, not to draw the conclusion that having a large number of firefighters present is the cause of a great deal of damage—the real underlying cause (of the number of firefighters and the amount of damage) is likely to be the size of the fire (Bock, Velleman, & DeVeaux, 2004). Situations involving spurious correlations help to illustrate that correlation and causation are not equivalent concepts. In such cases, there are lurking variables (e.g., the size of the fire) that are the real driving forces behind the patterns seen in the data. To help students sharpen their intuition regarding bivariate data analysis, teachers can present situations involving spurious correlations and then pose the task of identifying lurking variables. As students communicate their reasoning about the situations, they should be encouraged to make careful and deliberate use of terms such as *association* and *correlation*, avoiding misuses of the terms frequently encountered in everyday language.

STOP TO REFLECT

Provide at least five examples of situations that might lead students toward illusory correlation. Also provide at least five examples of situations involving spurious correlations. Use your examples to explain the difference between illusory and spurious correlation.

Understanding Correlation Coefficients

Correlation coefficients can be thought of as descriptive statistics that are helpful for measuring the strength of the linear relationship between two variables. When interpreting correlation coefficients, however, caution is in order. It is important for students to develop the habit of examining the associated data, not just the numerical value of the correlation coefficient. Anscombe (1973) provided classic examples of data sets for which examining the correlation coefficient in isolation is misleading (Figure 11.13). They have come to be known as **Anscombe's quartet**. These four data sets have many of the same descriptive and inferential statistics. The Pearson correlation coefficient and least-squares regression line are the same in each case. Nonetheless, the data sets differ drastically.

Teachers can help students gain several insights about bivariate data analysis by using Anscombe's quartet. The graph in the top left corner illustrates a situation where the correlation coefficient is likely to be useful for describing the strength of the linear relationship between two variables. Points on the scatterplot approximate a straight line, as one might expect when a moderately strong correlation is computed. The graph in the top right corner, however, exemplifies a situation where computing a correlation may be misleading. There appears to be a strong nonlinear relationship between the two variables, so attempting to measure the linear association between the variables or trying to fit a least-squares line to the data is not appropriate. The graph in the bottom left corner shows a situation in which a single data point strongly influences the direction of the regression line and the value of the correlation coefficient. In a real data analysis situation, it would be reasonable to inquire about the nature of the outlying point before drawing

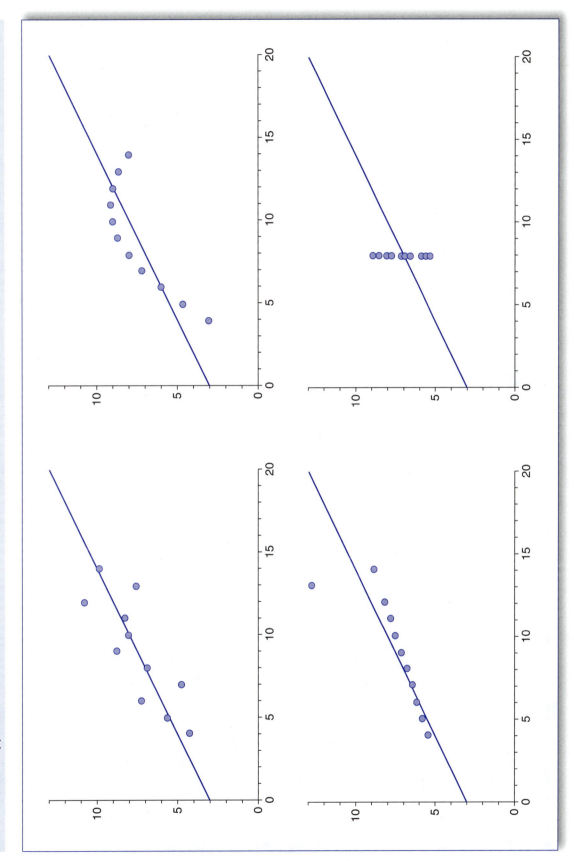

Figure 11.13 Data sets to illustrate the dangers inherent in interpreting summary statistics in isolation from graphs (Anscombe, 1973, pp. 19–20).

conclusions. Such outlying points may be caused by measurement error or observations from populations that are dramatically different from those in the main cluster of data. The graph in the bottom right corner again shows the effect of a point from outside the main cluster, and also shows a situation where the regression line is not reflective of the trend of most of the data points in the set. Asking students to critique the appropriateness of computing correlation coefficients and fitting least-squares lines in situations like these can help sharpen intuition regarding bivariate data analysis.

In addition to examining data sets like those shown in Figure 11.12, it is important for students to have opportunities to work with variables that are negatively correlated. Students often find it difficult to reason about covariation when there is a negative relationship (Garfield & Ben-Zvi, 2008). There is a subtle linguistic distinction between saying that there is a negative relationship and that there is no relationship. Some students, unsure of how to interpret correlation coefficients, believe that a correlation of -0.7 is weaker than a correlation of 0.1 (Shaughnessy, 2007). Hence, when selecting data sets and graphs for students to analyze, choosing some that exemplify negative relationships provides opportunities to address a frequently problematic aspect of students' thinking.

TECHNOLOGY CONNECTION

Tools for Bivariate Data Analysis

Online tools facilitate the study of bivariate data. One notable website is the Data and Story Library (DASL; http://lib.stat.cmu.edu/DASL/). Teachers can use DASL to select data sets that match students' interests and address patterns of thinking they exhibit. Some collections of online Java applets also help sharpen students' intuition about bivariate data analysis. The Rossman/Chance website (www.rossmanchance.com), for example, has a Guess the Correlation applet. The applet generates samples of bivariate data and asks students to estimate their correlations. After an estimate is provided, the applet shows the exact value of the correlation, and students can compare their estimates against it. The National Library of Virtual Manipulatives (NLVM; http://nlvm.usu.edu/en/nav/vlibrary.html) has an applet that allows the user to enter data on height versus hand span for several individuals. The applet shows how the correlation coefficient changes with each data point entered and also how the equation of the least-squares line changes. Observing how the correlation coefficient, graph, and least-squares line are linked can help improve students' judgment about correlation.

Implementing the Common Core

See Clinical Task 11 to engage students in a task that involves computing and interpreting the correlation coefficient in a technological environment (Content Standard S-ID.8).

Although least-squares is the most common method for fitting lines to data, students frequently do not understand how the least-squares regression line is produced. Once again, technology can help address the problem. Using the software program Fathom, students can fit a "movable line" to bivariate data to approximate a regression line, as shown in the first graph of Figure 11.14. Once the movable line is in place, the Show Squares function can be turned on (second graph). Doing so provides a visual of the squares involved in the least-squares calculation. One can then adjust the movable line to try to minimize the total area of the squares shown (third graph). When it appears that the collective area of the squares has been minimized as much as possible, Fathom can plot the graph of the actual least-squares line (fourth graph). Comparing the actual least-squares line to the estimated line can help students understand the behavior and meaning of the frequently used technique of least-squares regression. The technology enables students to produce a number of examples quickly, something that it is not feasible with paper and pencil alone.

Figure 11.14 Using Fathom to understand the meaning of "least-squares."

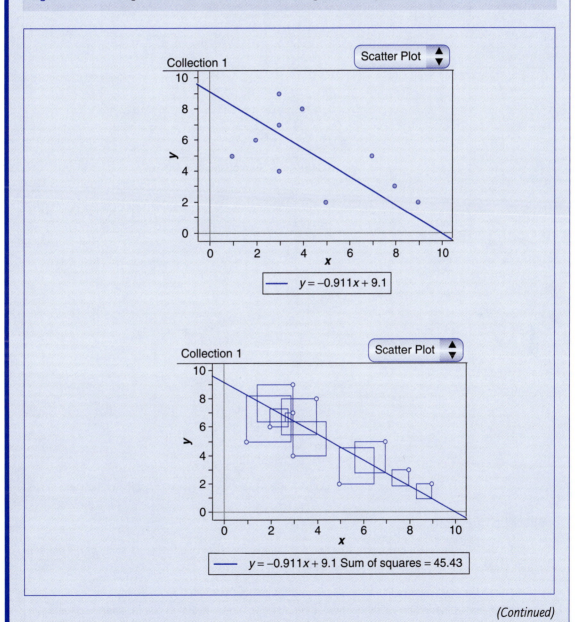

(Continued)

(Continued)

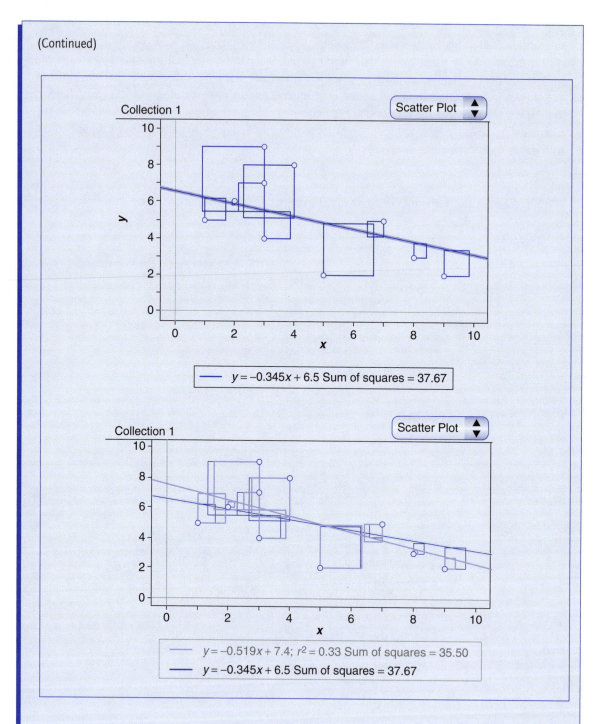

Collection 1 — Scatter Plot

$y = -0.345x + 6.5$ Sum of squares = 37.67

Collection 1 — Scatter Plot

$y = -0.519x + 7.4$; $r^2 = 0.33$ Sum of squares = 35.50
$y = -0.345x + 6.5$ Sum of squares = 37.67

Follow-up questions:

1. Explore the websites mentioned in this Technology Connection. How might you use some of the material on the websites in teaching your own students?

2. How might the least-squares line be misleading for some data sets? Provide an example of a data set for which the least-squares line would be misleading and give alternative methods for analyzing the data.

Theme 2: Sampling and Experimentation

Some aspects of sampling and experimentation were discussed in Chapter 9, and that discussion will not be repeated below. Instead, this chapter will concentrate on some of the more formal aspects of study design associated with the AP course. One of the major conceptual leaps students must make in AP Statistics is to decide when different types of studies, such as surveys and experiments, are necessary. They must also be able to understand and use formal tools for quality control in the design of surveys and experiments so that sound conclusions can be drawn from studies (College Board, 2008).

Designing and Conducting Surveys

The ability to design surveys and judge their quality does not always come naturally. When initially studying survey design, it is not always obvious why random samples are desirable. Students at times propose flawed methods such as voluntary Internet polls or unsystematic door-to-door visits for gathering data to learn about a population of interest (Groth, 2003). Students often understand that sample size is an important factor in judging the quality of a survey, but they tend to overlook other relevant factors. This phenomenon was illustrated by the results of a question on the National Assessment of Educational Progress (NAEP). When eighth graders were asked if a survey conducted at a baseball park would accurately reflect the overall population, only 45% recognized it as a problematic sampling method. However, 57% recognized that larger samples were more desirable than small ones in another scenario involving a survey (Tarr & Shaughnessy, 2007). These characteristics of student thinking suggest that techniques for survey design need careful, explicit instructional attention.

The AP Statistics course description and sample examinations provide a variety of survey design scenarios to discuss with students. One such scenario is shown in Figure 11.15. Exploring the possible responses to the multiple-choice question provides an opportunity to address many important aspects of survey design and interpretation. Response A presents an opportunity to discuss the meaning of the word *bias*. Bias generally arises from the design and administration of an instrument. Response B brings up the importance of independence. There is nothing in the scenario to suggest that the assumption of independence was invalidated. Likewise, there is nothing in the scenario to support the notion set forth in choice C. Choice D raises the issue of stratified random sampling versus simple random sampling. Nothing in the scenario suggests that stratification occurred at the outset of the survey, and simply receiving a low response rate does not stratify the sample. Response E, the best choice, acknowledges the possibility of bias arising from characteristics that led some to respond to the survey and others to ignore it. Students can be asked to brainstorm situations in which this type of bias may occur. A mailed questionnaire about perceptions of a political party, for example, may catch the attention of fervent supporters and critics of the party, while those who have more neutral feelings may not invest the time to respond. As a result, it would be difficult to draw reasonable inferences about perceptions that exist across the entire country.

Figure 11.15   AP Statistics item on interpretation of survey results (College Board, 2008, p. 23).

Each person in a simple random sample of 2,000 received a survey, and 317 people returned their survey. How could nonresponse cause the results of the survey to be biased?

(A) Those who did not respond reduced the sample size, and small samples have more bias than large samples.
(B) Those who did not respond caused a violation of the assumption of independence.
(C) Those who did not respond were indistinguishable from those who did not receive the survey.
(D) Those who did not respond represent a stratum, changing the simple random sample into a stratified random sample.
(E) Those who did respond may differ in some important way from those who did not respond.

Source: Copyright © 2008 The College Board. Reproduced with permission. http://apcentral.collegeboard.com

Designing and Conducting Experiments

As with surveys, the important elements of experimental study design are not necessarily intuitive. After studying random sampling in the context of administering surveys, students may confuse it with the idea of random assignment (Derry, Levin, Osana, Jones, & Peterson, 2000). Additionally, students may not see much difference between observational studies and experiments, thinking that similar conclusions can safely be drawn from each (Groth, 2003). Even students who see the importance of systematically comparing one group against another may not recognize the value of experimental controls such as random assignment to treatments or double-blinding in appropriate situations (Groth, 2003). Generally speaking, there is a great deal of room for growth in understanding experimental study design as students enter AP Statistics.

Figure 11.16 shows an item from an AP Statistics examination that can be discussed with students to help sharpen ideas about comparing treatments against one another. After reading the scenario in the item, one can raise questions about how an experimental study design may allow more accurate inferences than the survey design that is described. Each of the possible responses to the question (with the exception of response D) highlights problems with the study design. Before students are shown the possible responses, they can be encouraged to identify as many flaws in the study design as possible. They can also work on designing alternative studies that would allow reasonable conclusions about the quality of each cookie recipe. After students do some of their own thinking about the scenario, they can be allowed to examine the answer choices to reflect on the weaknesses they identified and the alternative study designs they proposed. Choice A brings out the important aspect of selecting participants for the study. Choices B and C describe how the recipe's quality may be confounded by other variables. Choice E emphasizes the importance of replication. Students can decide whether or not their own strategies take these important factors into account. As they do so, a question that was initially just a multiple-choice item on the AP Statistics exam can become a powerful tool for helping students reflect on and refine their own statistical thinking.

Figure 11.16   AP Statistics item on comparing cookie recipes (College Board, 2008, p. 26).

16. George and Michelle each claimed to have the better recipe for chocolate chip cookies. They decided to conduct a study to determine whose cookies were really better. They each baked a batch of cookies using their own recipe. George asked a random sample of his friends to taste his cookies and to complete a questionnaire on their quality. Michelle asked a random sample of her friends to complete the same questionnaire for her cookies. They then compared the results. Which of the following statements about this study is false?

(A) Because George and Michelle have a different population of friends, their sampling procedure makes it difficult to compare the recipes.
(B) Because George and Michelle each used only their own respective recipes, their cooking ability is comfounded with the recipe quality.
(C) Because George and Michelle each used only the ovens in their houses, the recipe quality is confounded with the characteristics of the oven.
(D) Because George and Michelle used the same questionnaire, their results will generalize to the combined population of their friends.
(E) Because George and Michelle each baked one batch, there is no replication of the cookie recipes.

Recognizing the difficulties students encounter in thinking about statistical study design, the American Statistical Association produced a set of guided investigations for classroom use (Peck & Starnes, 2009). In one of the investigations, students are taken through the process of designing an experiment to determine whether teenagers prefer blue soda or clear soda. The investigation begins by prompting students to formulate a blue soda recipe with food coloring. Once the recipe is done, the investigation worksheet raises critical questions to be answered in designing and conducting the experiment. The first question, for instance, asks why it is important to randomize the order in which the drinks are presented. Another question asks about the possibility of using blinding for the study. Students are also asked to think about the population to which the results of the experiment might generalize. A variety of additional important details about conducting the study are addressed, such as the precise wording of the question to be asked of all study participants after they have tasted the two sodas.

Once the results of the blue soda experiment have been gathered, the investigation prompts students to make graphical displays and decide if there was a preference for one drink over the other. Finally, students are asked to write a report for a soft drink company summarizing the results of the study and commenting on the advisability of manufacturing blue soda. Going through all of the steps of a study in this manner allows students to sharpen their study design skills in an interesting context while also drawing on exploratory analysis and formal inference skills developed in AP Statistics.

> **Implementing the Common Core**
>
> See Clinical Task 12 to assess students' ability to "make inferences and justify conclusions from sample surveys, experiments, and observational studies" when given the items shown in Figures 11.15 and 11.16 (Content Standard S-IC).

Theme 3: Anticipating Patterns in Random Phenomena

As discussed in Chapter 8, the behavior of random phenomena is often counterintuitive. Students have difficulty, for example, knowing how much faith to place in random samples. At times, they may have too much confidence that a sample represents a population well, and at other times they may have too little confidence (Rubin, Bruce, & Tenney, 1991). They may think that the proportion of the population sampled is more important than absolute sample size, not seeing value in a survey unless 10% or more of the population is included in the sample (M. H. Smith, 2004). Such notions are in conflict with sound intuitions about the behavior of random samples. Ideally, students should view a sample as a small-scale version of the population that produces statistics approximating a characteristic of interest. Such a view reflects an understanding of sample-to-sample variability and the need for a sampling distribution as the grounds for inference (Saldanha & Thompson, 2002).

Saldanha and Thompson (2007) described a teaching sequence aimed at helping students develop sound intuitions about sampling distributions. The initial activity was to draw from a bag that contained the same number of red and white candies. Students were not told that the bag contained the same number of reds and whites. Instead, they were directed to draw samples, record their composition, and describe sample-to-sample variability. Taking several samples helped students recognize limitations inherent in drawing just one random sample from a population. After drawing 10 samples of equal size from the population, students examined the distribution of the sample proportions and conjectured that there were more white candies than reds in the bag. When the instructor revealed that the number of whites was actually equal to the number of reds, discussion ensued about the likelihood of obtaining the results generated in class. Simulating the class sampling activity many times using computer software led to the conclusion that the results obtained by the class actually were not that unusual.

TECHNOLOGY CONNECTION

Simulation Software

In general, computer simulation software plays an indispensible role in teaching via simulation. It is advisable to have students begin by drawing samples by hand, as in the candy drawing example, but once that experience has occurred simulation can be automated by the software. One freely available, classroom-tested simulation package that helps build understanding of the concept of sampling distribution is Sampling Sim (download Windows and Mac versions from www.tc.umn.edu/~delma001/stat_tools/). It can produce simulations of the results obtained from computing the means and proportions of several samples drawn from a population.

Chance, delMas, and Garfield (2004) described their classroom experiences over the course of seven years of using Sampling Sim software to teach about sampling distributions. They found that one of the most effective activities consisted of three stages. First, students were shown a population distribution and asked to predict a correct sampling distribution. For example, given the population distribution shown in the first graph corner of Figure 11.17, they were to predict which of the graphs in the figure would represent the sampling distribution of sample means drawn from the population. After making a selection, students were to test their prediction by generating a simulated sampling distribution with the software. Finally, students

Figure 11.17 Predicting the shape of the sampling distribution of sample means, given a population distribution (Garfield, delMas, & Chance, 2002, n.p.).

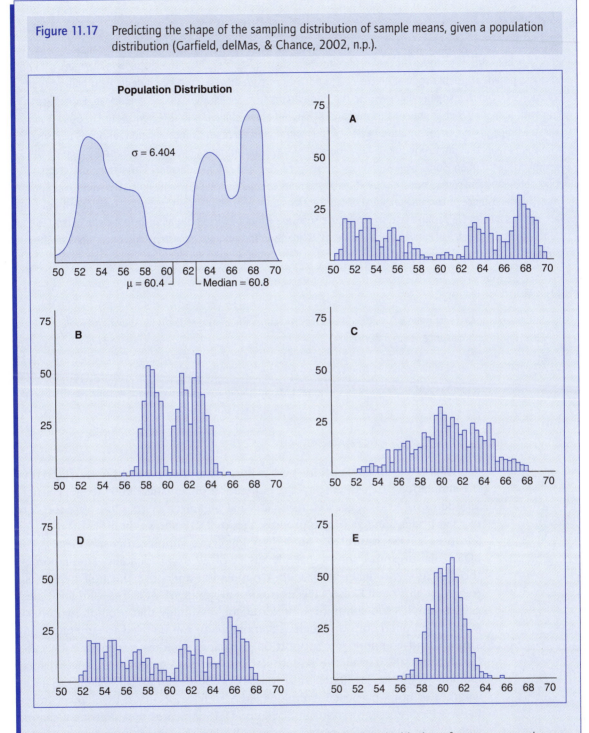

were to compare their initial conjectures against the results produced with the software, commenting on discrepancies or areas of agreement between the initial conjecture and the results obtained. The three-step process of predicting, testing, and evaluating caused students to address misconceptions they held at the outset of the activity.

(Continued)

(Continued)

Although predicting, testing, and evaluating conjectures about sampling distributions are valuable activities, teachers still need to be on the lookout for persistent misconceptions. The distracters in Figure 11.17 suggest potential trouble spots. Choices A, B, and D are frequently selected by students who believe that sampling distributions generally resemble the populations from which the samples are drawn. Experimenting with Sampling Sim addresses this misconception, but some still cling to it (Chance et al., 2004). Even those who begin to form an intuitive notion of the central limit theorem by recognizing C and E as plausible choices may exhibit confusion about how the size of each individual sample influences the shape of the sampling distribution. Some believe that as the size of individual samples increases, the sampling distribution becomes more spread out. This belief leads students to select choice C when choice E is more appropriate. The persistence of certain statistical misconceptions demonstrates that although the predict-test-evaluate cycle is a valuable pedagogical structure, the software cannot by itself put all incorrect ideas to rest. The teacher needs to actively converse with students to identify and address persistent misconceptions as they surface.

Follow-up questions:

1. Suppose your students believed that choice D was the correct response to the item shown in Figure 11.17. How might you convince them otherwise?

2. In what ways can Sampling Sim software be used as a metacognitive tool?

Simulation of Samples Model

Sometimes misconceptions persist because students do not understand how the simulation process leads to the generation of statistical knowledge. Garfield and Ben-Zvi (2008) suggested using a visual organizer to help illustrate the conceptual underpinnings of the process. Figure 11.18 shows their **Simulation of Samples (SOS) model** for an activity involving sampling Reese's Pieces. The SOS visual organizer shows three levels involved in a probability simulation. At the first level, students work with a population and pose the null hypothesis (e.g., that 45% of all Reese's Pieces candies are orange). At the second level, they gather samples and compute sample statistics assuming that the null hypothesis is true. Computer software such as Sampling Sim can be used to generate many samples and compute the proportion of orange candies drawn in each case. Finally, at the third level, the sample statistics are aggregated into a single display (Sampling Sim also facilitates this process). Students can use the display to judge whether or not a particular sample is likely to have occurred by chance. In the Reese's Pieces case, the bottom of Figure 11.18 indicates that drawing a sample with 40% orange candies would not be unlikely if the overall population proportion for orange is 45%. On the other hand, obtaining a random sample in which 67% were orange could lead one to suspect that the population proportion for orange is actually not 45%. The SOS visual organizer helps convey the logic of inference, the role of sampling distributions, and how simulation can be used to test hypotheses.

Figure 11.18   Simulation of Samples (SOS) visual organizer (Garfield, delMas, Zieffler, & Ben-Zvi, n.d., n.p.).

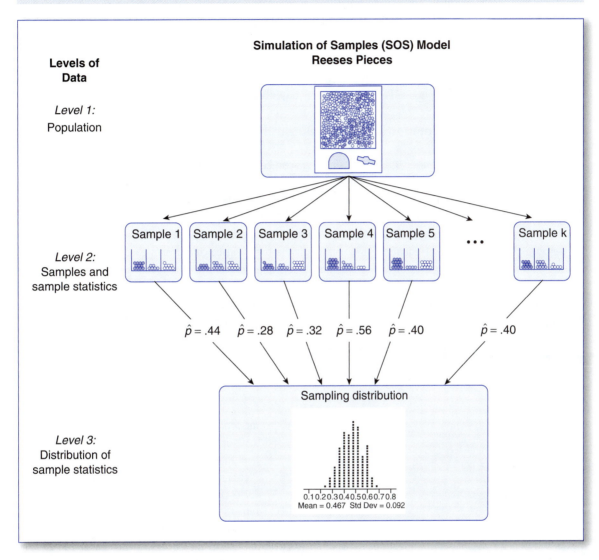

Write a simulation problem for which the Simulation of Samples (SOS) model would be applicable. Then design a visual organizer you would use to provide students an overview and summary of the simulation process to be used for the problem.

STOP TO REFLECT

Theme 4: Statistical Inference

As demonstrated in the SOS visual organizer and accompanying Reese's Pieces activity, reasoning about sampling distributions and random phenomena leads naturally into the topic of formal inference. Extensive attention to formal inference separates the content of AP Statistics from other statistical content students ordinarily encounter in secondary school. The AP course requires understanding and use of hypothesis tests and confidence intervals. Students must be able to use graphing calculators to perform hypothesis tests and construct confidence intervals. Most important, they need to interpret calculator output.

TECHNOLOGY CONNECTION

Interpreting Calculator Output for Hypothesis Tests

On the AP Statistics examination, graphing calculators can be used to carry out hypothesis tests. This minimizes potential difficulties with computational aspects of hypothesis testing. Figure 11.19 shows graphing calculator input and output related to the Reese's Pieces situation described earlier. Suppose that in a random sample of 30 candies drawn from the population, 20 were orange. The calculator can be used to test the null hypothesis that 45% of all Reese's Pieces manufactured are orange against the alternative that the percentage of orange is not equal to 45%. Although the graphing calculator can execute the necessary computations, interpreting the output remains the student's responsibility.

Figure 11.19 Graphing calculator input and output for a hypothesis test.

Graphing calculator input	Graphing calculator output
1-PropZTest	1-PropZTest
p_0: .45	prop $\neq$ .45
x: 20	z = 2.385421491
n: 30	p = .0170595426
Prop $\neq p_0$ < p_0 > p_0	$\hat{p}$ = .6666666667
Calculate Draw	n = 30

When interpreting calculator output, students may find it difficult to understand the meaning of the p-value and the firmness of the conclusions that can be drawn. One of the main misconceptions about p-value is that it provides the probability that the null hypothesis is true (or false; Garfield & Ben-Zvi, 2008). A student holding this type of misconception may interpret the p-value of .017 in Figure 11.19 to mean that the null hypothesis (i.e., that 45% of all Reese's Pieces manufactured are orange) has only a 1.7% chance of being true. It is more accurate to think of p-value as measuring "the probability of an outcome at least as extreme as the one we actually got" (Shaughnessy & Chance, 2005, p. 71). In the context of the Reese's Pieces example, this would mean that if it is actually true that 45% of all Reese's Pieces manufactured are orange, the probability of obtaining 20 or more orange in a random sample of 30 has a probability of only .017. This provides strong evidence against the null hypothesis, but it does not allow us to say with mathematical certainty that the null hypothesis is false. Students sometimes believe that hypothesis tests prove the null hypothesis to be true or false (Garfield & Ben-Zvi, 2008), perhaps expecting mathematical certainty because of previous experiences in school mathematics. If hypothesis testing is connected with previous experiences in building up and analyzing sampling distributions (Figure 11.18), students have firmer grounds for interpreting the p-value and the strength of conclusions that can be drawn from their calculations.

Follow-up questions:

1. What is the difference between "proving the null hypothesis to be false" and "having strong evidence against the null hypothesis?" How would you help students see the distinction?

2. See Figure 11.19. Explain the meaning of each piece of output displayed on the right side of the figure.

TECHNOLOGY CONNECTION

Interpreting Calculator Output for Confidence Intervals

As part of the study of formal inference in AP Statistics, students are to work with confidence intervals. The computational aspects of determining confidence intervals on the AP examination are facilitated by use of graphing calculators, just as with hypothesis testing. Figure 11.20 shows calculator input and output for a 95% confidence interval for estimating the proportion of manufactured Reese's Pieces that are orange, given that we have a random sample in which 20 out of 30 were orange. Again, although the graphing calculator computes the confidence interval, the interpretation of the output is up to the user.

Students often have considerable difficulty learning to interpret confidence intervals. Various misconceptions have been well-documented. Garfield and Ben-Zvi (2008) listed the following misconceptions as some of the most prevalent:

- There is a 95% chance the confidence interval includes the sample mean [or, in the case above, sample proportion].
- There is a 95% chance the population mean will be between the two values (upper and lower limits).
- 95% of the data are included in the confidence interval.
- A wider confidence interval means less confidence.
- A narrower confidence interval is always better (regardless of the confidence level). (p. 270)

Teachers should explicitly look for evidence of these misconceptions when assessing students.

Correctly interpreting the 95% confidence interval generated in Figure 11.20 hinges on understanding a confidence interval as a prediction about the location of the population proportion (Perkowski & Perkowski, 2007). The output shown in Figure 11.20 predicts that the proportion of all Reese's Pieces that are orange is somewhere between 49.8% and 83.5%. With repeated random sampling, we expect this interval to include the parameter 95% of the time. Of course, saying the population proportion is somewhere between 49.8% and 83.5% is a rather wide, imprecise estimate. We could obtain a narrower estimate by drawing a larger random sample. Additionally, if we wanted to produce a 99% confidence interval, it would be even wider than the 95% interval from the sample of 30, again underscoring the desirability of a larger random sample (assuming that time and money are available for obtaining one).

Figure 11.20   Graphing calculator input and output for a confidence interval.

Graphing calculator input	Graphing calculator output
1-PropZInt	1-PropZInt
x: 20	(.49798, .83535)
n: 30	$\hat{p} = .6666666667$
C-Level: .95	
Calculate	n = 30

Students should understand how the confidence level, sample size, and width of the confidence interval are related to one another. Graphing calculators and computer software packages can facilitate explorations leading to this type of understanding by allowing students to generate a variety of confidence intervals under a number of different conditions and then analyze the results. They can be asked to investigate, for instance, how the width of an 80% confidence interval compares to that of 95% and 99% confidence intervals generated using a given random sample. They can also investigate how changing the size of the random sample affects the width of each interval. Patterns observed in calculator output can be explained and verified by returning to the algebraic formulas used for computing confidence intervals and discussing how the selected confidence level and sample size influence the result.

(Continued)

(Continued)

Follow-up questions:

1. How would the width of an 80% confidence interval for a parameter differ from a 95% confidence interval? From a 99% confidence interval?

2. Why do students develop misconceptions about confidence intervals? Describe possible origins for each of the student misconceptions discussed in this Technology Connection.

CONCLUSION

Teaching an AP Calculus or Statistics course can be an exciting and challenging opportunity. The AP courses provide many high school students with their first glimpse of college-level mathematical and statistical thinking. Some of the core concepts addressed in AP Calculus are limits, differentiation, integration, and the fundamental theorem of calculus. In AP Statistics, exploratory analysis, study design, simulation, and formal inference form the conceptual structure of the curriculum. Students bring existing, sometimes unconventional, intuitions about the core concepts in the AP courses to the classroom. An important part of the teacher's job is to identify intuitions that are in conflict with conventional mathematical and statistical thinking and to help students form intuitions that will provide a solid foundation for further postsecondary study. This chapter provides a starting point for preparing to teach AP courses, but it is not exhaustive. Learning the finer details of teaching AP courses is a career-long process that can be pursued by taking advantage of professional development opportunities available through conferences, online discussions, AP exam grading sessions, and other means. The career-long journey of an AP teacher is marked by progressively deeper understanding of the content of the course, students' thinking, and the interaction between the two. A glimpse at teaching some of the AP topics not explicitly discussed in this chapter, including related rates, transformations of data sets, and chi-square statistics, is provided in the four-column lesson plans at the end of the chapter.

VOCABULARY LIST

After reading this chapter, you should be able to offer reasonable definitions for the following ideas (listed in their order of first occurrence) and describe their relevance to teaching Advanced Placement courses:

Advanced mathematical thinking 354	Local linearity 360
Calculus reform movement 356	Harmonic thinking 365
Harvard calculus 356	GAISE college report 370
Rule of four 356	Illusory correlation 371
External sources of conviction 357	Spurious correlation 372
Internal sources of conviction 357	Anscombe's quartet 372
Collapse metaphor for limit 357	Sampling Sim 380
Approximation metaphor for limit 357	Simulation of Samples (SOS) model 382

HOMEWORK TASKS

1. Visit the AP Central website (http://apcentral.collegeboard.com). After exploring the website, make a "top 10" list of the most useful resources available on the site. Write your list from the perspective of a teacher preparing to teach either AP Calculus or AP Statistics for the first time. Explain why each item on your list would be useful for establishing an AP Calculus or AP Statistics course.

2. Visit the AP Central website (http://apcentral.collegeboard.com). Download the most recent set of free-response questions from the AP Calculus AB or BC examination. Also download all available sample responses and scoring guidelines for one of the free-response questions. Use the scoring guidelines to assign a grade to each sample response. Then check the accuracy of your scoring by reading the scoring commentary accompanying the sample responses (do not read the scoring commentary before assigning your own grade to each sample response).

3. Visit the AP Central website (http://apcentral.collegeboard.com). Subscribe to the discussion group for AP Calculus. After following the conversation in the discussion group for a week and exploring the discussion archives, make a "top 10" list of the most important questions posed by AP Calculus teachers. Provide a satisfactory resolution to each question on your list.

4. Interview a mathematics professor at your university and ask what comes to mind when he or she hears the phrase "advanced mathematical thinking." Ask for specific examples of how the professor uses advanced mathematical thinking in his or her own teaching or research. Compare and contrast the professor's response to the description of advanced mathematical thinking in this chapter.

5. Write a calculus problem that is best solved by employing harmonic thinking. Then write a solution to the problem that demonstrates aspects of strong visual and analytic thinking. Also write at least three questions or prompts that would aid students who need help developing the harmonic thinking necessary for solving the task.

6. Explain how plotting the points from a table of values showing the areas of rectangles that approximate the area between the curve $f(t) = t^2$ and the horizontal axis can lead students to conjecture that $\int_0^x t^2\, dt = \frac{x^3}{3}$. Also explain how changing the lower boundary can lead to the broader conjecture that $\int_a^x t^2\, dt = \frac{x^3}{3} - \frac{a^3}{3}$.

7. Visit the AP Central website (http://apcentral.collegeboard.com). Download the most recent set of free-response questions from the AP Statistics examination. Also download all available sample responses and scoring guidelines for one of the free-response questions. Use the scoring guidelines to assign a grade to each sample response. Then check the accuracy of your scoring by reading the scoring commentary accompanying the sample responses (do not read the scoring commentary before assigning your own grade to each sample response).

8. Visit the AP Central website (http://apcentral.collegeboard.com). Subscribe to the discussion group for AP Statistics. After following the conversation in the discussion group for a week and exploring the discussion archives, make a "top 10" list of the most important questions posed by AP Statistics teachers. Provide a satisfactory resolution to each question on your list.

9. Download the AP Statistics course description (http://apcentral.collegeboard.com) and the GAISE College Report (www.amstat.org). Identify at least five areas in which the two documents express similar philosophies of teaching statistics. Describe how the documents complement one another in the five identified areas by using direct quotes from each source.

10. Download the software program Sampling Sim (www.tc.umn.edu/~delma001/stat_tools/). Also download at least one of its accompanying activities from the website. Using the AP Statistics course description (http://apcentral.collegeboard.com), identify the AP course objectives that can be addressed through use of the Sampling Sim activity you selected. Also complete the Sampling Sim activity you selected, showing all work and explaining your reasoning.

CLINICAL TASKS

1. Interview a high school principal to determine his or her perception of AP courses. How important are AP courses to the overall mission of the principal's high school? What are the advantages and disadvantages of offering AP courses? How do students from the high school tend to perform on AP examinations? Does the high school plan to offer more AP courses in the future? Why or why not?

2. Interview a teacher of AP Calculus or AP Statistics to determine what it is like to teach an AP course. In what kinds of professional development opportunities does the teacher engage? What kinds of unique challenges are associated with teaching an AP course? What reasons do students have for taking AP courses?

3. Observe an AP Calculus or AP Statistics class. Document and describe classroom events that have to do with building procedural knowledge. Also document and describe classroom events that have to do with building conceptual knowledge. Explain why you categorized each event as either procedural or conceptual. Then critique the lesson: Was there attention to both procedural and conceptual knowledge? What steps could you take to further improve the lesson?

4. Obtain an AP Calculus textbook from your clinical site. To what extent does it reflect the four principles underlying Harvard calculus? Provide specific examples for each of the four categories. In cases where the book does not have examples that fit one of the four categories, describe how one or more of the exercises could be modified to align with the category.

5. Interview an AP Calculus student and a student who has not yet taken calculus. Ask each student if the following statement is true: $.\overline{9} = 1$. Ask each student to explain his or her reasoning. Next, ask each student if the following statement is true: $.\overline{3} = \frac{1}{3}$. Again, ask each student to explain his or her reasoning. Compare and contrast the students' responses to one another. Also compare and contrast each individual student's justification for the first and second questions. Comment on the strengths and weaknesses in each student's reasoning.

6. Ask an AP Calculus class to solve the problems shown in Figures 11.2 and 11.3 in writing. Have them write an explanation of the reasoning supporting their answers. Then categorize the strategies students use to reason about the problems. Comment on the overall quality of the conceptual and procedural understanding exhibited by students in solving the two tasks.

7. Ask an AP Calculus class to solve the problem shown in Figure 11.4 in writing. Have them write an explanation of the reasoning supporting their answer. Then categorize the strategies students use to reason about the problem. Comment on the overall quality of the conceptual and procedural understanding exhibited by students in solving the task.

8. Ask an AP Calculus class to solve the problem shown in Figure 11.6 in writing. Have them write an explanation of the reasoning supporting their answer. Then categorize the strategies students use

to reason about the problem. Comment on the overall quality of the conceptual and procedural understanding exhibited by students in solving the task.

9. Write several composite functions that AP Calculus students should be able to differentiate (e.g., $y = \sin x^3$). Interview an AP Calculus student or a classmate who has taken calculus and ask him or her to differentiate the composite functions. Ask him or her to explain the rule applied in each case. Is there evidence that he or she sees the tasks as linked by the chain rule, or not? Explain.

10. Obtain an AP Statistics or AP Calculus syllabus from your clinical site. Also examine sample syllabi available online at http://apcentral.collegeboard.com. Then, write a syllabus for an AP Calculus or Statistics course you would teach. Explain how your syllabus draws on elements from each of the syllabi you examined. Also explain why you believe the syllabus would pass the AP audit process (see http://apcentral.collegeboard.com for the audit process and guidelines).

11. Interview at least two different students using the Guess the Correlation applet on the Rossman/ Chance website (www.rossmanchance.com). Ask each student to predict the correlations for at least five different sets of data. Record how close each student's prediction comes to the actual value. Also ask each student to explain the reasoning he or she used in estimating correlations. Describe strengths and weaknesses in each student's reasoning patterns.

12. Interview an AP Statistics student and a student who has not yet taken the course. Ask each student to respond to the items shown in Figures 11.15 and 11.16. Have each student explain his or her reasoning. Compare and contrast the students' responses to one another. Comment on the strengths and weaknesses in each student's reasoning.

13. Write a story to provide a context for the population and sampling distributions shown in Figure 11.17. In your story, include the situation of taking small random samples and also the situation of taking larger random samples. Share your story with a student and ask him or her to match the population distribution with a plausible sampling distribution for the case of small random samples and for the case of larger random samples. Ask the student to justify his or her answer. Describe the strengths and weaknesses in the patterns of thinking you observe in the student.

14. Share the calculator outputs from Figures 11.19 and 11.20 with an AP Statistics student. Explain how and why the output was generated, using the Reese's Pieces example from the chapter. Then ask the student to interpret the output. Does the student seem to have any misconceptions about hypothesis tests and confidence intervals? If so, which ones? If not, what did the student say in correctly interpreting the output in each case?

15. Ask a group of AP Calculus or AP Statistics students to respond to the following two writing prompts: "Why did you choose to enroll in this course?" "How does the course help support your future career goals?" Summarize the students' responses and draw conclusions about why students choose to enroll in the course.

16. Visit the Adapting and Implementing Innovative Material in Statistics (AIMS) website (www.tc.umn .edu/~aims/) with the AP Statistics teacher in your school. Together, select a lesson that would be feasible to coteach to a group of students. While teaching the lesson, have students generate work samples that demonstrate their understanding of the concepts in the lesson. Collect the work samples and write a report describing the important concepts students seemed to learn as well as important concepts with which students may still need more work.

Teaching and Learning AP Calculus

Items to Consider Before Reading the Vignette

1. State the chain rule and the power rule for differentiation. Give an example of the application of each rule. Explain how they are related to one another.

2. Suppose $f(7) = -5, f'(7) = 9, g(7) = 7$, and $g'(7) = -4$. What is $h'(7)$ if $h(x) = f(g(x))$? Explain your reasoning.

3. Explain how $y = \sin x^2$ differs from $y = \sin^2 x$. Then differentiate each function. Graph each one and its derivative. Explain why the derivative function graphs behave as they do.

Scenario

Mr. Matthews was in his second year of teaching and in his first year in a new school district. Part of the attraction of the new school district was the opportunity to teach AP Calculus. He had thoroughly enjoyed calculus as an undergraduate and was eager to share his knowledge with high school students seeking college credit through the AP examination. It did not take long, however, for his enthusiasm to be tempered by the reality of the demanding spot in which he found himself. Mr. Matthews was taking the place of a loved and admired teacher, Mr. Vandersetzig, who had just retired. Students in Mr. Vandersetzig's class had consistently earned high scores on the AP exam. It seemed that whenever Mr. Matthews talked to the school principal, Mr. Stanczyk, he would comment on how high the AP Calculus scores for the school had been in the past. None of the veteran teachers in the district wanted to teach AP Calculus because of the impossibly high bar that had been set. An additional pressure on Mr. Matthews was that as he reviewed the AP Calculus syllabus, many of the topics it contained seemed quite different from those in his undergraduate calculus syllabus. He found himself learning and relearning content as he progressed through the semester.

The Lesson

It was the designated time of the year for Mr. Stanczyk to observe one of Mr. Matthews's lessons and write a formal evaluation. Mr. Stanczyk liked to give his teachers freedom to choose the class period he would come to observe. Mr. Matthews was intent on proving that he was capable of filling Mr. Vandersetzig's shoes, so he invited Mr. Stanczyk to observe his AP Calculus class. Mr. Matthews also believed that his AP Calculus class would be better behaved than his basic math class, so he would not run the risk of being negatively evaluated on classroom management skills.

Mr. Matthews felt that his class was behind in preparing for the AP examination. He had not anticipated the high number of absences his students would have. Because the AP class contained some of the most active students in the school, extracurricular activities such as band trips and pep rallies would routinely wipe out half of his class attendance on some days. Fearing that students would not gain the knowledge they needed before the exam, he had decided to conduct his 50-minute class sessions mainly as lectures, despite the advice of much of the professional development he had received. On the day of Mr. Stanczyk's visit, Mr. Matthews planned to give a 50-minute lecture on the chain rule.

The lecture began with a few quick examples to review the power rule students had previously learned. Mr. Matthews began, saying, "Up to this point, we have differentiated functions like $y = x^3$ and $y = 2x^3 + 1$ by using the power rule. Today, we will learn how to handle functions like $y = (2x^3 + 1)^2$ by using the chain rule." He then showed the steps in differentiating the function and wrote: $y' = 2(2x^3 + 1) \cdot 6x^2 = 12x^3 (2x^3 + 1)$. Amy, a student who was deeply disappointed that Mr. Vandersetzig had retired before teaching her AP class, quickly raised her hand as soon as this computation was placed on the board. She stated, "I don't know how that can be right. I used the product rule on the function. I wrote $y = 2(2x^3 + 1) (2x^3 + 1)$. Then I wrote $f = 2x^3 + 1$, $f' = 6x^2$, $g = 2x^3 + 1$, and $g' = 6x^2$. So that means the derivative should be $6x^2(2x^3 + 1) + 6x^2(2x^3 + 1)$. That is $12x^2 (2x^3 + 1)$. Shouldn't you have used the product rule instead?" Mr. Stanczyk, whose teaching background was in social studies rather than mathematics, busily scribbled notes on his paper as Amy spoke. Mr. Matthews, flustered by being challenged about the simplest example in the lesson and seeing Mr. Stanczyk taking notes, needed a moment to realize what had happened. After floundering at the board for a minute, he found the algebra error he had made in simplifying his example. He explained, "OK, I see what happened. When I simplified my answer, I mistakenly wrote that $2(6x^2) = 12x^3$. It is $12x^2$, so our answers actually agree. Somehow I saw an extra x in there." Amy had an unconvinced and confused look on her face, but Mr. Matthews continued the lesson.

The next example in the lesson involved differentiating the trigonometric function $y = \sin x^2$. Mr. Matthews explained that the "inner function" in this case was x^2, and the "outer function" was sine. He showed that applying the chain rule would produce $2x(\cos x^2)$. It did not take long for Amy's hand to shoot up once again. She said, "OK, this time I know that we didn't get the same answer. I used the product rule on this one, too. Here's how I set it up: $f = \sin x$, $g = \sin x$, $f' = \cos x$, and $g' = \cos x$. That means that the derivative has to be $(\cos x)(\sin x) + (\sin x)(\cos x) = 2(\cos x)(\sin x)$. That is not even close to what you have." Mr. Matthews told Amy that her product rule solution was not correct because $\sin x^2$ is not the same as $(\sin x)^2$. Upon hearing this, Amy started whispering to her neighbors about how to do the problem. She stayed in conversation with the small group for the rest of the class period.

Sensing that students were starting to drift, Mr. Matthews decided to do some examples that looked nothing like problems that could be done with the product rule. He decided to demonstrate how to differentiate $y = \sqrt[7]{(2x-1)^2}$. He said, "The first step is to rewrite the radical expression as $(2x-1)^{\frac{2}{7}}$. Now, we can differentiate. Using the chain rule gives us $\frac{2}{7}(2x-1)^{\frac{-5}{7}} \cdot 2 = \frac{4}{7}(2x-1)^{\frac{-5}{7}}$. Now, converting back to radical form will give us our final answer of $\frac{4}{7^7\sqrt{(2x-1)^5}}$. Questions started pouring in. Jesse asked why it was permissible to simplify the radical in the first step. Emily wondered how the negative exponent became positive. Ryan asked why it was necessary to convert back into radical form. Amy interjected that in her algebra classes, one could never leave a radical in the denominator. Mr. Matthews was surprised at the number of questions about the example, and wondered if he had gone too fast in introducing this example right after some of the simpler ones.

Starting to feel as if his lesson was a disaster, Mr. Matthews decided to cut his planned lecture short and give students their homework assignment to do during the last 20 minutes of class. He said, "I can see there is a lot of confusion, so I will come around and help you on an individual basis as you work today's homework problems from the book." Many students began to open their books and work, although a few decided to use the time to socialize with friends. Mr. Matthews nervously tried to avoid Mr. Stanczyk while circulating about the room to help students. When the bell rang, Mr. Matthews approached Mr. Stanczyk and said, "I just don't know what happened today. This was absolutely the worst lesson I have taught all semester. I am sure that if you come back to observe another, it will not be such a disaster." Mr. Stanczyk nodded to Mr. Matthews, tried to be encouraging, and then set up an appointment to discuss the lesson in more detail. Mr. Matthews dreaded having to relive the lesson, wanting to simply put it behind him and forget about it.

Questions for Reflection and Discussion

1. What are some possible advantages and disadvantages of taking on an AP class as a beginning teacher?

2. Is lecturing an effective strategy for making sure that students are prepared for end-of-year high-stakes examinations? Why or why not?

3. Critique Mr. Matthews's introduction of the chain rule. Was the example he chose a good way to illustrate the concept? Why or why not?

4. How could a stronger connection have been made between students' understanding of the power rule and the new idea for the day, the chain rule?

5. How would you have responded to Amy's comments near the outset of the lesson?

6. The lesson involved mainly symbolic representations of the derivative. How might it be enhanced through the use of graphical and tabular representations?

7. How would you respond to the questions Jesse, Emily, Ryan, and Amy asked about Mr. Matthews's example with a radical?

8. How could a better balance be struck between procedural and conceptual knowledge in Mr. Matthews's lesson?

9. Put yourself in Mr. Stanczyk's position. What would you say to Mr. Matthews about the quality of the lesson? Remember, Mr. Stanczyk has no mathematics background.

RESOURCES TO EXPLORE

Books

Diefenderfer, C. L., & Nelsen, R. B. (Eds.). (2010). *The calculus collection: A resource for AP and beyond.* Washington, DC: Mathematical Association of America.

Description: This collection of articles contains ideas for engaging students in the study of introductory calculus. Teachers will find ways to explain difficult ideas, methods for incorporating technology in instruction, and insights about the structure of calculus.

Klymchuck, S. (2010). *Counterexamples in calculus.* Washington, DC: Mathematical Association of America.

Description: This book is unique in that it provides several incorrect mathematical statements, and students are to disprove them by designing counterexamples. The statements are grouped into the categories of functions, limits, continuity, differential calculus, and integral calculus.

Moore, T. L. (Ed.). (2000). *Teaching statistics: Resources for undergraduate instructors.* Washington, DC: Mathematical Association of America and American Statistical Association.

Description: This collection of articles captures the spirit and fundamental tenets of the statistics education reform movement. It provides ideas for teaching with data, using projects for active learning, choosing textbooks, using technology, and assessing students' learning.

Peck, R., & Starnes, D. (2009). *Making sense of statistical studies* (Teacher's module). Alexandria, VA: American Statistical Association.

Description: This book consists of 15 classroom investigations that help students design and analyze statistical studies. Each investigation is accompanied by an overview, a list of prerequisite knowledge needed, learning objectives, teaching tips, extensions, and suggested answers.

Websites

AP Central: **http://apcentral.collegeboard.com**

Description: AP Central contains a wealth of important information for AP Calculus and Statistics teachers. Resources include free-response items from past tests, course descriptions, sample syllabi, articles about teaching, the AP course audit, teacher discussion groups, and information about becoming an AP reader.

Consortium for the Advancement of Undergraduate Statistics Education (CAUSE): **www.causeweb.org**

Description: CAUSEweb.org provides information about resources, professional development, outreach, and research related to introductory statistics courses. Resources include lab activities, homework projects, teaching methods, data sets, and statistics cartoons for classroom use.

Mathematical Association of America: **www.maa.org**

Description: The Mathematical Association of America is dedicated to improving the quality of undergraduate mathematics courses. On its website, it has information about recent publications, meetings, competitions, funding opportunities, and research related to teaching and learning undergraduate mathematics.

Applets for teaching calculus: **http://www.ies.co.jp/math/products/calc/menu.html**

Description: This site contains several Java applets that can be used to help students attain a visual, conceptual understanding of ideas in introductory calculus. Topics addressed with applets include limits, derivatives, the chain rule, rectangle approximation of area under a curve, volumes of solids, and the fundamental theorem of calculus.

Applets for teaching statistics: **http://www.rossmanchance.com/applets**

Description: This site contains Java applets that can be used to help students attain a visual, conceptual understanding of ideas in introductory statistics. Applets deal with sampling distribution simulations, randomization distribution simulations, probability, inference, and mathematical models.

FOUR-COLUMN LESSON PLANS TO HELP DEVELOP STUDENTS' THINKING IN AP CALCULUS

Lesson Plan 1

Based on the following NCTM resource: Gordon, S. P., & Gordon, F. S. (2007). Discovering the fundamental theorem of calculus using data analysis. *Mathematics Teacher, 100,* 597–604.

Primary objective: To help students build an intuitive foundation for understanding the fundamental theorem of calculus

Materials needed: Graph paper; graphing calculator or spreadsheet

Steps of the lesson	Expected student responses	Teacher's responses to students	Assessment strategies and goals
1. Ask students to write an explanation for the meaning of the expression $\int_0^b \cos x \, dx$ without doing any calculations.	Some students may think of the integral only as an antiderivative and not mention the area interpretation. Some students may mention the area interpretation but be uncertain of how to deal with area below the *x*-axis and above the curve.	Elicit students' thinking until the area interpretation is mentioned. Explicitly attend to the meaning of the area below the *x*-axis and above the curve. Ask students how the desired area could be approximated, given the difficulty that we are dealing with a curved boundary.	Look for evidence that students are beginning to think about approximating the integral by using rectangles. If this idea is not given during class discussion, inject the idea into the conversation in preparation for the rest of the lesson.
2. Have students use rectangles to approximate $\int_0^{0.1} \cos x \, dx$. Direct them to do two approximations—one with a left sum and the other with a right sum—and then average the two.	Students may have difficulty determining the heights of the rectangle associated with the left sum and the right sum.	Help students sketch the two rectangles used to approximate the integral. Then have them identify the *y*-coordinate of the location at which each rectangle intersects the curve.	Check students' sketches for correct intersection points and rectangle widths. Understanding how to approximate area with rectangles is necessary throughout the lesson.
3. Have students extend the strategy introduced above to approximate the following expression: $\int_0^{0.2} \cos x \, dx$	Some students will use just one rectangle for the left sum and one rectangle for the right sum. Some students will begin to recognize that averaging the left and right sums is equivalent to using the trapezoidal rule.	Direct students to use two rectangles, each with a width of 0.1, to determine the left and right sums. Encourage students who recognize equivalence with the trapezoidal rule to continue to think about why the procedures are the same.	Ask students to discuss the advantages of using two small rectangles to approximate the integral rather than one large rectangle. Look for evidence of understanding that smaller rectangles result in less "leftover," or excess, area than larger ones.

Steps of the lesson	Expected student responses	Teacher's responses to students	Assessment strategies and goals
4. Use a graphing calculator or spreadsheet to approximate values for $\int_0^b \cos x\, dx$, letting $b = 0.3, 0.4, \ldots 12$. Use the procedure of averaging left- and right-hand estimates that was employed for the first two integrals.	Students should understand the idea of computing left- and right-hand sums at this point, but they may need help using a graphing calculator or spreadsheet to automate the computational process.	Have a technology-specific handout prepared to give students at this point in the lesson. It will describe how to use the calculator or spreadsheet used by students to automate the computation process. It will also describe how to graph the data.	Have students graph their estimates for each integral against the corresponding values of b. Do a quick visual inspection of the graphs to make sure that they approximate a sine curve. If not, check students' use of technology for "bugs."
5. Ask students to make and test a conjecture about an equation for a curve fitting the points produced by graphing $\int_0^b \cos x\, dx$ against the various values of b.	Some students will recognize the points on the graph as tracing out the function $y = \sin x$.	Have students superimpose the graph of $y = \sin x$ on the data. Encourage students to express their conjecture in formal terms: $$\int_0^b \cos x\, dx = \sin b$$	Ask students to express the meaning of $\int_0^b \cos x\, dx = \sin b$ in their own words to see if they connect the previous steps of the activity with the formal language of the conjecture.
6. Give the following task to students: "Make a conjecture about the value of $\int_a^b \cos x\, dx$."	Some students will not immediately see how changing the lower bound from 0 to a would change the conjecture made in the previous step of the lesson.	Encourage students to revisit the data they graphed in the last portion of the lesson. Have them experiment with different values of a and record the results obtained.	Ask students to express their conjectures for this portion of the lesson in writing. If they do not arrive at the conjecture that $\int_a^b \cos x\, dx = \sin b - \sin a$, have them discuss their conjecture with a classmate who did arrive at it.
7. Ask students to make a conjecture about the value of $\int_0^b x\, dx$.	Some students may recognize that it is possible to compute the value of the integral precisely by using the formula for the area of a triangle. Others may use the approximation method that was used for cosine.	Computing the integral using the triangle area formula and using the procedure developed earlier are both acceptable. If both strategies are used during the lesson, choose a student to present each strategy to the rest of the class.	Check to see that students arrive at the conjecture $\int_0^b x\, dx = \frac{1}{2}b^2 - \frac{1}{2}a^2$.
8. Have students make conjectures about the values of several additional integrals, including $\int_a^b x^2\, dx$ and $\int_a^b e^x\, dx$.	Some students may continue to use the approximation procedure outlined above, while others will look for patterns in the first two examples to try to make a conjecture about the value of each integral.	Encourage students who have different ways of forming their conjectures converse with one another so they can discuss the plausibility of the conjectures they are forming.	As a final assessment item, ask students to write and justify a conjecture about the value of $\int_a^b f(x)\, dx$ and the conditions under which it holds.

How This Lesson Meets Quality Control Criteria

- *Addressing students' preconceptions:* Students' previous school-based experiences of fitting functions to sets of data are foundational to the activity. The frequently encountered idea of area is drawn on throughout the lesson.
- *Conceptual and procedural knowledge development:* Students draw on and connect the ideas of area, antiderivative, and curve fitting to make conjectures about integrals. Students develop procedures to automate the computation of Riemann sum approximations.
- *Metacognition:* Students make conjectures about the values of several definite integrals and then revisit the conjectures to test and refine them by examining the data they produce.

Lesson Plan 2

Based on the following NCTM resource: Santulli, T. V. (2006). A deeper look at related rates in calculus. *Mathematics Teacher, 100,* 126–130.

Primary objective: To help students understand related rates problems and judge the reasonableness of their results

Materials needed: 20 to 30 balloons, ball of string, 15 to 20 rulers, graphing calculators

Steps of the lesson	Expected student responses	Teacher's responses to students	Assessment strategies and goals
1. Pose the following problem: "Air is being pumped into a spherical balloon at a rate of 200 cubic meters per second. 1. Find the rate of change of the radius of the balloon when the radius is 4 cm. 2. Find the rate of change of the radius of the balloon when the radius is 12 cm." (p. 126)	Students who know techniques for setting up and solving related rates problems may wish to proceed algebraically in solving the task. Students who are not familiar with solving related rates problems algebraically may try to estimate the solution by trying to imagine a reasonable result, given the context.	Encourage students to hold off on algebraic approaches to obtaining a precise solution to the problem at this point in the lesson. Instead, encourage them to think about why the answers for the two questions would not be the same, even though the rate of increase of volume is constant.	Assess students' ability to think about the situation in terms of the context in which it is set. Ask why the radius might increase more quickly at some points than at others. Ask for specific times at which they would expect the radius to increase quickly. Probe students' thinking about the change in radius at the very beginning of inflation and near the end of inflation.
2. Put students into pairs and give each pair a balloon, some string, and a ruler. Have one group member blow into the balloon and have another measure the circumference after each blow. Students should record the circumference data for each blow and then compute the radius after each blow as well as the increase in radius from one blow to the next.	Measurement error may hinder data collection at this point in the lesson. Some students may not blow the same amount of air into the balloon with each breath. The students measuring the circumference of the balloon may place the string too high or too low on the balloon to obtain an accurate measurement.	Monitor each pair of students as they gather data. To help minimize measurement error, remind them that the balloon is to be inflated at a constant rate, so the amount of air infused with each breath should be the same. If students measure the circumference incorrectly, help them with proper placement of the string on the balloon.	Check to see if students understand the conditions and assumptions of the problem by asking why it is not all right to put a different amount of air into the balloon with each breath. Check their understanding of the properties of three-dimensional objects by asking which cross sections of a sphere must be measured to determine circumference.

Steps of the lesson	Expected student responses	Teacher's responses to students	Assessment strategies and goals
3. Ask the class to set up and compute the solution to the related rates problem algebraically.	Students will likely begin with the formula for volume of a sphere, $V = \frac{4}{3}\pi r^2$, and then use implicit differentiation to obtain a rate of 0.995 cm/second for $r = 4$ and 0.111 cm/second for $r = 12$. Some may do a step of the implicit differentiation incorrectly and obtain different responses.	Direct students to compare the entries in their tables for $r = 4$ and $r = 12$ to the answers they computed using implicit differentiation. In particular, they should compute the ratio of the answer for $r = 4$ to the answer for $r = 12$ in both the table and the algebraic computations. Both ratios should be very close to 9.	Check the ratios that students compute. If they are not close to 9, have students check for an algebraic error or measurement error. Also ask students to write about why the two ratios they computed should be very close to one another. For an extra challenge, ask students to explain why the ratios should be approximately 9.
4. Pose a follow-up related rates problem to students: "A camera is 1200 feet away from the base of a rocket. The rocket is rising vertically at 500 ft./sec. 1. Find the rate of change in the distance from the camera to the rocket when the rocket is 600 feet high. 2. Find how fast the camera-to-rocket angle of elevation is changing when the rocket is 600 feet high." (p. 128)	Students will likely recognize both parts of the question as fairly routine related rates problems. They will likely solve the first by using the Pythagorean theorem and implicit differentiation. They will likely solve the second using the equation $y = 1{,}200 \tan(\theta)$ and implicit differentiation. Those who differentiate correctly will obtain $\frac{1}{3}$ radian/second as the answer to the second problem.	After answers have been obtained, lead students to check the reasonableness of their response to question 2. Suggest that they convert radians to degrees to obtain an answer of approximately 19.0986 degrees per second. Ask them why 19.0986 degrees per second is a reasonable answer for question 2, given the context of the situation.	As students explain why approximately 19.0986 is a reasonable response to question 2, monitor their explanations for connections to the actual context of the problem, and not just back to the computations they did to obtain the answer. Additionally, ask them if they would expect the camera-to-rocket angle to increase or decrease as the rocket rises. Again, monitor responses for evidence of connection to the context of the problem.
5. Ask students to produce a table that shows the rocket's height, the angle from the observer, and the change in angle measure from one observation to the next. Ask students to use 500-foot increments and to track progress from a minimum height of 0 feet to a maximum height of 6,000 feet.	Some students will struggle to find a formula that allows them to fill in the table requested. Others will recognize that the table can be completed by using the equation $y = \tan^{-1}\frac{x}{1{,}200}$. Some will use the graphing calculator to produce the desired table of values.	Ask students who use the equation $y = \tan^{-1}\frac{x}{1{,}200}$ in conjunction with the table feature of the graphing calculator to demonstrate their work for the rest of the class. They should explain their reasoning in setting up the equation as well as the mechanics of having the calculator produce the desired table of values.	Ask students to examine the table produced and describe what happens to the change in angle every 500 feet as well as what happens to the rate of increase. They should notice that the rate of change decreases as the angle increases. Ask students to explain why this occurs, again by referring back to the context of the problem.

(Continued)

(Continued)

Steps of the lesson	Expected student responses	Teacher's responses to students	Assessment strategies and goals
6. Pose a final question to consider in regard to the second related rates task: Why is 19 degrees per second a reasonable response to the second part of the second related rates task?	Students may try to calculate the angle for when the rocket is at 600 feet and then calculate it again for a second later. The difference would be 15.9 degrees rather than 19 degrees.	Encourage students to try to obtain a more precise estimate by looking at intervals smaller than 1 second. For instance, they can compute differences for increments of .01 sec.	Ask students to explain why it is reasonable to use intervals of less than 1 second for the problem. Explanations should include the rapid rate of increase one would expect to see in the given real-world situation.

How This Lesson Meets Quality Control Criteria

- *Addressing students' preconceptions:* The first related rates problem allows students to draw on out-of-school experiences with blowing up balloons to predict a reasonable response. The second related rates problem allows students to draw on out-of school experiences of watching objects rise. Each related rates problem allows students to draw on school-based knowledge of implicit differentiation.
- *Conceptual and procedural knowledge development:* The meanings of the related rates problems are drawn out by situating each in a real-world context. Students practice the algebraic procedures for solving each problem.
- *Metacognition:* Students check to see if the expected results of related rates problems align with their expectations. Students are encouraged to revisit their work to detect potential errors that lead to unrealistic responses.

FOUR-COLUMN LESSON PLANS TO HELP DEVELOP STUDENTS' THINKING IN AP STATISTICS

Lesson Plan 1

Based on the following NCTM resource: Fox, T. B. (2005). Transformations on data sets and their effects on descriptive statistics. *Mathematics Teacher, 99,* 208–217.

Primary objective: To help students examine the effects of data set translations and scale changes on descriptive statistics

Materials needed: Metric tape measure, computers with spreadsheets, graphing calculators, four worksheets accompanying the article on which the lesson plan is based

Steps of the lesson	Expected student responses	Teacher's responses to students	Assessment strategies and goals
1. Assemble students into groups of four. Give each group a metric tape measure and a set of four worksheets. Have students measure the height, wrist circumference, neck circumference, and arm length of each group member.	Most students will be able to complete the measurement portion of the activity with no problems, but measurement error may come into play when students gather data at the outset of the activity. Different individuals may use different techniques for measuring the same attribute.	Encourage students to consider how having different individuals measure the same attribute may impact the accuracy of the data. In cases where obvious measurement errors are being made, explicitly suggest a different measurement strategy.	Scan each group's corpus of data to see if there are any wildly inaccurate measurements. Ask students to produce a quick graphical display of any data sets that appear to be errant so that outliers due to measurement error are easier to spot.
2. Have students compute descriptive statistics for each attribute using the data collected. (The descriptive statistics to be computed are listed on the first worksheet and include mean, median, mode, standard deviation, and variance.)	Some students may recognize that the graphing calculator computes most of the descriptive statistics needed, while others will compute needed statistics by hand.	For computing descriptive statistics, students should be encouraged to take advantage of the calculator's capabilities, since hand computation is not the objective of the lesson.	Monitor students' use of the graphing calculator to ensure that they are using it as a tool to facilitate exploration and eliminate arduous computations to the fullest extent possible.
3. Ask students to predict how the descriptive statistics that were computed would change if each group member's height, wrist circumference, neck circumference, and arm length grew by the amounts specified on worksheet 2. The growth for each attribute is uniform (e.g., each person grew in height by 8 cm, each person	Some students will intuitively predict that the measures of center for each attribute would all increase by the amount of growth specified. Some of the students who predict correctly for measures of center (e.g., mean) will incorrectly predict that measures of spread (e.g., standard deviation) would also increase under these circumstances.	Ask students to use a graphing calculator or spreadsheet to check their predictions. If students do not understand why measures of center increase but measures of spread do not, have them revisit the algebraic formulas that are used to produce each measure. They can experiment by hand with a few data values to better understand what	Have students construct boxplots for the original and transformed data sets as specified on worksheet 2. In comparing the boxplots to one another, they should decide on the accuracy of their original conjectures. If their conjectures hold, they should discuss with their groups why they are true. If they do not hold, students should

(Continued)

(Continued)

Steps of the lesson	Expected student responses	Teacher's responses to students	Assessment strategies and goals
grew in wrist circumference by 1 cm, etc.).		each formula does with inputted data.	discuss possible revisions to the original conjectures with their groups.
4. Pose another scenario related to the measurement data collected at the outset of the lesson: "What if we wanted to convert from metric to the English system?" Ask students to predict how the descriptive statistics computed at the outset of the activity would change.	Some students may initially believe that the descriptive statistics will not change, since they are just converting from one unit of measure to another. Some students will determine a scale factor applicable to the conversion after experimenting with a few data points.	Have students check their conjectures by using the graphing calculator or a spreadsheet to carry out the calculations. Then have them compare the scaled statistics to the original ones, as specified on worksheet 3.	Ask students to describe what they see happening in the scaled data set compared to the original. Also ask how the change is related to the scale factor for converting metric units to English units, as specified on worksheet 3. They can use boxplots, if necessary, to check the accuracy of their conjectures for the median, max, min, and interquartile range.
5. Have each group present their final set of conjectures to the entire class. During the presentations, they should engage their classmates in a discussion of why they believe the conjectures are correct or incorrect.	Several groups are likely to have similar conjectures. Some groups will make conjectures that are still in need of revision and refinement. The difference between effects on measures and center and measures of spread may still need work in these cases.	If a number of groups have arrived at the same conjecture, it is not necessary to have various groups reiterate the same conjecture, unless one group's explanation is incomplete in some manner. Encourage students to share their work and boxplot representations with one another if discerning the difference between measures of center and spread remains problematic.	Ask students to put the descriptive statistics into the categories of "measures of center" and "measures of spread." Then ask them to explain how the two translations up to this point in the lesson change measures of center and also how they change measures of spread. Ask if the changes are uniform across all measures of center and all measures of spread.
6. Have students reassemble into groups of four and write a formal proof for an earlier conjecture—that increasing each value in a data set by a uniform, nonzero amount increases the mean by the same amount.	Some students will not see a need for proof because they were convinced by the examples provided earlier. Others will recognize the need for proof and begin approaching the problem algebraically.	Have students who do not see the need for proof converse with those who do see a need for it. The primary purpose of this segment of the lesson is to help students distinguish between conjecture and proof.	Assess students' written work for progress toward a formal proof. Formal proofs are not required on the AP exam, but doing this portion of the lesson helps give students a firmer foundation for later study.

How This Lesson Meets Quality Control Criteria

- *Addressing students' preconceptions:* Students' out-of-school knowledge of the dimensions of the human body and human growth help provide a motivating context for the activity. Students' school-based knowledge of computation of summary statistics is also foundational and is extended by the activities of the lesson.
- *Conceptual and procedural knowledge development:* Students draw on procedures for determining summary statistics in programming spreadsheets and/or working with a graphing calculator. Students connect the concept of transformation with the manner in which different summary statistics serve to describe a set of data.
- *Metacognition:* Students make conjectures about the effects of transformations on summary statistics, and then revisit, justify, and revise their conjectures, if necessary, after working with the data and discussing their thinking with one another.

Lesson Plan 2

Based on the following NCTM resource: White, W. (2001). Connecting independence and the chi-square statistic. *Mathematics Teacher, 94,* 134–136.

Primary objective: To develop students' understanding of the chi-square tests for goodness of fit and independence

Materials needed: Paper and pencil, chi-square tables, graphing calculators

Steps of the lesson	Expected student responses	Teacher's responses to students	Assessment strategies and goals
1. Review the chi-square test for goodness of fit to introduce the lesson. Ask students to imagine a situation in which a die is rolled 60 times. On the 60 rolls, one comes up 9 times, two comes up 13 times, three comes up 13 times, four comes up 9 times, five comes up 6 times, and six comes up 10 times. Ask, "Do you believe the die is balanced?"	From their previous experiences in class, some students will recognize the situation as requiring a chi-square test. They will calculate the chi-square statistic using the formula $\sum \frac{(observed - expected)^2}{expected} = \sum \frac{(O - E)^2}{E}$ and obtain a p-value between .50 and .70 using a chi-square table. Other students will make informal arguments about the possibility that the die is balanced.	Have students who do not see the desirability of formal inferential reasoning in this situation converse with those who do use formal inference. Emphasize that we have just a sample from the overall population, so it is desirable to have the ability to quantify how likely our observed results would be under the assumption that the die is balanced.	Assess students' procedural flexibility by asking if they can determine the p-value for this situation by using a graphing calculator as well as a table of values. Assess students' ability to identify the null hypothesis by writing a formal statement of it for this situation. Assess students' interpretation of p-value by asking them to write a sentence to explain its meaning in the context of this situation.
2. Provide a new set of dice rolling data for students to examine: On 60 rolls of a different die, one comes up 5 times, two comes up 11 times, three comes up 4 times, four comes up	Most students will do the formal chi-square computations, having been influenced by the problem from the first portion of the activity. In doing so, students will arrive at a p-value that is between .01 and .02.	Ask students to interpret the p-value they obtain within the context of the situation. Ask students to identify the null hypothesis in this situation and make a judgment	Assess students' interpretations of the p-value for evidence of interpretations connecting to the likelihood of the observed results. Assess students' claims about the null hypothesis for evidence of probabilistic

(Continued)

(Continued)

Steps of the lesson	Expected student responses	Teacher's responses to students	Assessment strategies and goals
19 times, five comes up 9 times, and six comes up 12 times. Ask, "Do you believe the die is balanced?"		about its feasibility in this case.	rather than deterministic thinking.
3. Ask students to say what comes to mind when they hear the word *independent*.	Some students will connect independence with the idea that if one event occurs, it does not change the probability of the other event. Other students will offer a previously studied formal definition of independence: that the joint probability is the product of the probability of each event.	Both ideas about independence will be somewhat useful in this activity, though the latter formal definition will allow for the precise probabilities. Since finding exact probabilities is a goal of this activity, the formal definition should be emphasized, and should be introduced if none of the students offers it.	Pose a quick, simple, straightforward probability exercise to assess students' ability to compute joint probability for independent events: "What is the probability of obtaining heads on the first flip of a coin and tails on the second flip?"

4. Introduce a problem to motivate consideration of the chi-square test for independence:
 	
Suppose we want to determine if a basketball team's performance depends on where it plays. We have the following data:
 	
      Wins Losses
Home 22   8   30
Away 13   7   20
      35   15   50
 	
These results serve as a sample.
	Some students will initially conclude that the team plays better at home than away, because they won 73% of their home games but only 65% of their away games. Other students will recognize that this situation, like the previous one, presents only a sample of the team's performance, and that they need to compute how likely this sample would be if it were true that the team's performance is independent of location.	Encourage students to discuss how this situation is similar to the last one, in that both require reasoning from a sample rather than from an entire population. To move students toward computing appropriate chi-square statistics, ask them to state how many games they would expect the team to win at home and away if it were true that the team's record is independent of the location where the game is played.	Assess students' computations of expected values for accuracy. Under the assumption that the team's record is independent of its location, one would expect the following data:
			
			      Wins Losses
			Home 21 9 30
			Away 14 6 20
			      35 15 50
			
			Students will likely note that these numbers are very close to the observed values.

5. Ask students if they are convinced that the location in which the team plays influences the result of the game, and ask them to justify their answers.	Many students will note that each cell of the expected values table differs only by one from each cell in the observed values table, and on that basis be convinced that location and	If students do not see a need for a chi-square test in this situation, ask them if they can quantify their certainty in the idea that performance and	Assess students' interpretations of the *p*-value that is calculated in this situation. Look for evidence that they see the *p*-value as implying that there is not much support

Steps of the lesson	Expected student responses	Teacher's responses to students	Assessment strategies and goals
	outcome of the game are independent. Some students will see a connection to the chi-square statistic and compute the p-value for the situation to be 0.529 using the graphing calculator.	location of the game are independent. Once students have done so by computing the p-value, push them to interpret the meaning of the p-value within the context of the situation.	for rejecting the null hypothesis that performance and location are independent. Also look for evidence that they do not interpret this to mean that the null hypothesis has deterministically been proven to be true.
6. Multiply each cell in the observed value table from Step 4 by 10 and reapply the chi-square test.	Students will find a very low p-value, approximately .046 for the modified situation.	Prompt students to interpret and explain the p-value within the context of the new situation.	Ask students to write about why this situation produces a lower p-value than the previous problem.

How This Lesson Meets Quality Control Criteria

- *Addressing students' preconceptions:* Students' out-of-school knowledge and beliefs about playing sports on a home court and away are activated by the context in which the activity is set. Students' previous in-school experiences with the concept of independence are drawn on to form a foundation for understanding the chi-square test for independence.

- *Conceptual and procedural knowledge development:* Students are introduced to a formula for computing chi-square statistics and also are asked to reflect on why the formula works as it does.

- *Metacognition:* Students are asked to examine the adequacy of their own informal reasoning about a situation and consider how their thinking could be extended and enhanced by using a formal inferential technique.

APPENDIX A

Common Core State Standards for Mathematical Practice*

MATHEMATICS | STANDARDS FOR MATHEMATICAL PRACTICE

The Standards for Mathematical Practice describe varieties of expertise that mathematics educators at all levels should seek to develop in their students. These practices rest on important "processes and proficiencies" with longstanding importance in mathematics education. The first of these are the NCTM process standards of problem solving, reasoning and proof, communication, representation, and connections. The second are the strands of mathematical proficiency specified in the National Research Council's report *Adding It Up*: adaptive reasoning, strategic competence, conceptual understanding (comprehension of mathematical concepts, operations and relations), procedural fluency (skill in carrying out procedures flexibly, accurately, efficiently and appropriately), and productive disposition (habitual inclination to see mathematics as sensible, useful, and worthwhile, coupled with a belief in diligence and one's own efficacy).

1. Make sense of problems and persevere in solving them.

Mathematically proficient students start by explaining to themselves the meaning of a problem and looking for entry points to its solution. They analyze givens, constraints, relationships, and goals. They make conjectures about the form and meaning of the solution and plan a solution pathway rather than simply jumping into a solution attempt. They consider analogous problems and try special cases and simpler forms of the original problem in order to gain insight into its solution. They monitor and evaluate their progress and change course if necessary. Older students might, depending on the context of the problem, transform algebraic expressions or change the viewing window on their graphing calculator to get the information they need. Mathematically proficient students can explain correspondences between equations, verbal descriptions, tables, and graphs or draw diagrams of important features and relationships, graph data, and search for regularity or trends. Younger students might rely on using concrete objects or pictures to help conceptualize and solve a problem. Mathematically proficient students check their answers to problems using a different method, and they continually ask themselves, "Does this make sense?" They can understand the approaches of others to solving complex problems and identify correspondences between different approaches.

2 Reason abstractly and quantitatively.

Mathematically proficient students make sense of quantities and their relationships in problem situations. They bring two complementary abilities to bear on problems involving quantitative relationships: the ability to decontextualize—to abstract a given situation and represent it symbolically and manipulate the representing symbols as if they have a life of their own, without necessarily attending to their referents—and the ability to contextualize, to pause as needed during the manipulation process in order to probe into the referents for the symbols involved. Quantitative reasoning entails habits of creating a coherent representation of the problem at hand; considering the units involved; attending to the meaning of quantities, not just how to compute them; and knowing and flexibly using different properties of operations and objects.

3 Construct viable arguments and critique the reasoning of others.

Mathematically proficient students understand and use stated assumptions, definitions, and previously established results in constructing arguments. They make conjectures and build a logical progression of statements to explore the truth of their conjectures. They are able to analyze situations by breaking them into cases, and can recognize and use counterexamples. They justify their conclusions,

---

*The full set of standards and illustrative examples are available online: http://www.corestandards.org.

communicate them to others, and respond to the arguments of others. They reason inductively about data, making plausible arguments that take into account the context from which the data arose. Mathematically proficient students are also able to compare the effectiveness of two plausible arguments, distinguish correct logic or reasoning from that which is flawed, and—if there is a flaw in an argument—explain what it is. Elementary students can construct arguments using concrete referents such as objects, drawings, diagrams, and actions. Such arguments can make sense and be correct, even though they are not generalized or made formal until later grades. Later, students learn to determine domains to which an argument applies. Students at all grades can listen or read the arguments of others, decide whether they make sense, and ask useful questions to clarify or improve the arguments.

4 Model with mathematics.

Mathematically proficient students can apply the mathematics they know to solve problems arising in everyday life, society, and the workplace. In early grades, this might be as simple as writing an addition equation to describe a situation. In middle grades, a student might apply proportional reasoning to plan a school event or analyze a problem in the community. By high school, a student might use geometry to solve a design problem or use a function to describe how one quantity of interest depends on another. Mathematically proficient students who can apply what they know are comfortable making assumptions and approximations to simplify a complicated situation, realizing that these may need revision later. They are able to identify important quantities in a practical situation and map their relationships using such tools as diagrams, two-way tables, graphs, flow-charts and formulas. They can analyze those relationships mathematically to draw conclusions. They routinely interpret their mathematical results in the context of the situation and reflect on whether the results make sense, possibly improving the model if it has not served its purpose.

5 Use appropriate tools strategically.

Mathematically proficient students consider the available tools when solving a mathematical problem. These tools might include pencil and paper, concrete models, a ruler, a protractor, a calculator, a spreadsheet, a computer algebra system, a statistical package, or dynamic geometry software. Proficient students are sufficiently familiar with tools appropriate for their grade or course to make sound decisions about when each of these tools might be helpful, recognizing both the insight to be gained and their limitations. For example, mathematically proficient high school students analyze graphs of functions and solutions generated using a graphing calculator. They detect possible errors by strategically using estimation and other mathematical knowledge. When making mathematical models, they know that technology can enable them to visualize the results of varying assumptions, explore consequences, and compare predictions with data. Mathematically proficient students at various grade levels are able to identify relevant external mathematical resources, such as digital content located on a website, and use them to pose or solve problems. They are able to use technological tools to explore and deepen their understanding of concepts.

6 Attend to precision.

Mathematically proficient students try to communicate precisely to others. They try to use clear definitions in discussion with others and in their own reasoning. They state the meaning of the symbols they choose, including using the equal sign consistently and appropriately. They are careful about specifying units of measure, and labeling axes to clarify the correspondence with quantities in a problem. They calculate accurately and efficiently, express numerical answers with a degree of precision appropriate for the problem context. In the elementary grades, students give carefully formulated explanations to each other. By the time they reach high school, they have learned to examine claims and make explicit use of definitions.

7 Look for and make use of structure.

Mathematically proficient students look closely to discern a pattern or structure. Young students, for example, might notice that three and seven more is the same amount as seven and three more, or they may sort a collection of shapes according to how many sides the shapes have. Later, students will see 7×8 equals the well remembered $7 \times 5 + 7 \times 3$, in preparation for learning about the distributive property. In the expression $x^2 + 9x + 14$, older students can see the 14 as 2×7 and the 9 as $2 + 7$. They recognize the significance of an existing line in a geometric figure and can use the strategy of drawing an auxiliary line for solving problems. They also can step back for an overview and shift perspective. They can see complicated things, such as some algebraic expressions, as single objects or as being composed of several objects. For example, they can see $5 - 3(x - y)^2$ as 5 minus a positive number times a square and use that to realize that its value cannot be more than 5 for any real numbers x and y.

8 Look for and express regularity in repeated reasoning.

Mathematically proficient students notice if calculations are repeated, and look both for general methods and for shortcuts. Upper elementary students might notice when dividing 25 by 11 that they are repeating the same calculations over and over again, and conclude they have a repeating decimal. By paying attention to the calculation of slope as they repeatedly check whether points are on the line through (1, 2) with slope 3, middle school students might abstract the equation $(y - 2)/(x - 1) = 3$. Noticing the regularity in the way terms cancel when expanding $(x - 1)(x + 1)$, $(x - 1)(x^2 + x + 1)$, and $(x - 1)(x^3 + x^2 + x + 1)$ might lead them to the general formula for the sum of a geometric series. As they work to solve a problem, mathematically proficient students maintain oversight of the process, while attending to the details. They continually evaluate the reasonableness of their intermediate results.

Connecting the Standards for Mathematical Practice to the Standards for Mathematical Content

The Standards for Mathematical Practice describe ways in which developing student practitioners of the discipline of mathematics increasingly ought to engage with the subject matter as they grow in mathematical maturity and expertise throughout the elementary, middle and high school years. Designers of curricula, assessments, and professional development should all attend to the need to connect the mathematical practices to mathematical content in mathematics instruction.

The Standards for Mathematical Content are a balanced combination of procedure and understanding. Expectations that begin with the word "understand" are often especially good opportunities to connect the practices to the content. Students who lack understanding of a topic may rely on procedures too heavily. Without a flexible base from which to work, they may be less likely to consider analogous problems, represent problems coherently, justify conclusions, apply the mathematics to practical situations, use technology mindfully to work with the mathematics, explain the mathematics accurately to other students, step back for an overview, or deviate from a known procedure to find a shortcut. In short, a lack of understanding effectively prevents a student from engaging in the mathematical practices.

In this respect, those content standards which set an expectation of understanding are potential "points of intersection" between the Standards for Mathematical Content and the Standards for Mathematical Practice. These points of intersection are intended to be weighted toward central and generative concepts in the school mathematics curriculum that most merit the time, resources, innovative energies, and focus necessary to qualitatively improve the curriculum, instruction, assessment, professional development, and student achievement in mathematics.

APPENDIX B

Common Core State Standards for Grades 6–8*

MATHEMATICS | GRADE 6

In Grade 6, instructional time should focus on four critical areas: (1) connecting ratio and rate to whole number multiplication and division and using concepts of ratio and rate to solve problems; (2) completing understanding of division of fractions and extending the notion of number to the system of rational numbers, which includes negative numbers; (3) writing, interpreting, and using expressions and equations; and (4) developing understanding of statistical thinking.

(1) Students use reasoning about multiplication and division to solve ratio and rate problems about quantities. By viewing equivalent ratios and rates as deriving from, and extending, pairs of rows (or columns) in the multiplication table and by analyzing simple drawings that indicate the relative size of quantities, students connect their understanding of multiplication and division with ratios and rates. Thus students expand the scope of problems for which they can use multiplication and division to solve problems, and they connect ratios and fractions. Students solve a wide variety of problems involving ratios and rates.

(2) Students use the meaning of fractions, the meanings of multiplication and division, and the relationship between multiplication and division to understand and explain why the procedures for dividing fractions make sense. Students use these operations to solve problems. Students extend their previous understandings of number and the ordering of numbers to the full system of rational numbers, which includes negative rational numbers, and in particular negative integers. They reason about the order and absolute value of rational numbers and about the location of points in all four quadrants of the coordinate plane,

(3) Students understand the use of variables in mathematical expressions. They write expressions and equations that correspond to given situations, evaluate expressions, and use expressions and formulas to solve problems. Students understand that expressions in different forms can be equivalent, and they use the properties of operations to rewrite expressions in equivalent forms. Students know that the solutions of an equation are the values of the variables that make the equation true. Students use properties of operations and the idea of maintaining the equality of both sides of an equation to solve simple one-step equations. Students construct and analyze tables, such as tables of quantities that are in equivalent ratios, and they use equations (such as $3x = y$) to describe relationships between quantities.

(4) Building on and reinforcing their understanding of number, students begin to develop their ability to think statistically. Students recognize that a data distribution may not have a definite center and that different ways to measure center yield different values. The median measures center in the sense that it is roughly the middle value. The mean measures center in the sense that it is the value that each data point would take on if the total of the data values were redistributed equally, and also in the sense that it is a balance point. Students recognize that a measure of variability (interquartile range or mean absolute deviation) can also be useful for summarizing data because two very different sets of data can have the same mean and

*The full set of standards and illustrative examples are available online: http://www.corestandards.org.

median yet be distinguished by their variability. Students learn to describe and summarize numerical data sets, identifying clusters, peaks, gaps, and symmetry, considering the context in which the data were collected.

Students in Grade 6 also build on their work with area in elementary school by reasoning about relationships among shapes to determine area, surface area, and volume. They find areas of right triangles, other triangles, and special quadrilaterals by decomposing these shapes, rearranging or removing pieces, and relating the shapes to rectangles. Using these methods, students discuss, develop, and justify formulas for areas of triangles and parallelograms. Students find areas of polygons and surface areas of prisms and pyramids by decomposing them into pieces whose area they can determine. They reason about right rectangular prisms with fractional side lengths to extend formulas for the volume of a right rectangular prism to fractional side lengths. They prepare for work on scale drawings and constructions in Grade 7 by drawing polygons in the coordinate plane.

GRADE 6 OVERVIEW

Ratios and Proportional Relationships

- Understand ratio concepts and use ratio reasoning to solve problems.

The Number System

- Apply and extend previous understandings of multiplication and division to divide fractions by fractions.
- Compute fluently with multi-digit numbers and find common factors and multiples.
- Apply and extend previous understandings of numbers to the system of rational numbers.

Expressions and Equations

- Apply and extend previous understandings of arithmetic to algebraic expressions.
- Reason about and solve one-variable equations and inequalities.
- Represent and analyze quantitative relationships between dependent and independent variables.

Geometry

- Solve real-world and mathematical problems involving area, surface area, and volume.

Statistics and Probability

- Develop understanding of statistical variability.
- Summarize and describe distributions.

Mathematical Practices

1. Make sense of problems and persevere in solving them.
2. Reason abstractly and quantitatively.
3. Construct viable arguments and critique the reasoning of others.
4. Model with mathematics.
5. Use appropriate tools strategically.
6. Attend to precision.
7. Look for and make use of structure.
8. Look for and express regularity in repeated reasoning.

MATHEMATICS | GRADE 7

In Grade 7, instructional time should focus on four critical areas: (1) developing understanding of and applying proportional relationships; (2) developing understanding of operations with rational numbers and working with expressions and linear equations; (3) solving problems involving scale drawings and informal geometric constructions, and working with two- and three-dimensional shapes to solve problems involving area, surface area, and volume; and (4) drawing inferences about populations based on samples.

(1)   Students extend their understanding of ratios and develop understanding of proportionality to solve single- and multistep problems. Students use their understanding of ratios and proportionality to solve a wide variety of percent problems, including those involving discounts, interest, taxes, tips, and percent increase or decrease. Students solve problems about scale drawings by relating corresponding lengths between the objects or by using the fact that relationships of lengths within an object are preserved in similar objects. Students graph proportional relationships and understand the unit rate informally as a measure of the steepness of the related line, called the slope. They distinguish proportional relationships from other relationships.

(2)   Students develop a unified understanding of number, recognizing fractions, decimals (that have a finite or a repeating decimal representation), and percents as different representations of rational numbers. Students extend addition, subtraction, multiplication, and division to all rational numbers, maintaining the properties of operations and the relationships between addition and subtraction, and multiplication and division. By applying these properties, and by viewing negative numbers in terms of everyday contexts (e.g., amounts owed or temperatures below zero), students explain and interpret the rules for adding, subtracting, multiplying, and dividing with negative numbers. They use the arithmetic of rational numbers as they formulate expressions and equations in one variable and use these equations to solve problems.

(3)   Students continue their work with area from Grade 6, solving problems involving the area and circumference of a circle and surface area of three-dimensional objects. In preparation for work on congruence and similarity in Grade 8, they reason about relationships among two-dimensional figures using scale drawings and informal geometric constructions, and they gain familiarity with the relationships between angles formed by intersecting lines. Students work with three-dimensional figures, relating them to two-dimensional figures by examining cross-sections. They solve real-world and mathematical problems involving area, surface area, and volume of two- and three-dimensional objects composed of triangles, quadrilaterals, polygons, cubes and right prisms.

(4)   Students build on their previous work with single data distributions to compare two data distributions and address questions about differences between populations. They begin informal work with random sampling to generate data sets and learn about the importance of representative samples for drawing inferences.

GRADE 7 OVERVIEW

Ratios and Proportional Relationships

- Analyze proportional relationships and use them to solve real-world and mathematical problems.

The Number System

- Apply and extend previous understandings of operations with fractions to add, subtract, multiply, and divide rational numbers.

Expressions and Equations

- Use properties of operations to generate equivalent expressions.
- Solve real-life and mathematical problems using numerical and algebraic expressions and equations.

Mathematical Practices

1. Make sense of problems and persevere in solving them.
2. Reason abstractly and quantitatively.
3. Construct viable arguments and critique the reasoning of others.
4. Model with mathematics.
5. Use appropriate tools strategically.
6. Attend to precision.
7. Look for and make use of structure.
8. Look for and express regularity in repeated reasoning.

Geometry

- Draw, construct and describe geometrical figures and describe the relationships between them.
- Solve real-life and mathematical problems involving angle measure, area, surface area, and volume.

Statistics and Probability

- Use random sampling to draw inferences about a population.
- Draw informal comparative inferences about two populations.
- Investigate chance processes and develop, use, and evaluate probability models.

MATHEMATICS | GRADE 8

In Grade 8, instructional time should focus on three critical areas: (1) formulating and reasoning about expressions and equations, including modeling an association in bivariate data with a linear equation, and solving linear equations and systems of linear equations; (2) grasping the concept of a function and using functions to describe quantitative relationships; (3) analyzing two- and three-dimensional space and figures using distance, angle, similarity, and congruence, and understanding and applying the Pythagorean Theorem.

(1) Students use linear equations and systems of linear equations to represent, analyze, and solve a variety of problems. Students recognize equations for proportions ($y/x = m$ or $y = mx$) as special linear equations ($y = mx + b$), understanding that the constant of proportionality (m) is the slope, and the graphs are lines through the origin. They understand that the slope (m) of a line is a constant rate of change, so that if the input or x-coordinate changes by an amount A, the output or y-coordinate changes by the amount $m \cdot A$. Students also use a linear equation to describe the association between two quantities in bivariate data (such as arm span vs. height for students in a classroom). At this grade, fitting the model, and assessing its fit to the data are done informally. Interpreting the model in the context of the data requires students to express a relationship between the two quantities in question and to interpret components of the relationship (such as slope and y-intercept) in terms of the situation.

Students strategically choose and efficiently implement procedures to solve linear equations in one variable, understanding that when they use the properties of equality and the concept of logical equivalence, they maintain the solutions of the original equation. Students solve systems of two linear equations in two variables and relate the systems to pairs of lines in the plane; these intersect, are parallel, or are the same line. Students use linear equations, systems of linear equations, linear functions, and their understanding of slope of a line to analyze situations and solve problems.

(2) Students grasp the concept of a function as a rule that assigns to each input exactly one output. They understand that functions describe situations where one quantity determines another. They can translate among representations and partial representations of functions (noting that tabular and graphical representations may be partial representations), and they describe how aspects of the function are reflected in the different representations.

(3) Students use ideas about distance and angles, how they behave under translations, rotations, reflections, and dilations, and ideas about congruence and similarity to describe and analyze two-dimensional figures and to solve problems. Students show that the sum of the angles in a triangle is the angle formed by a straight line, and that various configurations of lines give rise to similar triangles because of the angles created when a transversal cuts parallel lines. Students understand the statement of the Pythagorean Theorem and its converse, and can explain why the Pythagorean Theorem holds, for example, by decomposing a square in two different ways. They apply the Pythagorean Theorem to find distances between points on the coordinate plane, to find lengths, and to analyze polygons. Students complete their work on volume by solving problems involving cones, cylinders, and spheres.

GRADE 8 OVERVIEW

The Number System

- Know that there are numbers that are not rational, and approximate them by rational numbers.

Expressions and Equations

- Work with radicals and integer exponents.
- Understand the connections between proportional relationships, lines, and linear equations.
- Analyze and solve linear equations and pairs of simultaneous linear equations.

Functions

- Define, evaluate, and compare functions.
- Use functions to model relationships between quantities.

Mathematical Practices

1. Make sense of problems and persevere in solving them.
2. Reason abstractly and quantitatively.
3. Construct viable arguments and critique the reasoning of others.
4. Model with mathematics.
5. Use appropriate tools strategically.
6. Attend to precision.
7. Look for and make use of structure.
8. Look for and express regularity in repeated reasoning.

Geometry

- Understand congruence and similarity using physical models, transparencies, or geometry software.
- Understand and apply the Pythagorean Theorem.
- Solve real-world and mathematical problems involving volume of cylinders, cones, and spheres.

Statistics and Probability
- Investigate patterns of association in bivariate data.

APPENDIX C

*Common State Core Standards for High School**

MATHEMATICS STANDARDS FOR HIGH SCHOOL

The high school standards specify the mathematics that all students should study in order to be college and career ready. Additional mathematics that students should learn in order to take advanced courses such as calculus, advanced statistics, or discrete mathematics is indicated by (+), as in this example:

> (+) Represent complex numbers on the complex plane in rectangular and polar form (including real and imaginary numbers).

All standards without a (+) symbol should be in the common mathematics curriculum for all college and career ready students. Standards with a (+) symbol may also appear in courses intended for all students.

The high school standards are listed in conceptual categories:

- Number and Quantity
- Algebra
- Functions
- Modeling
- Geometry
- Statistics and Probability

Conceptual categories portray a coherent view of high school mathematics; a student's work with functions, for example, crosses a number of traditional course boundaries, potentially up through and including calculus.

Modeling is best interpreted not as a collection of isolated topics but in relation to other standards. Making mathematical models is a Standard for Mathematical Practice, and specific modeling standards appear throughout the high school standards indicated by a star symbol (*). The star symbol sometimes appears on the heading for a group of standards; in that case, it should be understood to apply to all standards in that group.

*The full set of standards and illustrative examples are available online: http://www.corestandards.org.

MATHEMATICS | HIGH SCHOOL—NUMBER AND QUANTITY

Numbers and Number Systems. During the years from kindergarten to eighth grade, students must repeatedly extend their conception of number. At first, "number" means "counting number": 1, 2, 3, . . . Soon after that, 0 is used to represent "none" and the whole numbers are formed by the counting numbers together with zero. The next extension is fractions. At first, fractions are barely numbers and tied strongly to pictorial representations. Yet by the time students understand division of fractions, they have a strong concept of fractions as numbers and have connected them, via their decimal representations, with the base-ten system used to represent the whole numbers. During middle school, fractions are augmented by negative fractions to form the rational numbers. In Grade 8, students extend this system once more, augmenting the rational numbers with the irrational numbers to form the real numbers. In high school, students will be exposed to yet another extension of number, when the real numbers are augmented by the imaginary numbers to form the complex numbers.

With each extension of number, the meanings of addition, subtraction, multiplication, and division are extended. In each new number system—integers, rational numbers, real numbers, and complex numbers—the four operations stay the same in two important ways: They have the commutative, associative, and distributive properties and their new meanings are consistent with their previous meanings.

Extending the properties of whole-number exponents leads to new and productive notation. For example, properties of whole-number exponents suggest that $(5^{1/3})^3$ should be $(5^{1/3})^3 = 5^1 = 5$ and that $5^{1/3}$ should be the cube root of 5.

Calculators, spreadsheets, and computer algebra systems can provide ways for students to become better acquainted with these new number systems and their notation. They can be used to generate data for numerical experiments, to help understand the workings of matrix, vector, and complex number algebra, and to experiment with non-integer exponents.

Quantities. In real world problems, the answers are usually not numbers but quantities: numbers with units, which involves measurement. In their work in measurement up through Grade 8, students primarily measure commonly used attributes such as length, area, and volume. In high school, students encounter a wider variety of units in modeling, for example, acceleration, currency conversions, derived quantities such as person-hours and heating degree days, social science rates such as per-capita income, and rates in everyday life such as points scored per game or batting averages. They also encounter novel situations in which they themselves must conceive the attributes of interest. For example, to find a good measure of overall highway safety, they might propose measures such as fatalities per year, fatalities per year per driver, or fatalities per vehicle-mile traveled. Such a conceptual process is sometimes called quantification. Quantification is important for science, as when surface area suddenly "stands out" as an important variable in evaporation. Quantification is also important for companies, which must conceptualize relevant attributes and create or choose suitable measures for them.

NUMBER AND QUANTITY OVERVIEW

The Real Number System

- Extend the properties of exponents to rational exponents
- Use properties of rational and irrational numbers.

Quantities

- Reason quantitatively and use units to solve problems

The Complex Number System

- Perform arithmetic operations with complex numbers
- Represent complex numbers and their operations on the complex plane
- Use complex numbers in polynomial identities and equations

Vector and Matrix Quantities

- Represent and model with vector quantities.
- Perform operations on vectors.
- Perform operations on matrices and use matrices in applications.

Mathematical Practices

1. Make sense of problems and persevere in solving them.
2. Reason abstractly and quantitatively.
3. Construct viable arguments and critique the reasoning of others.
4. Model with mathematics.
5. Use appropriate tools strategically.
6. Attend to precision.
7. Look for and make use of structure.
8. Look for and express regularity in repeated reasoning.

Mathematics | High School—Algebra

Expressions. An expression is a record of a computation with numbers, symbols that represent numbers, arithmetic operations, exponentiation, and, at more advanced levels, the operation of evaluating a function. Conventions about the use of parentheses and the order of operations assure that each expression is unambiguous. Creating an expression that describes a computation involving a general quantity requires the ability to express the computation in general terms, abstracting from specific instances.

Reading an expression with comprehension involves analysis of its underlying structure. This may suggest a different but equivalent way of writing the expression that exhibits some different aspect of its meaning. For example, $p + 0.05p$ can be interpreted as the addition of a 5% tax to a price p. Rewriting $p + 0.05p$ as $1.05p$ shows that adding a tax is the same as multiplying the price by a constant factor.

Algebraic manipulations are governed by the properties of operations and exponents, and the conventions of algebraic notation. At times, an expression is the result of applying operations to simpler expressions. For example, $p + 0.05p$ is the sum of the simpler expressions p and $0.05p$. Viewing an expression as the result of operation on simpler expressions can sometimes clarify its underlying structure.

A spreadsheet or a computer algebra system (CAS) can be used to experiment with algebraic expressions, perform complicated algebraic manipulations, and understand how algebraic manipulations behave.

Equations and inequalities. An equation is a statement of equality between two expressions, often viewed as a question asking for which values of the variables the expressions on either side are in fact equal. These values are the solutions to the equation. An identity, in contrast, is true for all values of the variables; identities are often developed by rewriting an expression in an equivalent form.

The solutions of an equation in one variable form a set of numbers; the solutions of an equation in two variables form a set of ordered pairs of numbers, which can be plotted in the coordinate plane. Two or more equations and/or inequalities form a system. A solution for such a system must satisfy every equation and inequality in the system.

An equation can often be solved by successively deducing from it one or more simpler equations. For example, one can add the same constant to both sides without changing the solutions, but squaring both sides might lead to extraneous solutions. Strategic competence in solving includes looking ahead for productive manipulations and anticipating the nature and number of solutions.

Some equations have no solutions in a given number system, but have a solution in a larger system. For example, the solution of $x + 1 = 0$ is an integer, not a whole number; the solution of $2x + 1 = 0$ is a rational number, not an integer; the solutions of $x^2 - 2 = 0$ are real numbers, not rational numbers; and the solutions of $x^2 + 2 = 0$ are complex numbers, not real numbers.

The same solution techniques used to solve equations can be used to rearrange formulas. For example, the formula for the area of a trapezoid, $A = ((b_1 + b_2)/2)h$, can be solved for h using the same deductive process.

Inequalities can be solved by reasoning about the properties of inequality. Many, but not all, of the properties of equality continue to hold for inequalities and can be useful in solving them.

Connections to Functions and Modeling. Expressions can define functions, and equivalent expressions define the same function. Asking when two functions have the same value for the same input leads to an equation; graphing the two functions allows for finding approximate solutions of the equation. Converting a verbal description to an equation, inequality, or system of these is an essential skill in modeling.

ALGEBRA OVERVIEW

Seeing Structure in Expressions

- Interpret the structure of expressions
- Write expressions in equivalent forms to solve problems

Arithmetic with Polynomials and Rational Expressions

- Perform arithmetic operations on polynomials
- Understand the relationship between zeros and factors of polynomials
- Use polynomial identities to solve problems
- Rewrite rational expressions

Creating Equations

- Create equations that describe numbers or relationships

Reasoning with Equations and Inequalities

- Understand solving equations as a process of reasoning and explain the reasoning
- Solve equations and inequalities in one variable
- Solve systems of equations
- Represent and solve equations and inequalities graphically

Mathematical Practices

1. Make sense of problems and persevere in solving them.
2. Reason abstractly and quantitatively.
3. Construct viable arguments and critique the reasoning of others.
4. Model with mathematics.
5. Use appropriate tools strategically.
6. Attend to precision.
7. Look for and make use of structure.
8. Look for and express regularity in repeated reasoning.

Mathematics | High School—Functions

Functions describe situations where one quantity determines another. For example, the return on $10,000 invested at an annualized percentage rate of 4.25% is a function of the length of time the money is invested. Because we continually make theories about dependencies between quantities in nature and society, functions are important tools in the construction of mathematical models.

In school mathematics, functions usually have numerical inputs and outputs and are often defined by an algebraic expression. For example, the time in hours it takes for a car to drive 100 miles is a function of the car's speed in miles per hour, v; the rule $T(v) = 100/v$ expresses this relationship algebraically and defines a function whose name is T.

The set of inputs to a function is called its domain. We often infer the domain to be all inputs for which the expression defining a function has a value, or for which the function makes sense in a given context.

A function can be described in various ways, such as by a graph (e.g., the trace of a seismograph); by a verbal rule, as in, "I'll give you a state, you give me the capital city"; by an algebraic expression like $r''(x) = a + bx$; or by a recursive rule. The graph of a function is often a useful way of visualizing the relationship of the function models, and manipulating a mathematical expression for a function can throw light on the function's properties.

Functions presented as expressions can model many important phenomena. Two important families of functions characterized by laws of growth are linear functions, which grow at a constant rate, and exponential functions, which grow at a constant percent rate. Linear functions with a constant term of zero describe proportional relationships.

A graphing utility or a computer algebra system can be used to experiment with properties of these functions and their graphs and to build computational models of functions, including recursively defined functions.

Connections to Expressions, Equations, Modeling, and Coordinates.

Determining an output value for a particular input involves evaluating an expression; finding inputs that yield a given output involves solving an equation. Questions about when two functions have the same value for the same input lead to equations, whose solutions can be visualized from the intersection of their graphs. Because functions describe relationships between quantities, they are frequently used in modeling. Sometimes functions are defined by a recursive process, which can be displayed effectively using a spreadsheet or other technology.

FUNCTIONS OVERVIEW

Interpreting Functions

- Understand the concept of a function and use function notation
- Interpret functions that arise in applications in terms of the context
- Analyze functions using different representations

Building Functions

- Build a function that models a relationship between two quantities
- Build new functions from existing functions

Linear, Quadratic, and Exponential Models

- Construct and compare linear, quadratic, and exponential models and solve problems
- Interpret expressions for functions in terms of the situation they model

Trigonometric Functions

- Extend the domain of trigonometric functions using the unit circle
- Model periodic phenomena with trigonometric functions
- Prove and apply trigonometric identities

Mathematical Practices

1. Make sense of problems and persevere in solving them.
2. Reason abstractly and quantitatively.
3. Construct viable arguments and critique the reasoning of others.
4. Model with mathematics.
5. Use appropriate tools strategically.
6. Attend to precision.
7. Look for and make use of structure.
8. Look for and express regularity in repeated reasoning.

MATHEMATICS | HIGH SCHOOL—MODELING

Modeling links classroom mathematics and statistics to everyday life, work, and decision-making. Modeling is the process of choosing and using appropriate mathematics and statistics to analyze empirical situations, to understand them better, and to improve decisions. Quantities and their relationships in physical, economic, public policy, social, and everyday situations can be modeled using mathematical and statistical methods. When making mathematical models, technology is valuable for varying assumptions, exploring consequences, and comparing predictions with data.

A model can be very simple, such as writing total cost as a product of unit price and number bought, or using a geometric shape to describe a physical object like a coin. Even such simple models involve making choices. It is up to us whether to model a coin as a three-dimensional cylinder, or whether a two-dimensional disk works well enough for our purposes. Other situations—modeling a delivery route, a production schedule, or a comparison of loan amortizations—need more elaborate models that use other tools from the mathematical sciences. Real-world situations are not organized and labeled for analysis; formulating tractable models, representing such models, and analyzing them is appropriately a creative process. Like every such process, this depends on acquired expertise as well as creativity,

Some examples of such situations might include

- estimating how much water and food is needed for emergency relief in a devastated city of 3 million people, and how it might be distributed.
- planning a table tennis tournament for 7 players at a club with 4 tables, where each player plays against each other player.
- designing the layout of the stalls in a school fair so as to raise as much money as possible.
- analyzing stopping distance for a car.
- modeling savings account balance, bacterial colony growth, or investment growth.
- engaging in critical path analysis, for example, applied to turnaround of an aircraft at an airport.
- analyzing risk in situations such as extreme sports, pandemics, and terrorism.
- relating population statistics to individual predictions.

In situations like these, the models devised depend on a number of factors: How precise an answer do we want or need? What aspects of the situation do we most need to understand, control, or optimize? What resources of time and tools do we have? The range of models that we can create and analyze is also constrained by the limitations of our mathematical, statistical, and technical skills, and our ability to recognize significant variables and relationships among them. Diagrams of various kinds, spreadsheets and other technology, and algebra are powerful tools for understanding and solving problems drawn from different types of real-world situations.

One of the insights provided by mathematical modeling is that essentially the same mathematical or statistical structure can sometimes model seemingly different situations. Models can also shed light on the mathematical structures themselves, for example, as when a model of bacterial growth makes more vivid the explosive growth of the exponential function.

The basic modeling cycle is summarized in the diagram. It involves (1) identifying variables in the situation and selecting those that represent essential features, (2) formulating a model by creating and selecting geometric, graphical, tabular, algebraic, or statistical representations that describe relationships between the variables, (3) analyzing and performing operations on these relationships to draw

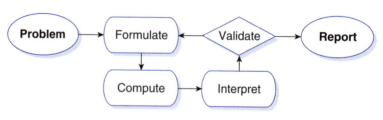

conclusions, (4) interpreting the results of the mathematics in terms of the original situation, (5) validating the conclusions by comparing them with the situation, and then either improving the model or, if it is acceptable, (6) reporting on the conclusions and the reasoning behind them. Choices, assumptions, and approximations are present throughout this cycle.

In descriptive modeling, a model simply describes the phenomena or summarizes them in a compact form. Graphs of observations are a familiar descriptive model— for example, graphs of global temperature and atmospheric CO_2 over time.

Analytic modeling seeks to explain data on the basis of deeper theoretical ideas, albeit with parameters that are empirically based; for example, exponential growth of bacterial colonies (until cut-off mechanisms such as pollution or starvation intervene) follows from a constant reproduction rate. Functions are an important tool for analyzing such problems.

Graphing utilities, spreadsheets, computer algebra systems, and dynamic geometry software are powerful tools that can be used to model purely mathematical phenomena (e.g., the behavior of polynomials) as well as physical phenomena.

Modeling Standards. *Modeling is best interpreted not as a collection of isolated topics but rather in relation to other standards. Making mathematical models is a Standard for Mathematical Practice, and specific modeling standards appear throughout the high school standards indicated by a star symbol (*).*

MATHEMATICS | HIGH SCHOOL—GEOMETRY

An understanding of the attributes and relationships of geometric objects can be applied in diverse contexts—interpreting a schematic drawing, estimating the amount of wood needed to frame a sloping roof, rendering computer graphics, or designing a sewing pattern for the most efficient use of material.

Although there are many types of geometry, school mathematics is devoted primarily to plane Euclidean geometry, studied both synthetically (without coordinates) and analytically (with coordinates). Euclidean geometry is characterized most importantly by the Parallel Postulate, that through a point not on a given line there is exactly one parallel line. (Spherical geometry, in contrast, has no parallel lines.)

During high school, students begin to formalize their geometry experiences from elementary and middle school, using more precise definitions and developing careful proofs. Later in college some students develop Euclidean and other geometries carefully from a small set of axioms.

The concepts of congruence, similarity, and symmetry can be understood from the perspective of geometric transformation. Fundamental are the rigid motions: translations, rotations, reflections, and combinations of these, all of which are here assumed to preserve distance and angles (and therefore shapes generally). Reflections and rotations each explain a particular type of symmetry, and the symmetries of an object offer insight into its attributes—as when the reflective symmetry of an isosceles triangle assures that its base angles are congruent.

In the approach taken here, two geometric figures are defined to be congruent if there is a sequence of rigid motions that carries one onto the other. This is the principle of superposition. For triangles, congruence means the equality of all corresponding pairs of sides and all corresponding pairs of angles. During the middle grades, through experiences drawing triangles from given conditions, students notice ways to specify enough measures in a triangle to ensure that all triangles drawn with those measures are congruent. Once these triangle congruence criteria (ASA, SAS, and SSS) are established using rigid motions, they can be used to prove theorems about triangles, quadrilaterals, and other geometric figures.

Similarity transformations (rigid motions followed by dilations) define similarity in the same way that rigid motions define congruence, thereby formalizing the similarity ideas of "same shape" and "scale factor" developed in the middle grades. These transformations lead to the criterion for triangle similarity that two pairs of corresponding angles are congruent.

The definitions of sine, cosine, and tangent for acute angles are founded on right triangles and similarity, and, with the Pythagorean Theorem, are fundamental in many real-world and theoretical situations. The Pythagorean Theorem is generalized to nonright triangles by the Law of Cosines. Together, the Laws of Sines and Cosines embody the triangle congruence criteria for the cases where three pieces of information suffice to completely solve a triangle. Furthermore, these laws yield two possible solutions in the ambiguous case, illustrating that Side-Side-Angle is not a congruence criterion.

Analytic geometry connects algebra and geometry, resulting in powerful methods of analysis and problem solving. Just as the number line associates numbers with locations in one dimension, a pair of perpendicular axes associates pairs of numbers with locations in two dimensions. This correspondence between numerical coordinates and geometric points allows methods from algebra to be applied to geometry and vice versa. The solution set of an equation becomes a geometric curve, making visualization a tool for doing and understanding algebra. Geometric shapes can be described by equations, making algebraic manipulation into a tool for geometric understanding, modeling, and proof. Geometric transformations of the graphs of equations correspond to algebraic changes in their equations.

Dynamic geometry environments provide students with experimental and modeling tools that allow them to investigate geometric phenomena in much the same way as computer algebra systems allow them to experiment with algebraic phenomena.

Connections to Equations. The correspondence between numerical coordinates and geometric points allows methods from algebra to be applied to geometry and vice versa. The solution set of an equation becomes a geometric curve, making visualization a tool for doing and understanding algebra. Geometric shapes can be described by equations, making algebraic manipulation into a tool for geometric understanding, modeling, and proof.

Geometry Overview

Congruence

- Experiment with transformations in the plane
- Understand congruence in terms of rigid motions
- Prove geometric theorems
- Make geometric constructions

Similarity, Right Triangles, and Trigonometry

- Understand similarity in terms of similarity transformations
- Prove theorems involving similarity
- Define trigonometric ratios and solve problems involving right triangles
- Apply trigonometry to general triangles

Circles

- Understand and apply theorems about circles
- Find arc lengths and areas of sectors of circles

Expressing Geometric Properties with Equations

- Translate between the geometric description and the equation for a conic section
- Use coordinates to prove simple geometric theorems algebraically

Geometric Measurement and Dimension

- Explain volume formulas and use them to solve problems
- Visualize relationships between two-dimensional and three-dimensional objects

Modeling with Geometry

- Apply geometric concepts in modeling situations

Mathematical Practices

1. Make sense of problems and persevere in solving them.
2. Reason abstractly and quantitatively.
3. Construct viable arguments and critique the reasoning of others.
4. Model with mathematics.
5. Use appropriate tools strategically.
6. Attend to precision.
7. Look for and make use of structure.
8. Look for and express regularity in repeated reasoning.

Mathematics | High School—Statistics and Probability

Decisions or predictions are often based on data—numbers in context. These decisions or predictions would be easy if the data always sent a clear message, but the message is often obscured by variability. Statistics provides tools for describing variability in data and for making informed decisions that take it into account.

Data are gathered, displayed, summarized, examined, and interpreted to discover patterns and deviations from patterns. Quantitative data can be described in terms of key characteristics: measures of shape, center, and spread. The shape of a data distribution might be described as symmetric, skewed, flat, or bell shaped, and it might be summarized by a statistic measuring center (such as mean or median) and a statistic measuring spread (such as standard deviation or interquartile range). Different distributions can be compared numerically using these statistics or compared visually using plots. Knowledge of center and spread are not enough to describe a distribution. Which statistics to compare, which plots to use, and what the results of a comparison might mean, depend on the question to be investigated and the real-life actions to be taken.

Randomization has two important uses in drawing statistical conclusions. First, collecting data from a random sample of a population makes it possible to draw valid conclusions about the whole population, taking variability into account. Second, randomly assigning individuals to different treatments allows a fair comparison of the effectiveness of those treatments. A statistically significant outcome is one that is unlikely to be due to chance alone, and this can be evaluated only under the condition of randomness. The conditions under which data are collected are important in drawing conclusions from the data; in critically reviewing uses of statistics in public media and other reports, it is important to consider the study design, how the data were gathered, and the analyses employed as well as the data summaries and the conclusions drawn.

Random processes can be described mathematically by using a probability model: a list or description of the possible outcomes (the sample space), each of which is assigned a probability. In situations such as flipping a coin, rolling a number cube, or drawing a card, it might be reasonable to assume various outcomes are equally likely. In a probability model, sample points represent outcomes and combine to make up events; probabilities of events can be computed by applying the Addition and Multiplication Rules. Interpreting these probabilities relies on an understanding of independence and conditional probability, which can be approached through the analysis of two-way tables.

Technology plays an important role in statistics and probability by making it possible to generate plots, regression functions, and correlation coefficients, and to simulate many possible outcomes in a short amount of time.

Connections to Functions and Modeling. Functions may be used to describe data; if the data suggest a linear relationship, the relationship can be modeled with a regression line, and its strength and direction can be expressed through a correlation coefficient.

STATISTICS AND PROBABILITY OVERVIEW

Interpreting Categorical and Quantitative Data

- Summarize, represent, and interpret data on a single count or measurement variable
- Summarize, represent, and interpret data on two categorical and quantitative variables
- Interpret linear models

Making Inferences and Justifying Conclusions

- Understand and evaluate random processes underlying statistical experiments
- Make inferences and justify conclusions from sample surveys, experiments and observational studies

Conditional Probability and the Rules of Probability

- Understand independence and conditional probability and use them to interpret data
- Use the rules of probability to compute probabilities of compound events in a uniform probability model

Using Probability to Make Decisions

- Calculate expected values and use them to solve problems
- Use probability to evaluate outcomes of decisions

Mathematical Practices

1. Make sense of problems and persevere in solving them.
2. Reason abstractly and quantitatively.
3. Construct viable arguments and critique the reasoning of others.
4. Model with mathematics.
5. Use appropriate tools strategically.
6. Attend to precision.
7. Look for and make use of structure.
8. Look for and express regularity in repeated reasoning.

REFERENCES

Albert, J. (2006). Interpreting probabilities and teaching the subjective point of view. In G. F. Burrill & P. C. Elliot (Eds.), *Thinking and reasoning with data and chance* (Sixty-eighth yearbook of the National Council of Teachers of Mathematics, pp. 417–433). Reston, VA: National Council of Teachers of Mathematics.

Aliaga, M., Cobb, G., Cuff, C., Garfield, J., Gould, R., Lock, R., . . . Witmer, J. (2010). *Guidelines for assessment and instruction in statistics education: College report.* Alexandria, VA: American Statistical Association. Retrieved May 19, 2010, from http://www.amstat.org/education/gaise/

American Association for the Advancement of Science. (2000). *Middle grades mathematics textbooks: A benchmark-based evaluation.* Retrieved December 23, 2009, from http://www.project2061.org/publications/textbook/mgmth/report/default.htm

Anscombe, F. J. (1973). Graphs in statistical analysis. *The American Statistician, 27,* 17–21.

Ardoin, S. P., Witt, J. C., Connell, J. E., & Koenig, J. L. (2005). Application of a three-tiered response to intervention model for instructional planning, decision making, and the identification of children in need of services. *Journal of Psychoeducational Assessment, 23,* 362–380.

Artigue, M., Batanero, C., & Kent, P. (2007). Mathematics thinking and learning at post-secondary level. In F. K. Lester, Jr. (Ed.), *Second handbook of research on mathematics teaching and learning* (pp. 1011–1049). Reston, VA: National Council of Teachers of Mathematics; Charlotte, NC: Information Age.

Artzt, A. F. (1999). Cooperative learning in mathematics teacher education. *Mathematics Teacher, 92,* 11–17.

Ashlock, R. B. (2005). *Error patterns in computation: Using error patterns to improve instruction* (8th ed.). Upper Saddle River, NJ: Pearson.

Asiala, M., Cottrill, J., Dubinsky, E., & Schwingendorf, K. E. (1997). The development of students' graphical understanding of the derivative. *Journal of Mathematical Behavior, 16,* 399–431.

Askey, R. (1999). Mathematical content. In S. G. Krantz (Ed.), *How to teach mathematics* (2nd ed., pp. 161–171). Providence, RI: American Mathematical Society.

Aspinwall, L., & Aspinwall, J. S. (2003). Investigating mathematical thinking using open writing prompts. *Mathematics Teaching in the Middle School, 8,* 350–353.

Aspinwall, L., Shaw, K. L., & Presmeg, N. C. (1997). Uncontrollable mental imagery: Graphical connections between a function and its derivative. *Educational Studies in Mathematics, 33,* 301–317.

Baker, B., Cooley, L., & Trigueros, M. (2000). A calculus graphing schema. *Journal for Research in Mathematics Education, 31,* 557–578.

Balacheff, N. (1988). Aspects of proof in pupils' practice of school mathematics. In D. Pimm (Ed.), *Mathematics, teachers, and children* (pp. 216–238). London, England: Hodder and Stoughton.

Ball, D. L. (1992). Magical hopes: Manipulatives and the reform of math education. *American Educator, 16*(2), 14–18, 46–47.

Ball, L., Pierce, R., & Stacey, K. (2003). Recognising equivalent algebraic expressions: An important component of algebraic expectation for working with CAS. In N. A. Pateman, B. J. Dougherty, & J. T. Zillox (Eds.), *Proceedings of the 27th PME International Conference, 4,* 15–22.

Barnsley, M. (1989). *Fractals everywhere.* Boston, MA: Academic Press.

Batanero, C., Estepa, A., Godino, J. D., & Green, D. R. (1996). Intuitive strategies and preconceptions about association in contingency tables. *Journal for Research in Mathematics Education, 27,* 151–169.

Batanero, C., Navarro-Pelayo, V., & Godino, J. D. (1997). Effect of the implicit combinatorial model on combinatorial reasoning in secondary school pupils. *Educational Studies in Mathematics, 32,* 181–199.

Batanero, C., & Serrano, L. (1999). The meaning of randomness for secondary school students. *Journal for Research in Mathematics Education, 30,* 558–567.

Battista, M. T. (1999). Fifth-graders' enumeration of cubes in 3D arrays: Conceptual progress in an inquiry-based classroom. *Journal for Research in Mathematics Education, 30,* 417–448.

Battista, M. T. (2003). *Shape makers.* Emeryville, CA: Key Curriculum Press.

Battista, M. T. (2007). The development of geometric and spatial thinking. In F. K. Lester, Jr. (Ed.), *Second handbook of research on mathematics teaching and learning* (pp. 843–908). Reston, VA: National Council of Teachers of Mathematics; Charlotte, NC: Information Age.

Battista, M. T. (2009). Highlights of research on learning school geometry. In T. V. Craine (Ed.), *Understanding geometry for a changing world* (pp. 91–108). Reston, VA: National Council of Teachers of Mathematics.

Baxter, J. A., Woodward, J., Olson, D., & Robyns, J. (2002). Blueprint for writing in middle school mathematics. *Mathematics Teaching in the Middle School, 8,* 52–56.

Bay-Williams, J. M. (2005). Poetry in motion: Using Shel Silverstein's works to engage students in mathematics. *Mathematics Teaching in the Middle School, 10,* 386–393.

Beaton, A. E., & Robitaille, D. F. (1999). An overview of the Third International Mathematics and Science Study. In G. Kaiser, E. Luna, & I. Huntley (Eds.), *International comparisons in mathematics education* (pp. 30–47). London, England: Falmer Press.

Beaton, T. (2004). Harry Potter in the mathematics classroom. *Mathematics Teaching in the Middle School, 10,* 23–25.

Behr, M., Erlwanger, S., & Nichols, E. (1980). How children view the equals sign. *Mathematics Teaching, 92,* 13–15.

Benacerraf, P., & Putnam, H. (Eds.). (1983). *Philosophy of mathematics: Selected readings.* New York, NY: Cambridge University Press.

Ben-Zvi, D., & Garfield, J. (2004). Statistical literacy, reasoning, and thinking: Goals and challenges. In J. Garfield & D. Ben-Zvi (Eds.), *The challenge of developing statistical literacy, reasoning, and thinking* (pp. 3–15). Dordrecht, The Netherlands: Kluwer.

Berkman, R. M. (2004). The chess and mathematics connection: More than just a game. *Mathematics Teaching in the Middle School, 9,* 246–250.

Bey, J. M., Reys, R. E., Simms, K., & Taylor, P. M. (2000). Bingo games: Turning student intuitions into investigations in probability and number sense. *Mathematics Teacher, 93,* 200–206.

Bezuidenhout, J. (1998). First-year university students' understanding of rate of change. *International Journal of Mathematical Education in Science and Technology, 29,* 389–399.

Bezuidenhout, J., & Olivier, A. (2000). Students' conceptions of the integral. In T. Nakahara & M. Koyama (Eds.), *Proceedings of the 24th PME International Conference, 2,* 73–80.

Biehler, R. (1994). Probabilistic thinking, statistical reasoning, and the search for causes—Do we need a probabilistic revolution after we have taught data analysis? In J. Garfield (Ed.), *Research papers from ICOTS-4* (pp. 20–37). Minneapolis: University of Minnesota.

Bloom, B. S. (Ed.), Englehart, M. D., Furst, E. J., Hill, W. H., & Krathwohl, D. R. (1956). *Taxonomy of educational objectives: The classification of educational goals. Handbook I: Cognitive domain.* New York, NY: Longman.

Blume, G. W., Galindo, E., & Walcott, C. (2007). Performance in measurement and geometry from the viewpoint of Principles and Standards for School Mathematics. In P. Kloosterman & F. K. Lester, Jr. (Eds.), *Results and interpretations of the 2003 mathematics assessment of the National Assessment of Educational Progress* (pp. 95–138). Reston, VA: National Council of Teachers of Mathematics.

Boaler, J. (1998). Open and closed mathematics: Student experiences and understandings. *Journal for Research in Mathematics Education, 29,* 41–62.

Boaler, J. (2008). *What's math got to do with it?* New York, NY: Penguin.

Boaler, J., & Staples, M. (2008). Creating mathematical futures through an equitable teaching approach: The case of Railside School. *Teachers College Record, 110,* 608–645.

Bock, D. E., Velleman, P. F., & DeVeaux, R. D. (2004). *Stats: Modeling the world.* Boston, MA: Pearson.

Booth, L. R. (1988). Children's difficulties in beginning algebra. In A. F. Coxford & A. P. Schulte (Eds.), *The ideas of algebra* (1988 yearbook of the National Council of Teachers of Mathematics, pp. 20–32). Reston, VA: National Council of Teachers of Mathematics.

Borasi, R. (1996). *Reconceiving mathematics instruction: A focus on errors.* Norwood, NJ: Ablex.

Borasi, R., & Siegel, M. (2000). *Reading counts: Expanding the role of reading in mathematics classrooms.* New York, NY: Teachers College Press.

Borenson, H. (2009). *Hands-on equations.* Retrieved May 9, 2009, from http://www.borenson.com

Boyer, C. B., & Merzbach, U. C. (1989). *A history of mathematics* (2nd ed.). New York, NY: Wiley.

Brandell, J. L. (1994). Helping students write paragraph proofs in geometry. *The Mathematics Teacher, 87,* 498–502.

Bransford, J. D., Brown, A. L., & Cocking, R. R. (Eds.). (1999). *How people learn: Brain, mind experience, and school.* Washington, DC: National Academy Press.

Breyfogle, M. L., & Herbel-Eisenmann, B. A. (2004). Focusing on students' mathematical thinking. *Mathematics Teacher, 97,* 244–247.

Britton, K. L., & Johannes, J. L. (2003). Portfolios and a backward approach to assessment. *Mathematics Teaching in the Middle School, 9,* 70–76.

Brown, S. A. (2005). You made it through the test; what about the aftermath? *Mathematics Teaching in the Middle School, 11,* 68–73.

Browning, C., & Garza-Kling, G. (2009). Conceptions of angle: Implications for middle school and beyond. In T. V. Craine (Ed.), *Understanding geometry for a changing world* (pp. 127–139). Reston, VA: National Council of Teachers of Mathematics.

Burns, M. (2000). *About teaching mathematics: A K–8 resource* (2nd ed.). Sausalito, CA: Math Solutions.

Buschman, L. (2001). Using student interviews to guide classroom instruction: An action research project. *Teaching Children Mathematics, 8,* 222–227.

Cahan, S., & Linchevski, L. (1996). The cumulative effect of ability grouping on mathematical achievement: A longitudinal perspective. *Studies in Educational Evaluation, 22*(1), 29–40.

Cai, J. (2000). Understanding and representing the arithmetic averaging algorithm: An analysis and comparison of U.S. and Chinese students' responses. *International Journal of Mathematical Education in Science and Technology, 31,* 839–855.

Cai, J., & Kenney, P. A. (2000). Fostering mathematical thinking through multiple solutions. *Mathematics Teaching in the Middle School, 5,* 534–539.

Calouh, L., & Herscovics, N. (1988). Teaching algebraic expressions in a meaningful way. In A. F. Coxford & A. P. Schulte (Eds.), *The ideas of algebra* (1988 yearbook of the National Council of Teachers of Mathematics, pp. 33–42). Reston, VA: National Council of Teachers of Mathematics.

Cannon, L., Dorward, J., Heal, B., & Edwards, L. (2001). *National library of virtual manipulatives for interactive mathematics.* MATTI Associates LLC. Retrieved April 17, 2009, from http://nlvm.usu.edu/EN/NAV/VLIBRARY.HTML

Carnegie Learning. (2009). Cognitive Tutor *software overview.* Retrieved December 26, 2009, from http://www.carnegielearning.com/software_features.cfm

Carpenter, T. P., Corbitt, M. K., Kepner, H. S., Jr., Lindquist, M. M., & Reys, R. E. (1981). *Results from the second mathematics assessment of the National Assessment of Educational Progress.* Reston, VA: National Council of Teachers of Mathematics.

Carpenter, T. P., Hiebert, J., Fennema, E., Fuson, K. C., Wearne, D., & Murray, H. (1997). *Making sense: Teaching and learning mathematics with understanding.* Portsmouth, NH: Heinemann.

Carpenter, T. P., Levi, L., Williams-Berman, P., & Pligge, M. (2005). Developing algebraic reasoning in the elementary school. In T. A. Romberg, T. P. Carpenter, & F. Dremock (Eds.), *Understanding mathematics and science matters* (pp. 81–98). Mahwah, NJ: Erlbaum.

Carraher, T. N., Carraher, D. W., & Schliemann, A. D. (1985). Mathematics in the streets and schools. *British Journal of Developmental Psychology, 3,* 21–29.

Caulfield, R., Harkness, S. S., & Riley, R. (2003). Surprise! Turn routine problems into worthwhile tasks. *Mathematics Teaching in the Middle School, 9,* 198–202.

Cavey, L. O., & Berenson, S. B. (2005). Learning to teach high school mathematics: Patterns of growth in understanding right triangle trigonometry during lesson plan study. *Journal of Mathematical Behavior, 24,* 171–190.

Chance, B., delMas, R., & Garfield, J. (2004). Reasoning about sampling distributions. In D. Ben-Zvi & J. Garfield (Eds.), *The challenge of developing statistical literacy, reasoning, and thinking* (pp. 295–323). Dordrecht, The Netherlands: Kluwer.

Chappell, M. F. (2003). Keeping mathematics front and center: Reaction to middle-grades curriculum projects research. In S. L. Senk & D. R. Thompson (Eds.), *Standards-based school mathematics curricula: What are they? What do students learn?* (pp. 285–298). Mahwah, NJ: Erlbaum.

Chappell, M. F., & Strutchens, M. E. (2001). Creating connections: Promoting algebraic thinking with concrete models. *Mathematics Teaching in the Middle School, 7,* 20–25.

Charles, R., Lester, F., & O'Daffer, P. (1987). *How to evaluate progress in problem solving.* Reston, VA: National Council of Teachers of Mathematics.

Chazan, D. (1993). High school geometry students' justification for their views of empirical evidence and mathematical proof. *Educational Studies in Mathematics, 24,* 359–387.

Chazan, D. (1996). Algebra for all students? *Journal of Mathematical Behavior, 15,* 455–477.

Chazan, D. (2000). *Beyond formulas in mathematics and teaching.* New York, NY: Teachers College Press.

Chazan, D., & Ball, D. (1999). Beyond being told not to tell. *For the Learning of Mathematics, 19*(2), 2–10.

Clark, J. M., Cordero, F., Cottrill, J., Czarnocha, B., DeVries, D. J., St. John, D., . . . Vidakovic, D. (1997). Constructing a schema: The case of the chain rule? *Journal of Mathematical Behavior, 16,* 345–364.

Clement, J., Lochhead, J., & Monk, S. (1981). Translation difficulties in learning mathematics. *American Mathematical Monthly, 88,* 286–290.

Clement, L. L., & Bernhard, J. Z. (2005). A problem-solving alternative to using key-words. *Mathematics Teaching in the Middle School, 10,* 360–365.

Cobb, P. (1994). Where is the mind? Constructivist and sociocultural perspectives on mathematical development. *Educational Researcher, 23,* 13–20.

Cobb, P. (1999). Individual and collective mathematical development: The case of statistical data analysis. *Mathematical Thinking and Learning, 1,* 5–43.

Cobb, P., & Yackel, E. (1996). Constructivist, emergent, and sociocultural perspectives in the context of developmental research. *Educational Psychologist, 31,* 175–190.

Cohen, K. S., & Adams, T. L. (2004). A primer for preproblem ponderings: Anticipating the answer. *Mathematics Teacher, 97,* 110–115.

Cole, K. A. (1999). Walking around: Getting more from informal assessment. *Mathematics Teaching in the Middle School, 4,* 224–227.

College Board. (2008). *AP Statistics course description.* Retrieved May 19, 2010, from http://apcentral.collegeboard.com/

College Board. (2009). *AP Calculus course description.* Retrieved April 27, 2010, from http://apcentral.collegeboard.com/apc/Controller.jpf

College Board. (2010). *AP Central.* Retrieved April 27, 2010, from http://apcentral.collegeboard.com/apc/Controller.jpf

College Preparatory Mathematics. (2010). *Sample problems for Algebra 2 Connections.* Retrieved June 29, 2010, from http://www.cpm.org/pdfs/information/CPM_Sample_Problems_2.pdf

Confrey, J., & Smith, E. (1991). A framework for functions: Prototypes, multiple representations, and transformations. In R. G. Underhill (Ed.), *Proceedings of the 13th PME-NA Conference, 1,* 57–63.

Confrey, J., & Smith, E. (1995). Splitting, covariation, and their role in the development of exponential functions. *Journal for Research in Mathematics Education, 26,* 66–86.

Connected Mathematics Project. (2009). *Connected Mathematics Project student materials.* Retrieved December 27, 2009, from http://connectedmath.msu.edu/components/student.shtml

Consortium for Mathematics and Its Applications. (2010). *Mathematics: Modeling our world* (Course 1, 2nd ed.). Bedford, MA: Author.

Cramer, K., Wyberg, T., & Leavitt, S. (2008). The role of representations in fraction addition and subtraction. *Mathematics Teaching in the Middle School, 13,* 490–496.

Crites, T. (1992). Skilled and less skilled estimators' strategies for estimating discrete quantities. *The Elementary School Journal, 92,* 601–619.

Cuoco, A., Goldenberg, E. P., & Mark, J. (1996). Habits of mind: An organizing principle for mathematics curricula. *Journal of Mathematical Behavior, 15,* 375–402.

Dancis, J. (n.d.). *Pattern MIS-recognition.* Retrieved November 11, 2009, from http://www-users.math.umd.edu/~jnd/Patterns.pdf

Davis, E. A., & Krajcik, J. S. (2005). Designing educative curriculum materials to promote teacher learning. *Educational Researcher, 34*(3), 3–14.

delMas, R. C. (2004). A comparison of mathematical and statistical reasoning. In J. Garfield & D. Ben-Zvi (Eds.), *The challenge of developing statistical literacy, reasoning, and thinking* (pp. 79–95). Dordrecht, The Netherlands: Kluwer.

Demby, A. (1997). Algebraic procedures used by 13- to 15-year-olds. *Educational Studies in Mathematics, 33,* 45–70.

Derry, S. J., Levin, J. R., Osana, H. P., Jones, M. S., & Peterson, M. (2000). Fostering students' statistical and scientific thinking: Lessons learned from an innovative college course. *American Educational Research Journal, 37,* 747–773.

Devaney, R. L. (1998). Chaos in the classroom. In R. Lehrer & D. Chazan (Eds.), *Designing learning environments for developing understanding of geometry and space* (pp. 91–104). Mahwah, NJ: Erlbaum.

de Villiers, M. (1994). The role and function of a hierarchical classification of quadrilaterals. *For the Learning of Mathematics, 14,* 11–18.

de Villiers, M. (1998). An alternative approach to proof in dynamic geometry. In R. Lehrer & D. Chazan

(Eds.), *Designing learning environments for developing understanding of geometry and space* (pp. 369–393). Mahwah, NJ: Erlbaum.

de Villiers, M., Govender, R., & Patterson, N. (2009). Defining in geometry. In T. V. Craine (Ed.), *Understanding geometry for a changing world* (pp. 189–203). Reston, VA: National Council of Teachers of Mathematics.

Donoghue, E. F. (2003). Algebra and geometry textbooks in twentieth-century America. In G. M. A. Stanic & J. Kilpatrick (Eds.), *A history of school mathematics* (Vol. 1, pp. 329–398). Reston, VA: National Council of Teachers of Mathematics.

Driscoll, M. (1999). *Fostering algebraic thinking: A guide for teachers, Grades 6–10.* Portsmouth, NH: Heinemann.

Educational Testing Service. (2009). *Mathematics: Content knowledge (0061) test at a glance.* Princeton, NJ: Educational Testing Service. Retrieved September 16, 2009, from http://www.ets.org/Media/Tests/PRAXIS/pdf/0061.pdf

Edwards, M. T. (2003). Visualizing transformations: Matrices, handheld graphing calculators, and computer algebra systems. *Mathematics Teacher, 96,* 48–56.

Eizenberg, M. M., & Zaslavsky, O. (2004). Students' verification strategies for combinatorial problems. *Mathematical Thinking and Learning, 6,* 15–36.

Ellis, M. W., & Pagni, D. (2008). Exploring segment lengths on the geoboard. *Mathematics Teaching in the Middle School, 13,* 520–525.

English, L. D. (2005). Combinatorics and the development of children's combinatorial reasoning. In G. A. Jones (Ed.), *Exploring probability in school: Challenges for teaching and learning* (pp. 121–141). New York, NY: Springer.

Ernest, P. (1996). Varieties of constructivism: A framework for comparison. In L. P. Steffe, P. Nesher, P. Cobb, G. A. Goldin, & B. Greer (Eds.), *Theories of mathematical learning* (pp. 335–350). Mahwah, NJ: Erlbaum.

Even, R. (1993). Subject matter knowledge and pedagogical content knowledge: Prospective secondary teachers and the function concept. *Journal for Research in Mathematics Education, 24,* 94–116.

Falk, R. (1986). Conditional probabilities: Insights and difficulties. In R. Davidson & J. Swift (Eds.), *Proceedings of the Second International Conference on Teaching Statistics* (pp. 292–297). Victoria, British Columbia, Canada: University of Victoria.

Falkner, K. P., Levi, L., & Carpenter, T. P. (1999). Children's understanding of equality: A foundation for algebra. *Teaching Children Mathematics, 6,* 232–236.

Fast, G. (1997). Using analogies to overcome student teachers' probability misconceptions. *Journal of Mathematical Behavior, 16,* 325–344.

Fendel, D., Resek, D., Alper, L., & Fraser, S. (2010). *Interactive mathematics program, year 2* (2nd ed.). Emeryville, CA: Key Curriculum Press.

Ferrini-Mundy, J., & Lauten, D. (1993). Teaching and learning calculus. In P. S. Wilson (Ed.), *Research ideas for the classroom: High school mathematics* (pp. 155–175). Reston, VA: National Council of Teachers of Mathematics.

Filloy, E., Rojano, T., & Solares, A. (2003). Two meanings of the "equal" sign and senses of comparison and substitution methods. In N. A. Pateman, B. J. Dougherty, & J. T. Zillox (Eds.), *Proceedings of the 27th PME International Conference, 4,* 223–229.

Fiore, G. (1999). Math-abused students: Are we prepared to teach them? *Mathematics Teacher, 92,* 403–406.

Forster, P. (2000). Process and object interpretations of vector magnitude mediated by use of the graphics calculator. *Mathematics Education Research Journal, 12,* 269–285.

Forster, P. A., & Taylor, P. C. (2000). A multi-perspective analysis of learning in the presence of technology. *Educational Studies in Mathematics, 42,* 35–59.

Franklin, C., Kader, G., Mewborn, D., Moreno, J., Peck, R., Perry, M., & Scheaffer, R. (2007). *Guidelines for assessment and instruction in statistics education* (GAISE) *report: A pre-K–12 curriculum framework.* Alexandria, VA: American Statistical Association.

Friel, S. N. (2008). The research frontier: Where technology interacts with the teaching and learning of data analysis and statistics. In G. W. Blume & K. M. Heid (Eds.), *Research on technology and the teaching and learning of mathematics: Vol. 2. Cases and perspectives* (pp. 279–331). Charlotte, NC: Information Age.

Friel, S. N., Curcio, F. R., & Bright, G. W. (2001). Making sense of graphs: Critical factors influencing comprehension and instructional implications. *Journal for Research in Mathematics Education, 32,* 124–158.

Fuller, F. F., & Brown, O. H. (1975). Becoming a teacher. In K. Ryan (Ed.), *Teacher education* (Seventy-fourth yearbook of the National Society for the Study of Education). Chicago, IL: University of Chicago Press.

Fuson, K. C., Kalchman, M., & Bransford, J. D. (2005). Mathematical understanding: An introduction. In M. S. Donovan & J. D. Bransford (Eds.), *How students learn: Mathematics in the classroom* (pp. 217–256). Washington, DC: National Academies Press.

Gal, I. (1998). Assessing statistical knowledge as it relates to students' interpretation of data. In S. P. Lajoie (Ed.), *Reflections on statistics: Learning, teaching, and assessment in Grades K–12* (pp. 275–295). Mahwah, NJ: Erlbaum.

Gallardo, A. (2002). The extension of the natural-number domain to the integers in the transition from arithmetic to algebra. *Educational Studies in Mathematics, 49,* 171–192.

Garfield, J., delMas, R., & Chance, B. (2002). *Tools for teaching and assessing statistical inference.* Retrieved May 31, 2010, from http://www.tc.umn.edu/~delma001/stat_tools/

Garfield, J., delMas, R., Zieffler, A., & Ben-Zvi, D. (n.d.). *AIMS project: Adapting and implementing innovative material in statistics.* Retrieved May 31, 2010, from http://www.tc.umn.edu/~aims/index.htm

Garfield, J. B. (2003). Assessing statistical reasoning. *Statistics Education Research Journal, 2*(1), 22–38.

Garfield, J. B., & Ben-Zvi, D. (2008). *Developing students' statistical reasoning: Connecting research and teaching practice.* New York, NY: Springer.

Gentner, D., & Holyoak, K. J. (1997). Reasoning and learning by analogy. *American Psychologist, 52,* 32–34.

Georgakis, P. (1999). Oh, good, it's Tuesday! *Mathematics Teaching in the Middle School, 5,* 224–226.

Glidden, P. L. (2008). Prospective elementary teachers' understanding of order of operations. *School Science and Mathematics, 108*(4), 130–136.

Glynn, S. M. (1991). Explaining science concepts: A teaching-with-analogies model. In S. M. Glynn, R. H. Yeany, & B. K. Britton (Eds.), *The psychology of learning science* (pp. 219–240). Hillsdale, NJ: Erlbaum.

Goldin, G., & Shteingold, N. (2001). Systems of representations and the development of mathematical concepts. In A. A. Cuoco (Ed.), *The roles of representation in school mathematics* (pp. 1–23). Reston, VA: National Council of Teachers of Mathematics.

Goldsby, D. S., & Cozza, B. (2002). Writing samples to understand mathematical thinking. *Mathematics Teaching in the Middle School, 7,* 517–520.

Gonzalez-Martin, A. S., & Camacho, M. (2004). What is first-year mathematics students' actual knowledge about improper integrals? *International Journal of Mathematical Education in Science and Technology, 35,* 73–89.

Good, T. L., Mulryan, C., & McCaslin, M. (1992). Grouping for instruction in mathematics: A call for programmatic research on small-group processes. In D. A. Grouws (Ed.), *Handbook of research on mathematics teaching and learning* (pp. 165–196). Reston, VA: National Council of Teachers of Mathematics.

Goodson-Espy, T. (1998). The roles of reification and reflective abstraction in the development of abstract thought: Transitions from arithmetic to algebra. *Educational Studies in Mathematics, 36,* 219–245.

Goos, M. (2004). Learning mathematics in a classroom community of inquiry. *Journal for Research in Mathematics Education, 35,* 258–291.

Gordon, S. P., & Gordon, F. S. (2007). Discovering the fundamental theorem of calculus using data analysis. *Mathematics Teacher, 100,* 597–604.

Grossnickle, D. R., & Sesko, F. P. (1990). *Preventive discipline for effective teaching and learning.* Reston, VA: National Association of Secondary School Principals.

Groth, R. E. (2003). *Development of a high school statistical thinking framework* (Doctoral dissertation, Illinois State University, 2003). *Dissertation Abstracts International, 64*(04), 1202.

Groth, R. E. (2005). An investigation of statistical thinking in two different contexts: Detecting a signal in a noisy process and determining a typical value. *Journal of Mathematical Behavior, 24,* 109–124.

Groth, R. E. (2006a). Analysis of an online case discussion about teaching stochastics. *Mathematics Teacher Education and Development, 7,* 53–71.

Groth, R. E. (2006b). Expanding teachers' understanding of geometric definition: The case of the trapezoid. *Teaching Children Mathematics, 12,* 376–380.

Groth, R. E. (2008). Navigating layers of uncertainty in teaching statistics through case discussion. In C. Batanero, G. Burrill, C. Reading, & A. Rossman (Eds.), *Joint ICMI/IASE study: Teaching statistics in school mathematics. Challenges for teaching and teacher education* (Proceedings of ICMI Study 18 and the 2008 IASE Round Table Conference). Retrieved February 7, 2012, from http://www.ugr.es/~icmi/iase_study/

Groth, R. E. (2010). Three perspectives on the central objects of study for grades preK–8 statistics. In B. J. Reys & R. E. Reys (Eds.), *Mathematics curriculum issues, trends, and further directions* (Seventy-second yearbook of the National Council of Teachers of Mathematics, pp. 157–170). Reston, VA: National Council of Teachers of Mathematics.

Groth, R. E., & Bergner, J. A. (2006). Preservice elementary teachers' conceptual and procedural knowledge of mean, median, and mode. *Mathematical Thinking and Learning, 8,* 37–63.

Groth, R. E., & Powell, N. N. (2004). Using research projects to help develop high school students' statistical thinking. *Mathematics Teacher, 97,* 106–109.

Gutiérrez, A. (1992). Exploring the links between van Hiele levels and 3-dimensional geometry. *Structural Topology, 18,* 31–48.

Gutiérrez, A., Jaime, A., & Fortuny, J. M. (1991). An alternative paradigm to evaluate the acquisition of the van Hiele levels. *Journal for Research in Mathematics Education, 22,* 237–251.

Gutstein, E. (2003). Teaching and learning mathematics for social justice in an urban, Latino school. *Journal for Research in Mathematics Education, 34,* 37–73.

Gutstein, E. (2005). *Reading and writing the world with mathematics: Toward a pedagogy of social justice.* New York, NY: Routledge.

Haciomeroglu, E. S., Aspinwall, L., & Presmeg, N. C. (2009). Visual and analytic thinking in calculus. *Mathematics Teacher, 103,* 140–145.

Hadas, N., & Hershkowitz, R. (1998). Proof in geometry as an explanatory and convincing tool. In A. Olivier & K. Newstead (Eds.), *Proceedings of the 22nd PME Conference* (Vol. 3, pp. 25–32). University of Stellenbosch, Stellenbosch, South Africa.

Hadas, N., Hershkowitz, R., & Schwarz, B. B. (2000). The role of contradiction and uncertainty in promoting the need to prove in dynamic geometry environments. *Educational Studies in Mathematics, 44,* 127–150.

Hammersley, M. (2002). *Educational research: Policymaking and practice.* London, England: Paul Chapman.

Harel, G., Fuller, E., & Rabin, J. M. (2008). Attention to meaning by algebra teachers. *Journal of Mathematical Behavior, 27,* 116–127.

Harel, G., Selden, A., & Selden, J. (2006). Advanced mathematical thinking: Some PME perspectives. In A. Gutiérrez & P. Boero (Eds.), *Handbook of research on the psychology of mathematics education* (pp. 147–172). Rotterdam, The Netherlands: Sense Publishers.

Harel, G., & Sowder, L. (1998). Students' proof schemes: Results from exploratory studies. In A. H. Schoenfeld, J. Kaput, & E. Dubinsky (Eds.), *Research in collegiate mathematics*

education III (pp. 234–283). Providence, RI: American Mathematical Society; Washington, DC: Mathematical Association of America.

Heaton, R. M., & Mickelson, W. T. (2002). The learning and teaching of statistical investigation in teaching and teacher education. *Journal of Mathematics Teacher Education, 5,* 35–59.

Heid, M. K. (2005). Technology in mathematics education: Tapping into visions of the future. In W. J. Masalski & P. C. Elliot (Eds.), *Technology-supported learning environments* (Sixty-seventh yearbook of the National Council of Teachers of Mathematics, pp. 345–366). Reston, VA: National Council of Teachers of Mathematics.

Heid, M. K., & Blume, G. W. (2008). Algebra and function development. In M. K. Heid & G. W. Blume (Eds.), *Research on technology and the teaching and learning of mathematics: Vol. 1. Research syntheses* (pp. 55–108). Charlotte, NC: Information Age and National Council of Teachers of Mathematics.

Heid, M. K., Blume, G. W., Hollebrands, K., & Piez, C. (2002). Computer algebra systems in mathematics instruction: Implications from research. *Mathematics Teacher, 95,* 586–591.

Henningsen, M., & Stein, M. K. (1997). Mathematical tasks and student cognition: Classroom-based factors that support and inhibit high-level mathematical thinking and reasoning. *Journal for Research in Mathematics Education, 28,* 524–549.

Herbst, P. G. (2002). Establishing a custom of proving in American school geometry: Evolution of the two-column proof in the early twentieth century. *Educational Studies in Mathematics, 49,* 283–312.

Herman, M., & Laumakis, P. (2008). Using the CBR to enhance graphical understanding. *Mathematics Teacher, 102,* 383–389.

Herscovics, N., & Linchevski, L. (1991). Pre-algebraic thinking: Range of equations and informal solution processes used by seventh graders prior to any instruction. In F. Furinghetti (Ed.), *Proceedings of the 15th PME International Conference, 2,* 173–180.

Hershkowitz, R., Bruckheimer, M., & Vinner, S. (1987). Activities with teachers based on cognitive research. In C. R. Hirsch & M. J. Zweng (Eds.), *The secondary school mathematics curriculum* (pp. 222–235). Reston, VA: National Council of Teachers of Mathematics.

Hershkowitz, R., Dreyfus, T., Ben-Zvi, D., Friedlander, A., Hadas, N., Resnick, T., . . . Schwarz, B. (2002).

Mathematics curriculum development for computerized environments: A designer-researcher-teacher-learner activity. In L. D. English (Ed.), *Handbook of international research in mathematics education* (pp. 657–694). Mahwah, NJ: Erlbaum.

Hiebert, J. (2000). What can we expect from research? *Mathematics Teacher, 93,* 168–169.

Hiebert, J., & Lefevre, P. (1986). Conceptual and procedural knowledge in mathematics: An introductory analysis. In J. Hiebert (Ed.), *Conceptual and procedural knowledge: The case of mathematics* (pp. 1–28). Hillsdale, NJ: Erlbaum.

Hirsch, C. R., Coxford, A. F., Fey, J. T., & Schoen, H. L. (1998). *Contemporary mathematics in context: A unified approach.* New York, NY: Glencoe/McGraw-Hill.

Hirsch, C. R., Fey, J. T., Hart, E. W., Schoen, H. L., & Watkins, A. E. (2009). *Core-Plus Mathematics: Contemporary Mathematics in Context, scope and sequence.* Columbus, OH: Glencoe/McGraw-Hill.

Hodgson, T., Simonsen, L., Lubek, J., & Anderson, L. (2003). Measuring Montana: An episode in estimation. In D. H. Clements (Ed.), *Learning and teaching measurement* (Sixty-fifth yearbook of the National Council of Teachers of Mathematics, pp. 197–207). Reston, VA: National Council of Teachers of Mathematics.

Hollebrands, K. F., & Smith, R. C. (2009). Using interactive geometry software to teach secondary school geometry: Implications from research. In T. V. Craine (Ed.), *Understanding geometry for a changing world* (pp. 221–232). Reston, VA: National Council of Teachers of Mathematics.

House, P. (2005, February). Hey, taxi! *Student Math Notes,* pp. 1–4.

Hughes-Hallet, D., Gleason, M., et al. (1994). *Calculus.* New York, NY: Wiley.

Interactive Mathematics Program. (2007). *Interactive Mathematics Program resource center.* Retrieved December 23, 2009, from http://www.mathimp.org/general_info/intro.html

Izen, S. P. (1999). Fearless learning. *Mathematics Teacher, 92,* 756–757.

Jackson, C. D., & Leffingwell, R. J. (1999). The role of instructors in creating math anxiety in students from kindergarten through college. *Mathematics Teacher, 92,* 583–586.

Jacobs, J. K., Hiebert, J., Givvin, K., Hollingsworth, H., Garnier, H., & Wearne, D. (2006). Does eighth-grade teaching in the United States align with the NCTM Standards? Results from the TIMSS 1995 and 1999 video studies. *Journal for Research in Mathematics Education, 37,* 5–32.

Jacobs, V. R. (1999). How do students think about statistical sampling before instruction? *Mathematics Teaching in the Middle School, 5,* 240–246, 263.

Jaime, A., & Gutiérrez, A. (1995). Guidelines for teaching plane isometries in secondary school. *Mathematics Teacher, 88,* 591–597.

Janvier, C., Girardon, C., & Morand, J. (1993). Mathematical symbols and representations. In P. S. Wilson (Ed.), *Research ideas for the classroom: High school mathematics* (pp. 79–102). Reston, VA: National Council of Teachers of Mathematics.

Jean Piaget Society. (2006). *Jean Piaget Society: Society for the study of knowledge and development.* Retrieved July 20, 2006, from http://www.piaget.org/

John Wiley & Sons. (2010). *About the Calculus Consortium based at Harvard University.* Retrieved May 8, 2010, from http://he-cda.wiley.com/WileyCDA/Section/id-100325.html

Johnson, D. R. (1982). *Every minute counts: Making your math class work.* Parsippany, NJ: Dale Seymour.

Johnson, D. R. (1994). *Motivation counts: Teaching techniques that work.* Parsippany, NJ: Dale Seymour.

Johnson, H. L. (2010). Investigating the fundamental theorem of calculus. *Mathematics Teacher, 103,* 430–435.

Jones, G. A., Langrall, C. W., & Mooney, E. S. (2007). Research in probability: Responding to classroom realities. In F. K. Lester, Jr. (Ed.), *Second handbook of research on mathematics teaching and learning* (pp. 909–955). Reston, VA: National Council of Teachers of Mathematics; Charlotte, NC: Information Age.

Jones, K. (2000). Providing a foundation for deductive reasoning: Students' interpretations when using dynamic geometry software and their evolving explanations. *Educational Studies in Mathematics, 44,* 55–85.

Kahneman, D., & Tversky, A. (1983). Extensional versus intuitive reasoning: The conjunction fallacy in probability judgment. *Psychological Review, 90*(4), 293–315.

Kaiser, G., Luna, E., & Huntley, I. (Eds.) (1999). *International comparisons in mathematics education.* London, England: Falmer Press.

Kastberg, S. E. (2003). Using Bloom's Taxonomy as a framework for classroom assessment. *Mathematics Teacher, 96,* 402–405.

Kieran, C. (1984). A comparison between novice and more-expert algebra students on tasks dealing with the equivalence of equations. In J. M. Moser (Ed.), *Proceedings of the sixth annual meeting of the North American Chapter of the International Group for the Psychology of Mathematics*

Education (pp. 83–91). Madison: University of Wisconsin.

Kieran, C. (1992). The learning and teaching of school algebra. In D. A. Grouws (Ed.), *Handbook of research on mathematics teaching and learning* (pp. 390–419). Reston, VA: National Council of Teachers of Mathematics.

Kieran, C., & Saldanha, L. (2005). Computer algebra systems as a tool for coaxing the emergence of reasoning about equivalence of algebraic expressions. In H. L. Chick & J. L. Vincent (Eds.), *Proceedings of the 29th PME International Conference, 3,* 193–200.

Kilpatrick, J. (2003). What works? In S. L. Senk & D. R. Thompson (Eds.), *Standards-based school mathematics curricula: What are they? What do students learn?* (pp. 471–488). Mahwah, NJ: Erlbaum.

Kilpatrick, J., Swafford, J., & Findell, B. (Eds.). (2001). *Adding it up: Helping children learn mathematics.* Washington, DC: National Academy Press.

Kim, O.-K., & Kasmer, L. (2006). Analysis of emphasis on reasoning in state mathematics curriculum standards. In B. J. Reys (Ed.), *The intended mathematics curriculum as represented in state-level curriculum standards: Consensus or confusion?* (pp. 89–110). Charlotte, NC: Information Age.

Kirshner, D., & Awtry, T. (2004). Visual salience of algebraic transformations. *Journal for Research in Mathematics Education, 35,* 224–257.

Kline, M. (1977). *Why the professor can't teach: Mathematics and the dilemma of university education.* New York, NY: St. Martin's Press.

Knuth, E. J. (2000). Student understanding of the Cartesian Connection: An exploratory study. *Journal for Research in Mathematics Education, 31,* 500–508.

Knuth, E. J. (2002). Fostering mathematical curiosity. *Mathematics Teacher, 95,* 126–130.

Knuth, E. J., Alibali, M. W., Hattikudur, S., McNeil, N. M., & Stephens, A. C. (2008). The importance of equal sign understanding in the middle grades. *Mathematics Teaching in the Middle School, 13,* 514–519.

Knuth, E. J., & Perressini, D. (2001). Unpacking the nature of discourse in mathematics classrooms. *Mathematics Teaching in the Middle School, 6,* 320–325.

Knuth, E. J., Stephens, A. C., McNeil, N. M., & Alibali, M. W. (2006). Does understanding the equal sign matter? Evidence from solving equations. *Journal for Research in Mathematics Education, 37,* 297–312.

Koedinger, K. R. (1998). Conjecturing and argumentation in high-school geometry students. In R. Lehrer & D. Chazan (Eds.), *Designing learning environments for developing understanding of geometry and space* (pp. 319–347). Mahwah, NJ: Erlbaum.

Kohn, A. (2007). *The homework myth.* New York, NY: De Capo Press.

Konold, C. (1995). Issues in assessing conceptual understanding in probability and statistics. *Journal of Statistics Education, 3*(1). Retrieved February 8, 2012, from http://www.amstat.org/publications/jse/v3n1/konold.html

Konold, C. (2002). Alternatives to scatterplots. In B. Phillips (Ed.), *Proceedings of the Sixth International Conference on Teaching Statistics.* Cape Town, South Africa. Retrieved July 12, 2009, from http://www.stat.auckland.ac.nz/~iase/publications/1/7f5_kono.pdf

Konold, C., & Pollatsek, A. (2002). Data analysis as the search for signals in noisy processes. *Journal for Research in Mathematics Education, 33,* 259–289.

Kounin, J. S. (1970). *Discipline and group management in classrooms.* New York, NY: Holt, Rinehart & Winston.

Kulm, G. (1999). Making sure that your mathematics curriculum meets standards. *Mathematics Teaching in the Middle School, 4,* 536–541.

Kunkel, P., Chanan, S., & Steketee, S. (2009). *Exploring Algebra 1 with the Geometer's Sketchpad.* Emeryville, CA: Key Curriculum Press.

Kutscher, B., & Linchevski, L. (1997). Number instantiations as mediators in solving word problems. In E. Pekonen (Ed.), *Proceedings of the 21st PME International Conference, 3,* 168–175.

Lach, T. M., & Sakshaug, L. E. (2004). The role of playing games in developing algebraic reasoning, spatial sense, and problem solving. *FOCUS on Learning Problems in Mathematics, 26,* 34–42.

Lakoff, G. (1987). *Women, fire, and dangerous things: What categories reveal about the mind.* Chicago, IL: University of Chicago Press.

Lambdin, D. V., & Preston, R. V. (1995). Caricatures in innovation: Teacher adaptation to an investigation-oriented middle school mathematics curriculum. *Journal of Teacher Education, 46,* 130–140.

Lamon, S. J. (2005). *Teaching fractions and ratios for understanding: Essential content knowledge and instructional strategies for teachers.* Mahwah, NJ: Erlbaum.

Lamon, S. J. (2007). Rational numbers and proportional reasoning. In F. K. Lester, Jr. (Ed.), *Second handbook of research on mathematics teaching and learning* (pp. 629–667). Reston, VA: National

Council of Teachers of Mathematics; Charlotte, NC: Information Age.

Langrall, C. W., & Swafford, J. (2000). Three balloons for two dollars: Developing proportional reasoning. *Mathematics Teaching in the Middle School, 6,* 254–261.

Lannin, J. (2004). Developing mathematical power by using explicit and recursive reasoning. *Mathematics Teacher, 98,* 216–223.

Lannin, J., Barker, D., & Townsend, B. (2006). Algebraic generalization strategies: Factors influencing student strategy selection. *Mathematics Education Research Journal, 18,* 3–28.

Lappan, G., Fey, J. T., Fitzgerald, W. M., Friel, S. N., & Phillips, E. D. (1998a). *Shapes and designs: Two-dimensional geometry.* Palo Alto, CA: Dale Seymour.

Lappan, G., Fey, J. T., Fitzgerald, W. M., Friel, S. N., & Philips, E. D. (1998b). *Variables and patterns: Introducing algebra.* Palo Alto, CA: Dale Seymour.

LeCoutre, M. P. (1992). Cognitive models and problem spaces in "purely random" situations. *Educational Studies in Mathematics, 23,* 557–568.

Leikin, R., & Zaslavsky, O. (1999). Cooperative learning in mathematics. *Mathematics Teacher, 92,* 240–246.

Leinhardt, G. (1989). Math lessons: A contrast of novice and expert competence. *Journal for Research in Mathematics Education, 20,* 52–75.

Leitze, A. R., & Kitt, N. A. (2000). Using homemade algebra tiles to develop algebra and prealgebra concepts. *Mathematics Teacher, 93,* 462–466, 520.

Leung, S. S., & Wu, R. (1999). Posing problems: Two classroom examples. *Mathematics Teaching in the Middle School, 5,* 112–117.

Linchevski, L., & Kutscher, B. (1998). Tell me with whom you're learning, and I'll tell you how much you've learned: Mixed-ability versus same-ability grouping in mathematics. *Journal for Research in Mathematics Education, 29,* 533–554.

Lloyd, G. M. (1999). Two teachers' conceptions of a reform-oriented curriculum: Implications for mathematics teacher development. *Journal of Mathematics Teacher Education, 2,* 227–252.

Lobato, J., Clarke, D., & Ellis, A. B. (2005). Initiating and eliciting in teaching: A reformulation of telling. *Journal for Research in Mathematics Education, 36,* 101–136.

Lobato, J., Ellis, A. B., & Muñoz, R. (2003). How "focusing phenomena" in the instructional environment support individual students' generalizations. *Mathematical Thinking and Learning, 5,* 1–36.

Lott, J. W., & Nishimura, K. (Eds.). (2005). *Standards & curriculum: A view from the nation.* Reston, VA: National Council of Teachers of Mathematics.

Lubienski, S. T. (2000). Problem solving as a means toward mathematics for all: An exploratory look through a class lens. *Journal for Research in Mathematics Education, 31,* 454–482.

Lubienski, S. T., & Crockett, M. D. (2007). NAEP findings regarding race and ethnicity: Mathematics achievement, student affect, and school-home experiences. In P. Kloosterman & F. K. Lester, Jr. (Eds.), *Results and interpretations of the 2003 mathematics assessment of the National Assessment of Educational Progress* (pp. 227–260). Reston, VA: National Council of Teachers of Mathematics.

Ma, L. (1999). *Knowing and teaching elementary mathematics.* Mahwah, NJ: Erlbaum.

MacGregor, M., & Stacey, K. (1993). Cognitive models underlying students' formulation of simple linear equations. *Journal for Research in Mathematics Education, 24,* 217–232.

Mack, N. K. (1990). Learning fractions with understanding: Building on informal knowledge. *Journal for Research in Mathematics Education, 21,* 16–32.

Malara, N. A., & Zan, R. (2002). The problematic relationship between theory and practice. In L. D. English (Ed.), *Handbook of international research in mathematics education* (pp. 553–580). Mahwah, NJ: Erlbaum.

Malisani, E., & Spagnolo, F. (2009). From arithmetical thought to algebraic thought: The role of the "variable." *Educational Studies in Mathematics, 71,* 19–41.

Mamona-Downs, J., & Downs, M. (2002). Advanced mathematical thinking with a special reference to reflection on mathematical structure. In L. D. English (Ed.), *Handbook of international research in mathematics education* (pp. 165–195). Mahwah, NJ: Erlbaum.

Manouchehri, A., & Lapp, D. A. (2003). Unveiling student understanding: The role of questioning in instruction. *Mathematics Teacher, 96,* 562–566.

Mariotti, M. A. (2001). Justifying and proving in the Cabri environment. *International Journal of Computers for Mathematical Learning, 6,* 257–281.

Marrades, R., & Gutiérrez, A. (2000). Proofs produced by secondary school students learning geometry in a dynamic computer environment. *Educational Studies in Mathematics, 44,* 87–125.

Martin, T. S., McCrone, S. S., Bower, M. L., & Dindyal, J. (2005). The interplay of teacher and student actions in the teaching and learning of geometric proof. *Educational Studies in Mathematics, 60,* 95–124.

Martino, A. M., & Maher, C. A. (1999). Teacher questioning to promote justification and generalization in mathematics: What research practice has taught us. *Journal of Mathematical Behavior, 18,* 53–78.

Maryland State Department of Education. (2000). *Mathematics brief constructed response rubric.* Retrieved May 24, 2009, from http://mdk12.0rg/share/rubrics/hsa/mathematics/pdf/HSA_Mathematics RubricV1.pdf

Mason, J., Burton, L., & Stacey, K. (1982). *Thinking mathematically.* London, England: Addison-Wesley.

Mathematically Correct. (n.d.). *Mathematics program reviews and information.* Retrieved December 26, 2009, from http://mathematicallycorrect.com/programs.htm

Matthews, M. E., Hlas, C. S., & Finken, T. M. (2009). Using lesson study and four-column lesson planning with preservice teachers. *Mathematics Teacher, 102,* 504–508.

McIntosh, A., Reys, B. J., & Reys, R. E. (1992). A proposed framework for examining basic number sense. *For the Learning of Mathematics, 12*(3), 2–8.

McIntosh, M. E., & Draper, R. J. (2001). Using learning logs in mathematics: Writing to learn. *Mathematics Teacher, 94,* 554–557.

McKnight, C. C., Crosswhite, F. J., Dossey, J. A., Kifer, E., Swafford, J. O., Travers, K. J., & Cooney, T. J. (1987). *The underachieving curriculum: Assessing U.S. school mathematics from an international perspective.* Champaign, IL: Stipes.

McLeod, D. B. (1989). Beliefs, attitudes, and emotions: New views of affect in mathematics education. In D. B. McLeod & V. M. Adams (Eds.), *Affect and mathematical problem solving* (pp. 245–258). New York, NY: Springer-Verlag.

McLeod, D. B. (1992). Research on affect in mathematics education: A reconceptualization. In D. B. Grouws (Ed.), *Handbook of research on mathematics teaching and learning* (pp. 575–596). Reston, VA: National Council of Teachers of Mathematics.

McLeod, D. B., & Ortega, M. (1993). Affective issues in mathematics education. In P. S. Wilson (Ed.), *Research ideas for the classroom: High school mathematics* (pp. 21–36). New York, NY: Macmillan.

McMurray, R. (1978). Flow proofs in geometry. *Mathematics Teacher, 71,* 592–595.

McShea, B., Vogel, J., & Yarnevich, M. (2005). Harry Potter and the magic of mathematics. *Mathematics Teaching in the Middle School, 10,* 408–414.

Meichenbaum, D., Burland, S., Gruson, L., & Cameron, R. (1985). Metacognitive assessment. In S. Yussen (Ed.), *The growth of reflection in children* (pp. 1–30). Orlando, FL: Academic Press.

Merseth, K. K. (2003). *Windows on teaching math: Cases of middle and secondary classrooms.* New York, NY: Teachers College Press.

Middleton, J. A., & Spanias, P. A. (2002). Findings from research on motivation in mathematics education: What matters most in coming to value mathematics. In J. Sowder & B. Schappelle (Eds.), *Lessons learned from research* (pp. 9–15). Reston, VA: National Council of Teachers of Mathematics.

Miller, L. D. (1992). Teacher benefits from using impromptu writing prompts in algebra classes. *Journal for Research in Mathematics Education, 23,* 329–340.

Miller, W. A. (1991). Recursion and the central polygonal numbers. *Mathematics Teacher, 84,* 738–746.

Mish, F. C. (Ed.). (1991). *Webster's ninth new collegiate dictionary.* Springfield, MA: Merriam-Webster.

Mitchelmore, M. C., & White, P. (2000). Development of angle concepts by progressive abstraction and generalization. *Educational Studies in Mathematics, 41,* 209–238.

Mokros, J., & Russell, S. J. (1995). Children's concepts of average and representativeness. *Journal for Research in Mathematics Education, 26,* 20–39.

Mooney, E. S. (2002). A framework for characterizing middle school students' statistical thinking. *Mathematical Thinking and Learning, 4,* 23–63.

Moses, R. P., & Cobb, C. E., Jr. (2001). *Radical equations: Math literacy and civil rights.* Boston, MA: Beacon Press.

Mower, P. (2003). *Algebra out loud: Learning mathematics through reading and writing activities.* San Francisco, CA: Jossey-Bass.

Moyer, P. S. (2001). Are we having fun yet? How teachers use manipulatives to teach mathematics. *Educational Studies in Mathematics, 47,* 175–197.

Myers, D. A. (2008). The teacher as a service professional. *Action in Teacher Education, 30,* 4–11.

National Center for Education Statistics. (2010). *Trends in International Mathematics and Science Study.* Retrieved June 19, 2010, from http://nces.ed.gov/timss/

National Council of Teachers of Mathematics. (1989). *Curriculum and evaluation standards for school mathematics.* Reston, VA: Author.

National Council of Teachers of Mathematics. (1991). *Professional standards for teaching mathematics.* Reston, VA: Author.

National Council of Teachers of Mathematics. (1995). *Assessment standards for school mathematics.* Reston, VA: Author.

National Council of Teachers of Mathematics. (2000). *Principles and standards for school mathematics.* Reston, VA: Author.

National Council of Teachers of Mathematics. (2006). *Curriculum focal points for prekindergarten through Grade 8 mathematics.* Reston, VA: Author.

National Council of Teachers of Mathematics. (2007). *Mathematics teaching today: Improving practice, improving student learning* (2nd ed.). Reston, VA: Author.

National Council of Teachers of Mathematics. (2009). *Focus in high school mathematics: Reasoning and sense-making.* Reston, VA: Author.

National Governor's Association for Best Practices & Council of Chief State School Officers. (2010). *Common core state standards for mathematics.* Retrieved February 9, 2012, from http://www.corestandards.org/

National Mathematics Advisory Panel. (2008). *Foundations for success: The final report of the National Mathematics Advisory Panel.* Washington, DC: U.S. Department of Education.

Nickerson, S. D. (2002). What provides support for students' understanding of systems of linear equations? In D. S. Mewborn et al. (Eds.), *Proceedings of the 24th PME-NA Conference, 1,* 473–482.

No Child Left Behind Act of 2001, Pub. L. No. 107–110, 115 Stat. 1425 (2002).

Norwood, K. S. (1994). The effect of instructional approach on mathematics anxiety and achievement. *School Science and Mathematics, 94,* 248–254.

Noss, R., Pozzi, S., & Hoyles, C. (1999). Touching epistemologies: Meanings of average and variation in nursing practice. *Educational Studies in Mathematics, 40,* 25–51.

Novick, L. R. (2004). Diagram literacy in preservice math teachers, computer science majors, and typical undergraduates: The case of matrices, networks, and hierarchies. *Mathematical Thinking and Learning, 6,* 307–342.

Nurnberger-Haag, J. (2003). Order of op hop. *Mathematics Teaching in the Middle School, 8,* 234–236.

Oehrtman, M. (2008). Layers of abstraction: Theory and design for the instruction of limit concepts. In M. P. Carlson & C. Rasmussen (Eds.), *MAA notes: Vol. 73. Making the connection: Research and teaching in undergraduate mathematics*

(pp. 65–80). Washington, DC: Mathematical Association of America.

Oehrtman, M. (2009). Collapsing dimensions, physical limitation, and other student metaphors for limit concepts. *Journal for Research in Mathematics Education, 40,* 396–426.

Orey, D. (2005). *The algorithm collection project.* Retrieved November 7, 2009, from http://www.csus.edu/indiv/o/oreyd/ACP.htm_files/Alg.html

Ormrod, J. E. (2006). *Essentials of educational psychology.* Upper Saddle River, NJ: Pearson.

Orton, A. (1983a). Students' understanding of differentiation. *Educational Studies in Mathematics, 14,* 235–250.

Orton, A. (1983b). Students' understanding of integration. *Educational Studies in Mathematics, 14,* 1–18.

Panaoura, A., Elia, I., Gagatsis, A., & Giatilis, G.-P. (2006). Geometric and algebraic approaches in the concept of complex numbers. *International Journal of Mathematical Education in Science and Technology, 37,* 681–706.

Pape, S. J. (2004). Middle school children's problem-solving behavior: A cognitive analysis from a reading comprehension perspective. *Journal for Research in Mathematics Education, 35,* 187–219.

Peck, R., & Starnes, D. (2009). *Making sense of statistical studies* (Teacher's module). Alexandria, VA: American Statistical Association.

Perkowski, D. A., & Perkowski, M. (2007). *Data and probability connections: Mathematics for middle school teachers.* Upper Saddle River, NJ: Pearson.

Perlwitz, M. D. (2005). Dividing fractions: Reconciling self-generated solutions with algorithmic answers. *Mathematics Teaching in the Middle School, 10,* 278–283.

Pesek, D. D., & Kirshner, D. (2000). Interference of instrumental instruction in subsequent relational learning. *Journal for Research in Mathematics Education, 31,* 524–540.

Pfannkuch, M. (2005). Probability and statistical inference: How can teachers enable learners to make the connection? In G. A. Jones (Ed.), *Exploring probability in school: Challenges for teaching and learning* (pp. 267–294). New York, NY: Springer.

Pfannkuch, M., & Wild, C. (2004). Towards an understanding of statistical thinking. In D. Ben-Zvi & J. Garfield (Eds.), *The challenge of developing statistical literacy, reasoning, and thinking* (pp. 17–46). Dordrecht, The Netherlands: Kluwer.

Philipp, R. A. (1996). Multicultural mathematics and alternative algorithms. *Teaching Children Mathematics, 3,* 128–133.

Piaget, J. (1983). Piaget's theory. In P. Mussen (Ed.), *Handbook of child psychology* (pp. 103–128). New York, NY: Wiley.

Polya, G. (1945). *How to solve it.* Princeton, NJ: Princeton University Press.

Pollatsek, A., Lima, S., & Well, A. D. (1981). Concept or computation: Students' understanding of the mean. *Educational Studies in Mathematics, 12,* 191–204.

Presmeg, N. C. (1992). Prototypes, metaphors, metonymies, and imaginative rationality in high school mathematics. *Educational Studies in Mathematics, 23,* 595–610.

Presmeg, N.C. (1998). Metaphoric and metonymic signification in mathematics. *Journal of Mathematical Behavior, 17,* 25–32.

Putnam, R. T. (2003). In S. L. Senk & D. R. Thompson (Eds.), *Standards-based school mathematics curricula: What are they? What do students learn?* (pp. 161–180). Mahwah, NJ: Erlbaum.

Rambhia, S. (2002). A new approach to an old order. *Mathematics Teaching in the Middle School, 8,* 193–195.

Randolph, T. D., & Sherman, H. J. (2001). Alternative algorithms: Increasing options, reducing errors. *Teaching Children Mathematics, 7,* 480–484.

Rauff, J. V. (1994). Constructivism, factoring, and beliefs. *School Science and Mathematics, 94,* 421–426.

Reys, B. J. (Ed.). (2006). *The intended mathematics curriculum as represented in state-level curriculum standards: Consensus or confusion?* Charlotte, NC: Information Age.

Reys, B. J., Dingman, S., Olson, T., Sutter, A., Teuscher, D., & Chval, K. (2006). Analysis of K–8 number and operation grade-level learning expectations. In B. J. Reys (Ed.), *The intended mathematics curriculum as represented in state-level curriculum standards: Consensus or confusion?* (pp. 15–57). Charlotte, NC: Information Age.

Reys, R. E., & Dossey, J. A. (Eds.). (2008). *U.S. doctorates in mathematics education: Developing stewards of the discipline.* Washington, DC: American Mathematical Society.

Reys, R. E., Reys, B. J., Nohda, N., & Emori, H. (1995). Mental computation performance and strategy use of Japanese students in Grades 2, 4, 6, and 8. *Journal for Research in Mathematics Education, 26,* 304–326.

Reys, R. E., Rybolt, J. F., Bestgen, B. J., & Wyatt, J. W. (1982). Processes used by good computational estimators. *Journal for Research in Mathematics Education, 13,* 183–201.

Reys, R. E., & Yang, D.-C. (1998). Relationship between computational performance and number sense among sixth- and eighth-grade students in Taiwan. *Journal for Research in Mathematics Education, 29,* 225–237.

Rohrer, D. (2009). The effects of spacing and mixing practice problems. *Journal for Research in Mathematics Education, 40,* 4–17.

Romberg, T. A. (2001). *Designing middle-school mathematics materials using problems set in context to help students progress from informal to formal mathematical reasoning.* Madison, WI: Wisconsin Center for Education Research.

Roschelle, J. (2003). Unlocking the learning value of wireless mobile devices. *Journal of Computer Assisted Learning, 19,* 260–272.

Rosnick, P., & Clement, J. (1980). Learning without understanding: The effect of tutoring strategies on algebra misconceptions. *Journal of Mathematical Behavior, 3,* 3–27.

Rossman, A., Chance, B., & Medina, E. (2006). Some important comparisons between statistics and mathematics, and why teachers should care. In G. F. Burrill (Ed.), *Thinking and reasoning with data and chance* (Sixty-eighth yearbook of the National Council of Teachers of Mathematics, pp. 323–333). Reston, VA: National Council of Teachers of Mathematics.

Roth, W.-M., & McGinn, M. (1997). Graphing: Cognitive ability or practice? *Science Education, 81,* 91–106.

Rowe, M. B. (1986). Wait time: Slowing down may be a way of speeding up! *Journal of Teacher Education, 37,* 43–50.

Rowe, M. B. (1987). Using wait time to stimulate inquiry. In W. W. Willen (Ed.), *Questions, questioning techniques, and effective teaching* (pp. 95–106). Washington, DC: National Education Association.

Rubenstein, R. N. (1998). Historical algorithms: Sources for student projects. In L. J. Morrow (Ed.), *The teaching and learning of algorithms in school mathematics* (pp. 99–105). Reston, VA: National Council of Teachers of Mathematics.

Rubenstein, R. N., & Schwartz, R. K. (2000). Word histories: Melding mathematics and meanings. *Mathematics Teacher, 93,* 664–669.

Rubin, A., Bruce, B., & Tenney, Y. (1991). Learning about sampling: Trouble at the core of statistics.

In D. Vere-Jones (Ed.), *Proceedings of the Third International Conference on Teaching Statistics* (Vol. 1, pp. 314–319). Voorburg, The Netherlands: International Statistical Institute.

Saldanha, L., & Thompson, P. (2002). Conceptions of sample and their relationship to statistical inference. *Educational Studies in Mathematics, 51,* 257–270.

Saldanha, L., & Thompson, P. (2007). Exploring connections between sampling distributions and statistical inference: An analysis of students' engagement and thinking in the context of instruction involving repeated sampling. *International Electronic Journal of Mathematics Education, 2,* 270–297. Retrieved May 26, 2010, from http://www.iejme.com/032007/d9.pdf

Saxon Publishers. (2009). *Saxon math for Grades 9–12.* Retrieved December 26, 2009, from http://saxonpublishers.hmhco.com/en/sxnm_secondary.htm

Schoenfeld, A. H. (1983). Episodes and executive decisions in mathematical problem solving. In R. Lesh & M. Landau (Eds.), *Acquisition of mathematical concepts and processes* (pp. 345–395). New York, NY: Academic Press.

Schoenfeld, A. H. (1988). When good teaching leads to bad results: The disasters of "well-taught" mathematics courses. *Educational Psychologist, 23*(2), 145–166.

Schoenfeld, A. H. (1992). Learning to think mathematically: Problem solving, metacognition, and sense making in mathematics. In D. A. Grouws (Ed.), *Handbook of research on mathematics teaching and learning* (pp. 334–370). Reston, VA: National Council of Teachers of Mathematics.

Schoenfeld, A. H. (2004). The math wars. *Educational Policy, 18,* 253–286.

Schoenfeld, A. H. (2006). What doesn't work: The challenge and failure of the What Works Clearinghouse to conduct meaningful reviews of studies of mathematics curricula. *Educational Researcher, 35*(2), 13–21.

Schoenfeld, A. H., & Arcavi, A. (1988). On the meaning of variable. *Mathematics Teacher, 81,* 420–427.

Schroeder, T. L., & Lester, F. K. (1989). Developing understanding in mathematics via problem solving. In P. R. Trafton (Ed.), *New directions for elementary school mathematics* (1989 yearbook of the National Council of Teachers of Mathematics, pp. 31–42). Reston, VA: National Council of Teachers of Mathematics.

Schwartz, D. L., Goldman, S. R., Vye, N. J., & Barron, B. J. (1998). Aligning everyday and mathematical reasoning: The case of sampling assumptions. In S. P. Lajoie (Ed.), *Reflections on statistics: Learning, teaching, and assessment in Grades K–12* (pp. 233–273). Mahwah, NJ: Erlbaum.

Schwarz, B. B., & Hershkowitz, R. (1999). Prototypes: Brakes or levers in learning the function concept? The role of computer tools. *Journal for Research in Mathematics Education, 30,* 362–389.

Seidel, C., & McNamee, J. (2005). Phenomenally exciting joint mathematics–English vocabulary project. *Mathematics Teaching in the Middle School, 10,* 461–463.

Senechal, M. (1990). Shape. In L. A. Steen (Ed.), *On the shoulders of giants: New approaches to numeracy* (pp. 139–182). Washington, DC: National Academies Press.

Senk, S. L. (1985). How well do students write geometry proofs? *Mathematics Teacher, 78,* 448–456.

Senk, S. L., & Thompson, D. R. (Eds.). (2003). *Standards-based school mathematics curricula: What are they? What do students learn?* Mahwah, NJ: Erlbaum.

Sfard, A., & Linchevski, L. (1994). The gains and pitfalls of reification—The case of algebra. *Educational Studies in Mathematics, 26,* 191–228.

Shaughnessy, J. M. (2003). Research on students' understandings of probability. In J. Kilpatrick, W. G. Martin, & D. Schifter (Eds.), *A research companion to Principles and Standards for School Mathematics* (pp. 216–226). Reston, VA: National Council of Teachers of Mathematics.

Shaughnessy, J. M. (2007). Research on statistics learning and reasoning. In F. K. Lester (Ed.), *Second handbook of research on mathematics teaching and learning* (pp. 957–1009). Reston, VA: National Council of Teachers of Mathematics; Charlotte, NC: Information Age.

Shaughnessy, J. M., & Bergman, B. (1993). Thinking about uncertainty: Probability and statistics. In P. S. Wilson (Ed.), *Research ideas for the classroom: High school mathematics* (pp. 177–197). Reston, VA: National Council of Teachers of Mathematics.

Shaughnessy, J. M., & Chance, B. (2005). *Statistical questions from the classroom.* Reston, VA: National Council of Teachers of Mathematics.

Shaughnessy, J. M., Ciancetta, M., & Canada, D. (2004). Types of student reasoning on sampling tasks. In M. J. Hoines & A. B. Fuglestead (Eds.),

Proceedings of the 28th meeting of the International Group for the Psychology of Mathematics Education (Vol. 4, pp. 177–184). Bergen, Norway: Bergen University College Press.

Shaughnessy, J. M., & Pfannkuch, M. (2002). How faithful is Old Faithful? Statistical thinking: A story of variation and prediction. *Mathematics Teacher, 95,* 252–259.

Sherin, M. G., & Van Es, E. A. (2003). A new lens on teaching: Learning to notice. *Mathematics Teaching in the Middle School, 9,* 92–95.

Shockey, T. L., & Snyder, K. (2007). Engaging preservice teachers and elementary-age children in transformational geometry: Tessellating T-shirts. *Teaching Children Mathematics, 14,* 82–87.

Silver, E. A. (1990). Contributions of research to practice: Applying findings, methods, and perspectives. In T. J. Cooney (Ed.), *Teaching and learning mathematics in the 1990's* (pp. 1–11). Reston, VA: National Council of Teachers of Mathematics.

Simon, M. A. (1995). Reconstructing mathematics pedagogy from a constructivist perspective. *Journal for Research in Mathematics Education, 26,* 114–145.

Sirotic, N., & Zazkis, R. (2007). Irrational numbers: The gap between formal and intuitive knowledge. *Educational Studies in Mathematics, 65,* 49–76.

Sjoberg, C. A., Slavit, D., & Coon, T. (2004). Improving writing prompts to improve student reflection. *Mathematics Teaching in the Middle School, 9,* 490–493.

Skemp, R. R. (1976). Relational understanding and instrumental understanding. *Mathematics Teaching, 77,* 20–26.

Slavin, R. E. (1984). Students motivating students to excel: Cooperative incentives, cooperative tasks, and student achievement. *The Elementary School Journal, 85,* 53–63.

Slavit, D. (1998). Three women's understandings of algebra in a precalculus course integrated with the graphing calculator. *Journal of Mathematical Behavior, 17,* 355–372.

Sliva, J. A. (2003). *Teaching inclusive mathematics to special learners, K–6.* Thousand Oaks, CA: Sage.

Small, M. (2009). *Good questions: Great ways to differentiate mathematics instruction.* New York, NY: Teachers College Press.

Smith, J. P., III. (1996). Efficacy and teaching mathematics by telling: A challenge for reform. *Journal for Research in Mathematics Education, 27,* 387–402.

Smith, M. H. (2004). A sample/population size activity: Is it the sample size or the sample as a fraction of the population that matters? *Journal of Statistics Education, 12*(2). Retrieved May 24, 2010, from http://www.amstat.org/publications/jse/v12n2/smith.html

Smith, M. S., & Stein, M. K. (1998). Selecting and creating mathematical tasks: From research to practice. *Mathematics Teaching in the Middle School, 3,* 344–350.

Sowder, J. T. (1992). Estimation and number sense. In D. A. Grouws (Ed.), *Handbook of research on mathematics teaching and learning* (pp. 371–389). Reston, VA: National Council of Teachers of Mathematics.

Sowder, J. T. (2000). Editorial. *Journal for Research in Mathematics Education, 31,* 1–4.

Sriraman, B., & English, L. D. (2004). Combinatorial mathematics: Research into practice. *Mathematics Teacher, 98,* 182–191.

Stamper, A. W. (1906). *A history of the teaching of elementary geometry with reference to present day problems.* Doctoral dissertation, Teachers College, Columbia University.

Steen, L. A. (Ed.). (1988). *Calculus for a new century: A pump, not a filter* (MAA Notes, No. 8). Washington, DC: Mathematical Association of America.

Steen, L. A. (1999). Theories that gyre and gimble in the wabe. *Journal for Research in Mathematics Education, 30,* 235–241.

Stein, M. K., & Bovalino, J. W. (2001). Manipulatives: One piece of the puzzle. *Mathematics Teaching in the Middle School, 6,* 356–359.

Stein, M. K., Grover, B. W., & Henningsen, M. (1996). Building student capacity for mathematical thinking and reasoning: An analysis of mathematical tasks used in reform classrooms. *American Educational Research Journal, 33,* 455–488.

Stein, M. K., Remillard, J., & Smith, M. S. (2007). How curriculum influences student learning. In F. K. Lester, Jr. (Ed.), *Second handbook of research on mathematics teaching and learning* (pp. 319–369). Reston, VA: National Council of Teachers of Mathematics; Charlotte, NC: Information Age.

Stevens, B. A. (2001). My involvement in change. *Mathematics Teaching in the Middle School, 7,* 178–182.

Stigler, J. W., & Hiebert, J. (1999). *The teaching gap.* New York, NY: Free Press.

Stigler, J. W., & Hiebert, J. (2004). Improving mathematics teaching. *Educational Leadership, 61*(5), 12–16.

Stigler, S. M. (1986). *The history of statistics: The measurement of uncertainty before 1900.* Cambridge, MA: Harvard University Press.

Strauss, S., & Bichler, E. (1988). The development of children's concepts of the arithmetic average. *Journal for Research in Mathematics Education, 19,* 64–80.

Strogatz, S. H. (1994). *Nonlinear dynamics and chaos with applications to physics, biology, chemistry, and engineering.* Reading, MA: Addison-Wesley.

Struchens, M. E., Martin, W. G., & Kenney, P. A. (2003). What students know about measurement: Perspectives from the NAEP. In D. H. Clements (Ed.), *Learning and teaching measurement* (2003 yearbook of the National Council of Teachers of Mathematics, pp. 197–207). Reston, VA: National Council of Teachers of Mathematics.

Stump, S. L. (2001a). Developing preservice teachers' pedagogical content knowledge of slope. *Journal of Mathematical Behavior, 20,* 207–227.

Stump, S. L. (2001b). High school precalculus students' understanding of slope as a measure. *School Science and Mathematics, 101*(2), 81–89.

Swafford, J. (2003). Reaction to high school curriculum projects research. In S. L. Senk & D. R. Thompson (Eds.), *Standards-based school mathematics curricula: What are they? What do students learn?* (pp. 457–470). Mahwah, NJ: Erlbaum.

Swafford, J. O., & Langrall, C. W. (2000). Grade 6 students' preinstructional use of equations to describe and represent problem situations. *Journal for Research in Mathematics Education, 31,* 89–112.

Swokowski, E. W. (1991). *Calculus* (5th ed.). Boston, MA: PWS-Kent.

Szydlik, J. E. (2000). Mathematical beliefs and conceptual understanding of the limit of a function. *Journal for Research in Mathematics Education, 31,* 258–276.

Tabach, M., & Friedlander, A. (2008). The role of context in learning beginning algebra. In C. E. Greenes (Ed.), *Algebra and algebraic thinking in school mathematics* (Seventieth yearbook of the National Council of Teachers of Mathematics, pp. 223–232). Reston, VA: National Council of Teachers of Mathematics.

Taber, S. B. (2005). The mathematics of *Alice's Adventures in Wonderland. Mathematics Teaching in the Middle School, 11,* 165–171.

Takahashi, A., & Yoshida, M. (2004). Ideas for establishing lesson-study communities. *Teaching Children Mathematics, 10,* 436–443.

Tall, D., & Vinner, S. (1981). Concept image and concept definition in mathematics with particular reference to limits and continuity. *Educational Studies in Mathematics, 12,* 151–169.

Tarr, J., Reys, B. J., Barker, D. B., & Billstein, R. (2006). Selecting high-quality mathematics textbooks. *Mathematics Teaching in the Middle School, 12,* 50–54.

Tarr, J. E., Reys, R. E., Reys, B. J., Chavez, O., Shih, J., & Osterlind, S. J. (2008). The impact of middle-grades mathematics curricula and the classroom learning environment on student achievement. *Journal for Research in Mathematics Education, 39,* 247–280.

Tarr, J. E., & Shaughnessy, J. M. (2007). Student performance in data analysis, statistics, and probability. In P. Kloosterman & F. K. Lester, Jr. (Eds.), *Results and interpretations of the 2003 mathematics assessment of the National Assessment of Educational Progress* (pp. 139–168). Reston, VA: National Council of Teachers of Mathematics.

Taylor-Cox, J., & Oberdorf, C. (2006). *Family math night: Middle school math standards in action.* Larchmont, NY: Eye on Education.

Thompson, A. D., & Sproule, S. L. (2000). Deciding when to use calculators. *Mathematics Teaching in the Middle School, 6,* 126–129.

Thompson, D. R., & Rubenstein, R. N. (2000). Learning mathematics vocabulary: Potential pitfalls and instructional strategies. *Mathematics Teacher, 93,* 568–574.

Thompson, P. W. (1994). Images of rate and operational understanding of the fundamental theorem of calculus. *Educational Studies in Mathematics, 26,* 229–274.

Thompson, P. W. (2008). Conceptual analysis of mathematical ideas: Some spadework at the foundations of mathematics education. In O. Figueras, J. L. Cortina, S. Alatorre, T. Rojano, & A. Sépulveda (Eds.), *Proceedings of the 32nd annual meeting of the International Group for the Psychology of Mathematics Education* (Vol. 1, pp. 45–64). Morélia, Mexico: PME.

Tirosh, D., Even, R., & Robinson, N. (1998). Simplifying algebraic expressions: Teacher awareness and teaching approaches. *Educational Studies in Mathematics, 35,* 51–64.

Tobias, S. (1993). *Overcoming math anxiety.* New York, NY: W. W. Norton.

Tobias, S., & Weissbrod, C. (1980). Anxiety and mathematics: An update. *Harvard Educational Review, 50,* 63–70.

Tobin, K. (1986). Effects of teacher wait time on discourse characteristics in mathematics and language arts classes. *American Education Research Journal, 23,* 191–200.

Tomlinson, C. A. (2005). *The differentiated classroom: Responding to the needs of all learners.* Upper Saddle River, NJ: Pearson.

Tsamir, P., Almog, N., & Tirosh, D. (1998). Students' solutions to inequalities. In A. Oliver & K. Newstead (Eds.), *Proceedings of the 22nd PME International Conference, 4,* 129–136.

Tsamir, P., & Bazzini, L. (2002). Algorithmic models: Italian and Israeli students' solutions to algebraic inequalities. In A. D. Cockburn & E. Nardi (Eds.), *Proceedings of the 26th PME International Conference, 4,* 289–296.

Tversky, A., & Kahneman, D. (1974). Judgment under uncertainty: Heuristics and biases. *Science, 185,* 1124–1131.

U.S. Department of Education. (1999). *Exemplary and promising mathematics programs.* Washington, DC: Author.

U.S. Department of Education, Institute of Education Sciences. (2009). *What Works Clearinghouse.* Retrieved December 26, 2009, from http://ies.ed.gov/ncee/wwc/

Usiskin, Z. (1980). What should not be in the algebra and geometry curricula of average college-bound students? *Mathematics Teacher, 73,* 413–424.

Usiskin, Z. (1982). *Van Hiele levels and achievement in secondary school geometry.* Retrieved from ERIC database. (ED220288)

Usiskin, Z. (1988). Conceptions of school algebra and uses of variables. In A. F. Coxford & A. P. Schulte (Eds.), *The ideas of algebra* (1988 yearbook of the National Council of Teachers of Mathematics, pp. 8–19). Reston, VA: National Council of Teachers of Mathematics.

Usiskin, Z. (1998). Paper-and-pencil algorithms in a calculator-and-computer age. In L. J. Morrow (Ed.), *The teaching and learning of algorithms in school mathematics* (pp. 7–20). Reston, VA: National Council of Teachers of Mathematics.

Utley, J., & Wolfe, J. (2004). Geoboard areas: Students' remarkable ideas. *Mathematics Teacher, 97,* 18–26.

Van Boening, L. (1999). Growth through change. *Mathematics Teaching in the Middle School, 5,* 27–33.

Van de Walle, J. A. (2001). *Elementary and middle school mathematics: Teaching developmentally* (4th ed.). New York, NY: Addison Wesley Longman.

Van Dooren, W., Verschaffel, L., & Onghena, P. (2002). The impact of preservice teachers' content knowledge on their evaluation of students' strategies for solving arithmetic and algebra word problems. *Journal for Research in Mathematics Education, 33,* 319–351.

Van Dooren, W., Verschaffel, L., & Onghena, P. (2003). Pre-service teachers' preferred strategies for solving arithmetic and algebra word problems. *Journal of Mathematics Teacher Education, 6,* 27–52.

van Hiele, P. M. (1986). *Structure and insight: A theory of mathematics education.* New York, NY: Academic Press.

Veenman, S. (1984). Perceived problems of beginning teachers. *Review of Educational Research, 54,* 143–178.

Veljan, D. (2000). The 2500-year-old Pythagorean theorem. *Mathematics Magazine, 73,* 259–272.

Vinner, S., & Dreyfus, T. (1989). Images and definitions for the concept of function. *Journal for Research in Mathematics Education, 20,* 356–366.

Vlassis, J. (2002). The balance model: Hindrance or support for the solving of linear equations with one unknown. *Educational Studies in Mathematics, 49,* 341–359.

Vollmer, H. M., & Mills, D. L. (Eds.). (1966). *Professionalization.* Englewood Cliffs, NJ: Prentice Hall.

von Glasersfeld, E. (1990). An exposition of constructivism: Why some like it radical. In R. B. Davis, C. A. Maher, & N. Noddings (Eds.), *Constructivist views on the teaching and learning of mathematics* (pp. 19–29). Reston, VA: National Council of Teachers of Mathematics.

Vygotsky Centennial Project. (2006). Lev Semenovich Vygotsky. Retrieved July 20, 2006, from http://www.massey.ac.nz/~alock/virtual/project2.htm

Wagner, S. (1983). What are these things called variables? *Mathematics Teacher, 76,* 474–479.

Walmsley, A. L. E., & Hickman, A. (2006). A study of note-taking and its impact on student perception of use in a geometry classroom. *Mathematics Teacher, 99,* 614–621.

Wang, D. B. (2004). Family background factors and mathematics success: A comparison of Chinese and U.S. students. *International Journal of Educational Research, 41,* 40–54.

Watson, A. (2002). Instances of mathematical thinking among low attaining students in an ordinary secondary classroom. *Journal of Mathematical Behavior, 20,* 461–475.

Watson, A., Spirou, P., & Tall, D. (2003). The relationship between physical embodiment and mathematical symbolism: The concept of vector. *The Mediterranean Journal of Mathematics Education, 1*(2), 73–97.

Watson, A., & Tall, D. (2002). Embodied action, effect and symbol in mathematical growth. In A. Cockburn & E. Nardi (Eds.), *Proceedings of the 26th conference of the International Group for the Psychology of Mathematics Education* (Vol. 4, pp. 369–376). Norwich, England: University of East Anglia.

Watson, J. M. (2000). Statistics in context. *The Mathematics Teacher, 93,* 54–58.

Watson, J. M., & Moritz, J. B. (1999). The beginning of statistical inference: Comparing two data sets. *Educational Studies in Mathematics, 37,* 145–168.

Watson, J. M., & Moritz, J. B. (2000). Developing concepts of sampling. *Journal for Research in Mathematics Education, 31,* 44–70.

Watson, J. M., & Moritz, J. B. (2002). School students' reasoning about conjunction and conditional events. *International Journal of Mathematical Education in Science and Technology, 33,* 59–84.

Wearne, D., & Hiebert J. (1988). A cognitive approach to meaningful mathematics instruction: Testing a local theory using decimal numbers. *Journal for Research in Mathematics Education, 19,* 371–384.

Weber, K. (2005). Students' understanding of trigonometric functions. *Mathematics Education Research Journal, 17,* 91–112.

Weber, K. (2008). Teaching trigonometric functions: Lessons learned from research. *Mathematics Teacher, 102,* 144–150.

Weiss, M., Herbst, P., & Chen, C. (2009). Teachers' perspectives on "authentic mathematics" and the two-column proof form. *Educational Studies in Mathematics, 70,* 275–293.

Wiest, L. R. (2000). Mathematics that whets the appetite: Student-posed project problems. *Mathematics Teaching in the Middle School, 5,* 286–291.

Wiggins, G., & McTighe, J. (2005). *Understanding by design* (2nd ed.). Upper Saddle River, NJ: Pearson.

Wild, C. J., & Pfannkuch, M. (1999). Statistical thinking in empirical enquiry. *International Statistical Review, 67,* 223–265.

Williams, K. M. (2003). Writing about the problem solving process to improve problem solving performance. *Mathematics Teacher, 96,* 185–187.

Williams, N. B., & Wynne, B. D. (2000). Journal writing in the mathematics classroom: A beginner's approach. *Mathematics Teacher, 93,* 132–135.

Williams, S. R. (1991). Models of limit held by college calculus students. *Journal for Research in Mathematics Education, 22,* 219–236.

Wood, T. (1998). Alternative patterns of communication in mathematics classes: Funneling or focusing? In H. Steinbring, M. G. Bartolini-Bussi, & A. Sierpinska (Eds.), *Language and communication in the mathematics classroom* (pp. 167–178). Reston, VA: National Council of Teachers of Mathematics.

Woolfolk, A. E. (1993). *Educational psychology* (5th ed.). Boston, MA: Allyn & Bacon.

Worrall, L. J., & Quinn, R. J. (2001). Promoting conceptual understanding of matrices. *Mathematics Teacher, 94,* 46–49.

Wu, H. (1999a). Basic skills vs. conceptual understanding: A bogus dichotomy in mathematics education. *American Educator, 23*(3), 14–19, 50–52.

Wu, H. (1999b). The joy of lecturing—with a critique of the romantic tradition in education writing. In S. G. Krantz (Ed.), *How to teach mathematics* (2nd ed., pp. 261–271). Providence, RI: American Mathematical Society.

Yerushalmy, M. (2006). Slower algebra students meet faster tools: Solving algebra word problems with graphing software. *Journal for Research in Mathematics Education, 37,* 356–387.

Yu, P., Barrett, J., & Presmeg, N. (2009). Prototypes and categorical reasoning: A perspective to explain how children learn about interactive geometry. In T. V. Craine (Ed.), *Understanding geometry for a changing world* (pp. 109–125). Reston, VA: National Council of Teachers of Mathematics.

Yun, J. O., & Flores, A. (2008). The Pythagorean theorem with jelly beans. *Mathematics Teaching in the Middle School, 14,* 202–207.

Zawojewski, J. S., & Shaughnessy, J. M. (2000). Data and chance. In E. A. Silver & P. A. Kenney (Eds.), *Results from the Seventh Mathematics Assessment of the National Assessment of Educational Progress* (pp. 235–268). Reston, VA: National Council of Teachers of Mathematics.

Zazkis, R. (1998). Odds and ends of odds and evens: An inquiry into students' understanding of odd and even numbers. *Educational Studies in Mathematics, 36,* 73–89.

Zazkis, R., (1999). Intuitive rules in number theory: Example of "the more of A, the more of B" rule implementation. *Educational Studies in Mathematics, 40,* 197–209.

Zazkis, R. (2005). Representing numbers: Prime and irrational. *International Journal of Mathematical Education in Science and Technology, 36,* 207–218.

Zazkis, R., & Campbell, S. (1996a). Divisibility and multiplicative structure of natural numbers: Preservice teachers' understanding. *Journal for Research in Mathematics Education, 27,* 540–563.

Zazkis, R., & Campbell, S. (1996b). Prime decomposition: Understanding uniqueness. *Journal of Mathematical Behavior, 15,* 207–218.

Zazkis, R., & Liljedahl, P. (2002). Arithmetic sequence as a bridge between conceptual fields. *Canadian Journal of Science, Mathematics, and Technology Education, 2*(1), 93–120.

Zazkis, R., & Liljedahl, P. (2004). Understanding primes: The role of representation. *Journal for Research in Mathematics Education, 35,* 164–186.

Zimmerman, G., & Jones, G. A. (2002). Probability simulation: What meaning does it have for high school students? *Canadian Journal of Mathematics, Science, and Technology Education, 2,* 221–236.

Zodik, I., & Zaslavsky, O. (2008). Characteristics of teachers' choice of examples in and for the mathematics classroom. *Educational Studies in Mathematics, 69,* 165–182.

Zumwalt, K., & Craig, E. (2005). Teachers' characteristics: Research on the demographic profile. In M. Cochran-Smith & K. M. Zeichner (Eds.), *Studying teacher education: The report of the AERA Panel on Research and Teacher Education* (pp. 111–156). Mahwah, NJ: Erlbaum.

GLOSSARY

Additive change. The total raw amount of change in a quantity.

Advanced mathematical thinking. Engaging in thinking processes involved in producing mathematics, and not just in studying the theoretical products that have been handed down by mathematicians (Mamona-Downs & Downs, 2002). Fostering AMT can be conceptualized as inducting students into the same sort of thinking processes that mathematicians use in their profession (Harel, Selden, & Selden, 2006).

Affect. "A wide range of feelings and moods that are generally regarded as something different from pure cognition" (McLeod, 1989, p. 245).

Algebra-for-all movement. A movement sparked by lack of equity in access to algebra and characterized by a search for pedagogical strategies to reach all students.

Algebraic habits of mind. Thinking processes that are necessary for performing a variety of algebraic tasks. Three central habits of mind are doing-undoing, building rules to represent functions, and abstracting from computation (Driscoll, 1999).

Algebra tiles. Tools for helping students learn polynomial multiplication and factoring conceptually. They are designed to help students think about polynomial factoring and multiplication in terms of dimensions and areas of rectangles.

Algorithm. A "finite, step-by-step procedure for accomplishing a task that we wish to complete" (Usiskin, 1998, p. 7).

Analytic rubric. A rubric that assigns scores to various phases or aspects of the problem solving process.

Anscombe's quartet. Four data sets that have many of the same descriptive and inferential statistics. The Pearson correlation coefficient and least-squares regression line are the same in each case. Nonetheless, the data sets differ drastically.

Approximation metaphor for limit. Students employing the approximation metaphor may describe using a series of secant lines to approximate the tangent line to a curve. They may also think of the secant as collapsing into the tangent.

Assessment Standards for School Mathematics. A document published by NCTM in 1995 that provided guidance on classroom assessment.

Attitudes. "The affective responses that involve positive or negative feelings that are relatively stable. Liking geometry, disliking story problems, being curious about topology, and being bored by algebra are all examples of attitudes" (McLeod & Ortega, 1993, p. 29).

Availability heuristic. Individuals using the availability heuristic to reason about a situation overestimate the likelihood of a given event because of their personal experiences.

Average as a data reducer. Characterization of the idea of average in situations where the average helps condense a set of numbers down to a single value.

Average as a fair share. When thought of as a fair share, the arithmetic mean describes the amount of a given quantity shared among a given number of individuals.

Average as a signal in noise. The signal-in-noise interpretation characterizes average as a way to make sense of data generated by a noisy process that has a distinct signal to be identified.

Average as a typical value. Typical value problems involve selecting statistical measures to concisely describe the characteristic values of data sets.

Backward design. Backward design consists of three stages: (1) Identify desired learning results, (2) decide on the kinds of evidence that will be acceptable for determining if students attained the desired results, and (3) plan corresponding learning experiences and select instructional strategies. The three stages emphasize "beginning with the end in mind" rather than just selecting activities that seem engaging (Wiggins & McTighe, 2005).

Balance scale model of equations. Model in which equations consist of two sides of equal weight, meaning that the operations performed on each side of the scale must be the same to maintain the balance.

Beliefs. Several types of beliefs are relevant to the learning of mathematics, including beliefs about mathematics, beliefs about self, and beliefs about mathematics teaching. Students' beliefs in all of these areas are important to understand, because they influence the manner in which students function in class.

Benchmark estimation. This strategy involves estimating a quantity by comparing it to one that is known (e.g., estimating the number of candies in a jar by comparing its size to that of a jar holding a known number of candies).

Blocked practice. Blocked practice assignments are those for which all exercises are drawn from the preceding lesson.

Bloom's Taxonomy. A framework consisting of categories that may be used to construct questions that elicit various cognitive levels.

Buggy algorithm. A persistent, flawed procedure for carrying out mathematical tasks.

Calculus reform movement. This movement began among university mathematicians in the 1980s and continues to the present. In the 1980s, widespread dissatisfaction arose with traditional calculus courses because they served as filters to block students from advanced mathematical and scientific study rather than pumps to propel students into such fields (Steen, 1988).

Cardinality. The number of objects in a set.

Cartesian connection. The idea that every point on a graph represents a solution to a corresponding equation.

Classical probability. Sometimes referred to as theoretical or a priori probability. Under this approach, probability is defined as "the ratio of the number of favorable outcomes to the total number of outcomes, where outcomes are assumed to be equally likely" (G. A. Jones, Langrall, & Mooney, 2007, p. 912).

Classroom response system. A device that affords teachers efficient ways to tabulate students' responses to classroom questions and quickly gauge their understanding.

Clinometer. A tool consisting of a protractor, straw, washer, and string that can be used to measure angles in real-world situations involving right triangle trigonometry.

Closed-form question. A question that requests a simple "yes" or "no" answer or a single numerical answer.

Cognition. "The various ways in which people think about what they are seeing, hearing, studying, and learning" (Ormrod, 2006, p. 20).

Cojoining terms. Combining algebraic terms that should remain separate.

Collapse metaphor for limit. Students thinking in terms of the collapse metaphor may describe the gradual collapse of the rise and run triangle showing the slope of a secant line to a graph. Some also use the collapse metaphor when discussing integrals, imagining the rectangles used to approximate the area under a curve collapsing into line segments.

Common Core State Standards. A standards document written and adopted by several states in the U.S. that identifies mathematics proficiencies students are to attain along with mathematical practices.

Compensation. An estimation strategy that involves making adjustments to numbers to account for the effects of reformulation or translation.

Computational estimation. Estimation of the result of a calculation (e.g., estimation of the answer to 72 ÷ 0.025).

Computer algebra system (CAS). Technology that can perform traditional symbol manipulation tasks such as polynomial multiplication and factoring in a few keystrokes.

Compartmentalization phenomenon. A conflict between the concept image and the concept definition held by a student.

Concatenation errors. Difficulties in interpreting chains of algebraic symbols.

Concept definition. Consists of the words used by the discipline to specify a concept.

Concept image. Consists of the mental pictures, properties, and processes students associate with a given concept.

Conceptual knowledge. "Conceptual knowledge is characterized most clearly as knowledge that is rich in relationships. It can be thought of as a connected web of knowledge, a network in which the linking relationships are as prominent as the discrete pieces of information. Relationships pervade the individual facts and propositions so that all pieces of information are linked to some network" (Hiebert & Lefevre, 1986, p. 3).

Confusion of the inverse. The belief that, in all cases, $P(A|B) = P(B|A)$.

Conjunction fallacy. Students exhibiting the conjunction fallacy believe that conjunctive probabilities are greater than individual probabilities when they are not (i.e., thinking that $P(A \cap B) > P(A)$ when $P(A \cap B) < P(A)$).

Cues. Verbal and nonverbal indicators that show students when and how transitions are to occur during a lesson.

Curriculum and Evaluation Standards for School Mathematics. A landmark document published by NCTM in 1989 describing a vision for the teaching and learning of mathematics differing sharply with much of conventional practice.

Curriculum focal point. "Curriculum focal points are important mathematical topics for each grade level, pre-K–8. These areas of instructional emphasis can serve as organizing structures for curriculum design and instruction at and across grade levels. The topics are central to mathematics: they convey knowledge and skills that are essential to educated citizens, and they provide foundations for further mathematical learning" (NCTM, 2006, p. 5).

Decomposition/recomposition estimation. This strategy involves dividing a problem into parts and then recombining the parts to produce an estimate (e.g., estimating the number of words on a page by looking at the number of words in a typical line and then multiplying by the number of lines on the sheet).

Deductive reasoning. Involves incorporating accepted statements such as theorems, postulates, and definitions into a logical argument.

Descriptive-analytic reasoning. The second van Hiele level, at which students begin to differentiate among shapes by analyzing their component parts.

DGE construction. An object created in a dynamic geometry software environment that maintains its properties when dragged.

DGE drawing. An object created in a dynamic geometry software environment that loses its properties when dragged.

Dialogic discourse. Teachers who engage students in dialogic discourse work to understand and build on students' ways of thinking and solving problems. Dialogic discourse provides the opportunity to emphasize that the most important thing in solving a problem is to find a reasonable solution, not to try to imitate the teacher's solution.

Differentiated instruction. Differentiated instruction is based on the premise that students come to the classroom with diverse learning needs and draws attention to varying instruction along the dimensions of content, process, and product.

Disequilibrium. A necessary degree of discomfort experienced during the learning process.

Distracter. Incorrect choice to a multiple-choice question designed to appeal to students with a common misunderstanding.

Distributed practice. Sometimes called mixed review, distributed practice involves having students practice skills learned in a given lesson at several different points during the school year. Distributed practice prompts students to reflect on the content that has been taught in previous lessons, not just the most recent one.

Distribution combinatorics problem. A combinatorics problem that involves placing n objects into m cells. The objects may or may not be identical. The same holds for the cells. In some cases order matters, and in others it does not. There may be zero, one, or more than one object per cell.

Downtime. Occasions when no mathematical activity is taking place in class.

Dynamic statistics software. Computer software that allows students to efficiently produce and manipulate multiple representations of data.

Emotions. Emotions "may involve little cognitive appraisal, and may appear and disappear rather quickly, as when the frustration of trying to solve a hard problem is followed by the joy of finding a solution" (McLeod, 1992, p. 579).

Enacted curriculum. The curriculum that actually unfolds in a teacher's classroom setting.

Engineering view of research. This engineering view holds that research should provide "specific and immediately applicable technical solutions to problems, in the manner that natural science or engineering research is assumed to do" (Hammersley, 2002, p. 38).

Equiprobability bias. The belief that all random events are equally likely.

Euclid's *Elements*. Euclid's compilation and extension of the work of his predecessors, characterized as the most influential textbook in history and likely second only to the Bible in terms of number of editions published.

Explicit reasoning. Reasoning that involves formulating a rule to describe the characteristics of any given term in a sequence without reference to previous terms.

External sources of conviction. Students with external sources of conviction look to authorities such as the teacher, the answers in the back of a text, a calculator, or some other external source to determine mathematical correctness.

Extrinsic rewards. Incentives provided to induce students toward a desired behavior. Such rewards include praise, high grades, and certificates of merit.

Falk phenomenon. Occurs when students believe it does not make sense to compute the probability of an event when the outcome of a second, related follow-up event is given.

***Focus in High School Mathematics*.** NCTM curriculum document describing the reasoning and sense-making processes students are to develop and use when working across all mathematical content areas.

Focusing. Discourse that occurs when teachers' questions concentrate on uncovering students' thinking; the information gained about students' thinking then guides instruction.

Formal deductive proof (van Hiele level). The fourth van Hiele level, at which students understand the importance of undefined terms, definitions, axioms, and theorems in deductive reasoning. They can construct proofs by drawing on given information and using previous results to build a deductive argument.

Formative assessment. Assessments done primarily to inform the path that future instruction should take.

Four-column lesson plan. Often used when teachers engage in lesson study. The four-column lesson plan is designed to emphasize the idea that lesson planning should go beyond a single column that is essentially a to-do list for the mathematics teacher during a class period.

Fractal. "Roughly speaking, fractals are complex geometric shapes with fine structure at arbitrarily small scales. Usually they have some degree of self-similarity. In other words, if we magnify a tiny part of a fractal, we will see features reminiscent of the whole. Sometimes the similarity is exact; more often it is only approximate or statistical" (Strogatz, 1994, p. 398).

Fraction circles. Manipulatives representing fractions, consisting of slices of a circle's interior.

Frequentist probability. Sometimes referred to as experimental or a posteriori probability. Under this approach, probability can be defined as "the ratio of the number of trials favorable to the event to the total number of trials" (G. A. Jones, Langrall, & Mooney, 2007, p. 912).

Functional contexts for slope. Contexts for slope that involve interpreting slope as a measure of change in situations such as distance versus time or cost per hour.

Functions-based approach to algebra. An approach to teaching algebra that makes functions from everyday life the central objects of study.

Fundamental counting principle. The idea that if there are x ways of doing one thing and y ways of doing another, there are xy ways of doing both together.

Fundamental theorem of arithmetic. "The Fundamental Theorem of Arithmetic claims that decomposition of a number into its prime factors exists and is unique except for the order in which the prime factors appear in the product" (Zazkis & Campbell, 1996b, p. 207).

Funneling. Discourse that occurs when a teacher uses questions to walk students through a series of steps necessary to solve a problem.

GAISE college report. *Guidelines for Assessment and Instruction in Statistics Education* endorsed by the American Statistical Association, offering recommendations for the reform of introductory statistics courses.

Geoboard. A pegboard on which rubber bands can be strung to create geometric objects.

Geometric habits of mind. Characteristics of normative geometric thinking that include using proportional reasoning, using several languages at once, using a single language for everything, reasoning about systems, studying change and invariance, and analyzing shapes (Cuoco, Goldenberg, & Mark, 1996).

Guided investigation. Guided investigation is based on the premise that students will better retain information if they have a hand in discovering it. Curricula that utilize guided investigation do not leave students to reinvent all of the mathematics they need to learn on their own, but instead ask carefully sequenced questions to lead students in the desired direction.

Harmonic thinking. Blending visual and analytic thinking when necessary to analyze a situation.

Harvard calculus. A shorthand phrase for a curriculum developed by the Calculus Consortium at Harvard University (Hughes-Hallet, Gleason, et al., 1994) with funding from the National Science Foundation.

Hierarchical definitions. Definitions that establish a system in which a hierarchy of concepts can be formed.

Holistic rubric. A rubric that assigns a single score to a student's work on a problem.

Illusory correlation. Exists when a student believes that two uncorrelated variables are actually correlated.

Individualized education plan (IEP). The IEP originates from federal legislation charging schools to meet students' individual learning needs. IEPs attempt to chart paths to success for special-needs students who spend time in the regular classroom.

Instrumental understanding. Skemp (1976) associated instrumental understanding with "learning rules without reasons."

Integration of content strands. Traditionally, algebra and geometry have been considered separate courses, particularly at the high school level. The integration of content strands represents a sharply different approach in that students may study algebra, geometry, data analysis, probability, and other subject areas in any given year.

Intended curriculum. Refers to a teacher's instructional plans related to a given curriculum.

Internal sources of conviction. Students with internal sources of conviction look to their own reasoning to establish mathematical correctness.

Intrinsic motivation. Motivation that comes from pure enjoyment of an activity rather than from receiving a reward for doing it.

Isometry. A transformation in the plane that preserves the distance between points (Jaime & Gutiérrez, 1995).

Judgment tasks. Tasks that involve drawing plausible conclusions about a situation through interpretation of associated data.

Law of small numbers. The law of small numbers, in contrast to the law of large numbers, is the belief that sample size is not likely to influence the accuracy with which a sample statistic reflects a given characteristic of the population.

Lesson study. An increasingly popular model for teacher research that consists of "collaborating with fellow teachers to plan, observe, and reflect on lessons" (Takahashi & Yoshida, 2004).

Levels of cognitive demand. A framework for evaluating challenges of tasks used in a classroom setting.

Literal symbols. Symbols used to represent unknowns, variables, and constants in algebraic expressions, formulas, identities, equations, and inequalities.

Local linearity. The idea that repeated zooming eventually "straightens out" a graph of a function that is differentiable at a given point.

Mathematics anxiety. "The panic, helplessness, paralysis, and mental disorganization that arises among some people when they are required to solve a mathematical problem" (Tobias & Weissbrod, 1980).

Mean as a balance point. With this metaphor, the mean is analogous to a scale fulcrum that is closer to some of the data values than others.

Measure interpretation of rational numbers. Interpretation of $\frac{x}{y}$ as x lengths of size $\frac{x}{y}$ each.

Measurement estimation. This strategy involves estimating geometric measurements or large quantities of items (numerosity estimation).

Measurement model of division. Division situation that involves determining the number of groups of a certain size in a given collection.

Metacognition. Individuals' "awareness of their own cognitive machinery and how the machinery works" (Meichenbaum, Burland, Gruson, & Cameron, 1985, p. 5).

Metaphor. "A figure of speech in which a word or phrase literally denoting one kind of object or idea is used in place of another to suggest a likeness or analogy between them" (Mish, 1991, p. 746).

Mixed review. Mixed review assignments include exercises drawn from a combination of previous lessons.

Moderate enlightenment view of research. The moderate enlightenment view holds that "research is one among several sources of knowledge on which practice can draw. Moreover, the use made of it properly depends on practical judgments about what is appropriate and useful" (Hammersley, 2002, p. 42).

Multiple representations of functions. These include graphs, tables, equations, and verbal representations.

Multiplicative change. The amount of change expressed as a percentage of the original quantity.

Naive empiricists. Students who rely entirely on the empirical evidence produced by a finite number of cases.

National Council of Teachers of Mathematics (NCTM). An organization with more than 90,000 members dedicated to improving mathematics education.

NCTM principles. The principles "describe particular features of high-quality mathematics education" (NCTM, 2000, p. 11).

NCTM process standards. The process standards describe aspects of mathematical teaching and learning that should cut across all content areas.

No Child Left Behind (NCLB). Federal legislation prompting states to identify mathematics learning expectations for students.

Normative geometric thinking. The thinking of mathematicians as they are engaged in doing geometry.

NSF-funded curricula. After the NCTM (1989) standards were published, the National Science Foundation (NSF) funded a number of curriculum projects for middle and high school mathematics to model NCTM's vision for mathematics instruction.

Number sense. "Refers to a person's general understanding of number and operations. It also includes the ability and inclination to use this understanding in flexible ways to make mathematical judgments and to develop useful strategies for handling numbers and operations. It reflects an inclination and an ability to use numbers and quantitative methods as a means of communicating, processing, and interpreting information. It results in an expectation that numbers are useful and that mathematics has a certain regularity" (R. E. Reys & Yang, 1998, pp. 225–226).

Object view of function. Interpretation of a function such as $y = f(x)$ as a single entity. The entity may be thought of graphically in terms of its overall shape, and one may visualize making modifications to the entire entity by altering its coefficients or constants.

Odometer strategy. This strategy consists of holding one term constant and then forming all possible combinations that include it. Once the possibilities have been exhausted for one constant term, students hold another term constant and then document all possible combinations that include it.

Opaque representations. Representations that do not readily reveal certain characteristics of a mathematical object.

Open-form question. A questions that allows students to express diverse patterns of thinking and justification.

Open number sentence. An equation with one or more blanks inserted in place of quantities.

Open questions. Questions that allow multiple solution strategies.

Operational view of the equal sign. Interpretation of the equal sign as a prompt to combine the two quantities to the left of it and write the result to the right.

Operator interpretation of rational numbers. Interpretation of $\frac{x}{y}$ as taking a fractional amount $(\frac{x}{y})$ of a given quantity.

Ordered case value bars. Informal data representation that depicts ordered pairs from bivariate data sets as bar lengths placed in order according to the first value of each ordered pair.

Ordinality. Placement of numbers in the appropriate order on the number line.

Outcome approach. Students exhibiting the outcome approach interpret a probability task in terms of whether or not something will happen rather than as a statement about its likelihood.

Parallel tasks. Tasks that deal with the same mathematics concepts, yet differ in ways that accommodate individual differences.

Parallelogram rule for vector addition. This rule involves placing vectors together at their tails and then placing segments parallel to the two vectors to create a parallelogram. The sum of the two vectors is represented by the vector whose tail is located at the intersection of the tails of the vectors and whose tip is located at the intersection of the parallel segments.

Partition combinatorics problem. A combinatorics problem that involves splitting n objects into m subsets.

Partitional definitions. Definitions that partition concepts into separate, mutually exclusive bins.

Partitive model of division. Division situation in which the central task is to determine the number of objects per group.

Part-whole interpretation of rational numbers. Interpretation of $\frac{x}{y}$ as x parts out of y equal-size parts.

Physical contexts for slope. Contexts for slope that involve determining the steepness of objects such as ramps and hills.

Polygonal numbers. Numbers that can be represented by dots arranged in the shapes of various polygons.

Portfolio. Portfolios provide students an opportunity to assemble their best mathematical work into a single collection. Teachers may ask students to reflect on their work samples and write about how the work they have selected illustrates their growth in thinking over a given period of time.

Preventive discipline. The act of stopping discipline problems before they occur.

Principles and Standards for School Mathematics. A document published by NCTM in 2000 that built on the foundation of the original standards documents and organized curriculum recommendations into four grade bands.

Procedural knowledge. "Procedural knowledge ...is made up of two distinct parts. One part is composed of the formal language, or symbol representation system, of mathematics. The other part consists of the algorithms, or rules, for completing mathematical tasks" (Hiebert & Lefevre, 1986, p. 5).

Process of statistical investigation. A four-component, generally nonlinear process consisting of (1) formulating questions, (2) collecting data, (3) analyzing data, and (4) interpreting results (Franklin et al., 2007).

Process view of function. Interpretation of a function such as $y = f(x)$ as a procedure to be carried out by substituting a value for x and computing the result. From this perspective, the symbol y represents the finished computation.

Product-and-factor model of division. Division situation in which one wishes to determine the value of x in the equation $ax = b$, where a and b are known constants. The value of x is the quotient b/a.

Professional Standards for Teaching Mathematics. A document published by NCTM in 1991 that helped clarify NCTM's vision for school mathematics reform by recommending five major shifts in mathematics classroom environments.

Progressive formalization. "Progressive formalization of the mathematics involves, first, having

students approach problems and acquire algebraic concepts and skills in an informal way. They use words, pictures, and/or diagrams of their own invention to describe mathematical situations, organize their own knowledge and work, solve problems, and explain their strategies. In later units, students gradually begin to use symbols to describe situations, organize their mathematical work, or express their strategies. At this level, students devise their own symbols or learn certain nonconventional notation (e.g., arrow language). Their representations of problem situations and explanations of their work are a mixture of words and symbols" (Romberg, 2001, p. 5).

Project 2061. An American Association for the Advancement of Science (AAAS) project (Kulm, 1999) that evaluated textbooks for their mathematics content and their methods of teaching it.

Proportional reasoning. "Proportional *reasoning* refers to detecting, expressing, analyzing, explaining, and providing evidence in support of assertions about proportional relationships. The word *reasoning* further suggests that one uses common sense, good judgment, and a thoughtful approach to problem solving, rather than plucking numbers from word problems and blindly applying rules and operations" (Lamon, 2007, p. 647).

Prototype. Prototypes can be described as mental images that exemplify categories (Lakoff, 1987). Those that capture many of the relevant aspects of a category can be helpful in learning mathematics, and those that are more limited can be detrimental (Presmeg, 1992).

Quantitative analysis of a task. Technique for solving word problems that involves acting out a problem with representations of the quantities involved by using tools such as drawings and manipulatives.

Quotient interpretation of rational numbers. Interpretation of $\frac{x}{y}$ as x divided by y.

Radical constructivism. A theory of cognition that emphasizes the role of individual construction of knowledge in the learning process.

Range estimation. This strategy involves choosing the center of an arbitrary range as an estimate (e.g., believing that the actual number of words on a page is somewhere between 100 and 150 and choosing a number in the middle of the range as an estimate).

Ratio interpretation of rational numbers. Interpretation of $\frac{x}{y}$ as a statement of comparison between the quantities x and y (e.g., x miles per y gallons of gasoline).

Reading to learn mathematics. The use of reading as a springboard for mathematical problem solving.

Reasoning. "In the most general terms, reasoning can be thought of as the process of drawing conclusions on the basis of evidence or stated assumptions" (NCTM, 2009, p. 4).

Recursive reasoning. Reasoning that involves examining individual terms in a sequence to determine subsequent terms.

Reformulation. An estimation strategy that involves altering the numbers in a problem to make them more manageable.

Reification. Occurs as students move from viewing functions as processes to viewing them as objects.

Relational-inferential reasoning. The third van Hiele level, at which inferences about the characteristics of shapes are generally made from observing many examples.

Relational understanding. Skemp (1976) associated instrumental understanding with "knowing both what to do and why."

Relational view of the equal sign. Interpretation of the equal sign as a symbol of equivalence.

Representativeness heuristic. Students exhibiting the representativeness heuristic may expect that even very small samples will represent the population from which they were drawn.

Response to intervention (RTI). RTI programs are designed to identify and remediate students in need of special learning assistance. The amount and type of assistance students are given is based on their demonstrated learning needs.

Reversal error. Placing variables on the wrong sides of an equation written to represent a situation (e.g., writing the equation $6S = P$ rather than $6P = S$).

Rigor (van Hiele level). The fifth van Hiele level, at which students are able to reason about alternative axiomatic systems. They can understand that more than one logically consistent system of geometry exists.

Rule of four. The idea that, "where appropriate, topics should be presented geometrically, numerically, analytically, and verbally." (John Wiley & Sons, 2010, n.p.).

Sampling Sim. A freely available, classroom-tested simulation package that helps build understanding of the concept of sampling distribution. It can produce simulations of the results obtained from computing the means and proportions of several samples drawn from a population.

Scaffolding. Teacher action that involves conveying expectations at the beginning of the school year and then gradually withdrawing support during problem solving activities.

Selection combinatorics problem. A combinatorics problem that involves drawing samples, either with or without replacement. The sample may be either ordered or unordered.

Sense-making. "Developing understanding of a situation, context, or concept by connecting it with existing knowledge" (NCTM, 2009, p. 4).

Service orientation. The idea that self-interest must at times be secondary to serving those in need of assistance.

Simulation of Samples (SOS) model. A visual organizer that shows three levels involved in a probability simulation.

Sliced scatterplot. A data representation that takes a conventional scatterplot and partitions it into vertical slices. The mean y-value of the data points within each slice may be computed to provide a better picture of the overall trend of the data.

Social constructivism. "Social constructivism regards individual subjects and the realm of the social as indissolubly connected. Human subjects are formed through their interactions with each other as well as with their individual processes" (Ernest, 1996, p. 342).

Social justice pedagogy. An approach to teaching distinguished by three main goals: to help students develop sociopolitical consciousness, a sense of agency, and positive social and cultural identities (Gutstein, 2005).

Social norms. Cobb and Yackel (1996) described social norms as the delineation of the participation structure for classroom discourse. Norms exist for things such as justifying solutions, making sense of solution strategies offered by others in class, and indicating agreement or disagreement with a given solution.

Sociomathematical norms. Sociomathematical norms are specific to the subject of mathematics. Cobb and Yackel (1996) mentioned several types of sociomathematical norms, including what counts as a different mathematical solution, an efficient solution, or an acceptable explanation.

Spurious correlation. A correlation between variables that are not causally related.

Static comparison method. A strategy characterized by treating literal symbols as labels rather than variables.

Statistical intuition. Intuitive ability to "construct representations by balancing quantitative and contextual aspects of the data and by determining the reasonableness of inferences" (Mooney, 2002, p. 49).

Statistical literacy. Includes "basic and important skills that may be used in understanding statistical information or research results" (Ben-Zvi & Garfield, 2004, p. 7).

Statistical reasoning. "The way people reason with statistical ideas and make sense of statistical information" (Ben-Zvi & Garfield, 2004, p. 7).

Statistical thinking. "An understanding of how and why statistical investigations are conducted and the 'big ideas' that underlie statistical investigations" (Ben-Zvi & Garfield, 2004, p. 7).

Strong enlightenment view of research. The strong enlightenment view "implies that policy-makers and practitioners are normally in the dark, and that research is needed to provide the light necessary for them to see what they are doing, and/or what they ought to be doing" (Hammersley, 2002, p. 39).

Student preconceptions. The school-based and experiential knowledge students bring to a lesson.

Student-professor problem. "Write an equation for the following statement: 'There are six times as many students as professors at this university.' Use S for the number of students and P for the number of professors" (J. Clement, Lockhead, & Monk, 1981, p. 288).

Subjective probability. "Describes probability as a degree of belief, based on personal judgment and information about an outcome" (G. A. Jones, Langrall, & Mooney, 2007, p. 913).

Summative assessment. Summative assessments measure a student's performance at the end of a lesson or unit of study. Often, summative assessments are used for the purpose of assigning grades. Achievement tests given to students at the end of the school year are also summative in nature.

Taxicab geometry. Non-Euclidean geometry that redefines the conventional concept of distance in plane geometry. The coordinate grid can be thought of in terms of city streets, and distance is measured by considering the shortest path a taxi could drive in commuting between two points on the grid. Counting the number of unit grid lengths traced out by the shortest path gives the distance.

Teaching about problem solving. Teaching about problem solving involves helping students learn general problem solving strategies. For example, some curricula introduce Polya's (1945) problem solving process with the assumption that students will adopt his thinking strategies as they solve problems of their own.

Teaching for problem solving. Teaching for problem solving consists of explicitly teaching students mathematical ideas that they are expected to use to solve problems later on.

Teaching through problem solving. Teaching through problem solving involves selecting problems containing important mathematics and helping students learn mathematical ideas as they solve the problems.

Tessellation. Tiling of a plane that does not contain any gaps or overlaps.

Thematic unit. A thematic unit consists of a series of problems that arise from a real-world context. Problems in such units are chosen not only to bring out important mathematics, but also to pertain to a given real-world situation.

Third International Mathematics and Science Study (TIMSS). A large, comprehensive investigation of mathematics teaching and learning in different parts of the world. Conducted in 1994–1995, it tested mathematics achievement in more than 40 nations. The TIMSS acronym now signifies "Trends in International Mathematics and Science Study."

Tiers of statistical literacy. A set of three competencies relevant to interpreting statistics critically: "1. A basic understanding of statistical terminology[;] 2. An understanding of statistical language and concepts when they are embedded in the context of wider societal discussion[;] 3. A questioning attitude that can apply more sophisticated concepts to contradict claims made without proper statistical foundation" (Watson, 2000, p. 54).

Transfer. The ability to "apply what was learned in new situations and to learn related information more quickly" (Bransford, Brown, & Cocking, 1999, p. 17).

Translation. An estimation strategy that involves altering the mathematical structure of a problem.

Transnumeration. Thinking characterized by three aspects: "capturing measures of the real system that are relevant, constructing multiple statistical representations of the real system, and communicating to others what the statistical system suggests about the real system" (Shaughnessy & Pfannkuch, 2002, p. 256).

Transparent representations. Representations that readily reveal certain characteristics of a mathematical object.

Triangle rule for vector addition. This rule involves placing vectors that are to be added tip to tail. The sum of the vectors is the vector starting at the tail of the first and ending at the tip of the second.

Unit circle. A circle with a radius of 1 and centered on the point (0, 0).

Univocal discourse. Discourse on the univocal side of the continuum generally involves teachers trying to guide students to solve a given problem in a prescribed way. The teacher places value not on trying to understand students' alternative solution strategies, but only on making sure that they received the "correct" message about the manner in which the problem should be solved.

Van Hiele levels. Levels of development through which students tend to pass in learning geometry (van Hiele, 1986).

Visual decoding tasks. Tasks that involve direct reading of data from a graph or table.

Visual-holistic reasoning. The first van Hiele level, at which students can name shapes when they are shown to them. However, the names are based on the general appearance of the shapes rather than on careful analysis of their properties.

Visual salience. Occurs when the left and right sides of an equation appear to be naturally related to one another.

Wait time. The amount of time teachers wait before speaking after asking a question or the amount of time teachers wait before speaking after an answer to a question has been provided.

What Works Clearinghouse. A publisher-independent curriculum evaluation clearinghouse that aims to produce user-friendly guides for educators about the effectiveness of curricula, assess research evidence on curricular effectiveness, and develop and implement standards for the review and synthesis of education research.

Withitness. Communicating to students an awareness of what is happening in different parts of the classroom at all times.

Word order matching. A strategy reflecting the belief that algebraic symbols should be written into an equation in the same order as they appear in a corresponding word problem.

Worthwhile mathematical tasks. "Good tasks are ones that do not separate mathematical thinking from mathematical concepts or skills, that capture students' curiosity, and that invite them to speculate and to pursue their hunches. Many such tasks can be approached in more than one interesting and legitimate way; some have more than one reasonable solution" (NCTM, 1991, p. 25).

Writing prompts. Questions or statements posed by teachers to reveal students' cognition or affect in regard to a given mathematical idea.

Written curriculum. The curriculum that exists in print on the pages of textbooks and curriculum materials.

Zone of proximal development. "A symbolic space created through the interaction of learners with more knowledgeable others and the culture that precedes them" (Goos, 2004, p. 262).

INDEX